By: George and Marion Branigan

FUNDAMENTALS OF
GEOTECHNICAL
ANALYSIS

FUNDAMENTALS OF GEOTECHNICAL ANALYSIS

Irving S. Dunn
Loren R. Anderson
Fred W. Kiefer

Department of Civil Engineering
Utah State University

John Wiley & Sons
New York   Chichester   Brisbane   Toronto

Library of Congress Cataloging in Publication Data:

Dunn, Irving S.   1923–
   Fundamentals of geotechnical analysis.

   Bibliography: p.
   Includes index.
   1. Soil mechanics.   I. Anderson, Loren Runar,
1941– joint author.   II. Kiefer, Fred William,
1925– joint author.   III. Title.

TA710.D87   624′.1513   79-13583
ISBN 0-471-03698-6

Printed in the United States of America

10 9 8 7 6 5 4 3 2 1

PREFACE

Geotechnical engineering developed from the science of soil mechanics, which grew largely through the work of Karl Terzaghi and his associates in the early part of the twentieth century. Both the science and the art of geotechnical engineering are changing rapidly, and many important advances have been made in the past decade.

This book is designed to provide an introduction to methods of analysis at an elementary level for undergraduate engineering students. In order to be successful, a practicing geotechnical engineer must combine a thorough knowledge of analysis with experience and common sense.

Generally, soil mechanics books contain a great deal of reference material and many subjects that are appropriately taught at the graduate level. This book attempts to separate the basic ideas that are needed for a good understanding of geotechnical analysis from those that might best be left for another text on advanced problems and to treat these subjects in a way designed for optimum understanding by a college student. Teaching is thus a primary objective of this book, and no attempt is made to provide a large amount of reference material for the practicing engineer.

The first nine chapters of the text are arranged to provide a suitable sequence for a two-quarter, three-credits-per-quarter course for junior or senior engineering students. In our class at Utah State University we cover all the topics in the first nine chapters in the undergraduate courses. The last two chapters contain supplementary material that, although still basic in nature and desirable for use by some instructors, is not necessary for an adequate elementary introduction to analysis.

The SI system of units is used exclusively in this text, except in a very few instances where values may be given in English units for the purpose of comparison. No attempt is made to change gradually from the English system to the SI system. We believe that complete and sudden immersion in the SI system will result in the student's ability to think in terms of these units more clearly and more rapidly than with a gradual change.

We would like to acknowledge and express our appreciation to those who helped in preparing this text. Dr. Michael L. DeBloois and Dr. Steven J. Soulier of the Instructional Media Department at Utah State University provided support and consultation during the initial stages of our work. Mr. Thomas J. Allen worked all of the problems at the end of each chapter for the solutions manual and Mr. David W. Nyby helped prepare the computer programs. Dr. Joseph Olsen and Mr. Richard E. Riker used a draft form of the text in their classes and provided helpful comments. Dr. James K. Mitchell, Dr. W. L. Schroeder, Dr. Robert M. Koerner and Dr. Louis J. Thompson reviewed the manuscript and provided valuable comments. For the past four years we have used the manuscript as the text for our soil mechanics classes and many improvements were made from student comments and from this opportunity to field test the product. Genevieve Fonnesbeck, Suzanne Wilson and Jolene Kiefer typed the manuscript.

Irving S. Dunn
Loren R. Anderson
Fred W. Kiefer

Logan, Utah
January, 1980

CONTENTS

Contents

LIST OF SYMBOLS

A = area (L^2) or constant

A_n = area of voids (L^2)

a = area (L^2) or length (L)

a_c = ratio of contact area to total area

B = width of footing (L)

b = distance

C, C_1, C_2 = constants

C_c = compression index

C_u = uniformity coefficient

C_r = cohesive force (F)

C_s = shape coefficient

c_v = coefficient of consolidation

C_w = wall adhesive force (F) or water table correction factor

C_z = coefficient of curvature

c = cohesion

c_1, c_2, c_3 = constants

D = depth or particle diameter (L)

D_m = representative particle diameter (L)

D_f = depth of footing (L)

D_{10} = particle diameter with 10% finer than

D_r = relative density (granular soils)

d = distance (L)

d_c, d_q = depth factors for bearing capacity

E = modulus of elasticity (F/L^2) or energy (FL)

E = horizontal soil force (F)

e = void ratio

e_o = initial void ratio

e_{max}, e_{min} = maximum or minimum void ratio

F = safety factor or percent passing the 0.075 mm sieve

F = force

F_g, F_s, F_T = gravitational force or particle surface force or surface tension force (F)

f_s = soil-sleeve friction (F/L^2)

G = specific gravity

H = height or depth (L)

H_o = initial sample height

H_s = height of soil solids

h = hydraulic head (L) or vertical distance (L)

h_e = hydraulic elevation head (L)

h_L = hydraulic head loss (L)

h_p = hydraulic pressure head

h_T = total hydraulic head (L)

h_v = velocity head (L)

Δh = change in hydraulic head (L)

I_B = Boussinesq pressure influence coefficient, area load

I_W = Westergaard pressure influence coefficient, area load

i = hydraulic gradient or angle

i_c = critical hydraulic gradient

i_q = inclination factor for bearing capacity

K_a = active earth pressure coefficient

K_f = principal stress ratio at failure

K_o = lateral earth pressure coefficient

K_p = passive earth pressure coefficient

k = coefficient of permeability (L/T) (hydraulic conductivity)

k' = coefficient of permeability of transformed section (L/T)

L = length (L)

LI = liquidity index

LL = liquid limit (%)

M = integer or constant

M_B = Boussinesq pressure coefficient (area load)

M_i = mass of soil retained (M)

M_s = mass of dry soil (M)

M_w = mass of water (M)

M_W = Westergaard pressure coefficient (area load)

m = integer or shear strength by Bishop's modified method of slices (F)

m_v = coefficient of volume change (L^2/F)

N = ratio, integer, or blow count for standard penetration test

N = normal force (F)

N_B = Boussinesq pressure coefficient (point load)

N_W = Westergaard pressure coefficient (point load)

N_q, N_γ, N_c = bearing-capacity factors

n = integer number or porosity

P = pile penetration resistance (F)

P_a = active pressure force (F)

P_n = normal force (F)

P_p = passive pressure force (F)

PI = plasticity index

PL = plastic limit

$p = \dfrac{\sigma_1 + \sigma_3}{2}\ (F/L^2)$

Q = flow volume (L^3)

Q = force or load (F)

q = volume flow rate (L^3/T)

q = load per unit area (F/L^2)

$q = \dfrac{\sigma_1 - \sigma_3}{2}\ (F/L^2)$

q_a = allowable bearing capacity (F/L^2)

q_c = cone bearing capacity (F/L^2)

q_u = unconfined compressive strength (F/L^2)

R = radius (L) or force (F)

R_f = friction ratio f_s/q_c%

r = radius (L)

r_u = pore pressure coefficient

S = shear strength (F/L^2)

S = distance or settlement (L)

S_r = degree of saturation

S_u = undrained shear strength (F/L^2)

SL = shrinkage limit (%)

s = pile tip movement, distance (L) or shear resistance (F/L^2)

s_q = shape factor for bearing

T = time factor or time

T = torque (FL), force (F), or time (T)

T_s = surface tension (F/L)

t = time (T)

U = water force (F) or degree of consolidation

u = water or pore pressure (F/L^2)

u_e = excess pore pressure (F/L^2)

V = total volume (L^3)

V_a = volume of air (L^3)

V_s = volume of solids (L^3)

V_v = volume of voids (L^3)

V_w = volume of water (L^3)

v = velocity (L/T)

v_n = actual or pore velocity (L/T)

W = total weight (F)

W_s = weight of dry soil solids (F)

W_w = weight of water (F)

WCR = weighted creep ratio

WCD = weighted creep distance (L)

w = water content (%)

x, y, z = coordinate distance (L)

α = angular measure

β = constant or angular measure

γ = unit weight (F/L^3)

γ_b = buoyant unit weight (F/L^3)

γ_d = dry unit weight (F/L^3)

γ_{sat} = saturated unit weight (F/L^3)

γ_w = unit weight of water (F/L^3) (9.81 kN/m^3)

$\gamma_{d\,max}, \gamma_{d\,min}$ = maximum and minimum dry unit weight (F/L^3)

γ_i = initial unit weight (F/L^3)

Δ = displacement (L)

δ = wall friction angle

ϵ = strain

η = structural viscosity (FT/L^2)

θ = angular measure

θ_p = angle of slip lines for passive retaining wall

λ = decimal portion of cross-sectional area occupied by water

μ = viscosity (FT/L^2)

ν = Poisson's ratio

ρ = mass density of matter (M/L^3) or radius of curvature (L)

Σ = summation

σ = total normal stress (F/L^2)

$\bar{\sigma}$ = effective normal stress (F/L^2)

$\bar{\sigma}_o$ = initial effective stress (F/L^2)

σ_v = total vertical normal stress (F/L^2)

$\sigma_x, \sigma_y, \sigma_\theta$ = normal stress on x, y, or θ plane (F/L^2)

$\sigma_1, \sigma_2, \sigma_3$ = principal stresses (F/L^2)

τ = shearing stress (F/L^2)

τ_{xy} = shearing stress associated with the x–y plane (F/L^2)

τ_θ = shearing stress on θ plane (F/L^2)

ϕ = friction angle or velocity potential

$\bar{\phi}$ = effective friction angle

2 groups of soil

1. transported
   A. alluvium   soil transported
                 by water
   B. Colluvium   gravity
   C. GLACIERS, WIND

2. RESIDUAL

PHYSICAL   WEATHERING (LARGE ROCK)

1. Thermal expansion + contraction
2. Crystal Growth
3. Organic activity (plants
4. Colloid Plucking
5. PHYsical abrasion (Glaseation)
6. Abrasion (wind)

CHEMICAL   WEATHERING (small)

SOLUTION ← ROCK → Residue
         Deposition

1. Chelation - where certain
   cations are picked up by acids
2. OXIDATION
3. HYDROLYSIS

End Product (chemical) = f (parent material,
                           time, temp, rainfall)

Weathering process

Feldspars ⇒ sand ⇒ silt ⇒ Clays
Micas ⇒   " ⇒ " ⇒   "
Pyroxene ⇒   " ⇒ " ⇒   " + Fe_2O_3
Dolomite ⇒   " ⇒ " ⇒ impurities
Quartz ⇒   "

PLANER STRUCTURES

Clay - identifying feature ⇒ plasticity

FOR CLASSIFICATION

$W_L - W_P = I_P$
W_L - Liquid Limit
W_P - Plastic "

symbolic octahedral sheet
true gibbsite sheet have no
O, just OH & Al

Kaolinite

found in warm
wet regions of the
temperate zone

1

SOIL PROPERTIES

*Subsurface engineering is an art; soil mechanics is an
engineering science. . . . We would do well to recall and
examine the attributes necessary for the successful practice of
subsurface engineering. There are at least three: knowledge of
precedents, familiarity with soil mechanics, and a working
knowledge of geology. . . . (Peck, 1962)*

The purpose of this text is to familiarize the reader with the fundamental
principles of soil mechanics. These principles involve applying an understanding
of the physical properties of soil to the following.

- Analysis and design of earth structures such as dams and embankments.
- Evaluation of the stability of artificial and natural slopes.
- Evaluation of the ability of soil deposits to provide support for various
  structures.
- Evaluation of the magnitude and distribution of earth pressures against
  various structures.
- Prediction of water movement through soil.
- Improvement of soil properties by chemical or mechanical methods.

After the principles of soil mechanics have been mastered, the engineer will
have the necessary analytical tools to predict theoretically the behavior of soil
under various conditions. Familiarity with the principles of soil mechanics will
provide the young engineer with the framework to interpret and evaluate
experience and thus begin to develop a knowledge of precedents.

Illustrations of several typical projects that involve the application of the
principles of soil mechanics are shown in Figs. 2-1, 3-1, 3-2, and 5-1. Circular
steel sheet pile cells are used to support the wharf structure shown in Fig. 2-1.
The sheet pile cells must be designed to resist rupture from soil pressure within

1

Figure 2-1 Circular steel sheet pile cell wharf structure. The sand-filled cells support the wharf deck and serve as a retaining structure.

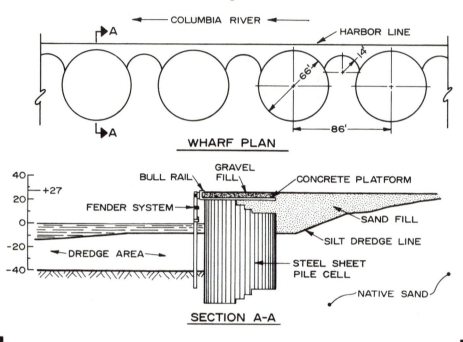

each cell as well as to resist overturning and sliding from the lateral pressures exerted by the soil behind each cell. Figure 3-1 illustrates the effect of the two failure criteria (shear failure, excessive settlement) used in the design of spread footings. The design of a zoned earth dam such as that shown in Figs. 3-2 and 4-1 must consider both the mass stability of the embankment and the amount of seepage that is likely to occur through the embankment and foundation. Figure 5-1 shows the test apparatus used to determine the design loads for a pile foundation.

Unlike many engineering disciplines, in geotechnical engineering the engineer must work with a material (soil) for which the physical properties are extremely variable and difficult to evaluate. The determination of the physical properties thus becomes part of the problem-solving process that is as difficult as, and equally important to, the other phases of the process.

The geotechnical engineer has a role in every project that involves earth structures, that requires a soil or rock foundation, or that is constructed below the ground surface. Although every project is different to some extent, there is a

Figure 3-1 A typical design chart used to proportion spread footings. The chart was developed from the results of a subsurface investigation at a particular site and using the principles of soil mechanics.

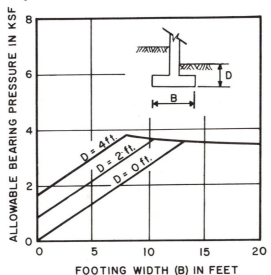

NOTE: Allowable bearing pressure is limited by failure by shear with a safety factor of 3 or by a total settlement of 1 inch.

ALLOWABLE BEARING PRESSURE
SPREAD FOOTINGS

Figure 3-2 Oroville Dam maximum section.

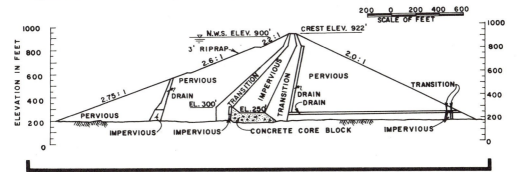

general procedure that is usually applicable in carrying out the soil and foundation work for most projects. This procedure is discussed here in order to give the beginning geotechnical engineering student a general feeling for the processes involved in the applications of soil mechanics. As the student's understanding of the principles of soil mechanics develops, it would be well to review this procedure from time to time. The procedure can be broken down into the following major parts (Worth, 1972).

1. *Define the Project Concept.* This involves establishing the purpose of the project; the schedule for design and construction; the number, type, and location of proposed structures; plans for future expansion of the project; potential future use of the facilities; and special or unusual performance requirements such as vibrating foundations.

1   Soil Properties

Figure 5-1 Pile load test that was conducted to determine allowable pile loads for a bridge foundation. (Courtesy of CH₂M HILL.)

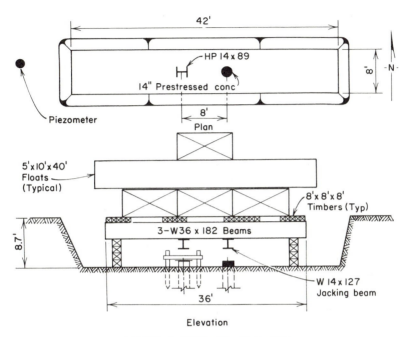

2. *Project Site Reconnaissance*. This involves a review of geologic literature, a review of existing subsurface information, and an inspection of the project site.

3. *Develop a "Working Hypothesis" of the Subsurface Conditions*. The "working hypothesis" (expected subsurface conditions) is developed from the definition of the project concept and on the information gained from the reconnaissance study.

4. *Plan a Field Investigation to Test the "Working Hypothesis."* A "working hypothesis" of subsurface conditions is necessary in order to plan intelligently the various features of a subsurface investigation such as the number and depth of test borings, the number and type of soil samples to be obtained, and the type of field tests to be performed.

Soil Properties   5

It is generally necessary to modify certain features of the subsurface investigation as information from the field activities develops.

5. *Develop a Model for Analysis.* The information obtained from the field investigation will probably reveal soil conditions that vary between borings. These conditions are generally much too complicated to analyze, and it will be necessary to develop an idealized model of the subsurface conditions that can be analyzed using the principles of soil mechanics. It is generally necessary to perform laboratory tests on selected samples from the borings in order to establish the physical parameters required to analyze the model.

6. *Evaluate Alternative Schemes.* The model of the subsurface conditions is used to analyze and evaluate (using the principles of soil mechanics) various alternative design and construction schemes. The evaluation of alternatives may involve comparing the serviceability and cost of various foundation types for a multistory building, comparing various designs of a zoned earth dam, comparing methods of dewatering a large open excavation below the groundwater table, comparing various types of retaining structures, etc.

7. *Make Specific Recommendations.* In making specific recommendations, consideration is generally given to cost, reliability, construction time, and environmental impact.

8. *Prepare Plans and Specifications.* Plans and specifications are prepared that reflect the conditions that were assumed during the evaluation of alternatives.

9. *Construction Inspection and Consultation.* It is important that the project is constructed in accordance with plans and specifications; therefore, detailed inspection of the construction operations is essential. Additionally, construction activities, particularly during excavation for the foundation, often reveal information about the subsurface conditions that were not evident at the time the subsurface investigation was made. If these conditions are radically different, it may be necessary to make design modifications during the construction stage.

10. *Performance Feedback.* All too often a soil and foundation engineer's involvement with a project ends at the time of completion of the foundation unless, of course, problems develop that can be related to foundation conditions. If soil engineers are to develop a knowledge of precedence (as Peck (1962) suggests, this is the most important attribute for one practicing the art of subsurface engineering), then it is very important to observe the long-range performance of all projects, not just those involved in catastrophic failures.

This text, in providing the reader with an understanding of the fundamental principles of soil mechanics, will provide a theoretical basis on which to carry out the procedure outlined above and will specifically introduce the tools to make the required analysis of alternatives in step 6 and to develop the physical parameters necessary to describe the simplified subsurface model of step 5. A soil engineer's ability to carry out a complete project will develop from first-hand experience and from reading the literature and thereby gaining from the experience of others.

A primary responsibility of the soil engineer is to design foundation systems, earth support systems, earth structures, etc., that will perform satisfactorily for the intended life of the structure. As in other areas of structural mechanics, the design must satisfy two specific criteria.

- There must be an adequate factor of safety against a failure by shear.
- Deformations must be within tolerable limits.

In a shear failure, stresses in a soil system exceed the strength of the soil, and this generally results in a collapse of the system. Landslides in natural and artificial slopes and overturned retaining walls are examples of this type of failure. These failures occur as a result of increasing the stresses along the failure plane or of decreasing the strength of the soil along the failure plane. Stresses can be increased from an external load or from a change in the stress distribution by some means, such as making a highway cut at the base of a natural slope. The strength of soil is often decreased during earthquakes, when the soil is subjected to a cyclic loading condition.

Excessive deformation of a foundation system can make a structure unusable. The amount of settlement that can be considered tolerable depends on the function of the structure. Undesirable deformations are caused by both expansion and compression of soil. Certain clay soils expand if the water content of the soil increases, and this can cause foundations and retaining walls to deform excessively. Many soils are very frost susceptible and expand during freezing temperatures, causing damage to highways, building foundations, retaining walls, and other structures. Sufficient measures must, therefore, be taken during design to prevent damage from expanding soils. If a foundation is not adequately designed, excessive settlement of the structure may occur as a result of compression of the underlying soil. The compression can be caused by the weight of the structure, by lowering the level of the groundwater table, or by vibrations from machines and earthquakes.

SOIL-FORMING PROCESS AND HISTORY

The definition of soil that is used by a civil engineer is rather arbitrary and is somewhat different from that used by a geologist, a soil scientist, or a lay person.

A civil engineer considers soil to include all the material, organic and inorganic, overlying bedrock. In interpreting and using the work of other disciplines, the engineer must keep in mind that there are many basic differences in the terminology and definitions used to classify and describe the physical and chemical behavior of soil. The terminology and definitions used in this text are those that are common to the engineering profession.

Based on origin, soils can be broadly classified as organic or inorganic. Organic soils are mixtures in which a significant part is derived from the growth and decay of plant life and in some instances from the accumulation of skeletons or shells of small organisms. Inorganic soils are derived from either chemical or mechanical weathering of rocks.

Inorganic soil that is still located at the place where it was formed is referred to as a residual soil. If the soil has been moved to another location by gravity, water, or wind, it is referred to as a transported soil.

A knowledge of the history of a given soil deposit can, in a general way, reveal much about its engineering properties. The engineering properties are essentially a function of the physical and chemical characteristics of the parent material, the type of weathering that formed the soil, whether the deposit is a residual or transported soil, the method of movement and deposition for a transported soil, the stress history of the soil deposit, the chemical history of the pore water, and the history of the position of the water table. Although detailed sampling and testing is required for a proper evaluation of the engineering properties of a soil, much information can be obtained from a knowledge of the soil type and its history. The following section gives examples of the general engineering characteristics of some soils.

GENERAL ENGINEERING CHARACTERISTICS OF SOME SOIL TYPES

The list of soil types and their general engineering characteristics as presented in this section is not an all-inclusive list. It is intended to show that, with some experience, a knowledge of the soil type and its history can provide preliminary information regarding the soil's general engineering characteristics. A description of the soil types in a given area is often available for many projects and can be obtained from geologic maps, soil maps, and reports of subsurface investigation made at adjacent sites. This kind of information should be considered preliminary. It is used by soil and foundation engineers only for planning subsurface investigations and laboratory testing programs and during feasibility studies to anticipate foundation requirements.

- Loose sand is simply a sand deposit with a low density. Vibratory loads tend to densify these deposits. Therefore, special measures must be taken in designing foundations for buildings that are to house machines, because the vibrations from operating the machine may induce intolerable settlements. Loose sands also present problems in high seismic risk areas, because seismic loading can induce liquefaction if the sand is saturated, as well as cause significant settlement.

- Loess is a deposit of relatively uniform, windblown silt. It has a relatively high vertical permeability but low horizontal permeability. Loess soils become very compressible when saturated. This often causes problems with hydraulic structures such as canals and earth dams that are constructed on loess foundations. Special design measures are necessary for these cases and may require prewetting the foundation to induce settlement before starting construction.

- Normally consolidated clays are clay soils that have never been subjected to a pressure greater than the existing pressure. These soils generally tend to be highly compressible, have a low ultimate bearing capacity, and, as with all clay soils, have a very low permeability. Because of the low ultimate bearing capacity and high compressibility, the soil is often not capable of supporting structures on shallow foundations. In these cases, other foundation types must be considered, such as pile foundations where the structural loads are transferred to a lower soil or rock strata that has a higher bearing capacity, or a floating foundation where the quantity of soil excavated for basement levels will equal the weight of the structure.

- Overconsolidated clays are clays that in the past have been subjected to a pressure greater than the existing pressure. Highly overconsolidated clays generally tend to have a rather high ultimate bearing capacity and are relatively incompressible.

- Bentonite is a highly plastic clay resulting from the decomposition of volcanic ash. It is an expansive soil that swells considerably when saturated. This can cause problems in the performance of foundations, sidewalks, concrete slabs, and other structural elements if the soil is subjected to seasonal changes in water content. On the other hand, bentonite is often used beneficially as an impermeable pond liner.

- Peat is fibrous, partly decomposed organic matter or a soil containing large amounts of fibrous organic matter. Peats have a very high void ratio and are extremely compressible. Su and Prysock (1972) reported that the settlement of an embankment 2.68 m high and underlain by 8.24 m of peat and 12.4 m of peaty clay was 2.13 m in 13 years. The ultimate settlement of the embankment was predicted to be 2.59 m after 25 years.

PARTICLE SIZE

The size of soil particles range from boulders more than 1 m in diameter to clay-size particles less than 0.001 mm in diameter. The table of Fig. 10-1 shows

Figure 10-1   Soil type based on particle size.

Soil Type	Particle-Size Range
Boulder	>0.3 m
Cobble	0.15–0.3 m
Gravel	2.0 mm–0.15 m
Sand	0.075–2.0 mm
Silt	0.002–0.075 mm
Clay	<0.002 mm

the common soil types (based on grain size) and their approximate particle-size range. In general, the principles of soil mechanics that are developed in this text apply to soils with particle sizes ranging from clay size to gravel.

PARTICLE FORCES AND BEHAVIOR

The behavior of individual soil particles and their interaction with other particles is influenced by body forces. Two body forces are generally associated with soil particles.

- Weight of the particle F_g.
- Particle surface forces F_s.

Weight is the result of gravitational forces and is a function of the volume of the particle. For equidimensional particles such as spheres of diameter D, the weight, F_g, is directly proportional to D^3. Particle surface forces are of an electrical nature. They are caused by unsatisfied electrical charges in the particle's crystalline structure. Surface forces, F_s, are directly proportional to the surface area and, hence, for equidimensional particles, to D^2.

The ratio of the weight of a particle to the particle's surface forces, F_g/F_s, is directly proportional to D. Thus, for large particle sizes, which include soil particles in the coarse fraction (> 0.075 mm), the weight of the particle is predominant over the surface forces. As the particle diameter decreases, the ratio, F_g/F_s, decreases; thus, for very small values of D, the surface forces predominate. This accounts for the cohesive nature of most fine-grained soils.

Characteristics of the Fine-Soil Fraction

Surface forces play a significant role in the behavior of clay soils and some silts. The crystalline structure of clays forms thin platy-shaped particles that carry a net negative charge on the flat surfaces of the particles. This charged surface attracts the cations that are in the pore water and even tends to orient water molecules into a somewhat structured arrangement. A basic understanding of clay particles and the aqueous solution that surrounds the particles helps in interpreting the engineering behavior of clays. The nature of clay particles can be illustrated by describing the three most common subgroups of the clay minerals: kaolinites, illites, and montmorillonites.

There are two basic crystalline units that form the clay minerals, and the manner in which they are combined differentiates between the three clay mineral subgroups mentioned above. One crystalline unit is the silicon-oxygen tetrahedron, as shown in Fig. 11-1a. The tetrahedrons combine to form the silica

Figure 11-1 Basic crystalline units that form the clay minerals. (*a*) Silicon-oxygen tetrahedron. (*b*) Silica Sheet. (*c*) Aluminium octahedron. (*d*) Gibbsite sheet. (After Lambe and Whitman.)

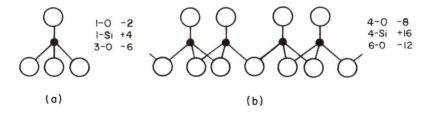

```
1-O   -2
1-Si  +4
3-O   -6
```

(a)

```
4-O   -8
4-Si  +16
6-O   -12
```

(b)

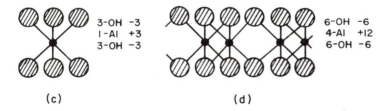

```
3-OH  -3
1-Al  +3
3-OH  -3
```

(c)

```
6-OH  -6
4-Al  +12
6-OH  -6
```

(d)

sheet, as shown in Fig. 11-1b. The second crystalline unit is the aluminium octahedron, Fig. 11-1c. Combining aluminium octahedrons forms the gibbsite sheet shown in Fig. 11-1d. The arrangement of the molecules in the silica and gibbsite sheets allows these sheets to fit together very closely. Relatively strong bonds hold the sheets in place. Two-layer and three-layer structures of silica and gibbsite sheets are shown in Figs. 12-1 and 13-1, respectively.

The clay mineral kaolinite is a basic two-layer unit (about 7.2 Å thick and indefinite in the other direction) formed by stacking a gibbsite sheet on a silica sheet, as shown in Fig. 12-1a and as shown symbolically in Fig. 12-1b. The actual mineral is formed by a number of these two-layer units stacked one on top of the other. The linkage between the units consists of hydrogen bonding and secondary valence forces.

Montmorillonite is a basic three-layer unit (about 9.5 Å thick and indefinite in the other directions) formed by placing one silica sheet on the top and one on the bottom of a gibbsite sheet, as shown symbolically in Fig. 13-1. Isomorphous

Figure 12-1 The structure of kaolinite. (a) Atomic structure. (b) Symbolic structure. (After Lambe and Whitman.)

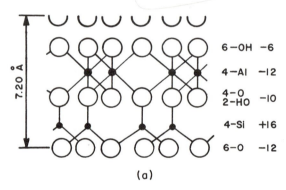

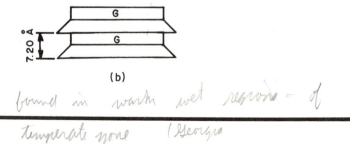

Figure 13-1 Symbolic structure of montmorillonite.

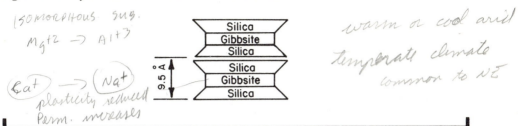

(handwritten notes, left:) ISOMORPHOUS SUB.
$Mg^{+2} \rightarrow Al^{+3}$

$Ca^+ \rightarrow (Na^+)$
plasticity reduced
Perm. increases

(center label:) Silica / Gibbsite / Silica / Silica / Gibbsite / Silica, 9.5 Å

(handwritten notes, right:) warm or cool arid
temperate climate
common to NE

substitution (replacement) of magnesium or iron for the aluminum in the gibbsite sheet is common and somewhat changes the characteristics of montmorillonite. A small amount of isomorphous substitution of aluminum for silicon in the silica sheet may also occur. The bonding between successive three-layer montmorillonite units is by secondary valence forces and exchangeable ion linkage, but it is very weak and water may enter between the units, causing the mineral to swell. Therefore, montmorillonite clays are very expansive, and this function must be considered in designing foundations.

Illite is a very stable three-layer unit that consists of the basic montmorillonite units bonded by secondary valence forces and potassium ions, as shown symbolically in Fig. 13-2. There is always substantial (about 20%) isomorphous substitution of aluminum for silicon in the silica sheet of illite. The mineral does not swell by the introduction of water between sheets, as does montmorillonite.

The above description of the three common clay minerals is very brief and simplified. For a more in-depth study of clay mineralogy, refer to texts that are devoted entirely to this subject, such as Grim (1953).

Figure 13-2 Symbolic structure of illite.

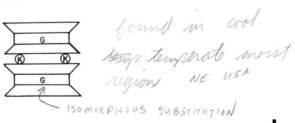

(center diagram labels:) G / K K / G

(handwritten notes, right:) found in cool
temp temperate moist
region NE USA

ISOMORPHOUS SUBSTITUTION

(handwritten notes, bottom:) periodically in the tetrahedron sheet
make it electrically deficient.
Potassium cations are attracted
to this structure.
$Al^{+3} \rightarrow Si^{+4}$

Individual clay particles consist of many basic clay mineral units stacked on one another. Because these basic units have a sheeted structure, the particles end up with a platy shape. Their surface dimensions are many times the particle thickness. The general shape of particles and the associated surface charge are shown in Fig. 14-1.

Most of the undisturbed clay deposits and the earth embankments of clay soil that are of concern to engineers are partially or fully saturated. Furthermore, the pore water generally contains dissolved cations that can interact with the negatively charged surface of the clay particles. The behavior of clay soils is, therefore, greatly affected by the nature of the individual clay particles and pore water.

Since the surfaces of the clay particles are negatively charged, they tend to adsorb (attract) the positively charged cations that are present in the pore water. The water molecules adjacent to the negatively charged surface may also undergo alteration and become structured. This water is also considered

Figure 14-1 Negatively charged clay particle and surrounding aqueous solution.

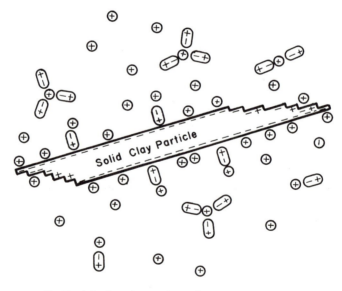

⊕ Positively charged cation

⊕⊝ Polar water molecule

— Net negative charge on the particle surface

adsorbed to the surface of the clay particle. The relationship between a negatively charged mineral particle and the associated aqueous solution that surrounds the particle is shown as a schematic sketch of a colloidal micella in Fig. 14-1.

The cations distribute themselves around the negatively charged surface of the clay particles with the greatest density near the surface and a decreasing density with increasing distance from the surface. The cations form a positively charged zone or layer which, together with a negatively charged surface of the clay particle, makes up the electric double layer. The generally accepted concept of the double layer was presented by Stern (1924) and is shown in Fig. 15-1.

Figure 15-1   Stern double layer.

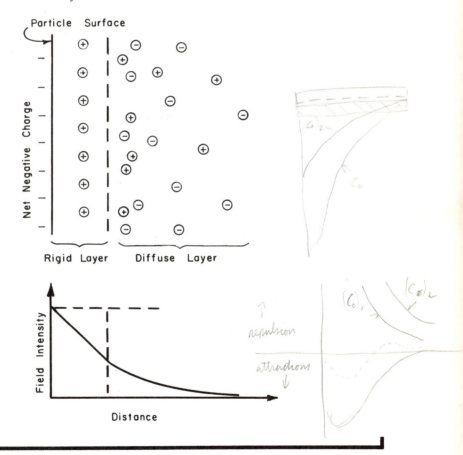

The nature of the electric double layer affects the structure of the aggregates of clay particles and, hence, the physical properties of the soil.

The cations are not permanently attached to the surface of the clay particles and can be replaced by other cations. This process of replacement is known as cation exchange. There is a known hierarchy of replaceable cations. For example, potassium ions (K+) tend to replace sodium ions (Na+). Thus, if a sodium soil is leached with a solution of potassium chloride (KCl), most of the Na+ ions will be replaced by K+ ions. This process of leaching will change the characteristics of the double layer and, hence, the physical properties of the soil.

Characteristics of the Coarse-Soil Fraction

The ratio of the volume to surface area for coarse-grained soils is large enough that electrical surface forces are negligible. Additionally, the pressures encountered in most soil mechanics applications are low enough that individual particles do not fracture. Hence, the aggregate physical properties of the coarse fraction (sand and gravel) are essentially functions of relative density (closeness of the particles) and particle shape (angular, rounded, etc.) and are relatively unaffected by the mineral composition of the soil.

WEIGHT-VOLUME RELATIONSHIPS

Figure 17-1 is a schematic representation of an element of soil, showing it as a three-phase system. An aggregate of solid particles, which are assumed to be incompressible under the loads normally encountered in engineering practice, makes up the solid phase of the system. The void spaces between the particles are filled with either liquid or gas or both and make up the remaining two phases.

The relationships defined in Fig. 17-1 between the volumes and weights of each phase are very useful in describing and evaluating the physical properties of soil. It is important that students of soil mechanics become familiar with these relationships and make them part of their technical vocabulary.

The total volume of the element of soil schematically represented in Fig. 17-1 is the sum of the volume of the voids (V_v) and the volume of the solids (V_s). The volume of the voids is the sum of the volume of gas (usually air) (V_a) and the volume of the liquid (usually water) (V_w). There are three useful relationships between the volumes of each phase.

- Void ratio (e) is the ratio between the volume of the voids and the volume of the solids; it is always expressed as a decimal. Void ratio is used extensively in soil mechanics to express various physical parameters as a function of the soil's density.

Figure 17-1 Schematic representation of an element of soil showing symbols, definitions, and useful weight-volume relationships.

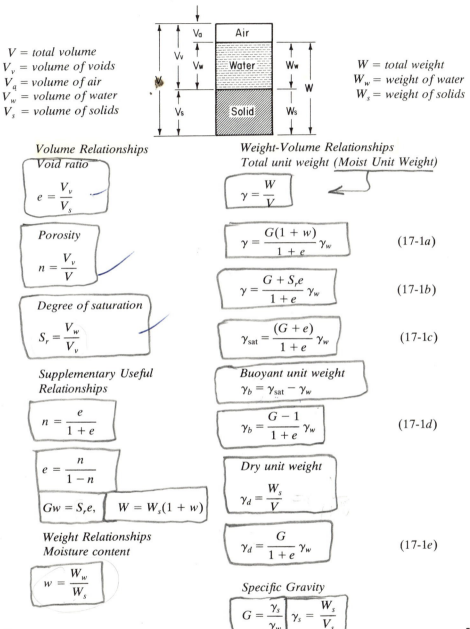

V = total volume
V_v = volume of voids
V_a = volume of air
V_w = volume of water
V_s = volume of solids

W = total weight
W_w = weight of water
W_s = weight of solids

Volume Relationships

Void ratio

$$e = \frac{V_v}{V_s}$$

Porosity

$$n = \frac{V_v}{V}$$

Degree of saturation

$$S_r = \frac{V_w}{V_v}$$

Supplementary Useful Relationships

$$n = \frac{e}{1+e}$$

$$e = \frac{n}{1-n}$$

$$Gw = S_r e, \quad W = W_s(1+w)$$

Weight Relationships
Moisture content

$$w = \frac{W_w}{W_s}$$

Weight-Volume Relationships
Total unit weight (Moist Unit Weight)

$$\gamma = \frac{W}{V}$$

$$\gamma = \frac{G(1+w)}{1+e}\gamma_w \qquad (17\text{-}1a)$$

$$\gamma = \frac{G+S_r e}{1+e}\gamma_w \qquad (17\text{-}1b)$$

$$\gamma_{\text{sat}} = \frac{(G+e)}{1+e}\gamma_w \qquad (17\text{-}1c)$$

Buoyant unit weight
$$\gamma_b = \gamma_{\text{sat}} - \gamma_w$$

$$\gamma_b = \frac{G-1}{1+e}\gamma_w \qquad (17\text{-}1d)$$

Dry unit weight

$$\gamma_d = \frac{W_s}{V}$$

$$\gamma_d = \frac{G}{1+e}\gamma_w \qquad (17\text{-}1e)$$

Specific Gravity

$$G = \frac{\gamma_s}{\gamma_w} \quad \gamma_s = \frac{W_s}{V_s}$$

17

Figure 18-1 Porosity, void ratio, and unit weight of typical soils in natural state. (After Peck, Hanson, and Thornburn, 1974.)

Description	Porosity (n)	Void Ratio (e)	Water Content $(w)^a$	Unit Weight			
				kN/m³		lb/cu ft	
				γ_d	γ_{sat}	γ_d	γ_{sat}
1. Uniform sand, loose	0.46	0.85	32	14.1	18.5	90	118
2. Uniform sand, dense	0.34	0.51	19	17.1	20.4	109	130
3. Mixed-grained sand, loose	0.40	0.67	25	15.6	19.5	99	124
4. Mixed-grained sand, dense	0.30	0.43	16	18.2	21.2	116	135
5. Windblown silt (loess)	0.50	0.99	21	13.4	18.2	85	116
6. Glacial till, very mixed-grained	0.20	0.25	9	20.7	22.8	132	145
7. Soft glacial clay	0.55	1.2	45	11.9	17.3	76	110
8. Stiff glacial clay	0.37	0.6	22	16.7	20.3	106	129
9. Soft slightly organic clay	0.66	1.9	70	9.1	15.4	58	98
10. Soft very organic clay	0.75	3.0	110	6.8	14.0	43	89
11. Soft montmorillonitic clay (calcium bentonite)	0.84	5.2	194	4.2	12.6	27	80

$^a w$ = water content when saturated, in percent of dry weight.

- Porosity (n) is the ratio of the volume of the voids to the total volume; it is expressed either as a decimal or as a percentage. Although porosity is used widely by some disciplines as a means of expressing the void volume, void ratio is used more commonly in soil mechanics.
- Degree of saturation (S_r) is the ratio of the volume of the water to the volume of the voids; it is expressed either as a decimal or as a percentage.

The total weight (W) of the soil element is the sum of the dry weight of the solids (W_s) plus the weight of the liquid phase (W_w). The gaseous phase is assumed to be weightless. Water or moisture content (w) is defined as the ratio of the weight of the water to the dry weight of the solids (W_s); it is usually expressed as a percentage. Some typical values of natural moisture content (for saturated soils) are given in the table of Fig. 18-1.

The unit weight of any substance (weight per unit volume) is the weight of the substance divided by its volume. In soil mechanics the unit weight of a soil mass can be described in terms of total unit weight (γ), dry unit weight (γ_d), and buoyant or submerged unit weight (γ_b). Total unit weight is defined as the total weight divided by the total volume, and dry unit weight is the weight of the solids (dry weight) divided by the total volume. Some typical values for total unit weight and dry unit weight are given in Fig. 18-1. When the soil is below the water table, it is often convenient to use its buoyant unit weight (γ_b), which is equal to the saturated unit weight minus the unit weight of water.

Specific gravity is the ratio of the unit weight of a substance to the unit weight of water (γ_w) at 4°C. In soil mechanics, specific gravity generally refers to

Figure 19-1 Specific gravity of some important soil constituents. (After Peck, Hanson, and Thornburn, 1974.)

Gypsum	2.32	Dolomite	2.87
Montmorillonite	2.65–2.80	Aragonite	2.94
Orthoclase	2.56	Biotite	3.0–3.1
Káolinite	2.6	Augite	3.2–3.4
Illite	2.8	Hornblende	3.2–3.5
Chlorite	2.6–3.0	Limonite	3.8
Quartz	2.66	Hematite, hydrous	4.3±
Talc	2.7	Magnetite	5.17
Calcite	2.72	Hematite	5.2
Muscovite	2.8–2.9		

the specific gravity of the solid particles (G), and is defined as the ratio of the unit weight of the solid particles to the unit weight of water $W_s / \gamma_w V_s$.

The value of the specific gravity can be determined from laboratory tests (ASTM D-854). Typical specific gravity values for some of the most important soil constituents are given in Fig. 19-1.

Unit weight can also be defined in terms of void ratio, specific gravity, the unit weight of water, and either moisture content or degree of saturation. These definitions are given as Eq. 17-1a to 17-1e.

The examples shown in Figs. 20-1 and 21-1 illustrate how these weight-volume relationships are often used.

Figure 20-1

Example Problem A sample of saturated clay from a consolidometer has a total mass of 1526 g and a dry mass of 1053 g. The specific gravity of the solid particles is 2.7. For this sample determine the water content, void ratio, porosity, and total unit weight.

Solution These quantities can be calculated using the relationships given on Fig. 17-1.

Water content

$$w = \frac{W_w}{W_s} = \frac{M_w}{M_s} = \frac{1526 - 1053}{1053}$$

M_w = mass of water

M_s = mass of dry soil

$w = 0.449$ or 44.9%

Void ratio

$$Gw = S_r e$$

$$e = \frac{wG}{S_r} = \frac{0.45(2.7)}{1}$$

$$e = 1.212$$

Porosity

$$n = \frac{e}{1 + e} = \frac{1.212}{1 + 1.212}$$

$$n = 0.548 \text{ or } 54.8\%$$

Total unit weight

$$\gamma = \left(\frac{G + S_r e}{1 + e} \right) \gamma_w = \frac{2.7 + 1\,(1.212)}{1 + 1.212} (9.81)$$

$$\gamma = 17.33 \text{ kN/m}^3$$

1   Soil Properties

Figure 21-1

Example problem A laboratory sample of silty clay has a volume of 14.88 cm³, a total mass of 28.81 g, a dry mass of 24.83 g, and a specific gravity of 2.7. For this sample determine the void ratio and the degree of saturation.

Solution These quantities can be calculated from the relationships given on Fig. 17-1.

Void ratio

$$e = \frac{V_v}{V_s}$$

$$V_s = \frac{W_s}{G\gamma_w} = \frac{24.83}{2.7\,(1)}$$

$$V_s = 9.2 \text{ cm}^3$$

$$V_v = V - V_s = 14.88 - 9.2$$

$$V_v = 5.68 \text{ cm}^3$$

$$e = \frac{5.68}{9.2} = 0.618$$

Degree of saturation

$$Gw = S_r e$$

$$w = \frac{W_w}{W_s} = \frac{28.81 - 24.83}{24.83} = 0.16$$

$$S_r = \frac{Gw}{e} = \frac{2.7\,(0.16)}{0.618} = 0.70$$

or $\quad S_r = 70\%$

Alternate method

$$\gamma_d = \frac{G}{1+e}\gamma_w = \frac{W_s}{V}$$

$$e = \frac{G\gamma_w V}{W_s} - 1$$

$$e = \frac{2.7\,(1)\,(14.88)}{24.83} - 1$$

$$e = 0.618$$

Figure 22-1 Relative density (granular soil only).

$$D_r = \frac{e_{max} - e}{e_{max} - e_{min}} \times 100\% \qquad (22\text{-}1a)$$

$$D_r = \frac{\gamma_{d\,max}}{\gamma_d} \times \frac{\gamma_d - \gamma_{d\,min}}{\gamma_{d\,max} - \gamma_{d\,min}} \times 100\% \qquad (22\text{-}1b)$$

e_{max} = maximum void ratio

e_{min} = minimum void ratio

e = void ratio of soil deposit

$\gamma_{d\,min}$ = minimum dry unit weight

$\gamma_{d\,max}$ = maximum dry unit weight

γ_d = dry unit weight of soil deposit

The unit weight of a given soil deposit can generally be related to certain engineering properties such as strength and compressibility. For granular soils, an especially useful relationship is relative density (D_r), which is defined by Eqs. 22-1a and 22-1b.

Since dry unit weight is much easier to measure than void ratio, the second relationship is generally used to calculate relative density. The maximum and minimum dry unit weights are determined in the laboratory using a standard procedure (ASTM D-2049). The minimum dry unit weight is determined by pouring dry soil (granular) into a container from a standard height without compaction. The maximum dry unit weight is determined by placing soil in a container and then densifying it on a vibrating table. A surcharge weight is placed on the soil while it is being vibrated. Samples are densified dry and fully saturated, and the greatest dry unit weight from the two methods is used in computing the relative density. The field value of γ_d for use in Eq. 22-1b is found by excavating a volume of soil (the dry weight is measured in the laboratory) and measuring the volume of the hole, using water in a rubber balloon or a premeasured volume of dry sand. Nuclear gauges are now being widely used to measure the dry unit weight in the field. The nuclear gauge has the advantage of providing an immediate evaluation of dry unit weight.

METHODS OF MECHANICAL ANALYSIS

The process by which soil particles are separated into the soil types as defined by various ranges of particle size (Fig. 10-1) is called mechanical analysis. Soil particles larger than approximately 0.075 mm (coarse fraction) are generally separated into particle-size ranges using a nest of sieves. A sample of soil is

Figure 23-1 Grain-size distribution curves for three soils.

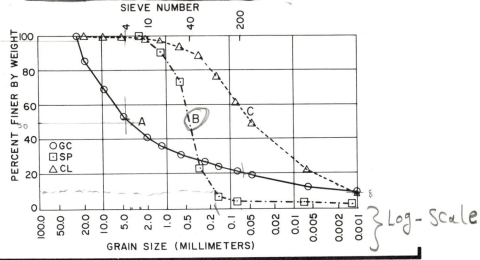

placed on the top sieve (largest openings are on the top sieve with progressively smaller openings on lower sieves), and the nest of sieves is vibrated to allow the individual particles to fall through the sieve openings. The grain-size distribution (particle-size distribution) of the soil is then determined by measuring the dry weight of material retained on each sieve. The results of the analysis are then plotted on semilog paper, as shown in Fig. 23-1.

Standard sieves are made from a woven wire fabric. Specifications have been established to control the size of the wire and the maximum permissible deviation from the average opening. The sieve opening is the distance between parallel wires of the fabric. U.S. Standard sieve-size numbers and their corresponding openings are given in Fig. 24-1 for several common sieve sizes. A complete list of sieve sizes is given in the ASTM Standards (E-11).

The procedure of using sieves to determine the grain-size distribution is referred to as dry mechanical analysis (ASTM D-422). Most soil particles of silt size or larger are usually rounded or angular and nearly equidimensional in shape; therefore, a sieve analysis gives a reasonably accurate measure of particle size. An exception is silt-size particles of mica, which are very flat and platy shaped. The results of a sieve analysis on this material must be used with caution.

The size of particles less than 0.075 mm (fine fraction) is generally determined by using a wet mechanical analysis procedure. In this method, a suspension of soil and water is prepared, and the relative size of particles is determined by measuring the time required for the particles to fall out of

Methods of Mechanical Analysis   **23**

Figure 24-1   Selected sieve sizes.

U.S. Standard Sieve Number	Opening (mm)
4	4.75
10	2.00
20	0.850
40	0.425
60	0.250
80	0.180
100	0.150
140	0.106
200	0.075

suspension. Stokes' law relates the terminal velocity of a sphere falling through a liquid to the square of its diameter and can be used to obtain the grain-size distribution. The time required for the particles to fall out of suspension can be measured by using a hydrometer to determine the specific gravity of the suspension at a given point. As all the particles of a certain size fall past the point, the specific gravity of that point decreases, and the value of the specific gravity of the suspension then becomes a measure of the grain-size distribution of the material. The development of this method is given in Fig. 24-2.

There are several inherent errors in using Stokes' law to determine the grain-size distribution of fine-grained soils.

Figure 24-2

Derivation   Develop the relationships necessary to determine the grain-size distribution of a fine grained soil by the hydrometer method.

Solution   A sample of soil of dry weight W_s and specific gravity G is placed into a cylinder of water. The soil–water solution is thoroughly mixed so that at the beginning of the test the solution has the same particle-size distribution at all points. After the solution is mixed, the cylinder is placed down and the particles begin to fall out of suspension.

The initial unit weight of the suspension (γ_i) can be expressed in terms of the dry weight of the soil (W_s), the weight of water (W_w) and the total volume (V) as

$$\gamma_i = \frac{W_s + W_w}{V}$$

This expression can also be stated in terms of the specific gravity of the solid particles G and the unit weight of water as

$$\gamma_i = \gamma_w + \left(\frac{G-1}{G}\right)\frac{W_s}{V}$$

Now consider a point at depth L in the soil–water solution and let T designate the time elapsed since the start of sedimentation. The size of particle D that would fall a distance L in time T can be computed from Stoke's law.

$$v = \frac{L}{T} = CD^2$$

where

$$v = \text{terminal velocity}$$
$$T = \text{elapsed time}$$
$$L = \text{distance}$$
$$C = \text{known constant}$$

Solving Stoke's law for D yields

$$D = \sqrt{\frac{L}{TC}} \qquad\qquad (25\text{-}1a)$$

At depth L and time T there will be no particles larger than D, and all particles smaller than D will be at the initial concentration. Let P equal the ratio of the dry weight of particles smaller than D to the total dry weight of the soil sample. The weight of particles per unit volume at depth L and time T then becomes

$$\frac{PW_s}{V}$$

and the unit weight of the suspension at this depth and time can be expressed in a manner similar to the initial unit weight as

$$\gamma = \gamma_w + \left(\frac{G-1}{G}\right)\left(\frac{PW_s}{V}\right)$$

Solving this expression for P yields

$$P = \frac{G}{(G-1)}\left(\frac{V}{W_s}\right)(\gamma - \gamma_w) \qquad\qquad (25\text{-}1b)$$

The ratio P represents the fraction of particles less than diameter D as given in Eq. 25-1a. Therefore, by measuring the unit weight γ of a soil water solution at a known depth L and at time T, Eqs. 25-1a and 25-1b can be used to furnish a point on the grain size distribution curve. A common method to determine γ is by using a hydrometer, and a specific test procedure is provided in ASTM D-422.

- The particles are never truly spherical. In fact, the shapes may bear little resemblance to spheres.
- The body of water is not indefinite in extent, and since many particles are present, the fall of any particle is influenced by the presence of other particles; similarly, particles near the side walls of the container are affected by the presence of the wall.
- The average value for specific gravity of grains is used; the values for some particles may differ appreciably from the average value. (Taylor, 1948)

The standard testing procedures that have been developed to determine the grain-size distribution by a sedimentation method, such as the hydrometer method (ASTM D-422), are such that the second and third errors listed above are minimized. The first item cannot be overcome, however, and the concept of an equivalent diameter must be accepted.

The grain-size distribution curves as determined from the wet and dry methods can be combined to produce the complete grain-size distribution for a soil that has both significant coarse and fine factions. However, in doing so, it must be realized that there is a difference in the definition of particle size as measured by the two methods. In a sieve analysis, a long cylindrical particle of diameter D would fit through the same sieve opening as a spherical particle of diameter D. These two particle shapes are very different, but their particle size as determined by a sieve analysis would be the same. Particle size as measured in a sedimentation method is an equivalent diameter equal to the diameter of a perfect sphere that would fall out of suspension at the same rate as the soil particle.

PARTICLE-SIZE GRADATION

The effective particle size of a given soil is defined as a particle size for which 10% of the material by weight is less than that size. Curve B in Fig. 23-1 has an effective particle size (D_{10}) equal to 0.17 mm. Other particle sizes are also frequently used in describing or classifying soils. The D_{50} of a soil is the median particle size. The D_{85} and D_{15} sizes are used to design filters for drainage systems in earth dams and other structures.

A soil that has a nearly vertical grain-size distribution curve (all particles of nearly the same size) is called a uniform soil. If the curve extends over a rather large range, the soil is called well graded. The distinction between a uniform and a well-graded soil can be defined numerically by the uniformity coefficient C_u and the coefficient of curvature C_z. The uniformity coefficient is defined by the ratio

$$C_u = \frac{D_{60}}{D_{10}}$$

1 Soil Properties

The coefficient of curvature is defined as

$$C_z = \frac{D_{30}^2}{D_{10}D_{60}}$$

Soils with C_u less than 4 are said to be uniform, and soils with C_u greater than 4 (6 for sands) are well graded provided that the grain-size distribution curve is smooth and reasonably symmetrical. The coefficient of curvature C_z is a measure of symmetry and shape of the gradation curve. For a well-graded soil, C_z will be between 1 and 3.

The soil of curve B in Fig. 23-1 has a uniformity coefficient of 2.8 and is, therefore, a uniform soil.

The uniformity coefficient and the coefficient of curvature are used as part of the Unified soil classification system shown in Fig. 37-1.

CONSISTENCY AND PLASTICITY
OF FINE-GRAINED SOILS

Remolded Soils

The consistency of clays and other cohesive soils is greatly influenced by the water content of the soil. If a clay slurry is slowly dried it will pass from a liquid state to a plastic state and finally into semisolid and solid states. The water content at which a soil passes from one state to another is different for different soils and can be used, in a qualitative way, to distinguish between, or classify, different fine-grained soil types.

Figure 28-1 shows a general relationship between volume and water content of a remolded soil. Superimposed in Fig. 28-1 are limits that divide the curve into four consistency states. The transition from one state to another occurs over a range of water contents, and the limits are, therefore, of an arbitrary nature. These limits were developed by Atterberg (1911) and are known as the Atterberg limits.

At very high water contents, the soil behaves as a viscous liquid in that it flows and will not hold a specific shape. The lowest water content at which the soil is in a liquid state is called the liquid limit (LL), and a specific test procedure (ASTM D-423) has been developed to determine this water content. The test is performed by placing a soil pat in a cup and making a standard-sized groove in the pat (Fig. 29-1). The cup is then dropped onto a hard surface from a height of 10 mm. The liquid limit is defined as the water content when the groove closes over 12.7 mm ($\frac{1}{2}$ in.) at 25 blows.

Soil is considered to be in a plastic state when it can be molded or worked into a new shape without crumbling. The lowest water content at which a soil is considered to be in a plastic state is the plastic limit (PL) of the soil. The plastic

Figure 28-1 States of soil consistency.

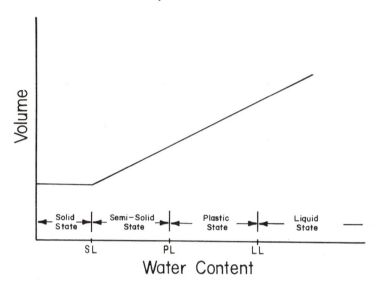

limit is determined by rolling a pat of soil into a thread. When the thread begins to crumble at a diameter of 3.18 mm ($\frac{1}{8}$ in.), the water content is the plastic limit (ASTM D-424).

Figure 28-1 shows that at some water content the soil will maintain a constant volume even if the water content decreases. This water content is defined as the shrinkage limit (SL), and a standard test procedure is used to determine its value (ASTM D-427).

The difference between the liquid limit and the plastic limit is the plasticity index (PI) and represents the range of water content over which the soil behaves in a plastic state.

Casagrande (1948) observed that many properties of clays and silts can be correlated with the Atterberg limits by means of the plasticity chart. The chart in Fig. 29-2 is divided into regions. By plotting the plasticity index and liquid limit of the soil, the soil may be classified. Inorganic clays lie above the *A* line, and classification symbols in Fig. 29-2 are defined on page 37. The liquid limit of the inorganic silts and organic silts and clays lie below the *A* line. The Unified soil can then be used to further classify the soil as having low plasticity or high plasticity. Organic soils are generally dark gray or black and have a distinctive odor.

28   **1**   Soil Properties

Figure 29-1   Liquid limit device.

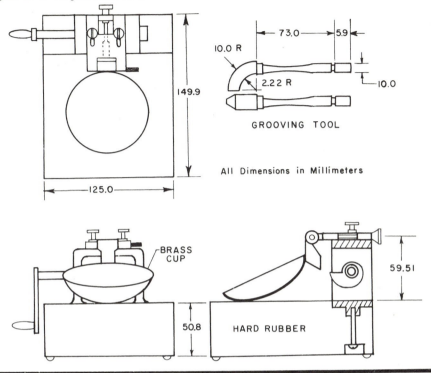

GROOVING TOOL

All Dimensions in Millimeters

Figure 29-2   Plasticity chart.

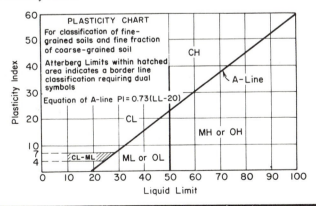

PLASTICITY CHART

For classification of fine-grained soils and fine fraction of coarse-grained soil

Atterberg Limits within hatched area indicates a border line classification requiring dual symbols

Equation of A-line $PI = 0.73(LL-20)$

29

Undisturbed Soils

Remolding changes the consistency of most undisturbed cohesive soils. The strength of a cylindrical soil sample in simple compression is referred to as the unconfined compressive strength and is often used to describe the consistency of undisturbed cohesive soils. Values of unconfined compressive strength are correlated with the consistency of the soil in Fig. 30-1.

The strength and other physical characteristics of a given fine-grained soil are functions of the soil's structure and the nature of the negatively charged mineral particle and the associated aqueous solution that surrounds the particle. Structure refers to the orientation and distribution of the individual particles. There are two extremes of soil structure, as shown in Fig. 31-1: flocculated structure and dispersed structure. In a flocculated structure the particles are edge to face and tend to attract each other; in a dispersed structure the particles are face to face (parallel) and tend to repel each other. In an actual soil deposit the soil structure is likely to be somewhere intermediate between flocculated and dispersed. In general, an element of flocculated soil will have a higher shear strength, a lower compressibility, and a higher permeability than the same element of soil at the same void ratio, but with a dispersed structure.

The structure of a specific soil deposit depends on how the deposit was originally formed, the characteristics of the pore water during the deposit's formation, and the history of the deposit since its formation. Lambe (1958) suggests that the soil structure of a compacted clay can tend to be essentially flocculated or dispersed, depending on the compactive effort and the remolding water content of the soil.

Figure 30-1 Consistency of soil in terms of unconfined compressive strength.

Consistency	Unconfined Compressive Strength, q_u	
	kN/m^2	kip/ft^2
Very soft	<25	<0.5
Soft	25–50	0.5–1
Medium	50–100	1–2
Stiff	100–200	2–4
Very stiff	200–400	4–8
Hard	Over 400	>8

Figure 31-1 Extremes of soil structure.

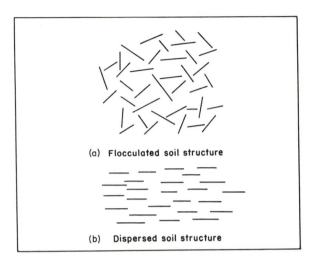

(a) **Flocculated soil structure**

(b) **Dispersed soil structure**

Remolding a cohesive soil generally causes a loss of strength. The loss of strength is probably caused by the following two conditions.

- Rearranging the particles toward a dispersed structure.
- Disturbing the chemical equilibrium of the particles and associated adsorbed ions and water molecules within the double layer.

The rearrangement of particles (change of the soil structure) is essentially an irreversible process, and the structure is not likely to improve with time. The soil may regain some strength with time, however, as a result of reestablishing a degree of chemical equilibrium. A regaining of strength with time is referred to as thixotropy.

The term sensitivity is used to describe the relative loss of strength of a soil after remolding, and it is defined as the ratio of the undisturbed unconfined compressive strength to the remolded unconfined compressive strength. Terzaghi and Peck (1967) group the sensitivity of clays as 2 to 4 for most clays, 4 to 8 for sensitive clays, and 8 to 16 for extrasensitive clays. The sensitivity of some clays, however, can be much higher. Bjerrum (1967) reported values for the sensitivity of a particular clay in the Drammen Valley, Norway, to range from 200 to 300. This very high sensitivity was caused by leaching the marine sediment clay with fresh water. The leaching process changed the characteristics of the double layer and thus the physical properties of the clay. The liquid limit of the soil was lowered to the point that it was less than the natural water content of the soil. Remolding this clay, because of the action of an external force, caused the consistency of the soil to approach that of a viscous liquid. Clays with these extremely high sensitivities are referred to as quick clays.

The liquidity index (LI) of a soil is defined as

$$LI = \frac{w - PL}{LL - PL} = \frac{w - PL}{PI}$$

It can be used to evaluate the behavior of a soil deposit if it is disturbed. The liquidity index of a soil will be greater than 1 if the natural water content (w) is greater than the liquid limit of the soil. Remolding this soil would then transform it into viscous liquid (thick slurry). When the natural water content is less than the plastic limit, the liquidity index is negative. For a negative liquidity index, a soil would be in the solid or semisolid state.

SOIL CLASSIFICATION SYSTEMS

Classification systems are used to group soils in accordance with their general behavior under given physical conditions. Soils that are grouped in order of performance for one set of physical conditions will not necessarily have the same order for performance under some other set of physical conditions. Thus, a number of classification systems have been developed depending on the intended purpose of the system.

Soil classification has proved to be very useful to the soil engineer. It can give general guidance through making available in an empirical manner the results of the field experience of others. However, classification systems must be used with caution. Blindly determining physical properties, such as compressibility, from empirical relationships and then using them in detailed calculations can lead to disastrous results.

Soil Classification Based on Grain Size

Grain size seems to be an obvious method by which to classify soils, and most of the earliest efforts to establish classification systems were based on grain size. Figure 33-1 shows several of these classification systems. The MIT system has probably been the most widely used. Since natural soil deposits generally contain a range of particle sizes, it is necessary to determine the grain-size distribution curve and then indicate the percentage of soil in each size range. The U.S. Department of Agriculture developed a grain-size classification system that specifically names a soil depending on the percentages of sand, silt, and clay.

1   Soil Properties

Figure 33-1 Soil classification based on grain size.

Classification System	Grain Size, mm			
	100 10	1 0.1	0.01 0.001	0.0001
MIT, 1931	Gravel	Sand	Silt	Clay
		2 0.06	0.002	
AASHO, 1970	Gravel	Sand	Silt	Clay [Colloids]
	75 2	0.05	0.002	
Unified 1953	Gravel	Sand	Fines (silt & clay)	
	75 4.75	0.075		

Figure 33-2 shows the triangular chart used to classify soil by this system. The percentages of sand, silt, and clay are plotted on the chart and the region where the point falls classifies the soil. The example in Fig. 34-1 classifies a soil based on its grain-size distribution.

Figure 33-2 Triangular soil classification chart. (U.S. Department of Agriculture.)

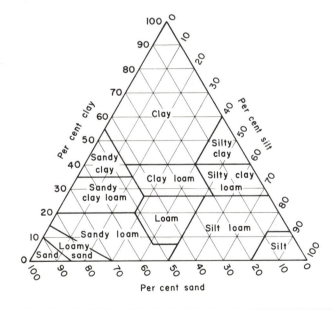

Figure 34-1

Example Problem Use the triangular soil classification chart (Fig. 33-2) to classify the soil with a grain-size distribution as shown on curve C of Fig. 23-1.

Solution The triangular chart is only applicable for material less than 2.0 mm, and curve C shows that 2% of the material is coarser than 2.00 mm. However, since this does not represent a significant quantity of the material, the soil will be classified on the basis of the fraction less than 2.00 mm.

Based on the MIT particle-size classification and adjusting the percentages of sand, silt, and clay from the distribution curve to include only the -2.0 mm material, the percentages are:

Sand	43%
Silt	44%
Clay	13%

Plotting these values on the triangular chart of Fig. 33-2 classifies the soil as loam.

If a significant portion of the material had been in the gravel-size range, the material might have been classified as a gravelly loam.

Although grain size seems to provide a very convenient means of classifying soil, it has a major shortcoming. There is very little relationship between grain size and physical properties for fine-grained soil. Other soil classification systems that include consistency and plasticity characteristics of the fine fraction have, therefore, been developed.

AASHTO Classification System

The AASHTO soil classification system was originally developed in the late 1920s by the U.S. Bureau of Public Roads for the classification of soils for highway subgrade use. The system originally classified soils into eight groups, A-1 to A-8. Group A-1 was considered to be the best suited for use as a highway subgrade. Soils of decreasing suitability are given higher group numbers. After some revision, this system was adopted by the American Association of State Highway Officials in 1945. The chart shown in Fig. 35-1 represents AASHTO Designation M145-73, adopted in 1973.

A soil is classified by proceeding from left to right on the chart to find the first group into which the soil test data will fit.

Soils containing fine-grained material are further identified by their group index. The group index is defined by the following equation.

$$\text{Group index} = (F - 35)[0.2 + 0.005(LL - 40)] + 0.01(F - 15)(PI - 10)$$

34

$PI = LL - PL$
$= (\text{liquid limit}) - (\text{plastic limit})$

Figure 35-1 AASHTO classification of soils and soil-aggregate mixtures.

AASHTO Classification of Soils and Soil-Aggregate Mixtures

General Classification	Granular Materials (35% or less passing 0.075 mm)							Silt-Clay Materials (More than 35% passing 0.075 mm)			
	A-1		A-3	A-2				A-4	A-5	A-6	A-7
Group Classification	A-1-a	A-1-b		A-2-4	A-2-5	A-2-6	A-2-7				A-7-5 / A-7-6
Sieve Analysis, Percent Passing:											
2.00 mm (No. 10)	50 max										
0.425 mm (No. 40)	30 max	50 max	51 min								
0.075 mm (No. 200)	15 max	25 max	10 max	35 max	35 max	35 max	35 max	36 min	36 min	36 min	36 min
Characteristics of Fraction Passing 0.425 mm (No. 40)											
Liquid limit				40 max	41 min	40 max	41 min	40 max	41 min	40 max	41 min
Plasticity index	6 max		N.P.	10 max	10 max	11 min	11 min	10 max	10 max	11 min	11 min[a]
Usual Types of Significant Constituent Materials	Stone Fragments Gravel and Sand		Fine Sand	Silty or Clayey Gravel Sand				Silty Soils		Clayey Soils	
General Rating as Subgrade	Excellent to Good							Fair to Poor			

[a]Plasticity index of A-7-5 subgroup is equal to or less than LL minus 30.
Plasticity index of A-7-6 subgroup is greater than LL minus 30.

A-7 ⇒ can either be A-7-5 or A-7-6

≤ 22

Example Problem Use the AASHTO soil classification system to classify the soil with a grain-size distribution as shown on curve A of Fig. 23-1. The liquid limit (LL) and plastic index (PI) of the material passing the 0.425-mm (No. 40) sieve for this soil are 39% and 19%, respectively.

Solution The AASHTO soil classification system is shown in Fig. 35-1. From curve A, the percentages of material passing the 2.0-mm, 0.425-mm, and 0.075-mm sieves are:

$$
\begin{array}{ll}
\text{2.0 mm (No. 10)} & 41\% \\
\text{0.425 mm (No. 40)} & 29.5\% \\
\text{0.075 mm (No. 200)} & 21\%
\end{array}
$$

Based on the preceding characteristics, the appropriate subgroup from Fig. 35-1 is A-2-6. Since the subgroup is A-2-6, only use the PI portion of the formula to calculate the group index.

The group index for the soil is:

Group index $= (F - 35)[(0.2 + 0.005(LL-40)] + 0.01(F - 15)(PI - 10)$

Group index $= 0.01(21 - 15)(19 - 10)$

Group index $= 0.54$ or Group index $= 1$

The final classification for this soil is A-2-6 (1).

where

 F = percent passing 0.075 mm (No. 200) sieve, expressed as a whole number
LL = liquid limit
 PI = plasticity index

The group index is always reported as a whole number unless it is negative, for which it is reported as zero. When calculating the group index of A-2-6 and A-2-7 subgroups, only the PI portion of the formula shall be used. The group index is reported as part of the AASHTO classification. If the group index of an A-7-6 soil is 15, the classification should be reported as A-7-6 (15). The higher the value of the group index, the less suitable the material for use as a highway subgrade. A group index of 0 indicates a "good" subgrade material, and a group index 20 or greater indicates a "very poor" subgrade material. The example shown in Fig. 36-1 classifies a soil according to the AASHTO classification system.

Unified Soil Classification System

The most popular soil classification system among soil and foundation engineers is the Unified soil classification system. This system was first developed by

40(20-10)

Figure 37-1 Unified soil classification system.

MAJOR DIVISIONS			GROUP SYMBOLS	TYPICAL NAMES	CLASSIFICATION CRITERIA
COARSE-GRAINED SOILS (More than 50% retained on 0.075 mm (No. 200) sieve)	GRAVELS (50% or more of coarse fraction retained on 4.75 mm (No. 4) sieve)	CLEAN GRAVELS	GW	Well graded gravels and gravel sand mixtures, little or no fines	$C_u = D_{60}/D_{10}$ Greater than 4 ; $C_z = \dfrac{(D_{30})^2}{D_{10} \times D_{60}}$ Between 1 and 3
			GP	Poorly graded gravels and gravel sand mixtures, little or no fines	Not meeting both criteria for GW
		GRAVELS WITH FINES	GM	Silty gravels, gravel-sand silt mixtures	Atterberg limits plot below "A" line or plasticity index less than 4 | Atterberg limits plotting in hatched area are borderline classifications requiring use of dual symbols
			GC	Clayey gravels, gravel-sand clay mixtures	Atterberg limits plot above "A" line and plasticity index greater than 7
	SANDS (More than 50% of coarse fraction passes 4.75 mm (No. 4) sieve)	CLEAN SANDS	SW	Well graded sands and gravelly sands, little or no fines	$C_u = D_{60}/D_{10}$ Greater than 6 ; $C_z = \dfrac{(D_{30})^2}{D_{10} \times D_{60}}$ Between 1 and 3
			SP	Poorly graded sands and gravelly sands, little or no fines	Not meeting both criteria for SW
		SANDS WITH FINES	SM	Silty sands, sand-silt mixtures	Atterberg limits plot below "A" line or plasticity index less than 4 | Atterberg limits plotting in hatched area are borderline classifications requiring use of dual symbols
			SC	Clayey sands, sand-clay mixture	Atterberg limits plot above "A" line and plasticity index greater than 7
FINE-GRAINED SOILS (50% or more passing 0.075 mm (No. 200) sieve)	SILTS AND CLAYS (Liquid limit 50% or less)		ML	Inorganic silts, very fine sands, rock flour, silty or clayey fine sands	
			CL	Inorganic clays of low to medium plasticity, gravelly clays, sandy clays, silty clays, lean clays	
			OL	Organic silts and organic silty clays of low plasticity	
	SILTS AND CLAYS (Liquid limit greater than 50%)		MH	Inorganic silts, micaceous or diatomaceous fine sands or silts, elastic silts	
			CH	Inorganic clays of high plasticity, fat clays	
			OH	Organic clays of medium to high plasticity	
Highly Organic Soils			PT	Peat, muck and other highly organic soils	

Classification on basis of percentage of fines:

Less than 5% Pass 0.075 mm sieve — GW, GP, SW, SP
More than 12% Pass 0.075 mm sieve — GM, GC, SM, SC
5% to 12% Pass 0.075 mm sieve — Border Classification requiring use of dual symbols

PLASTICITY CHART

For classification of fine-grained soils and fine fraction of coarse-grained soil

Atterberg Limits within hatched area indicates a border line classification requiring dual symbols

Equation of A-line $PI = 0.73(LL-20)$

Plasticity Index / Liquid Limit

A-Line, CH, MH or OH, CL, ML or OL, CL-ML

Visual-Manual Identification, See ASTM Designation D 2488.

37

Figure 38-1 Engineering-use chart based on the Unified soil classification. (USBR, 1963.)

TYPICAL NAMES OF SOIL GROUPS	GROUP SYMBOLS	IMPORTANT PROPERTIES			
		PERMEABILITY WHEN COMPACTED	SHEARING STRENGTH WHEN COMPACTED AND SATURATED	COMPRESSIBILITY WHEN COMPACTED AND SATURATED	WORKABILITY AS A CONSTRUCTION MATERIAL
WELL-GRADED GRAVELS, GRAVEL SAND MIXTURES, LITTLE OR NO FINES	G W	PERVIOUS	EXCELLENT	NEGLIGIBLE	EXCELLENT
POORLY GRADED GRAVELS, GRAVEL SAND MIXTURES, LITTLE OR NO FINES	G P	VERY PERVIOUS	GOOD	NEGLIGIBLE	GOOD
SILTY GRAVELS, POORLY GRADED GRAVEL-SAND-SILT MIXTURES	G M	SEMI PERVIOUS TO IMPERVIOUS	GOOD	NEGLIGIBLE	GOOD
CLAYEY GRAVELS, POORLY GRADED GRAVEL-SAND-CLAY MIXTURES	G C	IMPERVIOUS	GOOD TO FAIR	VERY LOW	GOOD
WELL-GRADED SANDS, GRAVELLY SANDS, LITTLE OR NO FINES	S W	PERVIOUS	EXCELLENT	NEGLIGIBLE	EXCELLENT
POORLY GRADED SANDS, GRAVELLY SANDS, LITTLE OR NO FINES	S P	PERVIOUS	GOOD	VERY LOW	FAIR
SILTY SANDS, POORLY GRADED SAND-SILT MIXTURES	S M	SEMI PERVIOUS TO IMPERVIOUS	GOOD	LOW	FAIR
CLAYEY SANDS, POORLY GRADED SAND-CLAY MIXTURES	S C	IMPERVIOUS	GOOD TO FAIR	LOW	GOOD
INORGANIC SILTS AND VERY FINE SANDS, ROCK FLOUR, SILTY OR CLAYEY FINE SANDS WITH SLIGHT PLASTICITY	M L	SEMI PERVIOUS TO IMPERVIOUS	FAIR	MEDIUM	FAIR
INORGANIC CLAYS OF LOW TO MEDIUM PLASTICITY, GRAVELLY CLAYS, SANDY CLAYS, SILTY CLAYS, LEAN CLAYS	C L	IMPERVIOUS	FAIR	MEDIUM	GOOD TO FAIR
ORGANIC SILTS AND ORGANIC SILT-CLAYS OF LOW PLASTICITY	O L	SEMI PERVIOUS TO IMPERVIOUS	POOR	MEDIUM	FAIR
INORGANIC SILTS, MICACEOUS OR DIATOMACEOUS FINE SANDY OR SILTY SOILS, ELASTIC SILTS	M H	SEMI PERVIOUS TO IMPERVIOUS	FAIR TO GOOD	HIGH	POOR
INORGANIC CLAYS OF HIGH PLASTICITY, FAT CLAYS	C H	IMPERVIOUS	POOR	HIGH	POOR
ORGANIC CLAYS OF MEDIUM TO HIGH PLASTICITY	O H	IMPERVIOUS	POOR	HIGH	POOR
PEAT AND OTHER HIGHLY ORGANIC SOILS	P T	——	——	——	——

Figure 39-1

GROUP SYMBOLS	RELATIVE DESIRABILITY FOR VARIOUS USES*									
	ROLLED EARTH DAMS			CANAL SECTIONS		FOUNDATIONS		ROADWAYS		
								FILLS		
	HOMO-GENEOUS EMBANK-MENT	CORE	SHELL	EROSION RESISTANCE	COMPACTED EARTH LINING	SEEPAGE IMPORTANT	SEEPAGE NOT IMPORTANT	FROST HEAVE NOT POSSIBLE	FROST HEAVE POSSIBLE	SURFACING
GW	—	—	1	1	—	—	1	1	1	3
GP	—	—	2	2	—	—	3	3	3	—
GM	2	4	—	4	4	1	4	4	9	5
GC	1	1	—	3	1	2	6	5	5	1
SW	—	—	3 If Gravelly	6	—	—	2	2	2	4
SP	—	—	4 If Gravelly	7 IF GRAVELLY	—	—	5	6	4	—
SM	4	5	—	8 if GRAVELLY	5 EROSION CRITICAL	3	7	8	10	6
SC	3	2	—	5	2	4	8	7	6	2
ML	6	6	—	—	6 EROSION CRITICAL	6	9	10	11	—
CL	5	3	—	9	3	5	10	9	7	7
OL	8	8	—	—	7 EROSION CRITICAL	7	11	11	12	—
MH	9	9	—	—	—	8	12	12	13	—
CH	7	7	—	10	8 VOLUME CHANGE CRITICAL	9	13	13	8	—
OH	10	10	—	—	—	10	14	14	14	—
PT	—	—	—	—	—	—	—	—	—	—

* Low number indicates preferred soil.

39

Casagrande (1948) and was known as the Airfield classification system. It was adopted with minor modifications by the U.S. Bureau of Reclamation and the U.S. Corps of Engineers in 1952. In 1969 the American Society for Testing and Materials (ASTM) adopted the Unified system as a standard method for classification of soils for engineering purposes (ASTM D-2487).

As shown in Fig. 37-1, the Unified system divides soils into three main groups: coarse grained, fine grained, and highly organic. Coarse-grained soils are those with more than 50 percent of the material retained on the No. 200 sieve (0.075 mm). Coarse-grained soils are divided into gravels (G) and sands (S). The gravels and sands are grouped according to their gradation and silt or clay content, as well graded (W), poorly graded (P), containing silt material (M), or containing clay material (C). A typical classification thus might be GP for a poorly graded gravel.

Fine-grained soils are those for which more than 50% of the material passes the No. 200 sieve. They are divided into silts (M), clays (C), and organic silts and clays (O), depending on how they plot on the plasticity chart (liquid-limit, plasticity-index relationship). The designation L and H are added to the fine-grained symbols to indicate low plasticity and high plasticity respectively (liquid limits below and above 50%). Highly organic soils (peats) are visually identified. The example shown in Fig. 40-1 classifies a soil in accordance with the Unified soil classification system. Figure 38-1 and Fig. 39-1 show a use chart, based on the Unified soil classification system, that can be used in a general way to characterize the important properties and the relative desirability of a soil for various uses.

Figure 40-1

Example Problem Use the Unified soil classification system to classify the soil with a grain-size distribution as shown on curve A of Fig. 23-1. The liquid limit (LL) and plastic index (PI) of the material passing the 0.425-mm (No. 40) sieve are 39% and 19% respectively.

Solution The Unified soil classification system is shown in Fig. 37-1. More than 50% of the material is retained on the 0.075-mm sieve, which initially groups the soil as a coarse-grained soil. Furthermore, since more than 50% of the material is retained on the 4.75-mm sieve it is a gravel soil. More than 12% of the coarse fraction passes the 0.075-mm sieve and, therefore, the soil is a gravel with fines. The nature of the fines can be determined from plotting the liquid limit and plastic limit on the plasticity chart that accompanies the Unified system. The plastic index is greater than 7 and the limits plot above the A-line; therefore, this soil is classified as GC (clayey gravel).

Soil engineers should commit to memory the details of the classification system that is most appropriate to their specific area of specialization. A continuing effort should then be made to field-classify soils according to that system and to check the field classification against laboratory results.

PROBLEMS

1-1 Briefly describe two national or international events that involved soil mechanics.

1-2 Comment on the responsibilities of a soil and foundation engineer with regard to the design and construction of a municipal sewage treatment plant.

1-3 Describe the type of foundation system that was used on a building you are familiar with. Describe any unusual subsurface conditions. Speculate on why that type of foundation was used.

1-4 Comment on the importance of "defining the project concept" before starting a subsurface investigation.

1-5 Comment on the importance of "performance feedback."

1-6 Describe a well-graded gravel and a uniform gravel.

1-7 Distinguish between the clay minerals kaolinite and illite.

1-8 Why do the flat surfaces of a clay mineral have a negative charge?

1-9 What is isomorphous substitution?

1-10 Explain why montmorillonite is an expansive clay.

1-11 Laboratory tests on a certain soil sample provided the following results: liquid limit = 60%, plastic limit = 30%, shrinkage limit = 25%. What is the plasticity index of the soil? Calculate the liquidity index for a natural water content of 36%.

1-12 The Atterberg limits of a particular soil are reported as liquid limit = 60%, shrinkage limit = 40%, and plastic limit = 35%. Are these values reasonable? Explain.

1-13 Define sensitivity and thixotropy.

1-14 Describe a clay particle and the environment that surrounds the particle.

1-15 What is the electric double layer?

1-16 Explain why electrical surface forces are important for fine-grained soils but have little effect on coarse-grained soils.

1-17 How many cubic meters of fill can be constructed at a void ratio of 0.7 from 191,000 m³ of borrow material that has a void ratio of 1.2?

1-18 The average void ratio of a given material is 0.67, and the specific gravity of the solid particles is 2.65. If the water content is 8%, calculate the total unit weight and the degree of saturation.

1-19 The wet unit weight of a soil is 18.80 kN/m³, the specific gravity of the solid particles of the soil is 2.67, and the moisture content of the soil is 12% by dry weight. Calculate dry unit weight, porosity, void ratio, and degree of saturation.

1-20 An undisturbed sample of clay soil is trimmed to fit into a consolidometer. The diameter of the consolidometer is 63.50 mm (2.5 in.) and the height is 25.4 mm (1.0 in.). The wet mass of the consolidation sample is 142.2 g, and the specific gravity of the solid particles is 2.7. At the completion of the test the soil sample is removed from the consolidometer and dried, and the oven-dried mass is determined to be 98.72 g. For this sample, determine the initial water content, void ratio, porosity, and degree of saturation.

1-21 Describe what is meant by dispersed and flocculated structure. What effect does structure have on the physical properties of clays?

1-22 Explain why clay soils lose strength as a result of remolding.

1-23 Explain why a soil classification system based on grain size alone is a poor method of classifying soils for engineering purposes.

1-24 Use the AASHTO soil classification system to classify the soils with grain-size distributions as shown on curves B and C of Fig. 23-1. The fines for the soil of curve B are nonplastic. The liquid limit (LL) and plasticity index (PI) of the material passing the 0.425 mm (No. 40) sieve for the soil of curve C are 32% and 12%, respectively.

1-25 Use the Unified soil classification system to classify the soils in problem 1-24.

1-26 A certain soil has 98% of the particles (by weight) finer than 1 mm, 59% finer than 0.1 mm, 24% finer than 0.01 mm, and 11% finer than 0.001 mm. Draw the grain-size distribution curve and determine the approximate percentage of the total weight in each of the various size ranges according to the MIT size classification. Determine the effective size (D_{10}) and the uniformity coefficient for this soil. Classify the soil using the U.S. Department of Agriculture triangular chart.

1-27 Tests run on a given soil produced the following results.

LL = 27%   PI = 15%

Grain Size (mm)	Sieve Size	Percent Passing
2.00	10	100
0.850	20	99
0.425	40	91
0.250	60	61
0.150	100	25
0.075	200	15
0.0065	—	3
0.001	—	0.5

For the given soil:

(a) Plot the grain-size distribution curve.
(b) Determine the effective particle size (D_{10}).
(c) Classify the soil by the U.S. Department of Agriculture's triangular chart.
(d) Classify the soil by the AASHTO classification system.
(e) Classify the soil by the Unified classification system.

1-19

$$\gamma_d = \frac{\gamma}{1+w} = \frac{18.8}{1+.12} = 16.8 \frac{kN}{m^3}$$

$$\gamma_d = \frac{G}{1+e}\gamma_w$$

$$S_r = \frac{Gw}{e} = \frac{2.67(.12)}{.56} = 57.2\%$$

$$e = \frac{G\gamma_w}{\gamma_d} - 1 = .56$$

$$n = \frac{e}{1+e} = \frac{.56}{1+.56} = .359$$

1-18 $e = .67$
       $G = 2.65$
       $W = .08$

$$S_r = \frac{Gw}{e} = \frac{2.65(.08)}{.67} = 31.6\%$$

$$\gamma = \frac{G(1+w)}{1+e}\gamma_w = \frac{(2.65)(1+.08)}{1+.67}(1) = 1.714 \ g/cm^3$$

2

SOIL WATER

The amount and distribution of moisture in soil have a very great effect on the physical properties of the soil. The ability of a soil to support foundation loads may vary over a range of several hundred percent, depending on the percent of moisture and the type of soil. Seepage flow through and below an earth dam must be considered in evaluating the dam's ability to impound water. The stability of artificial and natural slopes is greatly affected by the presence of water pressure. Because of these and many other reasons, it is, therefore, necessary to have an accurate and complete knowledge of the water conditions in the soil. The following sections are devoted to an understanding of this problem.

STATIC PRESSURES IN WATER

The pressure in a static body of water (see Fig. 46-1a) has a triangular distribution with a magnitude $\gamma_w y$. A pressure conforming to this definition is called hydrostatic pressure (Fig. 46-1a). A hydrostatic pressure distribution also exists in the pore water surrounding the soil particles in Fig. 46-1b. If, however, the particles are suspended in the water or are falling with a constant velocity (Fig. 46-1c), the magnitude of the pressure at any point below the surface may be found by the equation $u = \gamma y$ (see Fig. 46-1), where γ is the unit weight of the soil-water combination. The computation of static water pressure is illustrated in Fig. 46-2.

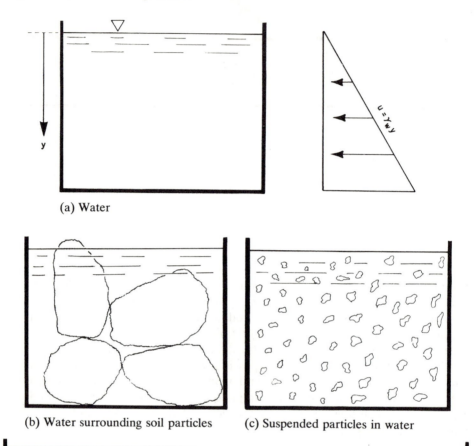

Figure 46-1 Static water pressure.

(a) Water

(b) Water surrounding soil particles   (c) Suspended particles in water

Figure 46-2

 Example Problem   Find the water pressure (u) at the bottom of the container in Fig. 46-1a (y = 8 m). Where u = water pressure and γ_w = Unit weight of water.

Solution
$$u = \gamma_w y = 9810 \text{ N/m}^3 \times 8 \text{ m}$$
$$= 78.5 \text{ kPa}$$

CAPILLARY MOISTURE

The water zone in a soil mass that has a water table may be divided into a saturated zone bélow the water table and a capillary zone above the water table. Some voids are filled with air in the saturated zone, but these air voids have little effect on the reaction of the soil to outside stresses. Above the water table, the number of air voids increases as the distance from the water table increases. The capillary zone may be divided into three different zones with somewhat arbitrary boundaries (see Fig. 47-1). The zone closest to the water table is called the zone of capillary saturation. The water content in this zone may be slightly less than 100%, but forces exerted on the soil structure by capillarity are very small and the soil reacts much as if it were in the saturated condition. Above the zone of capillary saturation is the zone of partial capillary saturation. In this zone water is connected through the smaller pores, but more of the larger pores are filled with air. The third zone is the zone of contact water. The water in this zone surrounds the points of contact between soil particles and also surrounds soil particles, but is disconnected through the pores. The water in the capillary zones

Figure 47-1 Capillary water system.

Figure 48-1 Height of rise in a capillary tube.

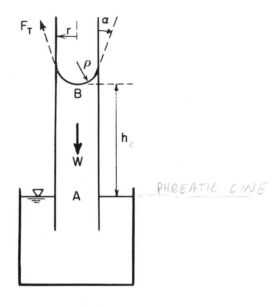

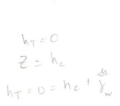

$W = F_T \cos \alpha$ (equilibrium of water column)

$$\gamma_w(\pi r^2 h) = T_s(2\pi r) \cos \alpha$$

where

$$T_s = \text{surface tension for water}$$

(Force/unit length) = 0.0735 N/m (water)

The height of capillary rise may be expressed as

$$h = \frac{2T_s}{\gamma_w r} \cos \alpha \qquad (48\text{-}1a)$$

and the pressure difference across the air water interface may be expressed as

$$u_B = \frac{-2T_s}{r} \cos \alpha \qquad (48\text{-}1b)$$

For water in a clean glass tube, $\alpha = 0$ and $\rho = r$.

Figure 49-1   Average capillary heights and tensions in soils.

Soil	Height (m)	Tension (kN/m^2)
Sand	0.05–1	0.5–10
Silt	1–10	10–100
Clay	>10	>100
Maximum	>35	>350

is held in place by capillary attraction and exerts relatively large stabilizing forces on the structure of the soil.

In fine-grained soils, the capillary zones reach to a considerable height above the water table (Fig. 49-1); in coarse sand, the capillary rise may be negligible. The capillary rise in soil is similar to capillary rise in a capillary tube, as shown in Fig. 48-1. The forces holding the column of water in the tube in equilibrium are complex, but an accurate expression for the height of rise in the tube may be obtained by assuming that the water surface acts as a membrane which may act in tension to support the weight of the column of water.

The pressure distribution in the water column of Fig. 48-1 is triangular in shape. The pressure is equal to zero at point A and $-\gamma_w h$ at point B. The curvature of the air-water interface is $1/\rho$, and the pressure difference across the interface is related to this curvature, as shown in Eq. 48-1b. In natural soil masses, as the height above the water table increases, the water content decreases, the curvature of the air-water interface increases, and the pressure difference across the interface increases. Negative pressures in the water film in soils with very low moisture content may become very large (Fig. 49-1). The application of Eqs. 48-1a and 48-1b is illustrated in the example problem of Fig. 49-2.

Figure 49-2

Example Problem   For the capillary tube of Fig. 48-1, determine the capillary height h and the negative pressure u_B at the interface for a 0.1-mm radius glass tube in water.

From Eq. (48-1a),

$$h = \frac{2(0.0735 \ N/m)}{(9810 \ N/m^3)(0.0001 \ m)} = 0.150 \ m$$

From Eq. 48-1b,

$$u_B = -\frac{2(0.0735 \ N/m)}{0.0001 \ m} = -1.47 \ kN/m^2$$

SATURATED FLUID FLOW

Darcy's Law

Water both below the water table and in the capillary zone is subject to forces that cause flow. In the zone below the water table, changes in pressure and elevation are the most important causes of flow. The property of soil that permits the passage of water under a gradient of force is called permeability.

In saturated soil, unless pores are very large, water flow is laminar. Reynolds (1883) has discussed a critical condition for water flow in pipes that describes a division between laminar and turbulent flow at a Reynolds number $vD\rho/\mu = 2100$. In soils a critical Reynolds number also exists, but is approximately in the range from 1 to 10.

For flow in the laminar range, energy losses are proportional to the first power of velocity, and Darcy (1856) has suggested Eq. 50-1a for velocity.

The head loss (h_L) in this equation is the name given to the energy loss per unit weight of fluid flowing and may be represented by a decrease in the Bernoulli

---

Figure 50-1

Darcy's equation for velocity and flow rate is

$$v = -k\frac{h_L}{L} = -ki \qquad (50\text{-}1a)$$

$$q = v_n A_n = vA \qquad (50\text{-}1b)$$

where

q = Volume flow rate

h_L = Energy loss per unit weight (head loss)

v = Darcy velocity (superficial velocity)

v_n = Actual velocity

A_n = Area of voids in cross section

A = Area of total cross section

L = Length of flow

i = Hydraulic gradient $\left(\dfrac{h_L}{L}\right)$

k = Coefficient of permeability

(*Note:* The negative sign in Darcy's equation indicates that the velocity is in the direction of the negative gradient. The equation is often written without the negative sign.)

---

head terms, which are (in units of length)

$$\text{velocity head} = h_v = (v^2/2g)$$
$$\text{pressure head} = h_p = (u/\gamma_w)$$
$$\text{elevation head} = h_e = (y)$$

where u is water pressure (pore pressure). The difference between the total Bernoulli heads at two points on the same flow line is the magnitude of the head loss h_L. The total Bernoulli head is equal to the sum of the head terms.

$$h_T = h_p + h_e + h_v$$

and

$$h_{T_1} = h_{T_2} + h_L$$

$$h_T = \frac{v^2 \approx 0}{2g} + z + \frac{u_w}{\gamma_w}$$

h_T = height of the permeameter point relative to datum

The head loss represents the amount of energy lost in heat through the viscous action between layers of flowing water.

It is more convenient in soil flow problems to use the total cross-sectional area as the area of flow than to find the area of voids [Eq. 50-1b]. Therefore, the flow q is usually represented by the equation $(q = vA)$, where A is the total cross-sectional area, including voids and solids, and v is a superficial velocity (also called Darcy's velocity) that would exist if the total section were a

---

Figure 51-1

Example Problem   Water flows through a soil mass that has a length of 4 m and a cross-sectional area (A) of 2 m^2. The fluid energy lost when 1 m^3 of water flows through the soil is 1500 N m. The void ratio of the soil is 0.64. The elapsed time for this flow is 30 hr. Find v, v_n, and k. From Eq. 50-1,

$$v = q/A = \frac{1 \text{ m}^3}{30(3600)2 \text{ m}^2} = 4.63 \times 10^{-6} \text{ m/sec}$$

$$v_n = v/n = v(1 + e)/e = 1.19 \times 10^{-5} \text{ m/sec}$$

$$h_L = \frac{\text{energy loss}}{\text{unit weight}} = \frac{1500 \text{ N m}}{1 \text{ m}^3(9810) \text{ N/m}^3} = 0.15 \text{ m}$$

$$k = v/i = vL/h_L = \frac{4.63 \times 10^{-6} \times 4}{0.15}$$

$$k = 1.23 \ 10^{-4} \text{ m/sec}$$

Saturated Fluid Flow

flow section. This equation is called Darcy's equation, and the term velocity, unless otherwise noted in this discussion, refers to superficial velocity.

In other disciplines the coefficient k, or constant of proportionality between velocity and hydraulic gradient, is termed a coefficient of conductivity, since it depends upon fluid properties as well as properties of the soil medium. However, through long-term use in geotechnical literature, k is popularly called the coefficient of permeability. Such usage is probably justified, since the temperature of water varies little in most soil application. (The kinematic viscosity of water decreases about 23% from 10° to 20°C.) Thus, in coarse-grained soils the primary factors affecting k are the soil particle size and gradation, particle shape and roughness, and the void ratio of the soil medium. In fine-grained soils, where particle surface forces predominate, other factors, such as the type of clay mineral and adsorbed ions, have an overriding influence on k. The application of Darcy's law is illustrated in the example problem of Fig. 51-1.

Hydraulic Gradient

The hydraulic gradient used in Darcy's equation, $i = h_L/L$, may be found quite easily in simple flow situations in the laboratory, in which reservoirs of water occur on each side of the soil sample. One such situation is shown in Fig. 53-1. The total head will be taken equal to the pressure head plus the elevation head. The velocity head in most soil flow situations is negligible compared to the pressure and elevation heads and is generally neglected.

The example in Fig. 53-1 gives the values of elevation head, pressure head, and total head at different points in the apparatus. The values of head at b and d may be found, after picking an arbitrary elevation datum, by adding the distance above the elevation datum and the distance below the surface of the water. Point d has a pressure head that may be found by using the principle that the pressure at equal elevations in static bodies of fluid is equal; therefore, the pressure at point e and the pressure at d are equal. It is true that the fluid in the tube is not exactly at rest due to the small amount of flow through the soil sample; however, these quantities of flow are small enough so that the head loss along the edge of the container in the tube is negligible, and the total head at any point in the tube may be considered constant.

The total head at point c in the middle of the soil sample may be found by examining the gradient of head through the sample. Since the sample is saturated and homogeneous and since the equation of continuity requires a constant velocity through the soil sample, the head loss per unit distance must also be constant; therefore, the head at point c is equal to the average of the heads at points b and d. The pressure head at c may then be found by using the relationship $h_T = h_p + h_e$.

Figure 53-1

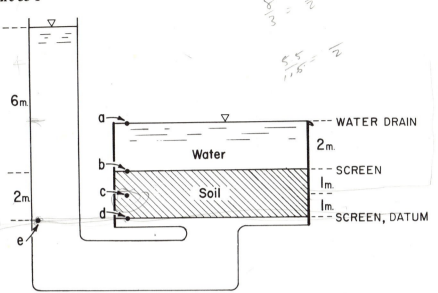

Example Problem Find the pressure head at point *c* by completing the following table.

Point	Pressure Head (h_p)	Elevation Head (h_e)	Total Head (h_T)
a	0	4	4
b	2	2	4
c	(5)	1	(6)
d	8	0	8

Determination of Permeability

The coefficient of permeability may be found by field tests or laboratory tests. If laboratory tests are used and the soil is fine grained, undisturbed samples of the soil must be taken to determine the in-situ values of *k*. For a sandy soil, *k* can be obtained accurately by remolding to the proper void ratio. Figure 54-1 shows a

Saturated Fluid Flow

Figure 54-1 Constant head permeameter.

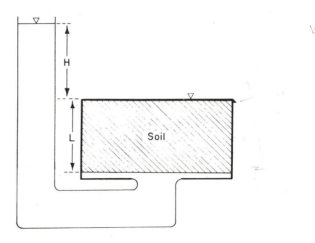

Darcy's law for Flow in a Constant Head Permeameter

$$q = \frac{Q}{t} = vA = kiA = \frac{kh_L A}{L}$$

from which,

$$k = \frac{QL}{th_L A} \qquad (54\text{-}1)$$

where

Q = flow volume in time t
A = area of soil sample (cross section)
L = length of soil sample
$h_L = H$

sample that has been placed into a constant head permeability apparatus. In this figure the water flow is upward. The coefficient of permeability may be found by applying Darcy's law to the flow (Eq. 54-1). From this equation, k may be computed by measuring Q for a given time interval. An example problem utilizing the data from a constant head permeability test is worked in Fig. 55-1.

Figure 55-2 shows a sample in a variable head permeability apparatus. The water level in the tube at the beginning of the test is H_1 and at the end of the test, H_2. At some intermediate time, the level is H and the change in level during a

Figure 55-1

Example Problem Given the following data from a constant head permeability test, compute the coefficient of permeability k.

$$Q = 0.034 \text{ m}^3$$
$$t = 500 \text{ sec}$$
$$H = 2.0 \text{ m}$$
$$L = 0.20 \text{ m}$$
$$A = 0.04 \text{ m}^2$$

From Darcy's law,

$$k = \frac{Q}{t} \frac{L}{h_L A} = \frac{(0.034 \text{ m}^3)(0.20 \text{ m})}{(500 \text{ sec})(2 \text{ m})(0.04 \text{ m}^2)} = 1.7 \times 10^{-4} \text{ m/sec}$$

Figure 55-2 Variable head permeameter.

$$Q = kiA = k\left(\frac{h}{t}\right)A$$

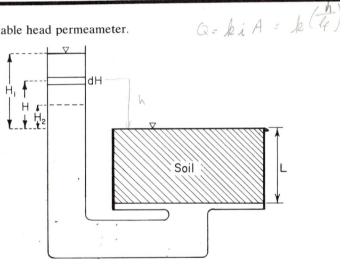

Darcy's Equation

$$-a \, dH = dQ = kA \frac{H}{L} \, dt; \quad \int_{H_1}^{H_2} -\frac{dH}{H} = \int_0^t \left(\frac{kA}{La}\right) dt; \quad k = \frac{aL}{At} \ln\left(\frac{H_1}{H_2}\right)$$

where

a = area of tube

A = area of soil sample (cross section)

dQ = flow volume in time dt

Saturated Fluid Flow

Figure 56-1   Permeability measured by the auger hole method.

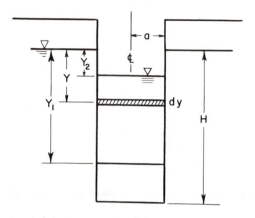

Hooghoudt Equation
Differential equation:

$$-dy(\pi a^2) = k(2\pi a H)\frac{y}{L}\,dt + k\pi a^2\,\frac{y}{L}\,dt$$

$$k = \frac{aL}{(2H + a)\,t}\,\ln\frac{y_1}{y_2}$$

L = An empirical length over which the head loss y occurs.

$$L = aH/0.19 \text{ (meters)}$$

Ernst Equation

$$k = \frac{40}{\left(20 + \dfrac{H}{a}\right)\left(2 - \dfrac{y}{H}\right)}\,\frac{a}{y}\,\frac{\Delta y}{\Delta t} \text{ (lengths in meters)}$$

Δy = rise in water level during time Δt.
 y = average y during increment.

small time interval, dt, is dH. The permeability k may again be computed by applying Darcy's equation as shown (water level, H, is equal to head loss, h_L).
   A field method for a preliminary estimate of the coefficient of permeability is the single auger hole method. An auger hole is drilled in the field and the water

is removed by bailing. Hooghoudt (1936) investigated the case of an auger hole in a homogeneous soil and assumed that the water table remained in a horizontal position and that water flowed horizontally into the sides of the auger hole and vertically through the bottom of the hole (see Fig. 56-1). The derivation is similar to a variable head laboratory test. Hooghoudt's differential equation for flow has an empirical constant, L, which represents some length over which the head loss, y, occurs (see Fig. 56-1). Experimentally, he found that L equals $aH/0.19$. His equation may be integrated as shown. Ernst's (1950) revised estimate of k from this same test is also shown in Fig. 56-1. The Hooghoudt and Ernst equations are illustrated in the example of Fig. 57-1.

Another field estimate of the coefficient of permeability may be obtained from wells by a pumping test. A steady-state flow is established and k is found

Figure 57-1

Example Problem A 3 m deep auger hole (diameter = 0.2 m) is drilled in a homogeneous soil. The hole extends 2.5 m below the water table. Readings on the water table are as follows:

Time (sec)	y (m)
0	2.5
600	1.6
1200	1.0
1800	0.5

Find the coefficient of permeability (k)
Hooghoudt Equation

$$1.3158$$

$$k = \frac{aL}{(2H + a)t} \ln \frac{y_1}{y_2} = \frac{(0.1 \text{ m}) \left(\dfrac{0.1 \text{ m} \times 2.5 \text{ m}}{0.19} \right)}{[2(2.5 \text{ m}) + 0.1 \text{ m}]1800 \text{ sec}} \ln \frac{2.5}{0.5}$$

$$5.1$$

$$k = 2.3 \times 10^{-5} \text{ m/sec}$$

Ernst Equation (use interval 600 to 1200 sec)

$$k = \frac{40}{\left(20 + \dfrac{2.5}{0.1} \right) \left(2 - \dfrac{1.3}{2.5} \right)} \left(\frac{0.1}{1.3} \right) \left(\frac{0.6 \text{ m}}{600 \text{ sec}} \right)$$

$$k = 4.62 \times 10^{-5} \text{ m/sec}$$

Saturated Fluid Flow

Figure 58-1

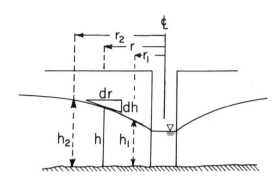

Well Equation—Unconfined Acquifer

$$q = kAi = kh(2\pi r)\frac{dh}{dr}; \quad \int_{r_1}^{r_2}\frac{q}{r}\,dr = k2\pi\int_{h_1}^{h_2}h\,dh; \quad k = \frac{q\,\ln(r_2/r_1)}{\pi(h_2^2 - h_1^2)} \quad (58\text{-}1)$$

from Eq. 58-1. Field measurements are made to determine h at two or more distances r from the well centerline.

In the absence of a measured permeability, the coefficient for clean granular soils can be estimated from the following equation.

$$k = \frac{2g}{C_s}\frac{\rho}{\mu}D^2\frac{e^3}{1+e} \quad (58\text{-}2)$$

where g is the accleration of gravity; μ/ρ is the kinematic viscosity of water; C_s is a particle shape factor, which varies from 360 for spherical particles to about 700 for angular particles; D is a weighted or characteristic particle diameter; and e is the void ratio. The characteristic diameter D is obtained from a grain-size analysis using the following equation.

$$D = \frac{\Sigma M_i}{\Sigma(M_i/D_i)}$$

where M_i is the mass retained between two adjacent sieves and D_i is the mean diameter of the adjacent sieves. (The kinematic viscosity of water varies approximately linearly from 1.31 mm²/sec at 10°C to 1.01 mm²/sec at 20°C.) The use of Eq. 58-2 is illustrated in the example problem of Fig. 59-1.

Figure 59-1

Example Problem Estimate the permeability for sand of the gradation shown when the dry unit weight is 14.2 kN/m³ and the grain shape is subangular. The water temperature is 15 C. Assume $G = 2.65$.

Sieve no.	Sieve Size (mm)	Mean Sieve Size $(D_i—mm)$	Mass Retained (M_i)	M_i/D_i
10	2.00	1.43	0.5	0.4
20	0.85	0.73	5.2	7.1
30	0.60	0.51	30.6	60.0
40	0.425	0.34	45.4	133.5
60	0.25	0.22	13.2	60.0
80	0.180	0.16	5.1	31.9
100	0.150			
		TOTALS	100.0	292.9

$$D_m = \frac{\Sigma M_i}{\Sigma M_i/D_i} = \frac{100}{293} = 0.34 \text{ mm} = 340 \ \mu m$$

$$e = G \frac{\gamma_w}{\gamma_d} - 1 = 2.65 \frac{9.81}{14.2} - 1 = 0.83$$

The kinematic viscosity of water is

$$\mu/\rho = 1.16 \text{ mm}^2/\text{sec} \text{ at } 15°C.$$

Estimate $C_s = 600$.

$$k = \frac{2g}{C_s} D^2 \frac{\rho}{\mu} \frac{e^3}{1+e} = \frac{2(9.81 \text{ m/sec}^2)(340 \times 10^{-6} \text{ m})^2}{600(1.16 \times 10^{-6} \text{ m}^2/\text{sec})} \frac{0.83^3}{1.83}$$

$$k = 1.0 \text{ mm/sec}$$

Soil Permeability

Some representative values of the coefficient of permeability for different soils are shown in Fig. 60-1. Note the extreme variation in permeability from gravel to clay.

Saturated Fluid Flow

Figure 60-1 Typical values of permeability.

Soil Type	Coefficient of Permeability (Range in mm/sec)
Gravel	>10
Sand	$10-10^{-4}$
Silt	$10^{-4}-10^{-6}$
Clay	$<10^{-6}$

Permeability of Nonhomogeneous Soils

In some cases such as that of the stratified bed of soil in Fig. 60-2, the proper value of the coefficient of permeability for the soil may not be found by taking the numerical average of the coefficient of permeability at several different locations. The coefficient of permeability to be used for flow through this type of soil deposit depends on the direction of flow. In some instances it is possible to compute effective coefficients of permeability. Development of effective permeability in stratified soils for horizontal flow and vertical flow are shown in Figs. 60-2 and 61-1.

Figure 60-2 Coefficient of permeability for horizontal flow through stratified nonisotropic soil.

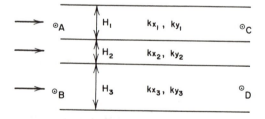

Horizontal Flow

$$q = q_1 + q_2 + q_3$$

$$k_x i \, \Sigma \, H_j = k_{x_1} i H_1 + k_{x_2} i H_2 + k_{x_3} i H_3 = \Sigma [k_x i H]_j$$

$$k_x i \, \Sigma \, H_j = i \, \Sigma (k_x H)_j$$

so

$$k_x = \frac{\Sigma (k_x H)_j}{\Sigma H_j} \qquad (60\text{-}2)$$

One-Dimensional Flow

Horizontal Flow (Fig. 60-2)

Assume a case in which the total head along line AB has a constant value, which would occur in hydrostatic pressure. Also, along line CD, a constant head exists, but at a value lower than along line AB. Flow will then be horizontal in each of the three layers shown. The effective coefficient of permeability in the x direction is, therefore, equal to the weighted average of the values of k_x in the three layers (Eq. 60-2).

Vertical Flow (Fig. 61-1)

If the gradient in head is in the vertical direction, then Eq. 61-1 may be used to estimate an effective coefficient of permeability through the horizontally stratified soil.

Figure 61-1 Coefficient of permeability for vertical flow through stratified, non-isotropic soil.

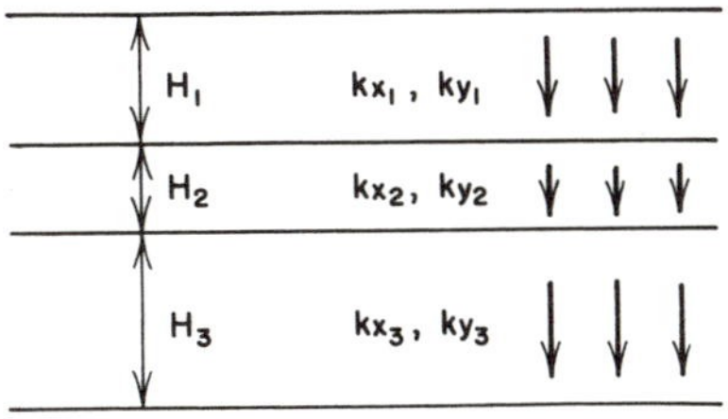

Vertical flow

$$q = q_1 = q_2 = q_3$$

$$k_y \frac{\Sigma(h_L)_j}{\Sigma H_j} A = k_{y1} \frac{h_{L_1}}{H_1} A = k_{y2} \frac{h_{L_2}}{H_2} A = k_{y3} \frac{h_{L_3}}{H_3} A$$

$$\Sigma(h_L)_j = \frac{q \, \Sigma H_j}{k_y A} = \frac{q H_1}{k_{y1} A} + \frac{q H_2}{k_{y2} A} + \frac{q H_3}{k_{y3} A}$$

$$\frac{\Sigma H_j}{k_y} = \left[\Sigma \frac{(H)}{k_y} \right]_j$$

and

$$k_y = \frac{\Sigma H_j}{\Sigma \left[\frac{(H)}{k_y} \right]_j} \qquad (61\text{-}1)$$

Figure 62-1

Example Problem The three layers of soil shown in Fig. 61-1 represent the soil profile beneath a reservoir in which the water depth is 10 m. Other data are as follows.

Layer	H (m)	$k_x \left(\dfrac{mm}{sec}\right)$	$k_y \left(\dfrac{mm}{sec}\right)$
1	2	2.0×10^{-4}	3.0×10^{-5}
2	1	1.5×10^{-5}	3.0×10^{-6}
3	3	1.0×10^{-5}	1.0×10^{-6}

A sandy layer lies below this profile. The sand has horizontal drainage, and the pore pressure in the sand is essentially zero.

Assume vertical flow through the layers shown and compute the water loss in three months for the reservoir ($A = 3000$ m²).

$$k_y = \frac{6 \text{ m}}{\left(\dfrac{2}{3.0 \times 10^{-5}} \text{ m/sec} + \dfrac{1}{3.0 \times 10^{-6}} + \dfrac{3}{1.0 \times 10^{-6}}\right)}$$

$$= 1.76 \times 10^{-6} \text{ mm/sec}$$

$$Q = k_y iAt = \left[\left(\frac{1.76 \times 10^{-6}}{1000} \text{ m/sec}\right)\left(\frac{16 \text{ m}}{6 \text{ m}}\right)(3000 \text{ m}^2)\right.$$

$$\left.(3 \times 30 \times 24 \times 3600 \text{ sec})\right] = 109 \text{ m}^3$$

Although the above effective values of coefficient of permeability may be used in flow situations where the direction of flow is horizontal or vertical (one dimensional), if the flow is curved it may be impossible to reduce this stratified soil to a simple homogeneous anisotropic soil mass by the method shown above. However, the example problem of Fig. 62-1 serves as an example of what may be done when the flow direction is known and the proper data determining permeability of the soil layers can be obtained.

Two-dimensional Flow — Flow Nets

Figure 63-1 shows the flow net for flow through the constant head permeameter in Fig. 54-1. The vertical lines in the soil sample represent the direction of flow

Figure 63-1 Flow net for permeameter.

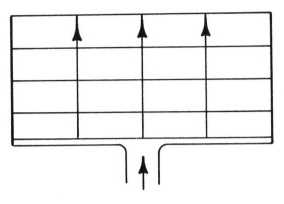

of water particles and are called flow lines. As the water progresses through the sample, head is lost at a constant rate and the horizontal lines represent lines of constant head. The two mutually perpendicular families of curves, flow lines and constant head lines, form a flow net. The flow net is a useful device in flow problems, because, once it is found, the values of the head at any point in the soil sample are uniquely determined.

The flow net of Fig. 63-2 has curved flow lines. This relatively simple laboratory experiment consists of an impermeable rectangular tank filled with

Figure 63-2 Curved flow lines.

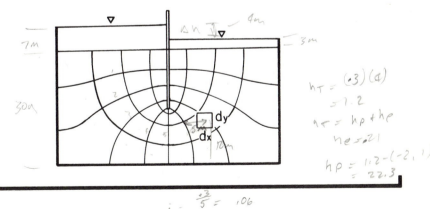

Saturated Fluid Flow

Figure 64-1 Derivation of the LaPlace equation.

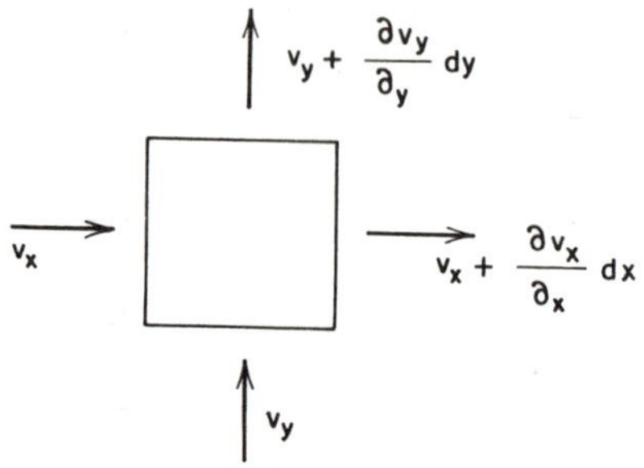

An element of soil in two-dimensional curved flow field.

Conservation of Mass

$$v_x\, dy\, dz - \left(v_x + \frac{\partial v_x}{\partial x}\, dx\right) dy\, dz + v_y\, dx\, dz - \left(v_y + \frac{\partial v_y}{\partial y}\, dy\right) dx\, dz = 0$$

where dz is the dimension of the element perpendicular to the two-dimensional flow.

From the preceding equation,

$$\frac{\partial v_x}{\partial x} + \frac{\partial v_y}{\partial y} = 0$$

Now, since

$$v_x = k_x i_x = k_x\,\frac{\partial h}{\partial x} \qquad \text{and} \qquad v_y = k_y\,\frac{\partial h}{\partial y}$$

then

$$k_x\,\frac{\partial^2 h}{\partial x^2} + k_y\,\frac{\partial^2 h}{\partial y^2} = 0 \qquad\qquad (64\text{-}1a)$$

For problems in which the coefficient of permeability in both directions is the same (isotropic), k may be cancelled and the equation becomes the LaPlace equation

$$\frac{\partial^2 h}{\partial x^2} + \frac{\partial^2 h}{\partial y^2} = 0 \qquad\qquad (64\text{-}1b)$$

The equation is sometimes written

$$\frac{\partial^2 \phi}{\partial x^2} + \frac{\partial^2 \phi}{\partial^2 h} = 0 \qquad\qquad (64\text{-}1c)$$

where $\phi = kh$ and is called the velocity potential.

2   Soil Water

soil that has a cutoff wall extending halfway through the soil. By maintaining a higher water level on the left-hand side of the cutoff wall, flow occurs from left to right. We may analyze the conditions of the flow at the element shown to determine the characteristics of curved flow. An enlarged picture of this element is shown in Fig. 64-1. Flow entering from the left and bottom of the element has velocities of v_x and v_y, respectively. The flow leaving the top of the element has a velocity equal to the velocity entering, plus the change in velocity in the distance dy. A similar incremental velocity has been added to v_x on the right-hand side of the sample. Conservation of mass specifies that if the volume of the sample remains the same and if the sample is completely saturated, then the net flow will equal zero. The derivation in Fig. 64-1 is based on this law. Either Eq. 64-1b or 64-1c is known as the LaPlace equation and describes the distribution of head in the interior of the homogeneous, isotropic soil region.

The solution to LaPlace's equation may be found by several different methods. Some of these are:

1. Direct mathematical solution.
2. Numerical solution.
3. Electrical analogy solution.
4. Graphical solution.

Direct Mathematical Solution
A demonstration of mathematical solutions can be made by picking functions that satisfy LaPlace's equation and choosing the boundary conditions that make the solution apply to a physical problem. Some examples of this method of solution follow.

Example 1
Consider the equation $h = 4y$, which satisfies LaPlace's equation. Values of h are plotted on the coordinate system in Fig. 66-1a. In Fig. 66-1b boundaries have been placed around a portion of the flow field to indicate a physical problem of which this equation is a solution. The values of h in Fig. 66-1b satisfy not only LaPlace's equation, but also the boundary conditions and therefore represent the proper solution.

Example 2
Consider the equation $h = 2xy$. This equation also satisfies LaPlace's equation. Values of h are plotted on the coordinate system in Fig. 67-1, and boundaries are placed in the flow field to show a physical problem.

Example 3
$h = C_1 \log C_2(x^2 + y^2)^{1/2}$. This equation satisfies LaPlace and is a solution for the problem shown in Fig. 68-1, in which water is being pumped from a well and is

Saturated Fluid Flow

Figure 66-1 Application of the LaPlace equation to one-dimensional flow.

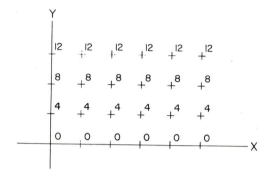

(a) Head distribution of $h = 4y$

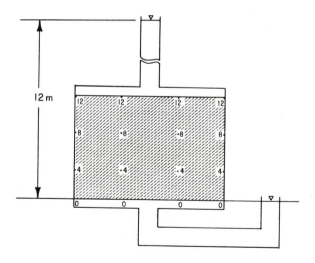

(b) Physical boundaries for $h = 4y$

Figure 67-1 Application of the LaPlace equation to two-dimensional flow.

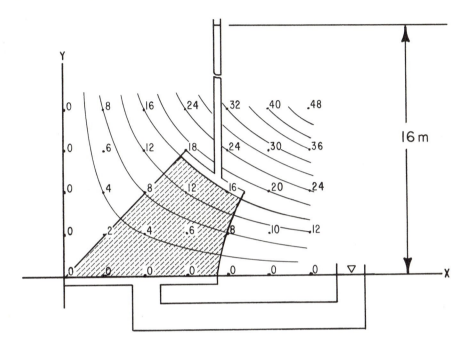

Physical problem for $h = 2xy$

flowing through a permeable aquifer bounded on the top and bottom by relatively impermeable soil.

The type of flows considered in this section are limited to two-dimensional flows. This theory may be extended to three-dimensional flow by including the direction perpendicular to the paper in the derivation of the differential equation. It is apparent that any number of possible solutions to problems may be analyzed, with the intent of working from the possible solution to the physical problem it represents. However, this method is not particularly suited for solving a problem with difficult boundary conditions. A more proper mathematical method is available using the theory of complex variables, but this method will not be discussed here. Numerical solutions to flow nets are discussed in Chapter 10.

Saturated Fluid Flow   **67**

Figure 68-1 Application of the LaPlace equation to axisymmetric flow.

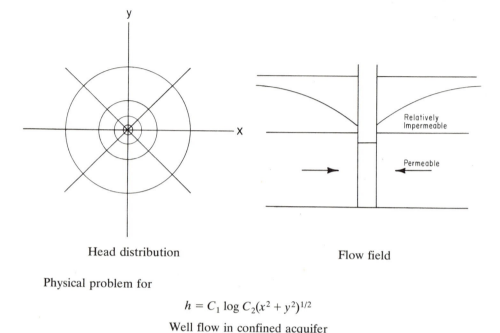

Head distribution   Flow field

Physical problem for

$$h = C_1 \log C_2 (x^2 + y^2)^{1/2}$$

Well flow in confined acquifer

Electrical Analogy Solution

A method of obtaining a solution to a two-dimensional flow problem using an electrical flow field analogy has proved very useful in some difficult flow problems. This method is based on the fact that steady-state electrical flow through a conductor of uniform thickness may also be represented by the LaPlace equation. For example, in Fig. 69-1 the flow field actually represents the flow through a permeable soil under a concrete dam. However, this problem may be solved by using a homogeneous electrical conducting medium in the shape of the flow field. (Conducting paper is often used.) Silver electrodes are then attached or painted on the medium along lines AB and CD. A difference in electrical potential is applied to these two electrodes, and the electrical flow occurs along the flow lines shown. The position of the constant head lines may be found by tracing lines of constant electrical potential (voltage). Flow lines may then be drawn perpendicular to the constant head lines. The flow net found in this manner is the same flow net that will exist in a hydraulic flow field.

Figure 69-1 Flow net for concrete dam on permeable foundation.

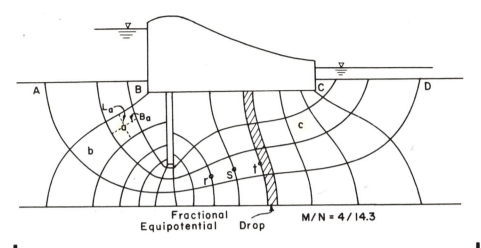

Fractional Equipotential Drop   M/N = 4/14.3

Graphical Solution

Flow nets such as those shown in Figs. 63-2 and 69-1 consist of two families of mutually perpendicular curves. These two families of curves represent the distribution of head within a flow field (constant head lines) and the path followed by water flowing through the field (flow lines). When the constant head lines are properly drawn, they describe the unique solution to the LaPlace equation (Eq. 64-1b) for the specific boundary conditions of the problem. A convenient way to obtain the flow net is by trial-and-error sketching of the flow lines and constant head lines. Several simple criteria must be satisfied in sketching the flow net.

1. The flow lines and constant head lines must be mutually perpendicular.
2. The areas formed by the intersection of the flow lines and constant head lines should form "square" figures.
3. The flow net must satisfy the boundary conditions of the flow field.

The criteria of square figures is developed in Fig. 70-1. The figure also demonstrates that the head drop across each square is constant and that the flow in any flow path equals the flow in any other flow path.

The graphical method for finding flow nets is most easily completed as follows.

Refer to Fig. 69-1. On a sheet of unlined paper draw to scale in ink the boundaries of the flow field, and then, using a soft-lead pencil, sketch four or five

Saturated Fluid Flow   **69**

Figure 70-1 Derivation of flow rate through two-dimensional flow field.

Consider the flow through squares a and c in Fig. 69-1. Since these squares are in the same flow path and since the flow is a steady flow, the value of q through the two squares will be the same. Therefore, from Darcy's law,

$$q_a = k \frac{\Delta h_a}{L_a} B_a = q_c = k \frac{\Delta h_c}{L_c} B_c$$

If the flow net is sketched so that B/L is constant, *the head drop across each square will be a constant.* The most convenient ratio to maintain is B/L equal to 1. This produces "square" figures.

Also
$$q_a = k \Delta h_a, \qquad q_b = k \Delta h_b$$

$\Delta h_a = \Delta h_b$, because these two squares lie between the same equal head lines. Therefore, $q_a = q_b$, *so the flow in any flow path equals the flow in any other path.* Now,

$$q = q_1 + q_2 + \ldots + q_m = q_1 M = k \Delta h_a M$$

$$q = k h_L \frac{M}{N} \tag{70-1}$$

where M is the number of flow paths, N is the number of equal head drops, and h_L is the total head loss through the net.

In the case of partial squares, either N or M may include a fractional square.

possible positions for flow lines. The first approximation for these lines should have a fairly uniform spacing. Then sketch constant head lines, making an attempt to keep all figures square and all intersections with flow lines at right angles. For the purpose of this method, a square figure may be defined as one in which the median flow line and the median equal head line through the center of the figure have the same length and that has 90° corners at all intersections. After drawing the complete family of constant head lines, erase all the flow lines and redraw them, attempting to keep the figures square and the intersections perpendicular. Then erase the constant head lines and redraw them. This procedure may be repeated until the figures all satisfy the conditions given. The flow net arrived at in this manner is the same flow net that would be produced by a mathematical solution if one were available for this problem — since there is only one pair of mutually perpendicular families of curves that will solve this problem, and that pair may be found by any of the methods suggested here.

During the sketching of the flow net, it may become obvious that an entire set of figures such as the shaded area of Fig. 69-1 cannot be drawn as squares but will turn out to be some fraction of a square. This will still be a proper flow net if the ratio of length to width of all the figures in such a line is constant. After the sketching procedure is completed, the flow through the net may be found, as shown in Fig. 70-1.

The water pressure at any point such as s in the flow net in Fig. 69-1 may be found from the flow net.

$$(h_T)_s = (h_e + h_p)_s = \left(h_e + \frac{u}{\gamma_w}\right)_s$$

The total head $(h_T)_s$ is equal to

$$h_{AB} - \frac{9}{14.3} h_L$$

and

$$u_s = \gamma_w (h_T - h_e)_s$$

Velocities in the flow net may be found, for example, at point s by taking the drop in head between r and t, dividing by the distance, and then applying Darcy's law.

$$v = -\frac{k \Delta h_{r-t}}{rt}$$

Two-dimensional Flow — Anisotropic Soil

A modification of the above methods is necessary in soils in which the horizontal and vertical coefficients of permeability are different. The permeabilities k_x and k_y then remain in Eq. 64-1a. Because this equation is not LaPlace's equation, the method of solution will not be the same. Consider a case in which the horizontal coefficient of permeability is greater than the vertical coefficient, as shown in Fig. 72-1. Let the area shown represent a mass of soil in which flow is occurring. The flow net in this situation will not be composed of mutually perpendicular families of curves. However, if we contract the x dimension, keeping the total resistance to flow in the x direction a constant, the resistance per unit distance increases and we may form a soil mass that has equal coefficients of permeability in both directions (transformed section). Let $x' = ax$ be the new dimension on the transformed section; then the LaPlace equation holds for the transformed section (Fig. 72-1) and the flow net in this section may be found in the usual manner. Intersections between flow lines and equal head lines may then be transformed back to the real section by multiplying the x' dimension by the factor $1/a$. In order to find the flow through the flow net in the transformed section in Fig. 72-1b, we may use the equation

$$q = k' h_L \frac{M}{N}$$

Figure 72-1 Transformed flow net for anisotropic soil.

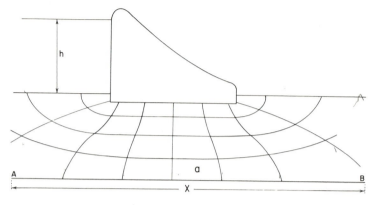

(*a*) Actual flow net for $k_x > k_y$

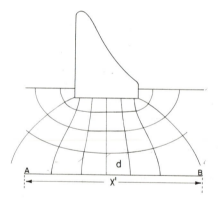

(*b*) Flow net—transformed section

Develop the transformation factor a.
   If $x' = ax$,

$$\frac{\partial h}{\partial x'} = \frac{\partial h}{\partial x}\frac{\partial x}{\partial x'} + \frac{\partial h}{\partial y}\frac{\partial y}{\partial x'}$$

but

$$\frac{\partial y}{\partial x'} = 0$$

$$\frac{\partial h}{\partial x'} = \frac{\partial h}{\partial x}\frac{\partial x}{\partial x'} = \frac{\partial h}{\partial x}\frac{1}{a}; \quad \frac{\partial^2 h}{\partial x'^2} = \frac{\partial(\partial h/\partial x')}{\partial x}\frac{\partial x}{\partial x'} = \frac{\partial^2 h}{\partial x^2}\frac{1}{a}\frac{1}{a}$$

Substituting into Eq. 64-1a,

$$k_x a^2 \frac{\partial^2 h}{\partial x'^2} + k_y \frac{\partial^2 h}{\partial y^2} = 0 \qquad (73\text{-}1)$$

which becomes the LaPlace equation if

$$a^2 = \frac{k_y}{k_x}$$

$$a = \sqrt{\frac{k_y}{k_x}}$$

(M is number of flow paths and N is number of head drops.) However, the value of the coefficient of permeability (k') will be different from either k_x or k_y. It may be found by analyzing the flow through two squares, one taken from the flow net of the real section and the corresponding one from the flow net of the transformed section (Fig. 74-1).

The example problem of Fig. 75-1 illustrates the computation of flow under an impermeable dam on an anisotropic soil deposit.

Flow in Earth Dams

In many soil problems involving flow nets, the boundaries of the flow net may not all be specified as in previous examples. One such case is in the flow through a homogeneous earth dam. The top flow line is a free surface and its exact position is not known. In order to approximate the position of this flow line, assume a dam having an upstream face in the shape of a parabola (Fig. 75-2) and an underdrain extending to a point directly under the intersection of the water surface and the upstream face of the dam. This is not a common design. However, if the dam were constructed in this manner and if the parabola had the focus at point F, the mutually perpendicular lines forming the flow net would be parabolas and the flow net would be as shown.

Changing the upstream face to a straight line and changing the position of the drain affects the position of the top flow line very little, and a parabola may generally still be used as an approximation to the top flow line (Fig. 76-1). This parabola is called the basic parabola. Studies of experimental flow nets have shown that the basic parabola intersects the water surface at a point A where AB is approximately 0.3 GB (Casagrande, 1937). The vertical line at E is called the

Figure 74-1 Flow rate through anisotropic soil. These squares are *a* and *d* in Fig. 72-1.

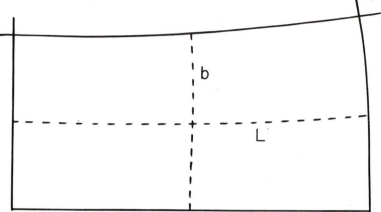

Square (a)—Real section

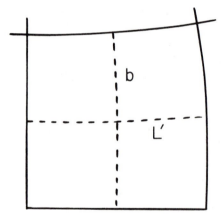

Square (d)—Transformed section

$$q_a = k_x \frac{\Delta h_a}{L} b = q_d = k' \frac{\Delta h_a}{L'} b;$$

$$k' = k_x \frac{L'}{L} = k_x a = \sqrt{k_x k_y} \tag{74-1a}$$

and

$$q = q' = k' h_L \frac{M}{N} \tag{74-1b}$$

Figure 75-1

Example Problem Use the dam shown in Fig. 72-1 (head upstream = 10 m, downstream = 0 m) $k_x = 2 \times 10^{-5}$ mm/sec, $k_y = 5 \times 10^{-6}$ mm/sec. Find the rate of flow beneath the dam.

$$a^2 = \frac{k_y}{k_x} = \frac{5 \times 10^{-6}}{2 \times 10^{-5}} = \frac{1}{4}$$

Note. The transformed section shown in Fig. 72-1 was drawn for this ratio.

$$k' = \sqrt{k_x k_y} = 1 \times 10^{-5} \text{ mm/sec}$$

$$q = k' h_L \frac{M}{N} = \left(\frac{1 \times 10^{-5}}{1000} \text{m/sec} \right)(10 \text{ m})\left(\frac{4}{8} \right)$$

$$q = 5 \times 10^{-8} \text{ (m}^3\text{/sec)/m}$$

Figure 75-2 Flow net with a free surface.

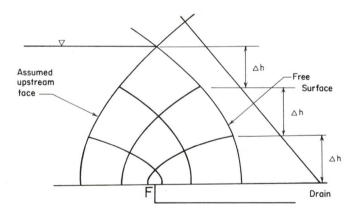

Flow net—confocal parabolas

Saturated Fluid Flow

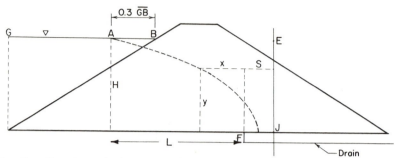

Figure 76-1 Flow net with a free surface.

Top flow line through A

directrix of the parabola, and point A is equidistant from E and F. Therefore,

$$\sqrt{H^2 + L^2} = L + S$$

where $S = \overline{FJ}$. S may be found from this equation and the directrix located. Points on the basic parabola may be found by using the geometric fact that each parabolic point is equidistant from the focus and the directrix; therefore,

$$\sqrt{x^2 + y^2} = x + S$$

The top flow line may be plotted from this equation, and the flow net sketched by the usual method (Fig. 77-1).

Since the parabola representing the top flow line does not intersect the upstream face of the dam at the water level, it is necessary to make a correction in the upper part of the top flow line. This correction is shown in Fig. 77-1, where the corrected top flow line meets the upstream face (which is an equal head line) at a 90° angle. In sketching the flow net, it is important to know that the intersections of equal head lines with the top flow line are equally spaced in the vertical direction, since head losses along this line are entirely losses in elevation head. The total flow through the net may be found by Eq. 70-1.

The total flow through the net may also be found by examining the symmetrical lower section of the net (see Fig. 77-1b). Note that $M = N$ in this section. Point R, directly above the focus, is equidistant from the focus and the directrix. The total head at point R is equal to S (so $h_L = S$), and $\overline{RT}$ is a constant head line; therefore

$$q = kh_L \frac{M}{N} = kS$$

Figure 77-1

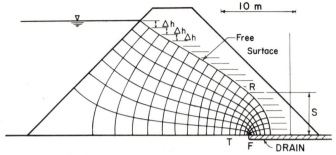

(*a*) Flow net for earth dam

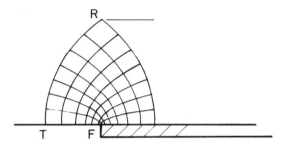

(*b*) Confocal parabolas

Figure 77-2

Example Problem For the dam of Fig. 77-1, $k = 2 \times 10^{-5}$ mm/sec, find the top flow line and draw the flow net. Find the rate of flow.

$$S = \sqrt{H^2 + L^2} - L = \sqrt{15.0^2 + 19.3^2} - 19.3 = 5.14; \quad \sqrt{x^2 + y^2} - S = x$$

The flow net is plotted in Fig. 77-1

$$q = kS = \left(\frac{2 \times 10^{-5} \text{ mm/sec}}{1000} \right) 5.14$$

$$1.03 \times 10^{-7} \text{ (m}^3\text{/sec)/m}$$

Alternate method From Eq. 70-1,

$$q = kh_L \frac{M}{N} = \left(\frac{2 \times 10^{-5}}{1000} \right) (15) \left(\frac{7}{19} \right)$$

$$q = 1.1 \times 10^{-7} \text{ (m}^3\text{/sec)/m}$$

The example problem of Fig. 77-2 illustrates two methods to compute the flow rate through an earth dam. The basic parabola may be a poor estimate of the flow line for some geometries ($L \ll H$), judgment should be exercised in its use.

Flow Net for Earth Dam on Impervious Foundation

In the homogeneous earth dam shown in Fig. 78-1a, the intersection of the downstream face and the impervious foundation is taken as the focus of the parabola for the top flow line. For slope angles greater than 30°, the table in Fig. 78-1b may be used to find the position of exit of the top flow line from the downstream face of the dam. This point is somewhat off the basic parabola forming the top flow line, and the flow line may be corrected to include this point. The corrected top flow line exits the downstream face along a direction tangent to the face, because the particles of water as they exit the dam tend to flow in the direction as close to vertical as the configuration of the dam allows. The equation $q = kS$ may be used to find the flow through the net. The example problem of Fig. 79-1 illustrates the method for locating the position of the top flow line. For

Figure 78-1

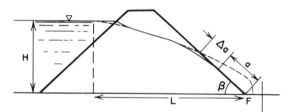

(a) Discharge face for earth dam on impervious foundation

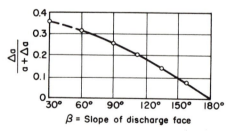

(b) Chart for locating exit point of top flow line

2   Soil Water

Example Problem From Fig. 78-1a, find the distance a to the intersection of the top flow line and the downstream face of the dam. From the figure: $\beta = 43°$, $H = 20$ m, and $L = 40$ m.

$$S = \sqrt{H^2 + L^2} - L = 4.72$$

The top flow line is drawn as shown in Fig. 78-1. $\Delta a + a$ is measured from the figure and is 16.7 m. From the graph in Fig. 78-1b.

$$\frac{\Delta a}{a + \Delta a} = 0.34$$

$$\Delta a = (0.34)(16.7) = 5.68$$

$$a = 16.7 - 5.7 = 11.0 \text{ m}$$

homogeneous dams in which β is less than 30°, Casagrande (1937) suggests the following equation for the distance a to the intersection of the top flow line.

$$a = \frac{L}{\cos \beta} - \sqrt{\frac{L^2}{\cos^2 \beta} - \frac{H^2}{\sin^2 \beta}}$$

For $\beta < 30°$, the quantity of flow may be found by

$$q = ka \sin \beta \tan \beta$$

SEEPAGE FORCES

Water flowing past a soil particle exerts a drag force on the particle in the direction of flow. The drag force is caused by pressure gradient and by viscous drag. To develop mathematical expressions for the drag forces, consider the free bodies in Fig. 80-1. If the height H of the water surface in the tube is raised, the pressure at the bottom of the soil sample is increased and the drag force on the soil particle becomes greater. When the height H reaches a certain critical value H_c, the drag forces and the buoyant weight of the particle are in balance and the soil particles will be washed out of the container. At this critical condition, the pressure force acting on the bottom of the soil sample will just equal the weight of the soil and water mass in the container. The hydraulic gradient existing at this time is called the critical gradient (i_c). This condition also occurs at the time at

Figure 80-1 For the one-dimensional vertical flow condition shown in Fig. 80-1a, determine expressions for the critical hydraulic gradient and for the critical seepage force.

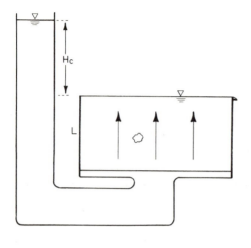

(a) Flow illustrating seepage forces

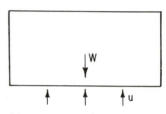

(b) Free-body of saturated soil

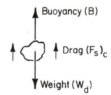

(c) Equilibrium forces

The determination of critical gradient from Fig. 80-1b is:

$$W = \left(\frac{G + e}{1 + e}\right)\gamma_w AL = uA = (H_c + L)\gamma_w A$$

from which

$$i_c = \frac{H_c}{L} = \frac{G - 1}{1 + e}$$

i_c is called the critical gradient.

The preceding result may be used to find the seepage force causing a critical condition. From Fig. 80-1c,

$$W_d = (F_s)_c + B$$

For the whole sample this becomes

$$(F_s)_c = \frac{G\gamma_w AL}{1+e} - \gamma_w V_s = \frac{G\gamma_w AL}{1+e} - \gamma_w(1-n)AL$$

$$= \left(\frac{G\gamma_w}{1+e} - \frac{\gamma_w}{1+e}\right) AL = i_c(\gamma_w)V$$

which individual soil particles are freely suspended in the flowing water and do not rest on adjacent particles. This equilibrium of forces is shown in Fig. 80-1c. From Fig. 80-1b, if H is less than H_c, the seepage force is proportionately less than $i_c\gamma_w$ and is always equal to $i\gamma_w \times$ volume in the direction of flow. The critical condition described above is called the quick condition, or boiling, and is responsible for the phenomenon called quicksand. It occurs in localities in which upward flow is occurring in fine sandy soils. The critical gradient is sufficient to cause boiling in sandy soils but not in cohesive soils in which the additional strength of the soil due to cohesion must be overcome before soil particles will be washed out of the soil mass.

PIPING AND BOILING

Erosion of soil particles along the contact surface between the soil foundation and concrete dams is responsible for many of the failures associated with concrete dams. Boiling and soil heave have been observed at the downstream point where flow emerges from the pervious foundation. This occurs when the exit hydraulic gradient approaches the critical hydraulic gradient. For the special case shown in Fig. 82-1, Terzaghi and Peck (1967) suggested analyzing the critical volume $ABCD$, where $\overline{CD}$ is one-half the depth $\overline{AD}$. The average vertical hydraulic gradient over the length $\overline{AD}$ is

$$i_{av} = \frac{(h_A + h_B)/2 - (h_C + h_D)/2}{\overline{AD}}$$

where h is the total head at the indicated points. If this gradient is equal to i_c in a granular soil, piping is imminent. The factor of safety against piping is

$$F = \frac{i_c}{i_{av}}$$

and is normally required to be three or greater. If used in cohesive soil, this method is conservative because of the added cohesive strength.

Figure 82-1 Piping analysis.

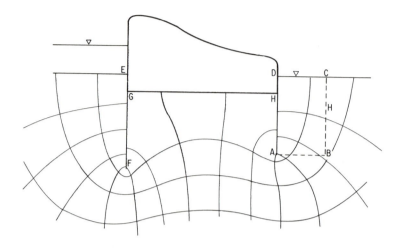

Lane (1935) investigated the problem of piping by using a different approach. He studied a large selection of concrete dams, some of which had failed due to piping, and compared distances of flow along the soil concrete contact surface. He found that as this distance increased, the piping failures occurred less frequently; he therefore devised the following criteria for design of these dams.

$$\text{WCR} = \frac{\text{WCD}}{h_L} = \frac{d_v + d_h/3}{h_L}$$

in which
    WCR = weighted creep ratio
    WCD = weighted creep distance
      d_v = the vertical distance along the contact path
      d_h = the horizontal distance along the contact path
      h_L = total head loss

The dam is considered safe with respect to erosion if WCR > $(\text{WCR})_{cr}$. See Fig. 83-1.

For concrete dams with upstream and downstream cutoff walls, such as shown in Fig. 82-1, the path of least resistance for water flow may be directly

Figure 83-1   Critical weighted creep ratios $(WCR)_{cr}$.

Material	Value
Very fine sand or silt to fine sand	8.5–7
Medium to coarse sand	6–5
Fine to coarse gravel	4–3
Boulders with some cobbles and gravel	2.5
Soft to medium clay	3–2
Hard to very hard clay	1.8–1.6

Figure 83-2
Example Problem   From Fig. 82-1, the soil is a silty sand, $k = 1 \times 10^{-3}$ mm/sec, $G = 2.65$, and $e = 0.60$. Check the design for safety against downstream erosion by the gradient method and the weighted creep method.
Solution *Gradient method.*   The water depth is 5.5 m. For an elevation datum at the ground surface,

$$h_C = h_D = 0, \qquad h_A = 1.38, \qquad h_B = 0.71$$

$$i_{av} = \frac{1.38 + 0.71 - 0 - 0}{2(15\ m)} = 0.070$$

$$i_c = \frac{G - 1}{1 + e} = \frac{2.65 - 1}{1 + 0.60} = 1.03$$

$$\frac{i_c}{i_{av}} = 14.7 = F$$

Weighted creep method

$$(WCD)_{EFGHAD} = 18.7 + 15.3 + \frac{29.0}{3} + 12.2 + 15.3 = 71.2$$

$$(WCD)_{EFAD} = 18.7 + 2(29.4) + 15.3 = 92.8$$

$$WCR = \frac{71.2}{5.5} = 12.95 > 8 \text{ is safe}$$

Piping and Boiling

through the soil from F to A. The two alternatives for weighted creep distance to be investigated in this case would, therefore, be

1. $\text{WCD}_1 = \overline{EF} + \overline{FG} + \dfrac{\overline{GH}}{3} + \overline{HA} + \overline{AD}.$

2. $\text{WCD}_2 = \overline{EF} + 2\,\overline{FA} + \overline{AD}.$

The least value is used in finding the weighted creep ratio. Lane recommended that the coefficient of the length $\overline{FA}$ in the equation for weighted creep distance have a value of 2, which indicates that he felt the resistance to flow along this line is twice as great as that along contact flow paths in the vertical direction. For this method, distances are considered to be vertical or horizontal if they are within 45° of that direction. Although the value of the factor of safety in Lane's method is not known precisely, the method itself is considered to be conservative. The example problem of Fig. 83-2 illustrates both the gradient method and the weighted creep method to check for safety against piping.

FILTER DESIGN

Internal and underdrain systems are used to control pore pressure buildup for a number of applications. One example is the blanket drain in the homogeneous earth dam of Fig. 77-1a. Drains must be designed to meet two basic performance criteria.

- The gradation of the drain material must be such that the fines of the ajacent soil will not migrate through the drain.
- The flow capacity of the drain must be high enough to convey all the seepage water without creating an excess hydrostatic head.

Considerable experimentation (U.S. Bureau of Reclamation) has shown that migration of particles from a fine soil into a coarse soil can be prevented if 15% (D_{85}) of the particles of the fine soil are larger than the effective pore size of the coarse soil. The effective pore size of the coarse soil corresponds to about one-fifth the D_{15} of the coarse soil. Criteria to prevent particle migration of the fine soil into the coarse soil can, therefore, be stated as

$$D_{85} \text{ (fine soil)} \geq \frac{1}{5} D_{15} \text{ (coarse soil)}$$

To convey water effectively, a drain must be considerably more permeable than the soil being drained. For uniform soils, the permeability is roughly proportional to the square of the effective grain-size diameter. Thus, for the drain material (coarse soil) to be twenty-five times (rule-of-thumb limit) as permeable as the material being drained (fine soil), the effective diameter of the coarse material must be five times the effective diameter of the fine material. This type of reasoning leads to the second design criteria.

$$D_{15} \text{ (coarse soil)} \geq 5\, D_{15} \text{ (fine soil)}$$

In addition to these criteria, the U.S. Bureau of Reclamation recommends that the grain-size distribution curves of the fine and coarse soils be roughly parallel.

Perforated drain pipes are frequently used in various internal and underdrain systems. Migration of fine material into the pipe can occur if the size of the perforations in the pipe are too large relative to the D_{85} of the soil. In order to prevent migration of fines into the pipe, the following criterion is often used.

$$\frac{D_{85} \text{ (soil)}}{\text{maximum diameter of perforations}} \geq 2$$

PROBLEMS

2-1 Discuss the distribution of water and water pressure in the capillary zone.

2-2 Is water pressure in the adsorbed layer transmitted equally in all directions? Discuss.

2-3 Find the height of rise in a capillary tube with a radius of 0.01 mm.

2-4 Draw a graph of water pressure versus height for the capillary tube in problem 2-3.

2-5 Discuss natural limitations on the height of rise of capillary water in fine-grained soils.

2-6 Find the pressure head, elevation head, and total head for points a through e in each of the following situations.

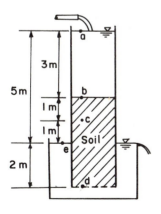

Prob. 2-6a

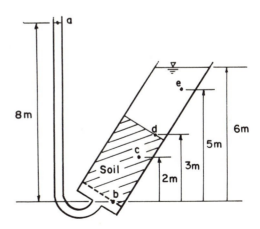

Prob. 2.6b

2-7 For the figure in problem 2-6a, the void ratio is 0.82, A (cross-sectional area) = 1 m², and Q = 0.5 liter in 15 min.
Find:

(a) v (velocity) in meters per second.
(b) v_n (actual velocity).
(c) k (coefficient of permeability) in meters per second.

2-8 If Fig. 53-1 represents a constant head permeameter and if k (coefficient of permeability) for the soil is 0.001 m/sec, find Q (flow volume) in 1 hr. (a, area of tube = 300 mm²; A, area of sample = 0.1 m²).

2-9 Let Fig. 53-1 represent a falling head permeameter. Use the data in problem 2-8. At a time when the head loss in the soil is 4 m, find q (rate of flow) in cubic meters per second.

2-10 The water depth in an auger hole (diameter = 0.15 m), which is drilled to a depth 3 m below the water table, rises from 0.2 m to 1.2 m (measured from the bottom of the hole) in 6 min. Estimate the coefficient of permeability of the soil. Use Hooghoudt and Ernst equations.

2-11 Plot on an x-y coordinate system the head distribution for the following equations. Draw constant head lines. Choose physical boundaries that describe a real problem in fluid flow. (Check the solutions to see if they satisfy LaPlace's equation.)

(a) $h = 4y + 2xy$.
(b) $h = x^2 + y^2$.

2-12 Water flows through a saturated silt formation at the rate of 0.1 m³/sec. What would be the rate if the head loss were increased 60% and the flow path doubled in length?

2-13 Refer to Fig. 63-2 (scale 1 mm = 1 m) $k = 1 \times 10^{-5}$ mm/sec at the position of the element. Find:

(a) Pressure head.
(b) Gradient.
(c) Velocity.
(d) Pore pressure.

2-14 Refer to Fig. 67-1. Choose a position on the constant head line labeled 12 and somewhere inside the soil mass. Find:

(a) Pressure head.
(b) Pore pressure.
(c) Gradient.

2-15   This homogeneous earth dam and its foundation are of the same soil down to the impervious layer. Draw the flow net and estimate the rate of flow through the dam.

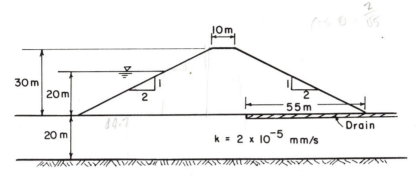

Prob. 2-15

2-16   (a)   Determine the flow rate (m^3/sec/m) through this permeable soil.
(b)   Determine the pore pressure at A.
(c)   Determine the hydraulic gradient at A.   2.5 m
(d)   Determine the seepage force per unit volume at A.

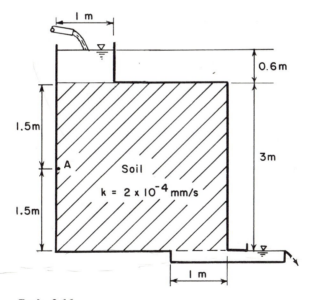

Prob. 2-16

Problems

2-17 Steady flow is established in this homogeneous earth dam ($k = 10^{-5}$ mm/sec). Draw the flow net. Find the seepage rate per meter.

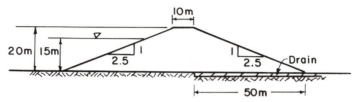

Prob. 2-17

2-18 Refer to problem 2-15. The permeable drain becomes clogged. Find the point at which the top flow line intersects the downstream surface of the dam. Find the seepage rate per meter.

2-19 Find the magnitude of the seepage force per unit volume at the center of the soil sample in Figs. 53-1 and 66-1b.

2-20 From the flow net shown below, find:

(a)   The flow (q) through the net.
(b)   The water pressure (u) in the middle of square A.
(c)   The actual velocity in the middle of square A.
(d)   The seepage force per unit volume (seepage pressure) at A.
(e)   The seepage force on the soil in square A.
(f)   The factor of safety against piping using the hydraulic gradient.

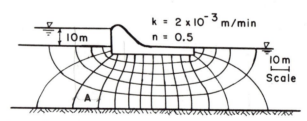

Prob. 2-20

Label all assumed dimensions.

2-21 Investigate the safety against downstream erosion for the dam in Fig. 82-1 (scale 1 mm = 2 m). The soil is a silty sand, $G = 2.65$, $e = 0.55$.

2-22 For the dam shown in problem 2-17, $k_x = 1 \times 10^{-5}$ mm/sec and $k_y = 2 \times 10^{-6}$ mm/sec. Draw the flow net for the transformed section and find q (flow rate).

3

SOIL STRESSES

Stresses imposed on the soil by the weight of the overburden or by structural loads may cause strengthening of the soil mass or failure, depending on the method of application of the load and the distribution of the stresses.

Methods that may be used to predict soil response to stress are discussed in Chapters 4 through 8 and 11. These methods all depend on a fairly accurate description of the stresses at different points in the soil. These stresses change greatly with time and position and are greatly affected by changes in moisture content and by previous stress history. This chapter deals with methods of estimating soil stresses under different loading conditions.

EFFECTIVE STRESS CONCEPT

Dry Soil

Soil pressure, also called effective stress or intergranular pressure, may be defined from Fig. 90-1a, which represents a cylindrical free body cut from a mass of dry homogenous soil with a level surface. The cross-sectional area of the free body is not plane but is taken in such a way that it passes through the points of contact between soil particles and does not cut through any particle. The free body is in vertical equilibrium under the action of the two forces shown; therefore,

$$\bar{\sigma} = W/A = \gamma H,$$

Figure 90-1 Vertical pressure in soils.

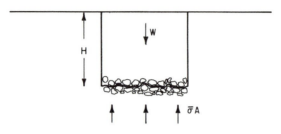

(*a*) Vertical pressure in dry soil

$$\sigma = \bar{\sigma} = \frac{W}{A} = \gamma H$$

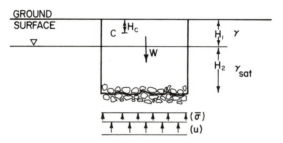

(*b*) Vertical pressure in saturated soil

Vertical Pressure in Saturated Soil

Total stress:

$$\sigma = \frac{W}{A} = \gamma H_1 + \gamma_{sat} H_2 \qquad (90\text{-}1a)$$

Pore pressure:

$$u = \gamma_w H_2$$

Effective stress:

$$\bar{\sigma} = \sigma - u \qquad (90\text{-}1b)$$

where $\bar{\sigma}$ is defined as the soil effective stress. It is apparent that the pressure $\bar{\sigma}$ is caused by forces acting through the contact points between soil particles. These contact points have very high concentrations of pressure. When the forces at all contact points are added and divided by the nominal cross-sectional area A, the effective stress $\bar{\sigma}$ is obtained. The effective stress represents an average stress over the whole area and is not a measure of the true intensity of stress at the contact points. However, $\bar{\sigma}$ as here defined, does give a good index that is convenient to use and is satisfactory for soil mechanics problems.

Saturated Soil

Now consider the same soil profile with the addition of a water table, as shown in Fig. 90-1b. The soil below the water table is assumed saturated and the soil above the water table contains capillary moisture. The total weight of the free body, in addition to the soil weight, now includes the weight of the capillary water above the water table, plus the water filling the soil pores below the water table. The total weight per unit area (total stress) is given in Eq. 90-1a, using the unit weight of the moist soil above the water table and the unit weight of saturated soil below the water table. The total stress is balanced by the stress in the water or pore pressure u at the lower surface of the free body and by the effective stress in the soil structure. The effective stress in Eq. 90-1b is then equal to the total stress minus the pore pressure. Equation 90-1b is known as the "effective stress principle." Recognition of this seemingly simple concept by Karl Terzaghi in the 1920s marked the beginning of a rational understanding of many soil phenomena. Terzaghi (1936) stated that "All the measurable effects of a change of the stress, such as compression, distortion and a change of the shearing resistance are exclusively due to changes in the effective stresses. . . ."

A further refinement to the effective stress equation has been suggested, assuming that the pore pressure does not act over that portion of the cross-section occupied by the soil contact points. Indications are that the portion of contact area $a_c A$ is small compared to the gross cross-sectional area A, and the factor a_c may be neglected.

$$\bar{\sigma} = \sigma - (1 - a_c)u = \sigma - u$$

The effective stress principle is illustrated in the example problem of Fig. 92-1.

Partially Saturated Soil

For a soil column above the water table at point C in Fig. 90-1b, a more rigorous analysis of equilibrium shows the total stress equal to

$$\sigma = \frac{W}{A} = \gamma H_c$$

Effective Stress Concept

Figure 92-1

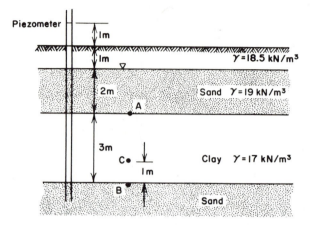

Example Problem An artesian pressure exists in the lower sand layer of the profile shown above, with a piezometric surface 1 m above ground surface. Determine the total stress, pore pressure and effective stress at points A, B and C.

Point A

$$\sigma = \gamma H_1 + \gamma_{sat} H_2$$
$$\sigma = 18.5(1) + 19.0(2) = 56.5 \text{ kPa}$$
$$u = \gamma_w H_2$$
$$u = 9.81(2) = 19.6 \text{ kPa}$$
$$\bar{\sigma} = \sigma - u$$
$$\bar{\sigma} = 56.5 - 19.6 = 36.9 \text{ kPa}$$

Point B

$$\sigma = 18.5(1) + 19.0(2) + 17.0(3)$$
$$\sigma = 107.5 \text{ kPa}$$

The pore pressure in the lower sand layer is determined by the height of water in the piezometer.

$$u = 9.81(7) = 68.7 \text{ kPa}$$
$$\bar{\sigma} = 107.5 - 68.7 = 38.8 \text{ kPa}$$

Point C

$$\sigma = 18.5(1) + 19.0(2) + 17.0(2)$$
$$\sigma = 90.5 \text{ kPa}$$

3 Soil Stresses

A steady-state seepage condition exists in the clay layer with water flowing upward. The pore pressure must, therefore, be evaluated by first calculating the pressure head at point C.

$$u = \gamma_w h_{p_C}$$

$$h_{p_C} = h_{T_C} - h_{e_C}$$

Establish a datum at the bottom of the clay layer.

$$h_{T_C} = h_{T_B} - i(y)$$

$$i = \frac{h_L}{L} = \frac{2}{3} = 0.667$$

$$h_{T_C} = 7 - 0.667(1)$$

$$h_{T_C} = 6.33 \text{ m}$$

$$h_{p_C} = 6.33 - 1.0 = 5.33 \text{ m}$$

$$u = 9.81(5.33) = 52.3 \text{ kPa}$$

$$\bar{\sigma} = 90.5 - 52.3 = 38.2 \text{ kPa}$$

and

$$\sigma = \bar{\sigma} + \lambda u + (1 - a_c - \lambda)u_a$$

where u_a is the pore air pressure and λ is the decimal portion of cross-sectional area occupied by water. If a_c and u_a are very small or zero, the equation for effective stress becomes $\bar{\sigma} = \sigma - \lambda u$.

In a study of the effective stress principle in saturated clay where u is negative, Evans and Lewis (1970) show that the preceding equation should read $\bar{\sigma} = \sigma - \beta u$ when u is negative, thus making the effective stress greater than the total stress. It is implied that β is due to a cause other than λ and that when u is negative, the full capillary tension is not effective in increasing the effective stress. The limitations of this correction and the application of the effective stress principle in partially saturated or unsaturated soils above the water table are not yet fully understood.

STRESSES DUE TO SURFACE LOADS

An increase in soil pressure is found in a soil mass that supports a point load in Fig. 94-1. If the vertical pressure were measured at different points along the

Figure 94-1 Vertical pressures due to overburden and point load at the surface.

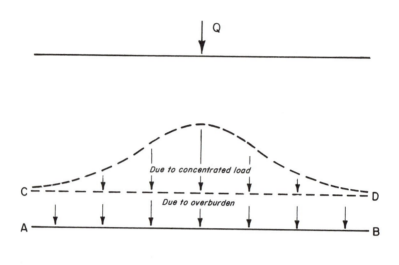

horizontal plane, *AB*, and each of these measured values were separated into two values—one representing the pressure caused by the weight of the soil above plane *AB* and another representing the effect of the concentrated load *Q*—the pressures could then be plotted on the diagram in Fig. 94-1 to yield the bell-shaped pressure curve shown. The maximum effect of the concentrated load is felt at a position directly beneath the load, and it becomes progressively less as the horizontal distance from the load increases. Values of the effect of the load *Q* on the vertical pressure at a point in the soil cannot be found by a simple free-body diagram, as in the previous section. The stress depends on the elastic properties of the soil and can be found only by an indeterminate analysis, which takes into account the displacements caused by the load.

Approximate Solutions

An approximate value of the maximum stress increase directly under a point load may be found, as in Fig. 95-1, by estimating the zone of influence of the concentrated load. This zone is represented here by the cone enclosed between the sloping lines. An angle of 51° between the sloping line and the horizontal line gives pressures that are reasonably close to the maximum stress provided by a more exact elastic analysis for the point load.

Figure 95-1 Approximate stress distribution under a point load.

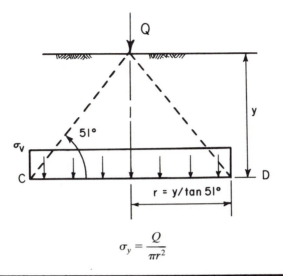

$$\sigma_y = \frac{Q}{\pi r^2}$$

For uniformly loaded circular or rectangular footings, an angle of 60° is often assumed, as in Fig. 96-1. The uniform pressure increase estimated by this means approximates the maximum pressure increase predicted by elastic methods (Fig. 101-1) when $1.5 < y/B < 5$. The approximate method of Fig. 96-1 is illustrated in Fig. 96-2.

Elastic Solution—Point Load

The analytical solution for stresses caused by a concentrated load on a semi-infinite elastic body is generally attributed to Boussinesq (1885), whose equations for stresses are defined in Fig. 97-1. Values of N_B in the equation for σ_y are given in Fig. 98-1.

Some soils are interspersed with thin layers of granular material that partially prevent lateral deformation of the soil. Westergaard (1938) found a solution for stresses in this type of material by considering an elastic medium in which the horizontal deformation was assumed to be zero; since most soils are nonhomogeneous, the Westergaard solution may be a better approximation of soil stresses than the solution proposed by Boussinesq. Equations and diagrams similar to those shown for the Boussinesq solution are presented for the Westergaard solutions in Figs. 98-1, 102-1, 103-1, and 105-1.

Stresses Due to Surface Loads

95

Figure 96-1 Approximate stress distribution under a uniformly loaded rectangle $B \times L$.

$$\sigma_v = \frac{Q}{(B + 2x)(L + 2x)}$$

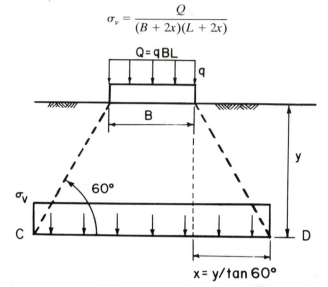

$$x = y/\tan 60°$$

Figure 96-2

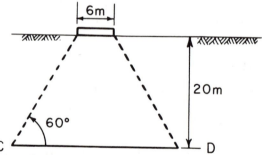

Example Problem Estimate the maximum vertical pressure 20 m below a square footing ($B = 6$ m) that exerts a uniform contact pressure of 20 kPa on the soil surface. Influence area:

$$CD = [6 + 2(20 \cot 60°)] = 29.09$$

$$\sigma_v = \frac{720}{CD^2} = 0.85 \text{ kPa}$$

Figure 97-1

(a) Stresses at a point due to a point load

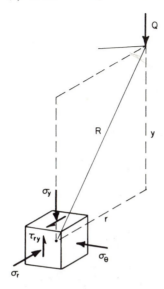

(b) Boussinesq equations

$$\sigma_y = \frac{3Qy^3}{2\pi R^5} = N_B Q/y^2 \qquad\qquad \sigma_\theta = \frac{Q}{2\pi}(1-2\nu)\left[\frac{y}{R^3} - \frac{1}{R(R+y)}\right]$$

$$\sigma_r = \frac{Q}{2\pi}\left[\frac{3r^2y}{R^5} - \frac{(1-2\nu)}{R(R+y)}\right] \qquad\qquad \tau_{ry} = \frac{3Qry^2}{2\pi R^5}$$

where ν is Poisson's ratio.

Stresses Due to Surface Loads

Figure 98-1 Boussinesq and Westergaard coefficients for a concentrated load.

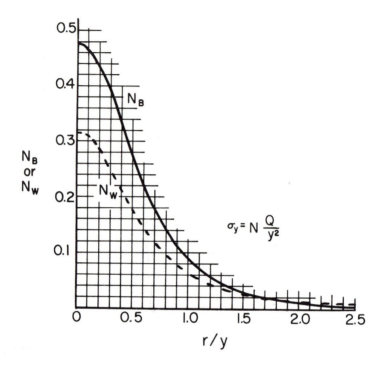

$$\sigma_y = N \frac{Q}{y^2}$$

$\dfrac{r}{y}$	N_B	N_w
0.0	0.477	0.318
0.1	0.465	0.308
0.3	0.385	0.248
0.6	0.221	0.141
1.0	0.084	0.061
1.5	0.025	0.025
2.0	0.008	0.012
2.5	0.003	0.006
3.0	0.0015	0.004

The Westergaard solutions are for a Poisson's ratio of zero, because this assumption gives the highest stresses. Stresses given by the Westergaard solutions range down to two-thirds those of the Boussinesq solutions.

Elastic Solution—Uniform Load

Stresses in a semi-infinite elastic mass due to a uniformly loaded area on the surface may be found by dividing the loaded area into small sections and considering the load on each section to be a concentrated load at the midpoint of the section. If the maximum dimension of the section is less than one-third the depth to the point where the stress is computed, this method will produce values for the stresses that are within approximately 5% of the correct stress.

A more exact expression for stresses caused by a uniformly loaded area may be found by integrating over the loaded area the stresses caused by a load dp on a differential portion of the loaded area. Newmark (1942) has performed this integration, and his solution is presented for the Boussinesq and Westergaard theories in Figs. 101-1 and 102-1, respectively. This chart may be used to find pressures under one corner of the loaded area. If stresses are desired at other points in the elastic medium such as directly beneath point A in Fig. 99-1, the area may be divided into four loaded rectangles and treated as the sum of four separate computations. See the example in Fig. 100-1b.

Influence Charts

Another method of integrating either the Boussinesq or the Westergaard relationship yields an influence chart as in Fig. 102-2 or Fig. 103-1, in which each

Figure 99-1 Stresses under point A for uniformly loaded retangle ($BCDE$).

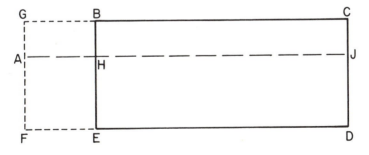

Figure 100-1

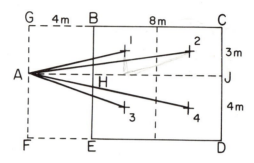

Example Problem Find the vertical pressure 20 m below point A by three methods (use Boussinesq theory).

(*a*) *Concentrated loads.*
(*b*) *Newmarks method, uniform loads.*
(*c*) *Influence chart.*

The contact pressure q is 20 kPa,

$$\overline{GB} = 4 \text{ m}, \overline{BC} = 8 \text{ m}, \overline{GA} = 3 \text{ m}, \overline{AF} = 4 \text{ m}.$$

(*a*) *Solution assuming concentrated loads*

Area	1	2	3	4
r/y	0.31	0.51	0.32	0.51
N_B	0.38	0.28	0.37	0.28
Q(kN)	240	240	320	320
N_BQ/y^2	0.23	0.17	0.30	0.22

$$\Sigma N_B Q/y^2 = 0.23 + 0.17 + 0.30 + 0.22$$

$$\sigma_y = 0.92 \text{ kPa}$$

(*b*) *Solution using stress under corner of uniformly loaded rectangles.*

Area	+AGCJ	+FAJD	−AGBH	−FAHE
L/B	4.0	3.0	1.33	1.0
B/y	0.15	0.2	0.15	0.2
M_B	+0.033	+0.043	−0.014	−0.018
qM_B	+0.66	+0.86	−0.28	−0.36

$$\sigma_y = q[M_{B(AGCJ)} + M_{B(FAJD)} - M_{B(AGBH)} - M_{B(FAHE)}] = 0.88 \text{ kPa}$$

(c) *Solution using the influence chart* A drawing of the loaded area scaled to the 20-m depth y and placed with point A at the centre of Fig. 102-2 covers about 19 "squares."

$$\sigma_y = I_B n q = (0.0025)(19)(20) = 0.95 \text{ kPa}$$

Figure 101-1 Boussinesq coefficient M_B for stress under the corner of a uniformly loaded rectangle. $L > B$

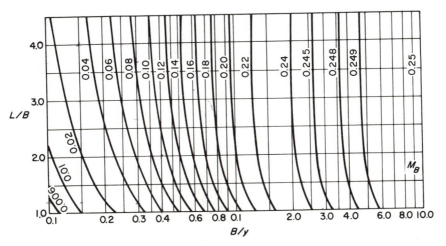

Figure 102-1 Westergaard coefficient M_W for stress under the corner of a uniformly loaded rectangle. $L > B$

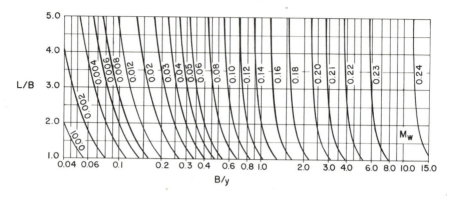

Figure 102-2 Influence chart for Boussinesq theory.

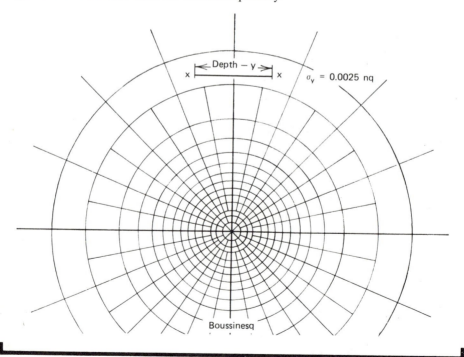

Figure 103-1 Influence chart for Westergaard theory.

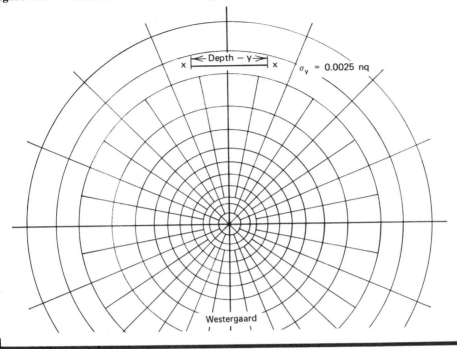

area on the chart, when loaded with a uniform stress, will produce the same increment of vertical stress at a depth y beneath the center of the diagram.

To use the influence chart, a plan of the uniformly loaded area is drawn to a scale such that the depth y at which the stress increase is to be determined equals the length x-x on the influence chart. The drawing of the loaded area is then positioned on the chart with the point where stresses are to be found located over the center of the influence chart. Count the number of squares covered (n) and then $\sigma_v = Inq$, where I is the influence value for the chart.

Graphical solutions for vertical stresses induced below a uniformly loaded square footing and a strip footing are shown in Figs. 104-1 and 105-1 for the Boussinesq and Westergaard solutions, respectively. Stresses below a circular footing of the same area would approximate the stresses below the square footing. Note that at a depth of $2B$ the vertical stress at the center of the square footing is less than $0.1q$. The line $0.1q$ defines a "pressure bulb" or zone of significant stresses below a footing. For the infinite strip the zone of influence is much deeper.

Stresses Due to Surface Loads

Figure 104-1 Vertical pressure due to uniformly loaded square or strip footings. Boussinesq solution.

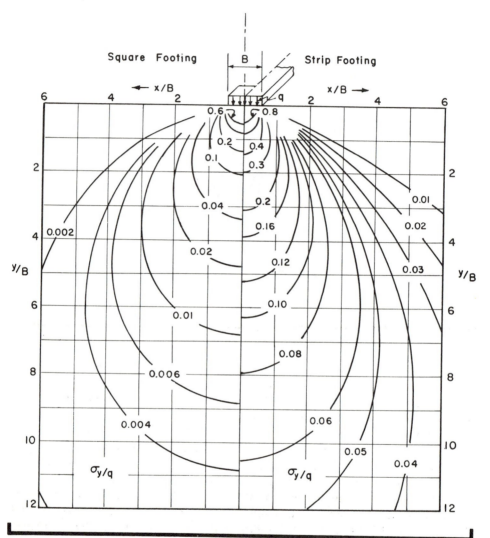

3 Soil Stresses

Figure 105-1 Vertical pressure due to uniformly loaded square or strip footings. Westergaard solution.

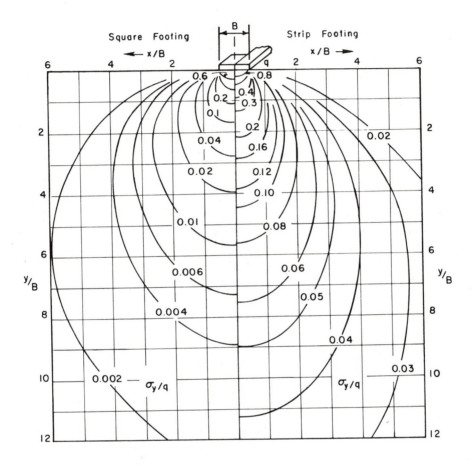

STRESS AT A POINT (MOHR CIRCLE)

To better understand the stress conditions in a soil, each of the infinite number of planes passing through a point in a stressed soil mass can be investigated (Fig. 107-1). In general, three of these planes have zero shear stress. The three planes are called principal planes and are mutually perpendicular to each other. Values of the normal stresses on these planes are called principal stresses—σ_1, the major principal stress; σ_2, the intermediate principal stress; and σ_3, the minor principal stress. In this analysis, compression is considered to be positive and tension negative, and σ_1 is the largest positive stress. A shear stress that produces a counterclockwise torque about the center of the free body is considered to be positive.

Assume that the plane containing the stress σ_2 has been found and that the stressed element is viewed from a direction perpendicular to this intermediate principal plane. Figure 107-2 shows this view of the element as oriented with respect to the x'-y' reference axes. A relationship between the stresses acting perpendicular to the intermediate principal stress may be found by applying the equations of equilibrium to the portion of this element enclosed by the triangle OPQ. All the stresses that have components in the plane and act on the surface of this free body are shown in Fig. 107-2. The area of face PQ is A.

Equations for the normal and shearing stresses on plane PQ are derived in Fig. 108-2. Equation 109-1c demonstrates the Mohr circle relationship for normal and shearing stresses at a point. This is shown graphically on Fig. 108-1. On the σ, τ axes, the Mohr circle has a radius equal to the square root of the right side of Eq. 109-1c, and is located at a center equal to

$$\frac{\sigma_x + \sigma_y}{2}$$

The graphical representation is a convenient aid in solving problems. Each point on the circle has coordinates (σ, τ) representing the normal and shearing stresses on a plane that makes an angle θ with respect to the plane on which σ_x acts, and that is also perpendicular to the intermediate principal plane. The point on the right side of the circle with coordinates $(\sigma_1, 0)$ represents the major principal stress which is greater than any other normal stress present on any plane through this point in the stressed body. The point on the left of the circle on the σ axis has coordinates $(\sigma_3, 0)$ and represents the stresses on the minor principal plane or least normal stress.

The stresses on plane PQ $(\sigma_\theta, \tau_\theta)$ are shown in Fig. 108-1 rotated an angle 2θ counterclockwise from the coordinates representing the stresses on plane PO (σ_x, τ_{xy}). The stresses σ_y and $-\tau_{xy}$ act on the plane 90° from PO and are represented on the Mohr circle 180° across the circle from the stresses σ_x and $+\tau_{xy}$. Stresses on any intermediate plane can be found by rotating the central angle of the Mohr circle in the same direction from the reference plane (plane PO in this example), an angle twice the actual rotation of the plane, and determining the σ, τ coordinates thus defined. Thus, the principal plane is located

106

Figure 107-1   Generalized stresses at a point.

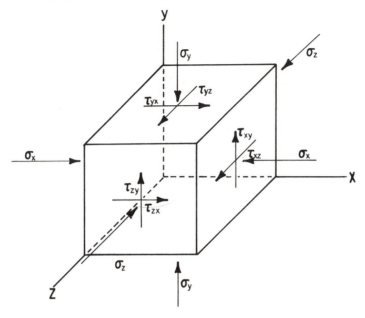

Figure 107-2   Stresses that act perpendicular to the intermediate principal stress at a point.

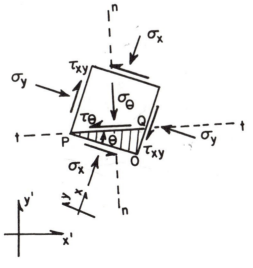

Figure 108-1 Mohr's circle for stress at a point.

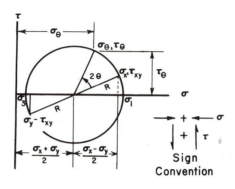

Figure 108-2 Development of equations of equilibrium for the free body *OPQ* of Fig. 107-2 and proof that the locus of points (σ_θ, τ_θ) forms a circle.

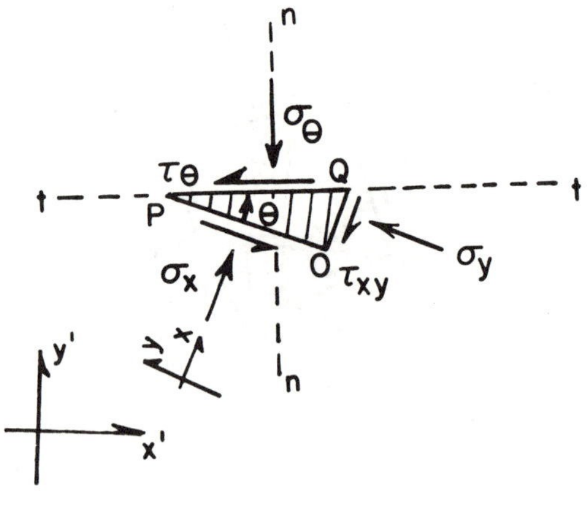

$\Sigma F_n = 0$

$\sigma_\theta A - \sigma_x(A \cos \theta) \cos \theta - \sigma_y(A \sin \theta) \sin \theta + \tau_{xy}(A \cos \theta) \sin \theta + \tau_{xy}(A \sin \theta)$
$+ \tau_{xy}(A \sin \theta) \cos \theta = 0$

Using the identities:

$$\sin \theta \cos \theta = \frac{\sin 2\theta}{2}$$

$$\cos^2\theta = \frac{1 + \cos 2\theta}{2} \quad \text{and} \quad \sin^2\theta = \frac{1 - \cos 2\theta}{2}$$

$$\sigma_\theta = \frac{\sigma_x + \sigma_y}{2} + \frac{\sigma_x - \sigma_y}{2} \cos 2\theta - \tau_{xy} \sin 2\theta \qquad (109\text{-}1a)$$

Summing forces in the *t-t* direction:

$\Sigma F_t = 0$

$$\tau_\theta A - \sigma_x(A \cos \theta) \sin \theta + \sigma_y(A \sin \theta) \cos \theta$$

$$- \tau_{xy}(A \cos \theta) \cos \theta + \tau_{xy}(A \sin \theta) \sin \theta = 0$$

$$\tau_\theta = \frac{\sigma_x - \sigma_y}{2} \sin 2\theta + \tau_{xy} \cos 2\theta \qquad (109\text{-}1b)$$

Write Eq. 109-1*a* as

$$\left[\sigma_\theta - \left(\frac{\sigma_x + \sigma_y}{2} \right) \right] = \left(\frac{\sigma_x - \sigma_y}{2} \right) \cos 2\theta - \tau_{xy} \sin 2\theta$$

Square both sides of the equation and add to the square of Eq. 109-1*b*.

$$\left[\sigma_\theta - \left(\frac{\sigma_x + \sigma_y}{2} \right) \right]^2 + \tau_\theta^2 = \left(\frac{\sigma_x - \sigma_y}{2} \right)^2 + \tau_{xy}^2 \qquad (109\text{-}1c)$$

which is the equation of a circle.

$$(\sigma_\theta - h)^2 + \tau_\theta^2 = R^2$$

Stress at a Point (Mohr Circle)

clockwise from plane *PO* at an angle equal to

$$\frac{1}{2}\tan^{-1}\frac{2\tau_{xy}}{(\sigma_x - \sigma_y)}$$

Origin of Planes

With the adopted sign convention for the normal and shearing stresses, a convenient orientation diagram can be superimposed on the Mohr diagram. If, through the coordinates σ_x, τ_{xy} on the Mohr circle, a line is drawn parallel to the plane on which these stresses act (see Fig. 110-1), this line intersects the Mohr circle at a unique point or origin of planes. Conversely, a line representing a parallel plane drawn through the origin of planes will intersect the Mohr circle at a point whose coordinates represent the normal and shearing stresses on the represented plane. The position of the origin of planes is dependent only on the orientation of the stressed element, whereas the Mohr diagram is dependent only upon the magnitude and sign of the normal and shearing stresses.

   In Fig. 110-1 the origin of planes is determined for the element previously given in Fig. 107-2. Using the origin-of-planes concept, stresses for three pairs of planes are determined as follows: (1) the θ plane and associated orthogonal

Figure 110-1   Use of the origin of planes.

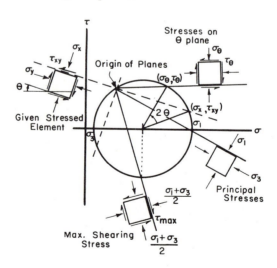

Figure 111-1

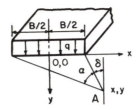

Example Problem Equations for the principal stresses in an elastic half-space below a uniformly loaded strip footing given by Paulos and Davis (1974) are as follows.

$$\sigma_1 = q/\pi(\alpha + \sin \alpha)$$

$$\sigma_3 = q/\pi(\alpha - \sin \alpha)$$

The direction of the major principal stress bisects the angle α.

Calculate the vertical stress σ_y, the horizontal stress σ_x, and τ_{xy} at point A if $x = 0.75B$ and $y = B/2$ using Mohr's diagram.

Solution

$$\alpha + \delta = \text{arc tan } 2.5 = 68.20°$$

$$\delta = \text{arc tan } 0.5 = 26.57°$$

$$\alpha = 68.20° - 26.57° = 41.63°$$

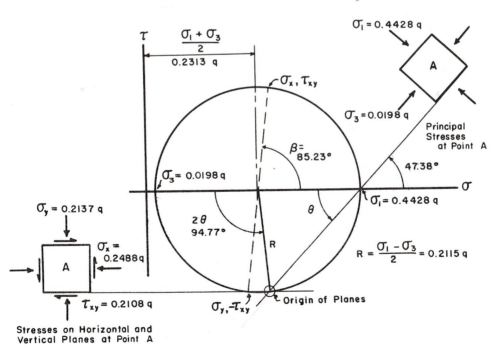

Stresses on Horizontal and
Vertical Planes at Point A

$$\delta + \alpha/2 = 47.38°$$

$$\sigma_1 = q/\pi(0.7267 + \sin 41.63) = 0.4428q$$

$$\sigma_3 = q/\pi(0.7267 - \sin 41.63) = 0.0198q$$

The stress σ_y is found by projecting the horizontal plane on which σ_y acts through the origin of planes. Evaluate the σ, τ coordinates where this plane intersects the circle. Compare the answer with Fig. 104-1.

$$\beta = 180° - 94.77° = 85.23°$$

$$\sigma_y = \frac{\sigma_1 + \sigma_3}{2} - R \cos \beta = 0.2313q - 0.0176q = 0.2137q$$

$$\tau_{xy} = R \sin \beta = 0.2115q \sin 85.23° = 0.2108q$$

Figure 112-1

Example Problem Given the stresses at a point in a soil, determine the principal stresses and show them on a properly oriented element.

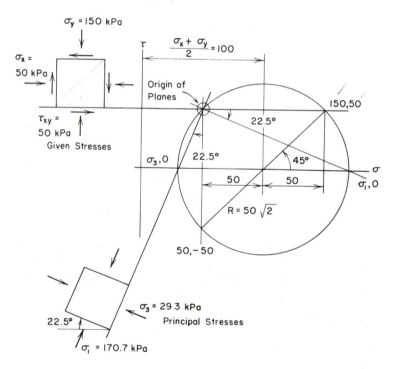

$$\sigma_1 = \frac{\sigma_x + \sigma_y}{2} + R = 100 + 70.7 = 170.7 \text{ kPa}$$

$$\sigma_3 = \frac{\sigma_x + \sigma_y}{2} - R = 29.3 \text{ kPa}$$

plane, (2) the principal planes, and (3) the planes for maximum shearing stresses. The planes and stresses acting on them are correctly oriented with respect to the given stress condition. The magnitude of the stresses can be conveniently determined from the geometry of the circle instead of relying on Eqs. 109-1a and 109-1b.

The application of the Mohr circle and origin-of-planes concept is illustrated in example problems in Figs. 111-1 and 112-1.

LATERAL PRESSURE IN NORMALLY CONSOLIDATED SOILS

A sedimentary soil deposit with a level surface is generally formed in such a way that, as the deposit is built up, the lateral strain is zero. For such soil deposits the horizontal and vertical stresses are the principal stresses, and the ratio of the principal stresses is the coefficient of earth pressure at rest.

$$\frac{\bar{\sigma}_3}{\bar{\sigma}_1} = \frac{\text{lateral effective stress}}{\text{vertical effective stress}} = \frac{\bar{\sigma}_h}{\bar{\sigma}_v} = K_o$$

From the generalized stress-strain equations, K_o can be shown to be a function of Poisson's ratio.

$$K_o = \frac{\nu}{1 - \nu}$$

Values of K_o computed from Poisson's ratio have been presented by Sowers and Sowers (1970) and are shown in Fig. 114-1. These values of K_o check well with those reported by Moore (1971), which were found experimentally.

The values of K_o are not appropriate for soils that have been subjected to lateral yielding. Soils that have been densified by artifical compaction or drying, or which have been affected by loads other than that of the present overburden, may have widely varying values of K_o.

The size of Mohr's circle for the at-rest case is, therefore, fixed by the type of soil. However, if the lateral strain is not restricted to zero, the ratio $\bar{\sigma}_3/\bar{\sigma}_1$ may vary over a considerable range. (See States of Equilibrium in Chapter 6.)

STRESS PATHS

The Mohr diagram can be useful for representing a series of stress states by drawing several Mohr circles showing progressive changes in the state of stress for a particular load application. Since a diagram with several complete stress circles would appear cluttered, it is convenient to plot only the point of maximum shearing stress (top of the circle). The complete circle can readily be constructed from this point if needed. A stress path is defined as the locus of points on the Mohr diagram whose coordinates represent the maximum shearing stress and the associated normal stress plotted for the entire stress history of a soil. Given the principal stresses σ_1 and σ_3, the coordinates of a point on the stress path are

$$p = \frac{\sigma_1 + \sigma_3}{2} \quad \text{and} \quad q = \frac{\sigma_1 - \sigma_3}{2} \tag{114-2}$$

When soil stresses are plotted in this manner, the resulting plot is called a p-q diagram.

The stress path for a soil where $\sigma_1 = \sigma_3$ initially and σ_1 is increased while σ_3 remains constant is a $45°$ line, as shown in Fig. 115-1a. Stress paths for other variations of σ_1 and σ_3 are readily constructed (Fig. 115-1b).

The stress path for a soil formed by a sedimentary process is a line through the origin at an angle α_o (Fig. 115-1c). The soil will be in the at-rest state, with

$$\sigma_h = K_o \sigma_v \quad \text{and} \quad \tan \alpha_o = \frac{\sigma_v - \sigma_h}{\sigma_v + \sigma_h} = \frac{\sigma_v(1 - K_o)}{\sigma_v(1 + K_o)} = \frac{1 - K_o}{1 + K_o}$$

Thus, lines through the origin represent stress states of constant principal stress ratio K.

Figure 115-1 Stress paths.

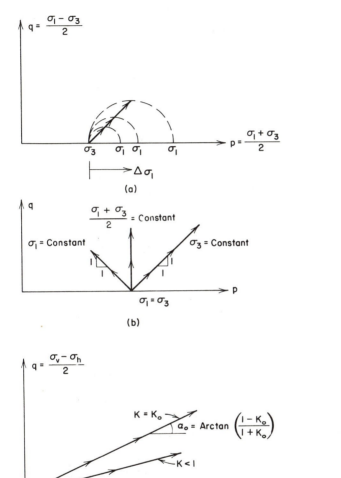

The stress path is a useful means of describing changes in states of stress in soil and of evaluating strength and compression (Lambe, 1967; Lambe and Whitman, 1969).

PROBLEMS

3-1 Calculate the effective stress at points c and d in the example problem of Fig. 53-1 ($e = 0.7$). Change the water height in the tube from 6 m to 3 m. ($G = 2.65$)

3-2 Calculate the effective stress at points c and d in problem 2-6a. ($e = 0.7$)

3-3 Complete the following table for the soil profile shown.

```
              O        Ground   Surface
                  ▽▽▽▽▽▽▽▽▽▽▽▽▽▽▽▽▽▽▽
                         S = 20%     Sand
              5 m   W.T. ▽           e = 0.7
                         S = 100%    G = 2.65
             10 m   ~~~~~~~~~~~~~~
                                     Clay
             20 m                    e = 0.9
                                     G = 2.65
```

Depth	σ Total Vertical Stress	$\bar{\sigma}$ Effective Stress	u Pore Pressure
5 m			
10 m			
20 m			

3-4 A 12-m thick layer of relatively impervious saturated clay lies over a gravel acquifer. Piezometers into the gravel show an artesian pressure condition with a piezometric surface 3 m above the surface of the clay. Clay properties are $e = 1.2$, $G = 2.7$.

(a) Determine the effective stress at the top of the gravel.
(b) How deep an excavation can be made in the clay layer without danger of bottom heave?

3-5 Plot the distribution of vertical stress σ_y on a horizontal plane 1 m below ground surface for a concentrated surface load of 10 kN (use Boussinesq). Compare this with the value of stress obtained by using the zone of influence method (Fig. 95-1).

3-6 Plot curves of vertical stress and shear stress on a vertical line extending from (0,0) to (0,4) due to a concentrated surface load of 10 kN at (1,0) (use Boussinesq).

3-7 Using Fig. 100-1, find the vertical stress 10 m below the midpoint of line AJ. Use data in the example problem in Fig. 100-1 and the Boussinesq equation. (Use three methods.)

3-8 Refer to Fig. 97-1. $Q = 10$ kN, $y = 1$ m, and $r = 0.5$ m. Use the Boussinesq equations to find the stresses. Plot Mohr's circle and find the normal stress and shear stress on a plane normal to the line between the load and the element ($\nu = 0.33$). Neglect the weight of the soil.

3-9 Plot Mohr's circle for problem 3-8 including soil weight, where $\gamma = 18$ kN/m³. The soil is dry. $K_o = 0.5$.

3-10 The principal stresses at a point in a soil are $\sigma_1 = 40$ kPa, $\sigma_2 = 28$ kPa, and $\sigma_3 = -8$ kPa. Draw Mohr's circle, considering each of the principal planes as the front plane. Place all three circles on the same drawing.

3-11 At a point in a soil layer the principal stresses are as shown in the figure.

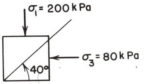

$\sigma_1 = 200$ k Pa

$\sigma_3 = 80$ k Pa

40°

Determine:

(a) The maximum shearing stress and associated normal stress, and show them on a properly oriented cube.

(b) The normal and shear stresses acting on the 40° plane as shown.

3-12 If the major principal stress in a soil is 360 kPa, find the minimum value of the minor stress if the shear stress is not to exceed 120 kPa.

3-13 If the stresses in the figure are as follows, determine:

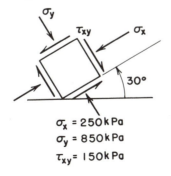

σ_y

τ_{xy}

σ_x

30°

$\sigma_x = 250$ k Pa

$\sigma_y = 850$ k Pa

$\tau_{xy} = 150$ k Pa

(a) The principal stresses shown acting on a properly oriented element.
(b) The normal and shearing stress on a horizontal plane.

3-14 Plot a stress path on a p-q diagram for the following conditions:

(a) Initially $\sigma_1 = \sigma_3 = 30$ kPa and $\Delta\sigma_3 = -0.3 \, \Delta\sigma_1$.
(b) Initially $\sigma_1 = 50$ kPa and $\sigma_3 = 30$ kPa, then σ_1 remains constant as σ_3 decreases.
(c) Initially $\sigma_1 = \sigma_3 = 30$ kPa and $\Delta\sigma_3 - 0.4 \, (\Delta\sigma_1)^2$.

Continue each stress path until the ratio q/p equals 2/3.

$\Delta\sigma_3 = 10$

$D\sigma_1 = -3$

$q = \frac{2}{3} p$

4

COMPRESSIBILITY, CONSOLIDATION, AND SETTLEMENT

Engineering structures with foundations on soil that are improperly designed may fail in one of two ways.

- Excessive settlement.
- Shear failure of the soil.

Soil deforms under load and if the deformation under different sections of the structure is enough to cause distress in structural members, then this condition constitutes a structural failure. A shear failure may be caused when structural loads exceed the ultimate bearing capacity or load-carrying capacity of the soil. This type of failure is characterized by a slippage along some critical plane of the soil foundation.

This chapter develops theories of compressibility and consolidation that may be used to design foundations on soil in such a way that differential settlements may be limited to an acceptable value.

COMPRESSIBILITY OF SANDS

Sandy soils deform less under static loads than under dynamic loads. However, a static load does cause a decrease in the void space in a sand mass. Figure 120-1 illustrates the change in height of two samples of the same sandy soil under a static load. The two samples have the same dry weight of soil, but sample 2 initially has a higher dry density and less void space than sample 1. As the loads are increased, it is apparent that the change in height of sample 1 is greater than that of sample 2.

Curves similar to those shown in Fig. 120-1 may be used to predict settlements of engineering structures under static loads. These curves may be obtained by field testing or by use of a simple compression test in the laboratory in which lateral deformations are prevented.

Laboratory tests may be used to predict settlement of soil layers in the field by subdividing the field layer of sand into several small layers and using either the Boussinesq or the Westergaard relationship to estimate the pressure increase in each of the sublayers. The vertical strain in each sand layer due to the pressure increase can be obtained from a laboratory compression curve of vertical strain versus pressure. The strain in each individual layer may then be used to compute total settlement.

Judgment must be exercised in using the results of laboratory tests to predict field settlement of sand layers. For small footings on sand, lateral bulging (which does not occur during a confined compression test) may contribute significantly to settlement, and thus the settlement predictions from the laboratory tests would be unreliable. The most common way to predict the settlement of spread footings on sand is by the use of semiempirical rules relating settlement to the results of field penetration tests. These methods are

Figure 120-1 Compression curves relating sample height and effective stress for two samples of sand at different initial densities.

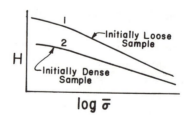

4 Compressibility, Consolidation, and Settlement

presented in Chapter 8 under settlement by standard penetration test and Dutch cone penetrometer test.

CONSOLIDATION OF CLAY

The settlement of a clay layer under a static load may still exhibit a curve similar to that in Fig. 120-1, but the nature of the settlement process is more difficult to analyze for clays than for sands. A static load on a sandy soil compresses the sand rapidly; a static load acting on a clay soil may settle very slowly.

Settlement Time Lag

There are two primary causes for the time lag in the settlement of clay soil.

- Hydrodynamic lag.
- Viscous lag.

Although the clay layer will exhibit some elastic compression due to a small change in volume in the soil and water particles, by far the greater amount of settlement must occur by expulsion of water from the void spaces. The static load produces a pressure gradient in the pore water and causes movement toward the drained surfaces. The movement, however, is slow because of the low permeability of the clay soil, and the rate of settlement becomes a function of permeability. The time lag in settlement caused by this phenomenon is known as hydrodynamic lag.

In order for clay particles to move closer together under a static load, the structured double layer of water surrounding the clay particles must deform. The deformation may be caused by loads that may tend to expel double-layer water and/or by shear loads that cause a shear deformation in the water surrounding the particle. Both actions have a viscous nature, and the speed of the deformation is a function of the magnitude of the load that causes the deformation. The time lag associated with this viscous resistance is called viscous lag.

The compression of saturated clay layers under a static load has been termed consolidation, and theories of consolidation that account for both hydrodynamic lag and viscous lag are available in the research literature in soil mechanics. At this time (1980), however, none of these more advanced theories has been accepted for common use in estimating clay consolidation. In general, these theories are complicated mathematically and have not been tested sufficiently to warrant confidence in their accuracy. The method of estimating the consolidation of clay commonly used at the present time is based on a one-dimensional theory of consolidation proposed by Terzaghi

Figure 122-1 Rheological model for consolidation.

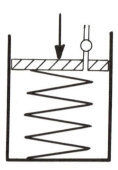

(1925), which only recognizes hydrodynamic lag as being responsible for time delays in settlement; for this reason, the theory must be used very carefully and according to standard procedures that have been tested over many years. Even though the Terzaghi theory does not consider viscous lag, at the time it was presented it represented a vast improvement in the procedures for estimating consolidation of clay and may be considered the forerunner of many of the more rigorous theories in soil mechanics today.

Rheological Consolidation Model

A rheological model of the consolidation process consisting of a container of water with a piston and spring is shown in Fig. 122-1. The excess water pressure in the container is initially zero and the spring is not compressed. If a load P is then applied to the piston with the valve closed, the water pressure increases to carry the load. The load in the spring remains zero, because it cannot compress until water drains from the container. When the valve is opened, water moves slowly through the piston and the spring depresses, carrying more and more of the load until the water pressure is finally reduced to zero and the spring carries the entire load. This is a condition of equilibrium. An element of saturated soil behaves in much the same way under the application of an additional load. The load is initially carried by the pore water because the soil structure must compress to take an additional load. Since the pore water is under an excess pressure, it will begin to seep out of the soil, allowing the soil to compress and begin to carry the additional load. At the end of the process the water pressure is zero and equilibrium is established.

4 Compressibility, Consolidation, and Settlement

Consolidation Test

The consolidation or oedometer test is performed on undisturbed samples of saturated clay soil. The sample is trimmed to fit a cylindrical container (Fig. 123-1) and is loaded with a small seating load, generally about 12 kPa. The sample may have porous plates on top and bottom, in which case water moves from the saturated soil, both upward and downward; or it may have a porous plate only on the top surface, in which case water moves upward through the entire sample. After equilibrium is reached under the seating load, the test proceeds by adding a new increment of load and allowing consolidation to occur until the new equilibrium point is reached. In the standard test, each increment of load is equal to the total preceding load and is double the preceding pressure increment. The test is generally run until the total load on the sample is about 1536 kPa. However, this will depend somewhat on the load anticipated in the field. The time required to reach a high percent of the total consolidation for a thin sample (20 to 30 mm), is about one day, and new increments may be added every 24 hr. An extensometer gage is mounted on the consolidometer and is used to measure the compression of the sample. A number of dial readings are taken during each load increment to measure the time rate of settlement. Time increments between dial readings are small at the beginning of the test, but increase as the test progresses.

Compressibility of Clay

Consider a soil sample dispersed in water. The system is then placed under a vertical compressive stress in a consolidometer. Water is allowed to drain

Figure 123-1   One-dimensional consolidation apparatus.

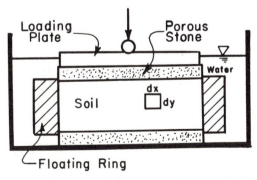

Figure 124-1 Compression curve for clay.

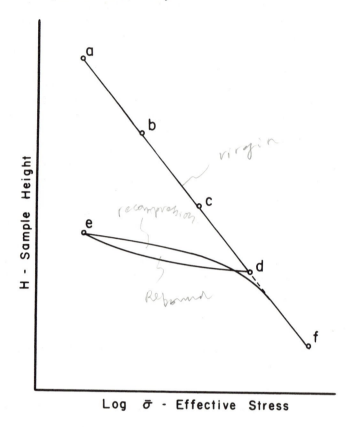

from the soil until the sample is in equilibrium under the compressive load. The height of the sample is measured and plotted against the effective stress on semilog paper, as shown on Fig. 124-1 (point *a*). The load is increased and again the equilibrium point (point *b*) is plotted on semilog paper. The load is increased several more times and the equilibrium points are plotted. A curve through these points on semilog paper is approximately straight in the range of pressures ordinarily encountered in soil mechanics problems, but tends to deviate from a straight line at very low and very high pressures. If the load on the sample is then decreased in increments and the equilibrium points are plotted, a rebound curve $\overline{de}$ will be produced. A large portion of the compression is irreversible and, therefore, the rebound curve is much flatter than the compression portion of the curve. A recompression curve can then be produced by again increasing the load and plotting the equilibrium points. This recompression curve is shown as $\overline{ef}$ in Fig. 124-1. Note that the recompression curve approaches a straight line after the maximum past pressure has been exceeded. The straight-line portion of the curve is referred to as the virgin compression curve.

A normally consolidated clay is a clay that has never been subjected to an effective stress that is greater than the existing stress. The compression characteristics of an element of normally consolidated clay in the field are similar to the straight-line portion of the curve in Fig. 124-1. Normally consolidated soils are generally compressible and are soft to great depths.

An overconsolidated or precompressed clay has been subjected to an effective stress that is greater than the existing stress. The ratio of the maximum past effective stress to the existing effective stress is the overconsolidation ratio. The compression characteristics of an element of overconsolidated clay in the field are represented by the recompression curve $\overline{ef}$ in Fig. 124-1. Overconsolidated clays are much less compressible than normally consolidated clays, as can be seen by the differences in the virgin compression curve and the recompression curve.

There are several processes by which clays become overconsolidated. Erosion of overburden material over many years will decrease the effective stress on underlying clay layers, causing them to rebound along a curve similar to $\overline{de}$ in Fig. 124-1. Loads from glaciers have also been responsible for overconsolidating clay deposits. Clays near the ground surface are often overconsolidated by cycles of wetting and drying. This process is referred to as overconsolidation by desiccation.

The compressibility of the soil may be evaluated from the results of a consolidation test by plotting the equilibrium points at the end of each load increment as a function of the logarithm of the effective stress, $\overline{\sigma}$, as shown in Fig. 124-1. The equilibrium points may be represented by the height of the sample, H, the change in height, ΔH, the strain, ϵ, or the void ratio, e. Void ratio versus logarithm of the effective stress has been the most common way of plotting the test results. However, strain is more direct than void ratio for presenting the results of one-dimensional consolidation tests and is used in this text (see Fig. 126-1). Strain is based on the original sample height H_o (height at the beginning of the test) and the change in height.

$$\epsilon = \frac{\Delta H}{H_o}$$

Figure 126-1 Finding field consolidation curve for normally consolidated clay.

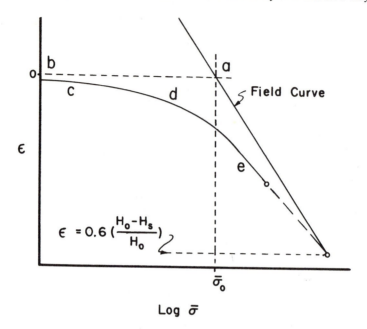

Normally Consolidated Clay

Undisturbed samples of soil taken by standard sampling procedures from a layer of normally consolidated clay soil in the field and then loaded in a consolidometer with standard pressure increments in the laboratory will exhibit a compression curve similar to $\overline{cde}$ in Fig. 126-1. The straight-line portion of the curve from the laboratory tests will be slightly different from the compression curve of the soil in its natural state in the field, because of disturbance of the soil sample during sampling and test preparation.

Since the field compression curve is required to estimate the settlement that will occur in the field, the laboratory compression curve must be adjusted. The following procedure is suggested for finding the field curve of a normally consolidated clay.

1. Plot the laboratory consolidation curve. This will appear as curve $\overline{cde}$ in Fig. 126-1. The sample in its natural state (point a in Fig. 126-1) is

4 Compressibility, Consolidation, and Settlement

in equilibrium under pressure $\bar{\sigma}_o$ and follows an unloading curve $\overline{ab}$ when it is removed from its natural state. The amount of volume rebound along curve $\overline{ab}$ is very small, because the tendency of the soil to expand produces capillary forces in the pore water. After the sample is trimmed, placed in the consolidometer, and tested in the laboratory, the reloading curve approaches a straight line as shown in Fig. 126-1. The straight-line portion of the reloading curve is displaced from the field curve due to disturbance in the soil structure caused from taking and trimming the sample.

2. One point on the field curve has the coordinates $(\bar{\sigma}_o, 0)$, which is the equilibrium condition in the field. This point is shown as point a in Fig. 126-1.

3. A second point on the field compression curve is found by extending the linear portion of the laboratory curve. The field curve is assumed to intersect the laboratory curve at a strain of approximately

$$\frac{0.6(H_o - H_s)}{H_o}$$

where
$$H_s = \frac{W_s}{G\gamma_w A},$$

A = sample area and H_o = initial height of sample (Schmertmann, 1955). This strain corresponds to a void ratio of $0.4\, e_o$, where e_o is the void ratio at zero strain.

4. The field curve is a straight line drawn between the two points identified in steps 2 and 3 and has the equation

$$\epsilon = C \log_{10} (\bar{\sigma}/\bar{\sigma}_o) \tag{127-1}$$

where $\bar{\sigma}_o$ is the initial effective stress.

Zero strain on the field curve corresponds to the equilibrium condition in the field for the sample that was tested. An element of the same soil at a greater depth would be under a larger equilibrium effective stress; therefore, the zero strain point for that element would correspond to the strain on the $\epsilon - \log \bar{\sigma}$ curve for the higher effective stress. The value of C (slope of the $\epsilon - \log \bar{\sigma}$ curve) for the deeper element would be the same as for the sample that was tested. An element of the same soil at a shallower depth would be under a smaller equilibrium effective stress, and the zero strain point for that element would correspond to a strain on the $\epsilon - \log \bar{\sigma}$ curve for a lower effective stress. Each element, therefore, has its own zero strain point. However, since the slope of the $\epsilon - \log \bar{\sigma}$ curve is the same for all elements of the same soil regardless of the depth, Eq. 127-1 may be used for the soil at any depth by using the appropriate value of $\bar{\sigma}_o$.

The field compression curve can also be plotted in terms of void ratio e, and effective stress, $\bar{\sigma}$. For normally consolidated clay this curve is also a straight line on semilog paper and has the equation

$$e = e_o - C_c \log_{10} \frac{\bar{\sigma}}{\bar{\sigma}_o} \tag{128-1}$$

where C_c is the compression index. The compression index has been empirically related to the liquid limit of clay soil (Terzaghi and Peck, 1967) and can be stated as

$$C_c \approx 0.009 \, (\text{LL} - 10\%) \tag{128-2}$$

This relationship was developed from tests on many different remolded soils and adjusted to reflect the difference in the behavior of remolded and undisturbed samples. It can provide a rough estimate of the compressibility of normally consolidated clay soils of low sensitivity when only the liquid limit of the soil is available.

The slope of the $\epsilon - \log \bar{\sigma}$ compression curve (C) is related to the compression index by

$$C = \frac{C_c}{1 + e_o} \tag{128-3}$$

Settlement Prediction for Normally Consolidated Clay

If the consolidation curve, as shown in Fig. 126-1, represents the compression characteristics of the soil for the entire field layer, then the strain caused in subdivisions of the layer can be found by evaluating the initial and final effective stresses in each division and calculating the strains from Eq. 127-1. The total settlement in the layer is then the sum of ϵH_o for each of the sublayers.

$$\Delta H = \Sigma(\epsilon H_o) = \Sigma \left(C \log_{10} \frac{\bar{\sigma}_o + \Delta \bar{\sigma}}{\bar{\sigma}_o} \right) H_o \tag{128-4}$$

Total settlement at a point on the ground surface is assumed to equal the settlement of the clay layer directly below the point in question. The increase in pressure in each sublayer can be evaluated using the methods discussed in Chapter 3.

Equation 128-4 is used in the example problem of Fig. 145-1 to predict the settlement of a layer of normally consolidated clay.

Settlement may also be calculated in terms of the change in void ratio from

$$\Delta H = \Sigma \left(\frac{\Delta e}{1 + e_o} H_o \right) \tag{128-5}$$

Overconsolidated Clay

Figure 124-1 demonstrates that the stress history of a soil has a great influence on its stress-strain behavior. As shown in the figure, the recompression curve is much flatter than the virgin compression curve. Since overconsolidated clays follow a recompression curve, their compressibility is generally much less than that of a normally consolidated clay.

Since the stress history of clay is important to its stress-strain characteristics, it is essential to be able to determine the maximum past effective stress (preconsolidation stress) for the clay. In some cases the maximum past effective stress may be estimated from geologic evidence; however, this is rarely the case. Casagrande (1936) developed a method for estimating the maximum past effective stress from the laboratory compression curve of an undisturbed sample. He developed this method from observations on slow cyclic compression tests on undisturbed clay samples. The basic steps in Casagrande's method are illustrated in Fig. 130-1 and described below.

1. Plot the laboratory compression curve on semilog paper.
2. By observation, find the point on the curve with the smallest radius of curvature.
3. At the point on the curve with the smallest radius of curvature, draw a horizontal line and a line tangent to the curve, and bisect the angle between these two lines.
4. Project the straight-line portion of the laboratory curve to its intersection with the line bisecting the angle from step 3. This point of intersection approximates the maximum past effective stress to which the soil has been subjected.

If the maximum past effective stress, as determined from Casagrande's method, is approximately equal to the existing effective stress, then the soil is normally consolidated and the field curve can be determined by the method discussed on page 126. If the maximum past effective stress is greater than the existing effective stress, then the soil is overconsolidated and the field compression curve should be determined by the method developed by Schmertmann (1955). This method is illustrated in Fig. 131-1 and described below.

1. Plot the laboratory consolidation curve and rebound curve on semilog paper.
2. One point on the field curve has the coordinates $(\bar{\sigma}_o, 0)$, which is the equilibrium condition in the field. This point is shown as point a in Fig. 131-1.
3. Rebound curves from any effective stress are nearly parallel. This can be used to establish the equilibrium condition at the time that the

Figure 130-1 Procedure for estimating the maximum past effective stress.

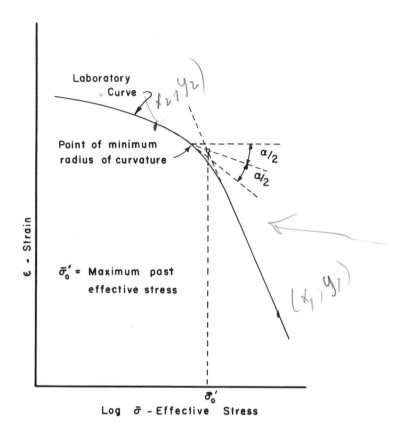

maximum past effective stress $\bar{\sigma}_o'$ acted on the sample. Draw a line parallel to the laboratory rebound curve from point a to the maximum past effective stress (point b). Point b also represents a point on the field curve.

4. Extend the straight-line portion of the laboratory curve to a strain equal to

$$\frac{0.6(H_o - H_s)}{H_o}$$

4 Compressibility, Consolidation, and Settlement

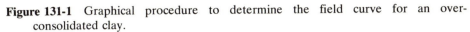

Figure 131-1 Graphical procedure to determine the field curve for an overconsolidated clay.

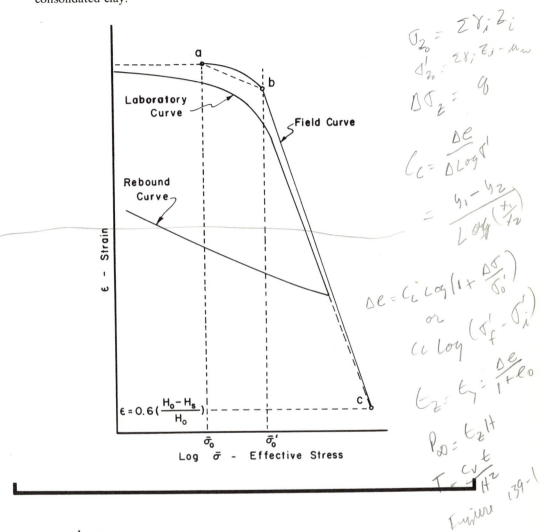

where

$$H_s = \frac{W_s}{G\gamma_w A}$$

This strain corresponds to a void ratio of 0.4 e_o, where e_o is the void ratio at zero strain; it is shown as point c in Fig. 131-1.

Consolidation of Clay

5. The virgin compression curve is a straight line from point b to point c.

6. The initial part of the field compression curve (overconsolidated region) is approximated by a curve from point a to point b that is nearly parallel to the laboratory curve in that stress region.

The field compression curve is $\overline{abc}$, and should be used to estimate the settlement of overconsolidated clays.

The use of high-quality undisturbed samples is extremely important in predicting the settlement of clay soils. Many clays are sensitive and their properties can be altered appreciably by poor sampling and testing techniques.

RATE OF CONSOLIDATION

Terzaghi (1925) presented a theory to describe the time rate of consolidation of clay soils.

Derivation of Differential Equation

The differential equation of consolidation as presented by Terzaghi can be developed by considering a typical differential element of clay soil from a one-dimensional consolidometer as shown in Fig. 133-1. This element is subjected to one-dimensional flow during the consolidation process. The time rate of change of volume of the element during consolidation will equal the difference in the flow into and out of the element.

$$\frac{dV}{dt} = q_{in} - q_{out}$$

From Fig. 133-1,

$$\frac{dV}{dt} = -\frac{\partial v_y}{\partial y}\, dx\, dy\, dz \tag{132-1}$$

From Darcy's law and assuming the coefficient of permeability, k, to be constant, the velocity can be expressed as

$$v_y = -k\frac{\partial h}{\partial y} \tag{132-2}$$

The pressure distribution in the clay sample at any time during consolidation is represented by Fig. 133-2. From the figure the total head can be expressed as

$$h = h_e + h_p = -y + \frac{(\gamma_w y + u_e)}{\gamma_w}$$

Figure 133-1 Typical element from a one-dimensional consolidometer.

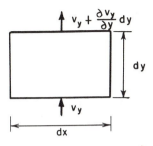

Figure 133-2 Pressure in soil during consolidation.

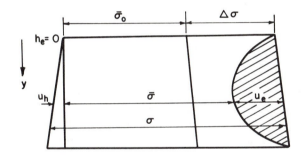

Therefore

$$\frac{\partial h}{\partial y} = \frac{1}{\gamma_w}\frac{\partial u_e}{\partial y}$$

and Eq. 132-2 becomes

$$v_y = -\frac{k}{\gamma_w}\frac{\partial u_e}{\partial y} \tag{133-3}$$

Rate of Consolidation

Substituting Eq. 133-3 into Eq. 132-1, the expression for the time rate of change in volume becomes

$$\frac{dV}{dt} = \frac{k}{\gamma_w} \frac{\partial^2 u_e}{\partial y^2} \, dx \, dy \, dz \qquad (134\text{-}1)$$

The total stress at any point in the soil is illustrated in Fig. 133-2 and can be expressed as,

$$\sigma = \bar{\sigma} + u_e + u_h$$

From this relationship the excess pore pressure is

$$u_e = \sigma - \bar{\sigma} - u_h$$

If $\sigma - u_h$ is a linear function of y, the second derivative with respect to y becomes

$$\frac{\partial^2 u_e}{\partial y^2} = - \frac{\partial^2 \bar{\sigma}}{\partial y^2} \qquad (134\text{-}2)$$

Substituting Eq. 134-2 into Eq. 134-1 yields an expression for the time rate of volume change in terms of effective stress

$$\frac{dV}{dt} = - \frac{k}{\gamma_w} \frac{\partial^2 \bar{\sigma}}{\partial y^2} \, dx \, dy \, dz \qquad (134\text{-}3)$$

This expression can be related to strain by assuming the effective stress-strain relationship to be linear, as shown in Fig. 135-1.
   From Fig. 135-1,

$$\epsilon = a + m_v \bar{\sigma}$$

where m_v is the coefficient of volume decrease.
   Solving for $\bar{\sigma}$ and taking the second derivative with respect to y yields

$$\frac{\partial^2 \bar{\sigma}}{\partial y^2} = \frac{1}{m_v} \frac{\partial^2 \epsilon}{\partial y^2} \qquad (134\text{-}4)$$

Substituting Eq. 134-4 into Eq. 134-3 yields

$$\frac{dV}{dt} = - \frac{k}{\gamma_w m_v} \frac{\partial^2 \epsilon}{\partial y^2} \, dx \, dy \, dz \qquad (134\text{-}5)$$

The time rate of volume change can also be stated in terms of strain by

$$\frac{dV}{dt} = - \frac{\partial(\epsilon \, dy) \, dx \, dz}{\partial t}$$

Figure 135-1 Assumed linear relationship between effective stress and strain.

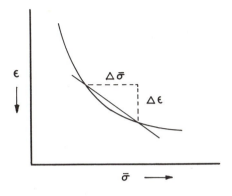

or

$$\frac{dV}{dt} = -\frac{\partial \epsilon}{\partial t} dx \, dy \, dz \qquad (135\text{-}2)$$

Equating the expressions for the time rate of volume change as given by Eqs. 134-5 and 135-2 yields

$$c_v \frac{\partial^2 \epsilon}{\partial y^2} = \frac{\partial \epsilon}{\partial t} \qquad (135\text{-}3)$$

where

$$c_v = \frac{k}{\gamma_w m_v}$$

and is called the coefficient of consolidation.

Equation 135-3 is the differential equation of consolidation. Its solution for appropriate boundary conditions yields the strain in the soil as a function of depth and time. This equation can also be derived in terms of excess pore pressure and takes the form

$$c_v \frac{\partial^2 u_e}{\partial y^2} = \frac{\partial u_e}{\partial t} \qquad (135\text{-}4)$$

Rate of Consolidation

Solution of Differential Equation

A solution of the differential equation of consolidation (Eq. 135-4) describes the excess pore pressure distribution as a function of depth and time.

The solution of Eq. 135-3 has the general form

$$\epsilon = c_1 + (c_2 \cos Ay + c_3 \sin Ay)e^{-c_v A^2 t} \tag{136-1}$$

where c_1, c_2, c_3, and A are constants.

The boundary conditions and the general shape of the solution for consolidation of a double-drained clay layer or for a clay sample in a double-drained consolidometer such as that shown in Fig. 123-1 is shown in Fig. 136-2. For this case a constant stress increase $\Delta\sigma$ has been applied to the full depth of the consolidating layer. The stress increase could have been the result, for example, of a large uniform load at the ground surface or from lowering the groundwater table. This constant stress would ultimately result in strain throughout the layer, as shown in Fig. 136-2. Consolidation at the drainage surfaces occurs almost instantly; therefore, the strain at these surfaces is equal to the equilibrium strain for all values of time. Flow of pore water during consolidation is toward the drainage surface and away from the

Figure 136-2 Shape of the solution surface and the boundary conditions for a constant pressure increase over the full height of a double-drained sample.

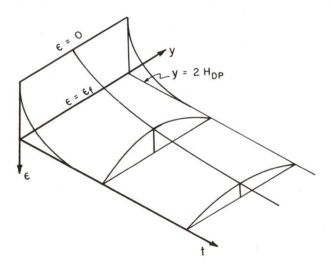

4 Compressibility, Consolidation, and Settlement

midplane. The length of the longest drainage path is H_{DP} and for a double-drained condition the total thickness of the sample is twice the length of the drainage path ($2H_{DP}$). The equilibrium strain is ϵ_f and the strain at the beginning of the consolidation process is defined as zero.

After applying the boundary conditions to Eq. 136-1, the solution of the differential equation of consolidation becomes

$$\epsilon = \epsilon_f - \sum_{n=1}^{\infty} \left[\frac{2\epsilon_f}{n\pi} (1 - \cos n\pi) \left(\sin \frac{n\pi y}{2H_{DP}} \right) e^{\frac{-c_v n^2 \pi^2 t}{4H_{DP}^2}} \right]$$

The degree of consolidation at any point in the sample is defined as

$$U_y = \frac{\epsilon}{\epsilon_f} \tag{137-1}$$

The solution in terms of the degree of consolidation becomes

$$U_y = 1 - \sum_{m=0}^{\infty} \frac{2}{M} \left(\sin \frac{My}{H_{DP}} \right) e^{-M^2 T} \tag{137-2}$$

where

$$M = \frac{\pi}{2} (2m + 1)$$

$$T = \frac{c_v t}{H_{DP}^2} \tag{137-3}$$

The dimensionless constant T is called the time factor. Equation 137-2 is represented in Fig. 138-1 and can be used to evaluate the progression of consolidation as a function of time and depth.

The degree of consolidation can also be stated in terms of excess pore pressure as

$$U_y = 1 - \frac{u_e}{\Delta\sigma} \tag{137-4}$$

where $\Delta\sigma$ is the pressure increment. From this definition of the degree of consolidation, Eq. 137-2 can be used to predict excess pore pressure as a function of time and depth.

The average degree of consolidation for the entire sample can be obtained by integrating over the thickness of the sample

$$U = \frac{\int_0^{2H_{DP}} U_y dy}{2H_{DP}}$$

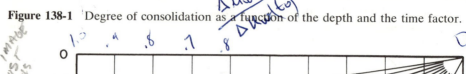

Figure 138-1 Degree of consolidation as a function of the depth and the time factor.

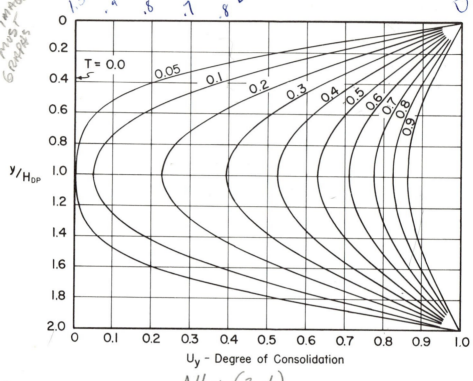

The average degree of consolidation becomes

$$U = 1 - \sum_{m=0}^{\infty} \frac{2}{M^2} e^{-M^2 T} \qquad (138\text{-}2)$$

Equation 138-2 is the solution for the boundary conditions shown in Fig. 136-2. If the applied pressure is not a linear function of y over the height of the sample, the solution may be somewhat different. Equation 138-2 is represented graphically in Fig. 139-1. The solution of the differential equation of consolidation for other boundary conditions has been presented by others, among them Taylor (1948) and Lambe and Whitman (1969).

The plotting of the theoretical settlement time curves depends on the choice of the value of the coefficient of consolidation, c_v, and total settlement.

4 Compressibility, Consolidation, and Settlement

Figure 139-1 Relationship between average degree of consolidation and time factor for a uniform load applied to the soil.

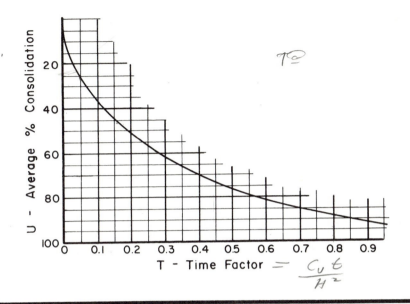

Four of the infinite number of possible curves are shown in Fig. 140-1, and it is apparent that a curve b, with the proper choice of c_v, may be made to fit the laboratory curve through a major portion of the consolidation process. If curve b is used as the solution, however, a portion of the total laboratory consolidation is left unexplained by the theory. This consolidation is called secondary consolidation. The portion of the consolidation explained by Terzaghi's theory is called primary consolidation. The difference between the two curves is due to neglecting viscous resistance in the derivation of the theoretical equation.

In using Terzaghi's theory to estimate consolidation, it is customary to find the value of c_v that will cause the theoretical curve to approximate the laboratory curve as closely as possible (curve b in Fig. 140-1), and then to use Terzaghi's theory with that value of c_v to explain the total consolidation in the field clay layer. This procedure neglects the occurrence of secondary consolidation in the field and, therefore, tends to overestimate the amount of consolidation occurring at a particular time. However, other factors influence the rate of consolidation in the field, and these tend to compensate for neglecting

Rate of Consolidation

Figure 140-1 Theoretical consolidation curves compared to the experimental curve. The theoretical curve depends on the selection of c_v.

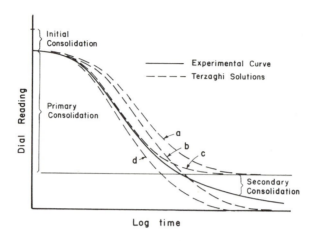

secondary consolidation. For example, radial drainage and small undetected drainage layers will tend to speed up the consolidation process. If the pressure increment in the field is very small compared to that used in the laboratory, the amount of secondary consolidation as opposed to primary consolidation increases in the field and the estimate for settlement at a particular time may greatly exceed the actual settlement.

Evaluation of c_v

The value of c_v is usually determined using curve-fitting procedures applied to the time-consolidation curves from one-dimensional consolidation tests on undisturbed samples. The curve-fitting methods that are commonly used are the logarithm of time-fitting method and the square root of time-fitting method (Taylor, 1948; Lambe and Whitman, 1969). Each method uses a procedure to approximate the time required to achieve a certain degree of primary consolidation of the laboratory sample. Equation 137-3 is then used to calculate c_v.

The logarithm of time-fitting method is described here. Plot on semilog paper the dial reading versus time results from one load increment of a one-dimensional consolidation test, as shown in Fig. 141-1. The dial readings are obtained from an extensometer gage mounted on top of the consolidation

Figure 141-1 Laboratory consolidation curve for one load increment showing the construction for the logarithm of time fitting method to evaluate c_v.

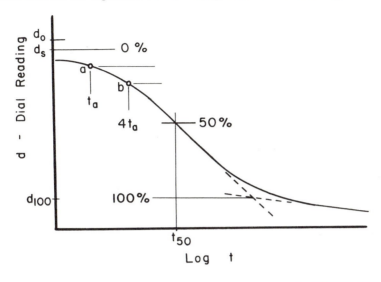

apparatus. This gage measures the compression of the sample during the test. Since some initial compression generally occurs from the compression of air in the voids of the sample, a correction must be applied to the consolidation curve to obtain the point of initial primary consolidation. The initial shape of the consolidation curve approximates a parabola, and the corrected initial point may be obtained by first choosing two points, a and b, on the early part of the curve for which the ratio of times $t_a/t_b = \frac{1}{4}$. The corrected initial dial reading, d_s, is then located a distance above point a equal to the difference in dial reading between points a and b. The dial reading d_{100} representing 100% primary consolidation is found by extending the straight-line portions at the middle of the curve and at the end of the curve to a point of intersection. Fifty percent consolidation then lies halfway between the dial readings d_s and d_{100}. Figure 139-1 shows the time factor at 50% consolidation to be 0.197. The value of c_v can, therefore, be calculated from Eq. 141-2.

$$c_v = \frac{0.197(H_{DP})^2_{50}}{t_{50}} \qquad (141\text{-}2)$$

where t_{50} is the time at 50% primary consolidation.

Rate of Consolidation **141**

The value c_v is different for each load increment and, therefore, must be calculated for all load increments used in the consolidation test. The values of c_v are generally plotted on semilog paper as a function of the average pressure for each load increment. Then the value of c_v for the field increment is read for the average field pressure. As pointed out earlier, standard load increments ($\Delta\sigma/\sigma$ equal to one) must be used.

Predicting Rate of Settlement

The rate of settlement of a clay layer subjected to an additional load can be predicted using the results of consolidation tests on undisturbed samples of the clay layer. The magnitude of settlement, ΔH, is first predicted using Eq. 128-4. An appropriate value of c_v for the average pressure during consolidation is then selected and the time rate of settlement is calculated using Eqs. 142-1 and 142-3.

$$t = \frac{T(H_{DP})^2}{c_v} \qquad (142\text{-}1)$$

where H_{DP} is the length of the drainage path of the field clay layer.
  Now

$$\epsilon_{av} = U\epsilon_f \qquad (142\text{-}2)$$

where

$$\epsilon_f = \frac{\Delta H}{H}$$

ϵ_{av} = average strain in the clay layer
ΔH = total calculated settlement
  H = height of field clay layer

and

$$\epsilon_{av} = \frac{S}{H}$$

where S = settlement of field layer at time t. Therefore, the degree of consolidation for the full clay layer can be expressed as

$$U = \frac{S}{\Delta H} \qquad (142\text{-}3)$$

The following table can be set up to find the settlement of the clay layer, S, at various values of time, t.

The time values are calculated from Eq. 142-1 and the settlement values from Eq. 142-3.

4 Compressibility, Consolidation, and Settlement

For very thick clay layers, tests on several undisturbed samples should be used to determine c_v. The usual procedure is to use a simple average of the values of c_v for each layer provided that there is not a wide variation in the values.

U	T	t	S
0.1	0.008		
0.2	0.031		
0.3	0.071		
0.4	0.126		
0.5	0.197		
0.6	0.287		
0.7	0.403		
0.8	0.567		
0.9	0.848		

For very complex problems involving several consolidating layers of various thicknesses and properties and with a common drainage surface, the preceding procedure cannot be used. Computer programs have been developed to handle the problems; these are discussed in Chapter 11.

Construction Period Correction

Real loads that cause consolidation of clay layers are added to the soil over a period of time and not instantaneously, as is the load on the laboratory sample. A method of correcting for the estimated settlement due to a gradually added load increment is shown in Fig. 144-1. The beginning point of consolidation is taken at the time when the net load returns to zero after the excavation period, and the loading period lasts from that time to time t at the end of construction. S_i is the instantaneous settlement curve that shows the consolidation that would occur if the load W_c had been added instantaneously. The correct settlement at time t is assumed to be the same as if the total load had been added at $t/2$ and may be found by dropping a vertical line from $t/2$ to the instantaneous settlement curve S_i and moving horizontally to the time t. The same rule may be applied to points between 0 and t. For example, at point $0.8t$, the corrected settlement is found by dropping a vertical line from $0.4t$ to the curve S_i and then horizontally to the line $0.8t$. This settlement is reduced by a factor of 0.8 because the load at $0.8t$ is $0.8 W_c$. This follows the same rule used for time t in that it is assumed that a load added gradually over the time $0.8t$ would have the same effect as the same load added instantaneously at $0.4t$. Points beyond the time t are found by displacing the curve S_i a horizontal distance equal to $t/2$. The adjusted time-settlement curve is shown by the dashed line.

Figure 144-1 Correction for gradual loading.

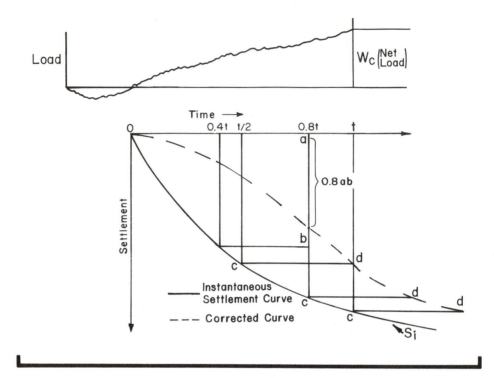

Discussion of Example Problem of Fig. 145-1.

The results of this problem are summarized by the corrected time settlement curve of Fig. 150-1. The importance of correcting the curve for the construction period is demonstrated by the difference in the instantaneous settlement curve and the corrected settlement curve.

In this problem, lowering the groundwater table increased the effective stress in the clay layer, which leads to consolidation of the clay. An increase in effective stress may also be caused by loads from earth embankments or structures. The methods discussed in Chapter 3 are used to calculate the stress increase. Such loads are always over a finite area and, in many cases, the increase in effective stress in a particular soil stratum will not be constant with depth. In these cases the soil stratum is generally divided into sublayers and the settlement of each sublayer is computed for the average increase in stress for that sublayer. Total settlement is the sum of the settlements in each

4   Compressibility, Consolidation, and Settlement

Figure 145-1

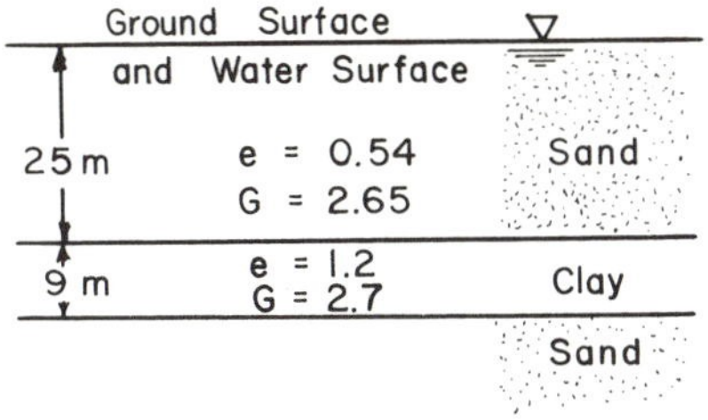

```
        Ground   Surface        ▽
         and   Water  Surface
    ↑
   25 m          e  =  0.54        Sand
    ↓            G  =  2.65
    ↑          e  = 1.2
   9 m          G  = 2.7           Clay
    ↓
                                 Sand
```

Example Problem   It is anticipated that the ground-water table in a certain area will be lowered 15.5 m over a period of 5.7 years. The groundwater table is currently at the ground surface and the soil profile for the area is shown. An undisturbed sample was obtained from the midpoint (depth of 29.5 m) of the clay layer. The results of a consolidation test on the undisturbed sample are shown in the tables. A double drained consolidation test was performed on a sample 2.5 in. (63.5 mm) in diameter. The initial height was 25.4 mm and the dry mass was 98.72 g. Predict, as a result of lowering the groundwater, table 15.5 m,

1. The ultimate settlement of the clay layer.
2. The time rate of settlement of the clay layer.

Solution   Compute the ultimate settlement from the strain at the midpoint of the clay layer.

Initial effective stress

$$\bar{\sigma}_o = \sigma - u$$

$$\bar{\sigma}_o = \frac{2.65 + 0.54}{1 + 0.54}(9.81)(25)$$

$$+ \frac{2.70 + 1.2}{1 + 1.2}(9.81)(4.5) - 9.81(29.5)$$

$$\bar{\sigma}_o = 296.9 \text{ kPa}$$

Final effective stress   Assume the average moisture content of the soil above the groundwater table is 9%.

$$\bar{\sigma}_f = \frac{2.65(1 + 0.09)}{1 + 0.54}(9.81)(15.5)$$

$$+ \frac{(2.65 + 0.54)}{1 + 0.54}(9.81)(9.5)$$

$$+ \left(\frac{2.70 + 1.2}{1 + 1.2}\right)(9.81)(4.5) - 9.81(14.0)$$

$$\bar{\sigma}_f = 419.2 \text{ kPa}$$

Rate of Consolidation

Field compression curve   The data from the consolidation test are given in the table and shown graphically on Fig. 146-1. The maximum past effective stress as determined by Casagrande's method is approximately equal to the existing effective stress; therefore, the clay is normally consolidated.

Compression data

$\bar{\sigma}$(kPa)	H(mm)	ϵ
12	25.34	0.0025
24	25.32	0.0030
48	25.30	0.0040
96	25.27	0.0050
192	25.08	0.0125
384	24.50	0.0355
768	23.32	0.0820
1536	22.11	0.1295

Figure 146-1   Compression curves.

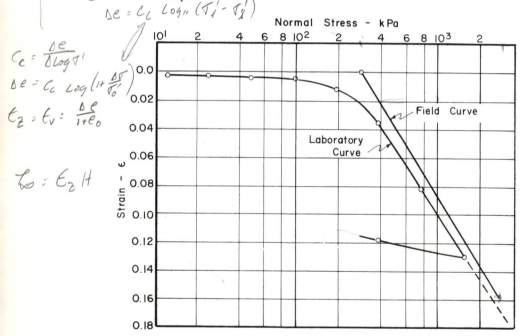

The field compression curve is a straight line between the existing equilibrium point (296.9, 0.0) and a point on the extended straight line portion of the

4   Compressibility, Consolidation, and Settlement

lab curve with a strain of

$$\epsilon = \frac{0.6(H_o - H_s)}{H_o}$$ *eq in p. 130-131*

$$H_s = \frac{W_s}{GA\gamma_w} = \frac{98.72(4)}{2.70(\pi)(63.5)^2(0.001)}$$

$$H_s = 11.55 \text{ mm}$$

$$\epsilon = \frac{0.6(25.4 - 11.55)}{25.4}$$

$$\epsilon = 0.327$$

The field curve is plotted on Fig. 146-1.

Equation of field curve

C'_c slope on $e-\sigma$ curve

$$\epsilon = C \log \frac{\bar{\sigma}}{\sigma_o} \qquad (127\text{-}1)$$

$$C = 0.161 \quad \sim \quad \epsilon - \sigma \text{ curve} \qquad C = \frac{C'}{1+e_o}$$

Note that C was evaluated as the difference in strain over one log cycle.

Strain at midpoint of clay layer

$$\epsilon = 0.161 \log \frac{419.2}{296.9} \quad \text{— from p. 145}$$

$$\epsilon = 0.0242$$

Ultimate settlement of clay layer The ultimate settlement can be computed from Eq. 128-4.

$$\Delta H = \epsilon H_o \quad \text{depts of clay layer}$$

$$\Delta H = 0.0242(9)$$

$$\Delta H = 0.217 \text{ m}$$

Rate of settlement The settlement-time curve can be established for the clay layer by first assuming the load is applied instantly and then using the relationships in Fig. 139-1, and Eqs. 142-1 and 142-3. The value of c_v can be determined from Fig. 149-1 which is a plot of computed values of c_v against pressure increment. The value of c_v for the increment 384 to 768 kPa must be computed to complete the curve. Fig. 148-1 shows the necessary graphical construction for this computation.

C

Rate of Consolidation

147

Figure 148-1   Time-compression curve illustrating logarithm of time fitting method.

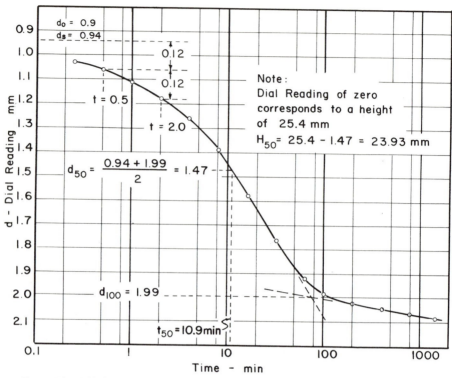

From Fig. 148-1,

$$t_{50} = 10.9 \text{ min}$$

The height of the consolidation sample at this time was

$$H_{50} = 23.93 \text{ mm}$$

Since the consolidation sample was double drained, the length of the longest drainage path in the test sample was

$$H_{DP} = \frac{23.93}{2} = 11.97 \text{ mm}$$

The coefficient of consolidation c_v can now be calculated from Eq. 141-2.

$$c_v = \frac{0.197(11.97)^2}{10.9}$$

$$c_v = 2.59 \text{ mm}^2/\text{min}$$

or $$c_v = 0.00373 \text{ m}^2/\text{day}$$

Other values of c_v have already been computed as follows.

Pressure increment (kPa)	c_v (mm²/min)
96–192	2.70
192–384	2.60
384–768	
768–1536	2.75

Figure 149-1 c_v versus normal stress.

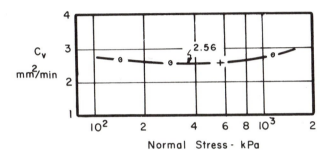

Normal Stress - kPa

From the curve choose $c_v = 2.56$ for the field increments 296.9 to 419.2 kPa.

$$(2.56 \text{ mm}^2/\text{min} = 0.00369 \text{ m}^2/\text{day})$$

The settlement-time curve for the field case can now be evaluated for an instantaneous loading from Fig. 139-1 and Eqs. 142-1 and 142-3.

$$t = \frac{TH_{DP}^2}{c_v}$$

$$S = \Delta H(U)$$

The field clay layer is double drained. Therefore,

$$H_{DP} = \frac{9}{2} = 4.5 \text{ m}$$

$$t = \frac{T(4.5)^2}{0.00369} = 5488 \, T \text{ days}$$

$$S = 0.217 \, U \text{ m}$$

Rate of Consolidation

Time-Settlement Tabulation

U	T	t (days)	$S = \Delta H(U)$ (m)
0.1	0.008	44	0.022
0.2	0.031	170	0.043
0.3	0.071	390	0.065
0.4	0.126	691	0.087
0.5	0.197	1081	0.108
0.6	0.287	1575	0.130
0.7	0.403	2212	0.152
0.8	0.567	3112	0.174
0.9	0.848	4654	0.195

Figure 150-1 Time-settlement diagrams.

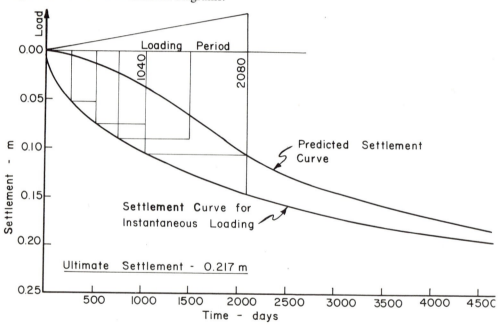

These time-settlement results have been plotted on Fig. 150-1. Since the load was applied over a period of 5.7 years (2080 days) instead of instantaneously, the time settlement curve must be adjusted. The adjustment is shown on Fig. 150-1 and was developed by assuming that the load increased linearly over the 5.7-year period and used the procedure for construction period correction.

4 Compressibility, Consolidation, and Settlement

sublayer. For an overconsolidated clay, the settlement analysis procedure is essentially the same except that the field compression curve is often not a straight line on a semilog plot and the strain would be taken directly from the field compression curve rather than calculated from Eq. 127-1.

Very often the consolidation sample will not be obtained from the mid-point of the clay layer. When this is the case, appropriate adjustments will have to be made that will require some judgment. Refer to problems 4-8, 4-17, and the example problem of Fig. 287-1.

Engineers should always keep in mind that the predicted settlement curve is an approximation of what is likely to happen in the field. Field experience has demonstrated that the method provides a reasonable estimate of the total settlement but that the rate of settlement often proceeds faster than predicted. Lateral drainage, particularly for smaller loaded areas, and the fact that the horizontal permeability of most soils is greater than the vertical permeability are factors that can account for the consolidation proceeding faster than predicted by one-dimensional drainage. The volume of samples tested are extremely small compared to the volume of field soils being consolidated, and test results are not necessarily representative of the entire soil stratum. Furthermore, all samples are disturbed to some extent during the sampling and testing activities. These problems need to be remembered and evaluated in establishing a level of confidence in settlement predictions.

PROBLEMS

4-1 Explain why there is a significant time lag in the settlement of clay soils but not of sandy soils. What causes the time lag in the settlement of clay?

4-2 Define normally consolidated clay and overconsolidated clay.

4-3 Develop the equation for ultimate settlement that is based on a change in void ratio.

$$\Delta H = \frac{\Delta e}{1 + e_o} H_o$$

4-4 Will lowering the water level in a lake from a depth of 50 m to a depth of 10 m cause consolidation settlement of the lake-bed sediments? Explain why or why not.

4-5 For the soil profile of the example problem in Fig. 145-1, calculate the magnitude of the ultimate settlement of the clay layer as a result of lowering the piezometer surface 15 m in the bottom sand layer. Assume that for the upper sand layer, the water table remains at the surface.

Problems   **151**

4-6 An undisturbed sample was obtained from the middle of a clay layer 6 m thick. The top of the clay layer is at depth 4.6 m, and there is a sandy soil both above and below the clay layer. The total unit weight of the upper sand is 17.3 kN/m³. The groundwater table is at a depth of 4.6 m. The results of a consolidation test on the undisturbed sample are shown below.

Consolidation Test Data

Initial sample thickness = 19.05 mm

Diameter = 75 mm

Wet mass = 164.7 gm

Dry mass = 126.6 gm

$G = 2.75$

Pressure versus Dial-Reading Data

σ (kPa)	Dial Reading (mm)	ϵ
24	0.021	.0011
48	0.025	.0613
96	0.050	.0026
192	0.203	.0196
384	0.686	.0360
768	1.238	.0650
1536	1.778	.0933
384	1.638	.086
96	1.457	.0765
24	1.307	.0686

(a) Plot the $\epsilon - \log \bar{\sigma}$ curve.
(b) Determine the maximum past pressure.
(c) Determine the ultimate settlement of the clay layer if a load of 200 kPa is applied to the surface over a large area.

4-7 A uniform load of 200 kPa is applied over a 12 m × 22 m area at the surface of the soil profile described in problem 4-6. Determine the ultimate settlement profile along the longitudinal centerline of the area. The settlement should be calculated under the center, at the edge, and at a point 4 m past the edge of the loaded area.

4-8 Given the laboratory $\epsilon - \log \bar{\sigma}$ curve from a consolidation test on an undisturbed sample of overconsolidated clay, explain how to find the field curve for an element of the same soil but at a shallower depth than where the sample was obtained. Explain how to find the field curve for the same soil at a deeper depth.

4 Compressibility, Consolidation, and Settlement

4-9 List the assumptions that were made in deriving Terzaghi's differential equation of consolidation.

4-10 For the example problem of Fig. 145-1 determine, by Terzaghi's theory, the pore pressure distribution in the clay layer after 500 days, 5000 days. Assume the load was applied instantaneously.

4-11 A certain clay layer 9 m thick is expected to have an ultimate settlement of 406 mm. If the settlement was 102 mm in four years, how much longer will it take to obtain a total of 150 mm?

4-12 How should initial excess pore pressures (residual pore pressures from a previous loading that have not yet dissipated) be accounted for in a settlement analysis?

4-13 A saturated clay stratum 8 m thick is located between two thick sand layers. In addition to the thick sand strata, there is a thin continuous sand layer (drainage layer) 2 m below the top of the clay layer. Consolidation tests on undisturbed samples of the clay show it to have a coefficient of consolidation, c_v, of 5×10^{-2} mm^2/sec. Application of a uniform load at the ground surface will cause a uniform increase in vertical stress throughout the clay layer.
(a) Determine the number of days required to obtain one-half the ultimate settlement.
(b) How many days would be required to obtain one-half the ultimate settlement without the thin drainage layer?
(c) If the increase in stress is 50 kPa throughout the clay layer, determine the pore pressure distribution in the clay for both drainage cases 175 days after adding the load.

4-14 Measured values of the time rate of settlement in the field may not agree with the predicted values. Explain why.

4-15 What is secondary consolidation and what causes it?

4-16 Given the following data from one loading increment of a consolidation test:
(a) Determine the coefficient of consolidation c_v by the logarithm of time-fitting method.
(b) Show on the laboratory consolidation curve the regions of initial, primary, and secondary consolidation.

Test Data

Double-drained sample, Loading increment = 48 to 96 kPa
Height = 30.86 mm (when dial reads 6.270)

Time (min)	Dial Reading (mm)
0	6.270
0.25	6.144
1.00	6.033
2.25	5.936
4.0	5.812
6.25	5.687
9.0	5.563
12.25	5.441
16.0	5.329
20.25	5.192
25.0	5.113
30.25	5.001
36.0	4.920
42.25	4.839
60	4.666
100	4.420
200	4.166
400	4.026
1440	3.810

4-17 The soil profile at the site of a proposed warehouse consists of 12.3 m of sand overlaying a clay stratum 10.0 m thick. Dense sand underlays the clay to a great depth. The groundwater table is at a depth of 5.5 m. The total unit weight of the upper sand layer is 19.2 kN/m³ above the water table and 19.8 kN/m³ below the water table. A consolidation test was performed on an undisturbed sample that was obtained from a depth of 14.7 m. The results of the test are shown below.

Test Data

Initial sample height = 25.4 mm

Diameter of sample = 63.5 mm

Wet mass = 153.78 g

Dry mass = 115.19 g

$G = 2.75$

4 Compressibility, Consolidation, and Settlement

$\bar{\sigma}$ (kPa)	Dial Reading (mm)	c_v (mm²/sec)
12.5	0.010	0.5
25	0.030	0.5
50	0.043	0.35
100	0.127	0.25
200	0.455	0.20
400	1.257	0.21
800	2.220	0.23
1600	3.175	
400	2.45	
100	2.20	
25	1.93	

Note. Zero dial reading corresponds to a sample height of 25.4 mm.

The floor dimensions shown in the plan of the proposed warehouse are 50 m × 40 m. It is anticipated that it will take about 13 months to construct the warehouse and put it into full operation. Initial construction will involve excavation and backfilling, so the net surface load will be less than zero for the first two months of construction. The final net surface load (dead load plus sustained live load) will average 50 kPa over the entire floor area of the warehouse.

Estimate:

(a) The total settlement under the center of the warehouse.
(b) The total settlement under a corner of the warehouse.
(c) The settlement time curve for a point at the center of the warehouse.
(d) If the maximum settlement is greater than 50 mm, recommend two alternative methods that could be used to correct the settlement problem.

5

SHEAR STRENGTH

The Mohr strength theory of failure is used to evaluate such soil mechanics problems as slope stability, ultimate bearing capacity, and lateral pressures. These types of analysis involve determination of stress along an assumed failure plane and comparing these stresses with the shear strength of the soil. If the strength of the soil is greater than the computed stresses, then the soil mass is safe against failure along the assumed failure plane. If the stresses are greater than the strength, then failure will occur. This type of analysis is referred to as a limiting equilibrium analysis and is a form of plastic analysis. This chapter introduces the Mohr strength theory, the concept of shear strength, and methods used to evaluate shear strength.

Consider a soil sample dispersed in water. The system is then placed under a compressive load, and water is allowed to drain until the sample is in equilibrium under the compressive load. If the sample is then subjected to a stress condition that causes a shear failure on some plane, the measured values of shear stress (τ_a) and normal stress (σ_a) on the failure plane plot as point A in Fig. 158-1. Other samples handled in the same way and tested under increasing consolidation pressures produce failure states represented by points B, C, and D. The curve $ABCD$, which passes through the origin of the diagram, is called the Mohr strength envelope of the soil. Point E, which falls below the failure envelope, has a shear stress that is less than the shear strength for that particular consolidation pressure and therefore has a factor

Figure 158-1 Shear strength envelope.

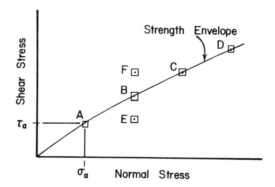

of safety against failure. Point *F*, which lies above the failure envelope, is not a possible condition, because the shear stress is greater than the shear strength for that particular normal stress.

The Mohr strength theory has been found to be the most practical method of describing soil strength. This theory states that failure occurs in a soil mass not on the plane where the shearing stress is maximum, but on a plane where there is a critical combination of normal and shearing stresses as defined by a strength envelope.

MEASUREMENT OF SHEAR STRENGTH

Shear strength is measured both in the laboratory and in the field. Laboratory tests are made on representative soil samples and must be done in a way that simulates the conditions that will exist in the field. The shear strength of sands and gravels can generally be made on disturbed samples that are remolded in the laboratory to field densities. However, as discussed in Chapter 1, remolding significantly affects the physical properties of clay, and laboratory tests on clay soils must therefore be made on undisturbed samples if the strength of a natural soil deposit is to be determined. The strength of proposed compacted earth embankments is often required, and for such cases the laboratory samples must be prepared to duplicate the density, water content, and compaction method of the field soil. The two most commonly used laboratory shear test methods are the direct shear test and the triaxial

Figure 159-1 Direct shear test apparatus.

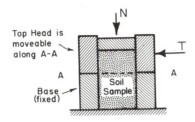

shear test. The unconfined compression test, which is widely used for saturated clay soils, is a special case of the triaxial shear test in which no confining pressure is used.

The vane shear test is a field test used to measure the shear strength of saturated clay. The field strength of both granular and clay soils is also measured through empirical relationships with penetrometer tests such as the standard penetration test and the Dutch cone penetrometer. See Chapters 8 and 11 for further discussion of these tests.

Direct Shear Test

The direct shear test is performed using a shear box similar to that shown in Fig. 159-1. A soil sample is placed in the shear box, and the load T required to shear the sample along the horizontal plane A-A under a given normal load N is measured. The average normal and shear stresses on the horizontal plane at failure represent a point on the strength envelope similar to that of Fig. 158-1. By repeating the test for different normal loads, the strength envelope for the soil can be developed. The relationship between the strength envelope and Mohr circle for the direct shear test is shown in the example problem in Fig. 160-1.

The direct shear test is a simple test to perform and generally provides good results for the strength of granular soil. It can also be used to determine the strength properties of silts and clays; however, it does not offer the flexibility of the triaxial shear test.

Triaxial Shear Test

Triaxial shear tests are performed on cylindrical soil samples encased in flexible membranes. A sample is subjected to a confining (all-around)

Measurement of Shear Strength **159**

Figure 160-1

Example Problem Results from a series of direct shear tests on a sandy soil are as follows.

Number	$\bar{\sigma}$ (kPa)	S (kPa)
1	50	34
2	150	103
3	250	172

Find the principal stresses on the failure plane for test 2.

Solution The Mohr circle is tangent to the strength envelope at point 2

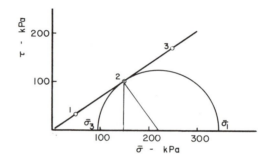

$$\bar{\sigma}_1 = 346 \text{ kPa}$$

$$\bar{\sigma}_3 = 96 \text{ kPa}$$

pressure by placing it in a pressure chamber similar to that shown in Fig. 161-1, and is then tested by increasing the axial load until the sample fails. The procedure is then repeated on other samples at other confining pressures. The test results are interpreted by plotting Mohr's circle for the stress conditions of each sample at the time of failure. This can be done by recognizing that the horizontal and vertical planes are the principal planes for which the principal stresses are the confining pressure, σ_3 (pressure in the pressure chamber), and σ_1 (σ_3 plus the deviator stress), as shown in Fig. 161-2.

5 Shear Strength

Figure 161-1 Section of triaxial shear test apparatus.

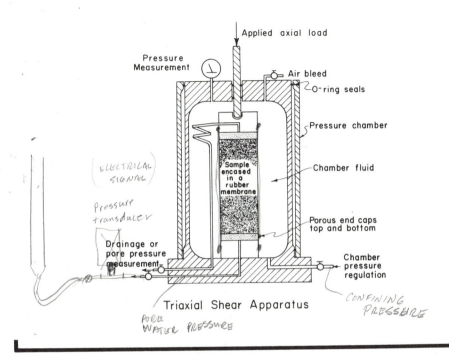

Applied axial load

Pressure
Measurement

Air bleed

O-ring seals

Pressure chamber

Chamber fluid

Sample
encased
in a
rubber
membrane

Porous end caps
top and bottom

Drainage or
pore pressure
measurement

Chamber
pressure
regulation

Triaxial Shear Apparatus

*ELECTRICAL
SIGNAL*

*Pressure
transducer*

*PORE
WATER PRESSURE*

*CONFINING
PRESSURE*

Figure 161-2 Differential element of a soil sample during a triaxial shear test.

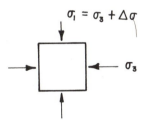

$$\sigma_1 = \sigma_3 + \Delta\sigma$$

σ_3

Measurement of Shear Strength

Figure 162-1 Strength envelope shown tangent to Mohr's circles from three triaxial shear tests.

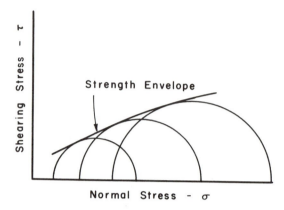

Mohr circles for three tests are shown in Fig. 162-1. The strength envelope is a curve that is tangent to the Mohr circles as shown. The point of tangency to Mohr's circle represents the stress conditions on the plane of failure for that specific sample. The orientation of the failure plane can be obtained from Mohr's circle by locating the origin of planes and drawing a line from that point to the point representing the stress condition on the failure plane (point of tangency).

The unconfined compression test is a special case of the triaxial shear test in which the confining pressure is zero. This test can be run only on soils with some cohesive strength.

Vane Shear Test

The vane shear test (ASTM D2573-72) is used to determine the in-situ shear strength of clay soil. The testing apparatus consists of four vanes on the end of a rod, as shown in Fig. 163-1. A boring is made to the depth at which the vane shear test is to be performed. The vane shear apparatus is placed in the hole and forced into the soil. The torque required to rotate the vane is measured, and the shear strength of the soil is evaluated by relating the maximum torque to the shear resistance provided by the soil around the perimeter and at the top and bottom of the cylinder. The expression for shear strength from the results of the vane shear test is given in Fig. 163-1.

Figure 163-1 Vane shear device.

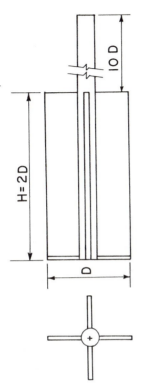

Rod diameter 1/2 in.

$$S = \frac{T}{\pi D^2(H/2 + D/6)}$$

where

S = shear strength
T = maximum torque
H = height of vane
D = diameter of vane

Recommended Dimensions (AASHTO T223)

Casing Size	Diameter (mm)	Blade Thickness (mm)
AX	38.1	1.6
BX	50.8	1.6
NX	63.5	3.2
4 in. (101.6 mm)	92.1	3.2

Sample Drainage

There are three basic types of shear test procedures as determined by the sample drainage condition. Although these three types apply to both the direct shear test and triaxial shear test, they are explained below for the triaxial test only.

The unconsolidated undrained (UU) test is performed by placing the sample in the pressure chamber and increasing the confining pressure without allowing the sample to consolidate (drain) under the confining pressure. The axial load is then applied without allowing drainage of the sample, and thus additional pore pressure develops. The shear strength determined from this testing procedure is referred to as the undrained strength, and the failure

Measurement of Shear Strength

envelope is a total stress failure envelope. Neither consolidation nor drainage is allowed during an unconfined compression test or a vane shear test. Both these tests would, therefore, theoretically yield results that are the same as the results from a UU triaxial shear test.

The consolidated undrained (CU) test is performed by placing the sample in the pressure chamber and increasing the confining pressure. The sample is then allowed to consolidate under the all-around confining pressure, σ_3 by leaving the drain line open (see Fig. 161-1). The drain lines are then closed and the axial stress is increased without allowing further drainage. Since no volume change is allowed during application of the deviator stress, an excess pore pressure normally develops in the sample. The shear strength determined is the consolidated undrained strength of the soil, and the resulting shear strength envelope is a total stress failure envelope.

The consolidated drained (CD) test is similar to the CU test except that the sample is allowed to drain as the axial load is added so that high excess pore pressures do not develop. Since the permeability of clay soil is very low, the axial load must be added very slowly during tests on these soils so that the pore pressures can dissipate. The shear strength determined from this testing procedure is referred to as the drained strength, and the resulting failure envelope is an effective stress failure envelope.

The UU test can be run rather quickly, because the sample is not required to consolidate under the confining pressure or to drain during application of the axial load. Because of the short time required to run the test, it is often referred to as the quick, or Q, test. The CD test may take a considerable period of time to run because of the time required both for consolidation under the confining pressure and for drainage during the application of the axial load. Since the time requirement is long for low-permeability soils, it is referred to as the slow, or S, test. The time required to run the CU test is between that for the UU and the CD tests and, therefore, the CU test is often referred to as the R test, because the letter R falls between Q and S in the alphabet.

The specific type of shear test that should be used will depend on the field conditions to be simulated. This topic is discussed in the following sections. Lowe (1967) suggests that anisotropic consolidation of the soil sample may be required in order to better simulate certain field conditions. This involves consolidating the sample with vertical pressure greater than the lateral pressure before increasing the deviator stress to failure.

SHEAR STRENGTH OF GRANULAR MATERIAL

The shear strength of granular material is affected largely by the initial void ratio of the soil and to a lesser extent by the particle shape, the surface roughness, and the grain-size distribution. Particle shape, surface roughness, and grain-size distribution are characteristics of a specific soil deposit, and their effect on shear strength leads to the differences in the strength characteristics of various deposits. For a given granular soil, the strength characteristics will depend on the void ratio or dry unit weight to which the soil is compacted. The higher the dry unit weight, the higher the shear strength, as represented by the ordinate of the Mohr envelope.

Figure 165-1 Stress versus strain and volume change versus strain for samples of dense and loose sand.

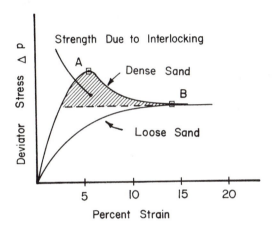

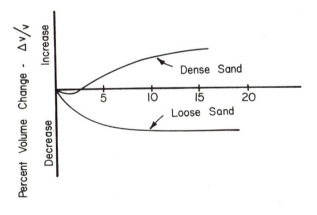

The results from a triaxial shear test can be represented by a plot of deviator stress, $\Delta\sigma$, versus percent strain. Figure 165-1 shows stress-strain curves for two drained tests on the same sand, using the same confining pressure but with different initial void ratios. There is a definite peak in the stress-strain curve for the dense sample, followed by a decrease in the deviator stress with increasing strain until it levels out to a residual value. Failure of the dense sample could thus be defined either by the peak-point stress conditions (point A) or by the ultimate stress conditions (point B). These two failure criteria would produce strength envelopes similar to those shown in Fig. 166-1. The appropriate strength envelope for a given problem will depend on the field conditions that are to be simulated. However, the peak-point shear strength is used for most applications.

As shown in Fig. 165-1, the stress-strain curve for the loose sand does not have a definitive peak point. The failure condition is generally defined by the stress condition at the maximum deviator stress or by the stress condition at a specific value of strain, such as 10 or 20%.

Plotted directly below the stress-strain curves in Fig. 165-1 are curves of percent volume change versus percent strain. The loose sand decreases in volume as the deviator stress is increased. The dense sample shows a slight decrease in volume during the earliest stages of loading, but with increased strain the sample increases in volume until the volume of the sample is larger than the initial volume, even though it has decreased in length. At some density between the loose and dense states, the sample changes very little in volume on application of the deviator stress. The void ratio at this density is referred to as the critical void ratio.

Figure 166-1   Strength envelope based on peak strength and ultimate strength.

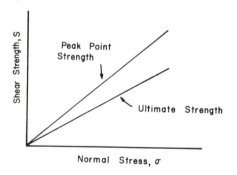

Figure 167-1 Sliding block on a rough plane as an illustration of frictional shear strength.

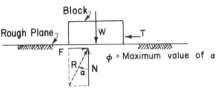

The differences in the volume change characteristics described above for the loose and dense samples is a result of the different particle arrangements at the time of application of a disturbing shearing stress. The particles in a loose sample tend to seek a more compact arrangement on application of the shearing stress, whereas the volume of the dense sample tends to increase, because the particles must either fracture or raise out of their positions to pass by one another, thus leading to an increase in volume.

A comparison of the stress-strain curves and the associated volume change characteristics of the two samples at different densities suggests two components of the shearing strength of granular soils. The shearing resistance of the loose sample is due essentially to the friction between the soil particles. The friction component is illustrated in Fig. 167-1 by a block on a rough plane. As the force T is increased, the angle of obliquity, α, between the normal force N and the resultant R increases until it reaches a maximum value ϕ, at which time motion of the block is impending. The maximum friction force F can then be expressed as

$$F = N \tan \phi$$

This relation can be defined in terms of stress by dividing by the contact area of the block, yielding

$$S = \sigma \tan \phi \qquad (167\text{-}2)$$

where

S = frictional shear strength at the block-plane interface
σ = normal stress
ϕ = maximum angle of obliquity (friction angle)

Equation 167-2 indicates that the frictional shear strength, S, between the block and the plane is a linear function of the normal stress, σ, as shown in Fig. 166-1. The shear strength of a loose sand is similarly the result of friction between the particle-to-particle contacts, and is thus a function of the inter-granular stress, $\bar{\sigma}$, and the coefficient of friction, $\tan \phi$.

Shear Strength of Granular Material **167**

In addition to the frictional component of shear strength, the dense sand has a component referred to as interlocking and is a result of the arrangement of the soil particles. The area between the stress-strain curves as shown in Fig. 165-1 is the result of the interlocking component. At strains greater than the peak point on the stress-strain diagram, the interlocking stress is overcome and the curve for the dense samples at high strain tends to the same level as the loose sample.

The permeability of most granular materials is very high, and thus sufficient time is generally available to allow complete drainage of the sample when subjected to a shear stress. The appropriate shear strength for most applications is, therefore, the drained strength. An exception to this is dynamic loads applied to fine sands and silty sands. Fine sands and silty sands have a rather low permeability, and excess pore pressures cannot dissipate under the rapid loading conditions from earthquakes and other dynamic loads, such as those from pile-driving operations. The excess pore pressures for loose sands are positive and can lead to a complete loss of strength under a long-duration dynamic load, such as that from a high-magnitude earthquake. This loss of strength is referred to as spontaneous liquefaction, and has resulted in many catastrophic failures (Seed, 1968). Dense, fine sand deposits may actually realize a gain in strength as a result of dynamic shock loads, because the excess pore pressure will generally be negative. Because of the differences in the strength characteristics of fine sands under static and dynamic loads, the static strength characteristics cannot be used to evaluate the dynamic strength. Special cyclic shear tests have been developed to determine the dynamic strength (Seed and Lee, 1966).

SHEAR STRENGTH OF CLAY

The shear strength of clay soil is composed of the following components.

- A stress component, $S_{\bar{\sigma}}$, which is a function of the effective stress, $\bar{\sigma}$. As $\bar{\sigma}$ is increased, contact pressures between particles become greater and the resistance to motion of one particle relative to another becomes greater. This component is intimately related to the frictional characteristics of the material.
- A void ratio component, S_e, which varies in a linear fashion with the log of the void ratio. As the void ratio becomes smaller, the electrical and molecular forces surrounding the soil particles play a greater part in resisting the relative movement of the particles. The viscosity of the structured water molecules close to the surface of the soil particles also assumes a more prominent role in influencing shear strength as the soil particles approach each other.

- A rheological component, S_ρ, which is a viscous phenomenon; its magnitude depends on the rate of motion between soil particles. This rate has been characterized by the following equation.

$$S_\rho = \frac{\eta \, d\epsilon}{dt}$$

where $d\epsilon/dt$ is the rate of change of axial strain and η is the structural viscosity of the soil. This equation represents a linearly viscous material, and a more correct relationship would probably consider a nonlinear viscosity such as

$$S_\rho = c_1 \left(\frac{d\epsilon}{dt} \right)^{c_2}$$

- A dilation component, S_d, which is caused by the necessity of the soil particles to move up and down as they pass by one another during failure. If the soil exhibits a tendency to increase in volume as failure occurs, then an additional shear force is necessary to overcome the attraction of the soil particles for one another and the S_d component and shear strength would be greater. A tendency for a volume decrease during failure decreases the force necessary to cause failure, and hence the S_d component and the shear strength would be less.

Because of the very low permeability and the high compressibility (under static loads) of clays and plastic silts, the drainage conditions, rate of loading, and the stress history of the soil greatly affect the shear strength properties. Shear strength in this section is therefore discussed in terms of the history of soil stress and the drainage conditions during a triaxial shear test, and the strength is related to the strength components just discussed.

As discussed in Chapter 4, a soil that has never been subjected to an effective pressure greater than the existing effective pressure is considered a normally consolidated soil. One that has been subjected to a greater effective pressure is an overconsolidated soil. These terms are used to describe qualitatively the stress history of the soil.

Normally Consolidated Clay—CD Test

Consider a saturated clay sample placed in a triaxial pressure chamber and allowed to consolidate under a confining pressure ($\overline{\sigma}_c$) that is equal to or greater than its maximum past pressure. If the axial load on the sample is then slowly increased with drainage at each end of the sample, a deviator stress ($\overline{\sigma}_1 - \overline{\sigma}_3$) versus strain diagram similar to that in Fig. 170-1a will be obtained. The percent volume change versus strain is plotted on Fig. 170-1b. This is referred to as a consolidated drained (CD) test on a normally

Figure 170-1 Typical curves of stress and volume change versus axial strain for a CD triaxial shear test on a normally consolidated clay soil.

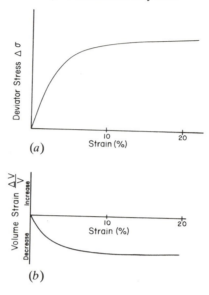

(a)

(b)

consolidated clay. Figure 170-1 shows that the sample consolidates during the shearing process. If the load is applied too fast, the sample would not fully consolidate during the test and excess pore pressures would develop in the sample. This would tend to reduce the effective stress and hence the strength.

The Mohr circles for stress conditions at failure are plotted for three chamber pressures in Fig. 171-1. The stress at failure in each case is defined either by the maximum deviator stress or by the deviator stress at a specified strain, such as 10 or 20%. The effective strength envelope is determined by drawing a line tangent to the three effective stress circles.

The points of tangency represent the stress conditions on the failure plane in each sample. The orientation of the failure plane is indicated by a line from the origin of planes $(\sigma_3, 0)$ to the point of tangency with the effective strength envelope as shown for the second circle in Fig. 171-1.

The strength components $S_{\bar{\sigma}}$, S_e, S_ρ, and S_d would all have some effect on the strength of a normally consolidated soil subject to drained shear. The component S_d will be small for a drained test on a normally consolidated soil, because the volume of the sample tends to decrease as a result of applying a

Figure 171-1 Typical effective strength envelope from a CD triaxial shear test on a normally consolidated clay.

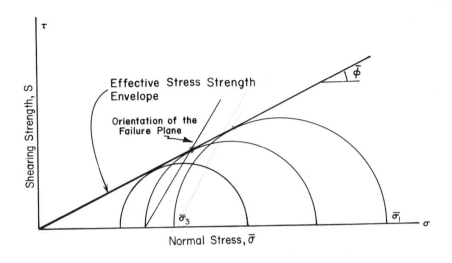

Normal Stress, $\bar{\sigma}$

shearing stress. Since the soil is allowed to consolidate under the confining pressure, the void ratio will be smaller and the effective stress larger for higher confining pressures, and thus the strength components S_e and $S_{\bar{\sigma}}$ will be greater. Furthermore, as the deviator stress is applied, the sample continues to consolidate so that S_e and $S_{\bar{\sigma}}$ are even greater at the end of the test. The increased values of the S_e and $S_{\bar{\sigma}}$ components are evident on the strength envelope of Fig. 171-1 in that the strength increases as the effective normal stress increases. The effect of S_ρ is small in the relatively slow CD test.

Normally Consolidated Clay—CU Test

Assume that the saturated clay sample, after consolidating under the confining pressure, is maintained at a constant volume during the application of the deviator stress and that the samples may be loaded to failure at a relatively rapid rate. The resulting deviator stress-versus-strain curve will have the general shape of that shown in Fig. 172-1a. However, the volume change will be zero throughout the test. The volume is maintained constant during a test on

Shear Strength of Clay

171

Figure 172-1 Typical stress and pore pressure versus strain curves from a CU triaxial shear test on a normally consolidated clay.

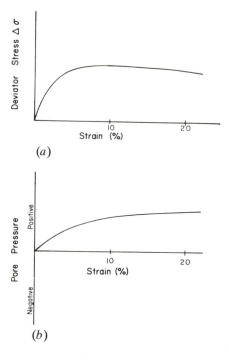

(a)

(b)

a saturated soil by closing the drain lines, hence the name consolidated undrained (CU) shear test. Since the sample is not allowed to consolidate during application of the deviator stress, there will be a buildup of excess pore pressure, as illustrated in Fig. 172-1b. It is necessary to measure the pore pressure during the test in order to obtain the effective stresses $\bar{\sigma}_1$ and $\bar{\sigma}_3$ that are used to plot Mohr's circle. The pore pressure is generally measured by connecting a pressure-measuring device, such as a strain gauge pressure transducer, to the drain lines that come from each end of the sample. The strength components $S_{\bar{\sigma}}$, S_e, S_ρ, and S_d would all contribute to the strength on the failure plane. Depending on the magnitude of S_ρ, the effective strength envelope from the CU test would fall slightly above or slightly below the strength envelope from the CD test (Fig. 171-1).

The failure envelope shown in Fig. 171-1 is generally referred to as the effective stress strength envelope because it is based on the effective stresses

Figure 173-1 Total and effective strength envelopes from CU triaxial shear tests on a normally consolidated clay.

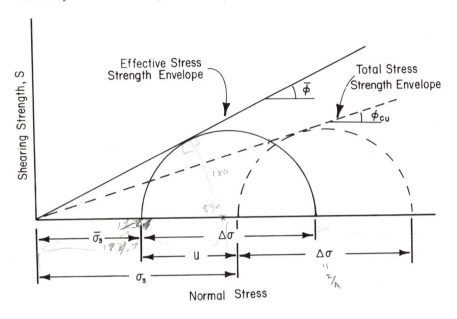

at the time of failure. Another common method of presenting the results from a CU test is to plot Mohr's circle in terms of total stress. The strength envelope formed by the curve that is tangent to the Mohr's circles plotted from the total stresses at the time of failure is referred to as the total stress strength envelope. A comparison of the two envelopes obtained from a CU triaxial shear test is shown in Fig. 173-1. Note that Mohr's circle has the same diameter for total stresses as for effective stresses but that the effective stress circle is displaced to the left an amount equal to the pore pressure at the time of failure. The orientation of the failure plane in a CU test is obtained as in Fig. 171-1 by drawing a line from the origin of planes to the point of tangency of the effective stress circle and envelope.

Normally Consolidated Clay—UU Test

An unconsolidated undrained (UU) shear test is performed by placing a sample in a pressure chamber and applying a confining pressure without allowing

Shear Strength of Clay

Figure 174-1 Typical strength envelope from a UU test on a normally consolidated clay.

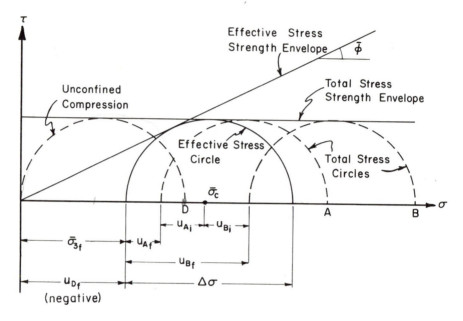

any volume change to occur. If the applied confining pressure (σ_3) is equal to the pressure to which the sample had been consolidated ($\overline{\sigma}_c$), then the initial excess pore pressure in the sample will be zero. If the confining pressure is greater or less than $\overline{\sigma}_c$, then the initial excess pore pressure will be equal to the confining pressure minus $\overline{\sigma}_c$. Thus, if the confining pressure is greater than $\overline{\sigma}_c$, the initial pore pressure will be positive; if it is less than $\overline{\sigma}_c$, the initial pore pressure will be negative (see Fig. 174-1). Following application of the confining pressure, the deviator stress is applied without allowing the sample to drain, and a deviator stress-versus-strain curve with a shape similar to that shown in Fig. 172-1a will be obtained. Mohr's circle at failure for the UU test is plotted in terms of total stress. Mohr's circles A, B, D for three different UU tests on a normally consolidated saturated clay are plotted on Fig. 174-1. Note that all circles have the same diameter and that a line drawn tangent to them is a horizontal strength envelope. This envelope represents

174

the undrained strength S_u. If pore pressures are measured during the test and then subtracted from the total pressures, approximately the same effective stress circle would be obtained at failure for all three tests. The minor principal stress at failure is identified as $\overline{\sigma}_{3f}$ in Fig. 174-1. For circles A and B the initial confining pressure was greater than $\overline{\sigma}_{3f}$; hence, the final pore pressures u_{Af} and u_{Bf} are positive. The point of tangency of the effective stress strength envelope to the effective stress circle is again used to locate the orientation of the failure plane. The components S_e and $S_{\overline{\sigma}}$ would both remain the same for each test, because the void ratio and effective stress on the failure plane are the same. If all tests were run at the same strain rate, then S_ρ would be the same for each test. If the strain rate is not the same for each test, the size of Mohr's circle would be slightly smaller or larger.

Circle D in Fig. 174-1 is a special case of the UU test, the unconfined compression test. As the name implies, the test is run without using a confining pressure. The pore pressure in the sample is, therefore, always negative. The minor principal stress of the total stress circle will be zero. The diameter of the circle, which is equal to the applied axial load at failure, is referred to as the unconfined compressive strength q_u. Since the circle is tangent to a horizontal strength envelope, the undrained strength, S_u, is approximately one-half the unconfined compressive strength.

$$ S_u = \frac{q_u}{2} $$

The example problem of Fig. 176-1 illustrates test results for CD, CU, and UU shear tests.

Overconsolidated Clay

As previously defined, an overconsolidated clay is one that has been subjected to a pressure greater than the existing pressure. The strength behavior of an intact overconsolidated clay can be illustrated by considering a sample that has previously been consolidated under a pressure of $\overline{\sigma}_c$ on the plane of failure. (This is not referring to the chamber pressure in a triaxial shear test.) If the normal pressure on the failure plane is then reduced to $\overline{\sigma}_a$ and the sample is allowed to expand, the void ratio will increase. However, since the rebound curve is much flatter than the loading curve, the void ratio will be much less than it would be for a normally consolidated clay at pressure $\overline{\sigma}_a$, as shown in Fig. 177-1. The value of the void ratio component of strength, S_e, will therefore be greater than the value of S_e for a normally consolidated sample under a normal pressure $\overline{\sigma}_a$. The corresponding points on the strength envelope for normally consolidated and overconsolidated samples that have an effective stress of $\overline{\sigma}_a$ on the failure plane are shown as points A and A',

Figure 176-1

Example Problem Two consolidated, undrained triaxial shear tests, with pore pressure measurements, were performed on saturated samples of a given clay soil. The specimens were loaded to failure after consolidation under confining pressures of 200 and 400 kPa. The results follow.

Specimen number	P_c σ_3 (kPa)	$\Delta\sigma$ (at Failure) (kPa)	Δu (at Failure) (kPa)
1	200	150	140
2	400	300	280

1. Plot the total stress failure envelope.
2. Plot the effective stress failure envelope.
3. Determine the orientation of the failure plane for sample 1.
4. A sample of this soil was consolidated under a 400-kPa confining pressure and then tested in an unconfined compression apparatus. Plot the Mohr circle and determine the unconfined compressive strength.

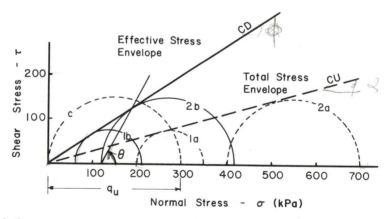

Solution
1. Draw a tangent to circles 1a and 2a.
2. Draw a tangent to circles 1b and 2b.
3. (Use the effective stress envelope.)

$$\theta = 45 + \frac{\phi}{2} = 45 + \frac{32°}{2} = 61° \text{ (see drawing)}$$

176

5   Shear Strength

4. The lateral pressure is zero. Draw circle c.

$$q_u = \text{diameter of circle}$$
$$= 300 \text{ kPa}$$

Figure 177-1 Void ratio versus consolidating pressure from a consolidation test on an undisturbed sample of clay.

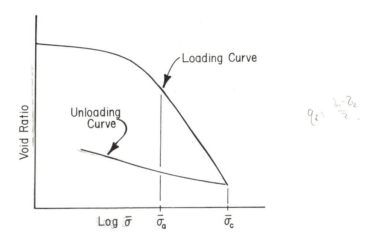

respectively, in Fig. 178-1. Line *ABCD* in the figure is the effective stress strength envelope for a normally consolidated clay and is the same as the strength envelope shown in Fig. 171-1. Point *B'* represents the stress condition at failure on a sample that had been consolidated to $\bar{\sigma}_c$ on the plane of failure and then allowed to expand before testing such that the effective normal stress at the time of failure was equal to $\bar{\sigma}_b$. Line *A'B'C* in Fig. 178-1 represents the strength envelope for an overconsolidated soil that has been initially consolidated to a normal stress of $\bar{\sigma}_c$ on the plane of failure. Figure 178-1 shows that for normal stresses greater than $\bar{\sigma}_c$, the strength envelope for the overconsolidated soil is the same as for the normally consolidated soil.

Shear Strength of Clay   **177**

Figure 178-1 Strength envelope for an overconsolidated clay.

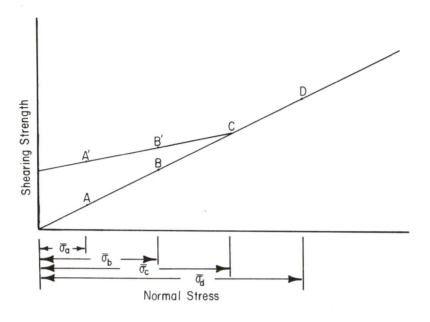

A typical stress-strain relationship from a CD triaxial shear test on a highly overconsolidated saturated clay soil is shown in Fig. 179-1*a*. The shape of the curve has a definite peak followed by a decrease in the deviator stress with increasing strains. The curve representing the volume change is plotted directly below the stress-strain curve and indicates a slight decrease in volume at the beginning of the test, followed by an increase in volume until the volume of the sample is much greater than the initial volume. This volume change characteristic was also observed for dense sands. The explanation for this phenomenon in clays is more complex than that for sands but still has to do with the soil grains being in a very compact position such that a disturbing force will tend to cause an increase in volume as the sample is distorted.

In a consolidated undrained triaxial shear test, the sample is maintained at constant volume during the application of the deviator stress. Since a highly overconsolidated clay has a tendency to increase in volume (take on water) during a drained test, the pore pressure in the CU test will become

Figure 179-1 Typical curves of stress and volume change versus strain from a CD triaxial shear test on a highly overconsolidated clay.

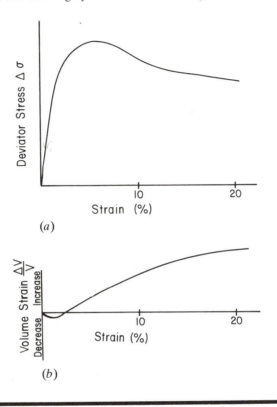

(a)

(b)

negative when drainage of the sample is not permitted. Typical stress-strain and pore pressure-strain relationships from a CU triaxial shear test on a saturated highly overconsolidated clay are shown in Fig. 180-1.

Because the pore pressure becomes negative during an undrained test on a highly overconsolidated clay (overconsolidation ratios greater than about 4 to 8, Terzaghi and Peck, 1967), the effective stress circle will be displaced to the right of the total stress circle, and hence the undrained strength will be greater than the drained strength. In the field the negative pore pressures over a long period will draw water into the pores of the soil matrix, causing an increase in void ratio and a decrease in the effective stress. This results in a decrease in the $S_{\bar{\sigma}}$ and S_e strength components and hence a decrease in the

Shear Strength of Clay

Figure 180-1 Typical curves of stress and pore pressure versus strain from a CU triaxial shear test on a highly overconsolidated clay.

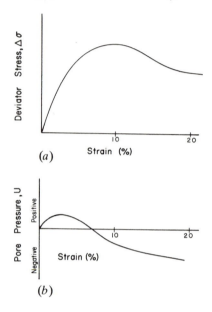

(a)

(b)

long-term strength of the soil. For this reason the undrained strength (both CU and UU) cannot be depended on. Terzaghi and Peck (1967) recommend that except for overconsolidation ratios as low as 2 to 4, the UU strength envelope should not be used for overconsolidated clays under long-term loads.

The somewhat idealized relationship between the CD, CU, and UU strength envelopes for a typical saturated inorganic clay are shown in Fig. 181-1. The UU envelope is horizontal, the slope of the CD envelope $\bar{\phi}$ (normally consolidated) ranges from more than 30° for clays with low plasticity index to less than 15° for clays with higher plasticity index. Stress circles to the right of $\bar{\sigma}_c$ represent normally consolidated conditions, and circles left of $\bar{\sigma}_c$ represent overconsolidated conditions.

Many highly overconsolidated clays contain a network of very small cracks. Clays of this type are referred to as fissured overconsolidated clays, and their shear strength characteristics are often time dependent and are very difficult to describe. Terzaghi and Peck (1967) discuss the problem of evaluating the stability of open cuts in fissured overconsolidated clays.

Figure 181-1 Typical strength envelopes for CD, CU and UU tests on clay.

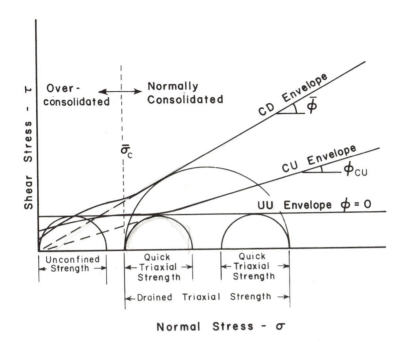

Compacted Clays

The design of compacted earth embankments such as earth dams requires the determination of the strength of the compacted material. The strength of compacted granular material is straightforward, because it is primarily a function of the relative density of the material. The strength of granular material is little affected by the moisture content at the time of compaction or by changes in the embankment water content with time. The strength of compacted clay soils, on the other hand, is a much more complex problem, because the remolding moisture content and subsequent changes in the moisture content have a great influence on strength.

The general behavior of compacted clays can be characterized as being similar to a moderately overconsolidated clay. A total stress analysis is generally used because the pore pressures of unsaturated soils are very difficult to predict. High-excess pore pressures can develop during compaction, especially if the remolding moisture content is high (much greater than the plastic limit).

On saturation, compacted clays generally experience a significant loss of strength. The final strength characteristics depend a great deal on the dry density and the moisture content at which the material was compacted.

Because of the sensitivity of the strength of compacted clays to initial and long-term moisture conditions, it is extremely important to assure that the preparation and testing of compacted clay samples are done in a way that simulates the expected moisture and stress history of the soil in the field. Considerable experience is needed to select the appropriate test procedure and to interpret the results. The properties of compacted expansive clays are discussed in Chapter 9.

MOHR-COULOMB STRENGTH ENVELOPE

Measured strength envelopes are generally slightly curved but, for convenience, they are usually represented by a straight line over the applicable stress range, as shown in Figs. 183-1 and 184-1.

Figure 183-1 shows a Mohr strength envelope for a general case. The equation of the envelope is given by Eq. 183-1a and is referred to as Coulomb's equation. The values of the parameters c and ϕ depend on the material and the testing conditions. The effective stress strength envelope for granular material and normally consolidated clays generally passes through the origin and therefore c is zero. Hence, Coulomb's equation (Eq. 183-1a) for sand or normally consolidated clay in drained shear is expressed as

$$S = \bar{\sigma} \tan \bar{\phi} \qquad \text{or} \qquad S = (\sigma - u) \tan \bar{\phi}$$

where $\bar{\phi}$ indicates the slope of the effective stress strength envelope.

As pointed out earlier, the strength of granular material (including nonplastic silts) is primarily a function of the material density. The table of Fig. 183-2 gives some typical values of ϕ for granular materials.

The total stress strength envelope from a consolidated undrained test on a normally consolidated clay can be described by the relationship

$$S = \sigma \tan \phi_{cu}$$

The normal stress for this case is expressed in terms of the total stress, σ, and ϕ_{cu} is the slope of the total stress envelope from a CU test (Fig. 173-1).

The failure envelope from an unconsolidated undrained (UU) shear test on a saturated clay is approximately a horizontal line (Fig. 174-1) and, therefore, the value of ϕ is zero. For this case the equation of the envelope reduces to

$$S_u = c$$

Figure 183-1 Mohr-Coulomb strength envelope.

$$S = c + \sigma \tan \phi \qquad\qquad (183\text{-}1a)$$

where

S = strength

c = cohesion intercept

σ = normal stress

ϕ = slope of the strength envelope in degrees.

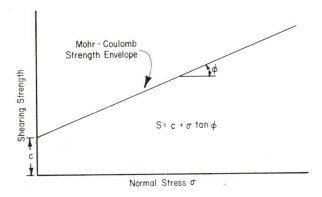

Figure 183-2 Representative values of $\bar{\phi}$ for sands and silts. (After Terzaghi and Peck, 1967.)

Material	Degrees	
	Loose	Dense
Sand, round grains, uniform	27.5	34
Sand, angular grains, well graded	33	45
Sandy gravels	35	50
Silty sand	27–33	30–34
Inorganic silt	27–30	30–35

Figure 184-1

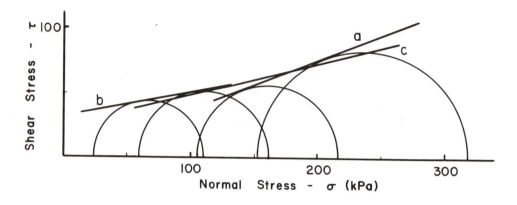

Example Problem Consolidated, drained, triaxial tests on saturated samples of preconsolidated clay give the following results.

Sample	$\bar{\sigma}_3$ (kPa)	$\Delta\bar{\sigma}$ at Failure (kPa)
1	25	85
2	60	101
3	105	113
4	154	163

1. Write Coulomb's equation for a problem in which the pressure range is 120 to 220 kPa.
2. Write Coulomb's equation for a problem in which the pressure range is up to 120 kPa.
3. Write Coulomb's equation for a problem in which the pressure range is 60 to 200 kPa.
4. Estimate the preconsolidation pressure.

Solution

1. Line a: $S = \bar{c} + \bar{\sigma} \tan \bar{\phi}$

$\qquad\qquad = 0 + \bar{\sigma} \tan 20°$

2. Line b: $S = 30 + \bar{\sigma} \tan 11°$

3. Line c: $S = 24 + \bar{\sigma} \tan 14°$

4. The preconsolidation pressure seems to be about 120 kPa.

Note: The Mohr-Coulomb envelope is the best-fit straight line that represents the soil strength over a range in pressures that may exist in a field problem. It may be used whenever a single straight line is not too inaccurate. It is especially useful whenever a single expression for shear strength is necessary in a mathematical derivation or analysis.

and it is generally referred to as the undrained strength. The strength envelope from a UU test on an unsaturated soil will have a slight slope at lower confining pressures, and it can be described by Eq. 183-1a.

The strength parameters for overconsolidated clays can be described in terms of both an effective stress strength envelope or a total stress strength envelope. Equations for each follow.

$$S = \bar{c} + \bar{\sigma} \tan \bar{\phi} \quad \text{(effective stress)}$$

$$S = c + \sigma \tan \phi \quad \text{(total stress)}$$

As pointed out earlier, the undrained strength of a highly overconsolidated clay can be greater than the drained strength, and care must be taken in using strength parameters for overconsolidated clays for long-term conditions that are based on total stress.

It must be emphasized that c and ϕ are not unique values for a given soil. Shear test results that give only values of c and ϕ are, therefore, meaningless without a description of the type of test (drainage conditions), type of strength envelope, and the normal stress range.

Selection of the appropriate strength parameters for a specific application is an important part of the analysis process. Field conditions will dictate whether short-term or long-term strength is required, and the specific method of analysis to be used. The selection of strength parameters for specific cases is presented in the following chapters.

The figures that have been used to illustrate the concept of a strength envelope must be considered as "textbook examples." In actual practice some judgment will be required to establish the strength envelope because experimental errors are likely to occur in the sampling and laboratory testing procedures. The ability to interpret laboratory results correctly is gained from an understanding of the basic concepts of shear strength, from the way in which the strength parameters are to be applied, and from experience.

Figure 186-1 *p—q* Diagrams.

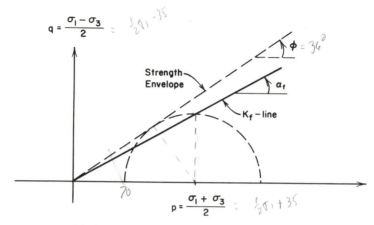

(*a*) Cohesionless soils

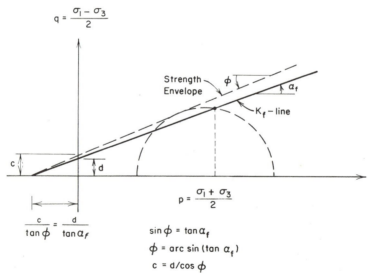

(*b*) Cohesive soils

p-q Diagram and Stress Paths

In solving the problem in Fig. 184-1, c and ϕ were determined by plotting the Mohr circles and drawing strength envelopes tangent to the circles for the respective pressure ranges of interest. An alternative is to plot a p-q diagram as described by Eq. 114-2. It is easier to draw a best-fit line through a series of points than tangent to a series of circles. The line through the p-q points is called a K_f line.

The relationships between slope and intercept of the K_f line (α_f and d) are related to the slope and intercept of the strength envelope (ϕ and c) in Fig. 186-1. The factor K_f represents the principal stress ratio at failure.

$$K_f = \left(\frac{\sigma_h}{\sigma_v}\right)_f = \frac{1 - \tan \alpha_f}{1 + \tan \alpha_f} = \frac{1 - \sin \phi}{1 + \sin \phi} = \tan^2\left(45 - \frac{\phi}{2}\right)$$

The use of this method is demonstrated in the example problem of Fig. 188-1, using the data from the example of Fig. 184-1. When the stress path extends to the K_f line, it indicates a state of plastic slip or failure. Stress paths may be drawn for either total or effective stress. The horizontal distance on a p-q diagram between the effective stress path and total stress path is the pore pressure u. The pore pressure equals the p coordinate for the total stress path minus the p coordinate for the effective stress path. If the effective stress path lies to the right of the total stress path the pore pressures are negative. In Fig. 189-1 are shown total and effective stress paths for CU triaxial shear tests on a normally consolidated clay and an overconsolidated clay. The samples were first subjected to a cell pressure of σ_c. As the deviator stress was applied, the effective stress path progressed to the K_f line as shown. The pore pressures at failure are shown in the figure. In a heavily overconsolidated clay the effective stress path could be to the right of the total stress path because of the generation of negative pore pressures.

Figure 188-1

Example Problem Using the data from the example problem of Fig. 184-1, plot a *p-q* diagram and fit K_f lines through points for the following pressure ranges.

1. 120 to 220 kPa
2. 0 to 120 kPa
3. 60 to 200 kPa

Sample	$\bar{\sigma}_3$	$\Delta\sigma$	$\bar{\sigma}_1$	$p = \dfrac{\bar{\sigma}_1 + \bar{\sigma}_3}{2}$	$q = \dfrac{\bar{\sigma}_1 - \bar{\sigma}_3}{2}$
1	25	85	110	67.5	42.5
2	60	101	161	110.5	50.5
3	105	113	218	161.5	56.5
4	154	163	317	235.5	81.5

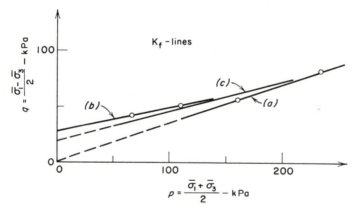

1. $\alpha_f = 18.7°$ $\phi = 19.8°$
   $d = 0$ $c = 0$ kPa
2. $\alpha_f = 10.5°$ $\phi = 10.5°$
   $d = 29.9$ $c = 30.5$ kPa
3. $\alpha_f = 13.1°$ $\phi = 13.5°$
   $d = 20.0$ $c = 20.6$ kPa

5 Shear Strength

Figure 189-1 Typical stress paths.

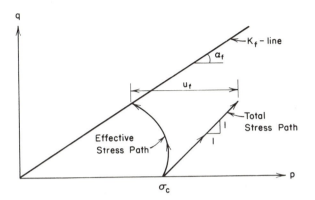

(*a*) Normally consolidated clay

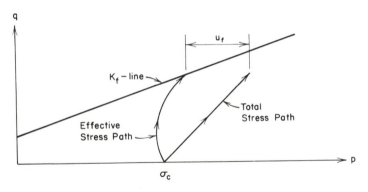

(*b*) Overconsolidated clay

Mohr-Coulomb Strength Envelope

PROBLEMS

5-1 Describe Mohr's theory of failure.

5-2 Derive the equation used to evaluate the shear strength from a vane shear test. Assume that the vanes have been pushed into the soil so that the soil must fail at both the top and the bottom surfaces.

(*Hint.* Take the summation of moments about the axis of rotation, and remember that the shear stress at failure along the failure surfaces is constant and equal to the shear strength of the soil.)

5-3 What is liquefaction? Why does it occur?

5-4 In your own words describe the friction and interlocking components of the shear strength of sand.

5-5 Use appropriate sketches to define for sand:

(a) Peak-point shear strength.
(b) Ultimate shear strength.
(c) Critical void ratio.

5-6 A fully saturated cohesionless soil specimen is to be tested in shear. Will its shearing resistance in an undrained test be the same as, greater than, or less than in a drained test:

(a) If the soil is initially dense?
(b) If the soil is initially very loose?

Explain.

5-7 The following data were obtained from a series of drained triaxial shear tests on samples of a clean uniform sand at two different void ratios.

$e_{min} = 0.42, e_{max} = 0.87$

Test Number	1	2	3	4	5	6
Confining Pressure (kPa)	34.5	34.5	69.0	69.0	138.0	138.0
Void Ratio	0.44	0.80	0.43	0.81	0.45	0.79

Axial Strain (%)	Deviator Stress (kPa)					
0	0	0	0	0	0	0
0.5	62.0	21.0	69.0	21.0	102.9	69.0
1.0	93.0	28.0	138.0	35.0	165.0	99.5
1.5	121.0	41.1	186.0	51.4	234.3	125.2
2.0	138.0	49.0	220.1	62.0	310.2	143.9
3.0	121.0	52.0	266.3	76.0	406.1	163.0
4.0	115.0	55.1	258.1	87.0	434.5	186.0

$$e_{min} = 0.42, e_{max} = 0.87$$

Test Number	1	2	3	4	5	6
Confining Pressure (kPa)	34.5	34.5	69.0	69.0	138.0	138.0
Void Ratio	0.44	0.80	0.43	0.81	0.45	0.79

Axial Strain (%)			Deviator Stress (kPa)			
6.0	100.0	62.0	228.0	110.1	427.5	207.1
8.0	86.0	65.0	217.0	125.0	390.2	214.0
10.0	82.9	65.0	214.1	131.1	349.8	216.8
12.0	78.0	65.0	210.1	131.1	324.3	216.8
14.0	76.0	65.0	206.3	129.0	296.0	216.8

(a) Plot the stress-strain curves for each test.
(b) Plot Mohr's circles and draw the failure envelope for the dense and loose samples. (Indicate the relative density of each sample.)
(c) Graphically measure the orientation of the failure plane for each test.

5-8 In your own words explain why the pore pressure becomes negative during a CU triaxial shear test on a highly overconsolidated clay soil.

5-9 Sketch typical curves of deviator stress ($\Delta\sigma$) versus axial strain, and pore pressure (u) versus axial strain for a consolidated-undrained test on:

(a) Loose sand or normally loaded clay.
(b) Dense sand or highly overconsolidated clay.

5-10 Sketch a possible strength envelope for an overconsolidated clay soil. Clearly indicate the maximum past consolidation pressure.

5-11 Should the unconfined compression test be used to determine the shear strength of a highly overconsolidated clay? Explain.

5-12 In a direct-shear test on a medium-dense sand, the normal stress on the failure plane was 140 kPa, and a shearing stress of 81.1 kPa was recorded at failure.

(a) What is the value of $\bar{\phi}$?
(b) Determine the major and minor principal stresses at failure.

5-13 In a direct-shear test on a medium-dense sand, a shearing stress of 58.6 kPa was recorded at failure. If the friction angle, $\bar{\phi}$, of the soil was 38°, what was the normal stress on the failure plane? What were the values of the major and minor principal stresses at failure?

5-14 A confining pressure of 70 kPa is applied to a sample of sand during a consolidated drained triaxial shear test. If the friction angle of the sand is 36°, what will be the deviator stress at the time of failure?

5-15 A vane shear test was made in a deposit of soft clay. The vane was 114 mm long and 75 mm in diameter, and it was pressed into the clay so that both the top and bottom surfaces of the vane were fully in the clay. Torque was applied and gradually increased to 44.8 Nm when failure occurred.

(a) Find the shear strength of the clay on the basis of the above vane shear test.
(b) What type of triaxial shear test would give the same value for shear strength if a perfectly undisturbed sample could be obtained?

5-16 In a consolidated drained triaxial test on a normally consolidated clay soil, a deviator stress ($\Delta\sigma$) of 152 kPa was required to fail the sample when the confining pressure was 70 kPa.

(a) What was the angle of shearing resistance $\bar{\phi}$?
(b) What was the orientation of the failure plane?
(c) Suppose that after consolidation under the 70 kPa confining pressure, the sample is placed in an unconfined compression apparatus and tested. Determine the resulting shear strength if $\phi_{cu} + 10° = \bar{\phi}$.
(d) During the unconfined compression test, what will be the value of the pore pressure at failure?

5-17 Two consolidated undrained triaxial shear tests with pore pressure measurements were performed on saturated samples of a given clay soil. The specimens were loaded to failure after consolidation under confining pressures of 200 and 400 kPa. The results are shown here.

Specimen Number	σ_3 (kPa)	$\Delta\sigma$ (at failure) (kPa)	Δu (at failure) (kPa)
1	200	150	140
2	400	300	280

(a) Plot the total stress failure envelope and write Coulomb's equation for the failure envelope.
(b) Plot the effective stress failure envelope on the same diagram as in (a) and write Coulomb's equation for the failure envelope.
(c) Comment on whether the clay appears to be normally consolidated or over consolidated.
(d) Determine the orientation of the failure plane for specimen No. 1.

5-18 A series of consolidated drained tests on a normally consolidated clay

5 Shear Strength

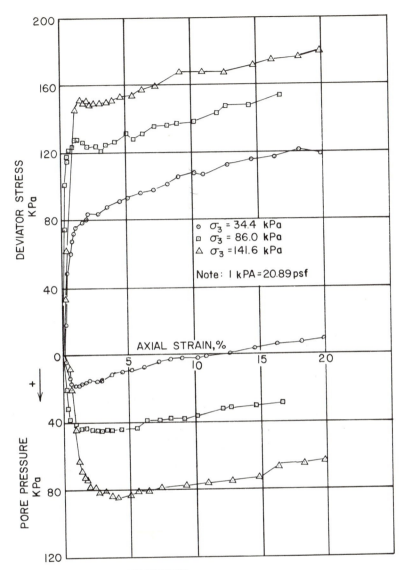

SAMPLE: RED, SILTY CLAY W/GRAVEL
         TEST PIT TP-8
         DEPTH 2.2' TO 3.3'
         1.4" DIAMETER SAMPLE

CONSOLIDATED-UNDRAINED
TRIAXIAL SHEAR TEST
STRESS-STRAIN CURVES

REMOLDED TO 95% OF DWR S-10 METHOD A
MAXIMUM DRY DENSITY, AT 1.5% WET OF
OPTIMUM MOISTURE. SAMPLE SATURATED
BEFORE TESTING.

Problems

indicated a friction angle $\bar{\phi}$ of 31°. A consolidated undrained triaxial shear test on the same sample yields the following results.

$$\sigma_3 = 200 \text{ kPa}$$
$$\Delta\sigma = 180 \text{ kPa}$$

Estimate the pore pressure at the time of failure in the CU test.

5-19 The unconfined compressive strength of a normally consolidated clay is determined to be 200 kPa. If the friction angle $\bar{\phi}$ of the clay is 25°, estimate the pore pressure at failure during the unconfined compression test.

5-20 For the data of problem 5-7, determine Coulomb's equation for the loose and dense samples. Estimate the value of $\bar{\phi}$ for a relative density of 60%.

5-21 Prove that the orientation of the failure plane in a triaxial shear test on a sand or a normally consolidated clay is at $45 + \bar{\phi}/2$ from the horizontal plane.

5-22 The stress-versus-strain and pore pressure-versus-strain-curves from consolidated undrained triaxial shear tests on a compacted silty clay are shown below. Determine Coulomb's equation for the total and effective stress failure envelopes for the compacted material. Consider failure at 10% strain.

5-23 Determine the effective strength parameters $\bar{c}$ and $\bar{\phi}$ for the data of problem 5-17, using a p-q diagram.

5-24 Two consolidated undrained triaxial shear tests with pore pressure measurement on undisturbed clay samples gave the following results. The saturated samples were initially consolidated at the given $\bar{\sigma}_3$ pressures.

Sample Number	σ_3 (kPa)	σ_1 (kPa)	u_f (kPa)
1	220	480	78
2	420	760	216

(a) Using a p-q diagram, determine the effective strength parameters $\bar{c}$ and $\bar{\phi}$.
(b) Is the clay normally consolidated or overconsolidated?

6

LATERAL PRESSURES AND RETAINING STRUCTURES

The design of many structures such as retaining walls, anchored bulkheads, buried pipes, basement walls, braced excavations, thrust blocks, and others, require the determination of lateral earth pressures. Lateral pressures can be grouped into three states.

- Active.
- At rest.
- Passive.

The at-rest lateral pressure is the lateral pressure that exists in soil deposits that have not been subject to lateral yielding. The active and passive states of lateral pressure are limiting conditions and represent states of plastic equilibrium or at least partial plastic equilibrium. A state of plastic equilibrium exists when all parts of a soil mass are on the verge of failure. A state of active stress occurs when the soil deposit yields in such a manner that the deposit tends to stretch horizontally—for example, a retaining wall moving away from its backfill. A state of passive stress occurs when the movement is such that the soil tends to compress—for example, when a thrust block moves against the soil. The yield required to develop the passive state is much greater than for the active case.

This chapter deals with the determination of lateral pressures for design purposes and the stability requirements for various retaining structures.

STATES OF EQUILIBRIUM

At-Rest State

Consider element A of Fig. 197-1 in a natural or artificially placed soil deposit that has not been subject to lateral yielding. The vertical intergranular pressure, $\bar{\sigma}_v$, is a function of the depth of the element, the unit weight of the material, and the position of the water table. The horizontal intergranular pressure, $\bar{\sigma}_h$, is related to the vertical intergranular pressure by Eq. 196-1.

$$\bar{\sigma}_h = K_o \bar{\sigma}_v \qquad (196\text{-}1)$$

where K_o is the coefficient of earth pressure at rest. The value of K_o depends upon the process by which the soil deposit was formed, the relative density of the soil for sands, and the overconsolidation ratio of the soil for clays. Typical values of K_o for normally loaded soil deposits are given in Fig. 114-1. These K_o values were computed on the basis of common values of Poisson's ratio and the assumption of zero lateral yielding during deposition. Terzaghi and Peck (1967) suggest that the value of K_o for a sand that was not deposited by artificial compaction (tamped in layers) ranges from about 0.4 for dense sand to 0.5 for loose sand. Tamping the sand in layers may increase the value of K_o to about 0.8. Figure 197-2 shows a range in values of K_o as a function of the overconsolidation ratio. This range includes data found by Brooker and Ireland (1965) for soils with a PI of 0-80, and data on sands found by Hendron (1963). The PI exerts a secondary effect on K_o, and Alpan (1967) has suggested the following equation for clays with an overconsolidation ratio of one (normally consolidated).

$$K_o = 0.19 + 0.233 \log \text{PI} \qquad (196\text{-}2)$$

where PI is the plasticity index in percent. The coefficient of earth pressure at rest relates the vertical and horizontal intergranular pressures. Pore pressure, u, has the same magnitude in all directions. The total lateral pressure in a saturated soil, therefore, is the sum of the intergranular lateral pressure and the pore water pressure as given by Eq. 196-3.

$$\sigma_h = \bar{\sigma}_h + u \qquad (196\text{-}3)$$

Figure 197-1 At-rest state of stress.

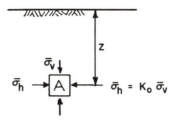

Figure 197-2 K_o as a function of overconsolidation ratio.

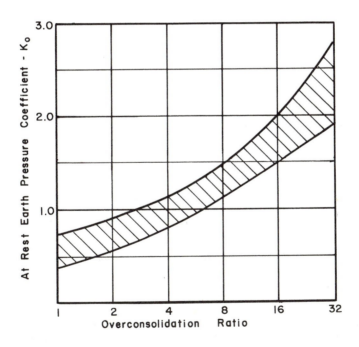

Figure 198-1

Example Problem Estimate the total at-rest lateral earth pressure at a depth of 5 m in a dense sand deposit with a water table much deeper than 5 m and also for the case of the water table at the ground surface. Assume the saturated unit weight of the sand to be 20.5 kN/m³ and the unit weight of the sand above the water table to be 18.4 kN/m³.

Solution Consider the water table below 5 m.

$$u = 0$$

Therefore

$$\sigma_h = \bar{\sigma}_h = K_o \bar{\sigma}_v$$

$$\bar{\sigma}_v = \gamma z = 18.4(5) = 92.0 \text{ kPa}$$

for dense sand $K_o = 0.4$

$$\sigma_h = 0.4(92.0)$$

$$\sigma_h = 36.8 \text{ kPa}$$

Consider the water table at the ground surface.

$$\sigma_h = \bar{\sigma}_h + u = K_o \bar{\sigma}_v + u$$

$$\bar{\sigma}_v = \gamma_b z$$

$$\bar{\sigma}_v = (20.5 - 9.81)(5) = 53.5 \text{ kPa}$$

$$u = \gamma_w z$$

$$u = 9.8(5) = 49.0 \text{ kPa}$$

$$\sigma_h = 0.4(53.5) + 49.0$$

$$\sigma_h = 70.4 \text{ kPa}$$

Figure 198-1 illustrates the computation of at-rest lateral earth pressures and shows the effect of a water table on the magnitude of the total lateral pressure.

Active State

Consider an element of sand below a level soil surface. The initial relationship between the vertical and horizontal intergranular pressures is given in Fig. 197-1 and is shown by Mohr's circle in Fig. 199-2. If the horizontal effective pressure is then decreased (by lateral stretching under drained conditions) while the vertical intergranular pressure is held constant, the minor principal stress moves to the limiting value, $\bar{\sigma}_a$, at which time Mohr's circle will be tangent to the effective stress failure envelope, as shown in Fig. 199-2. The soil is then in the active Rankine state of plastic equilibrium (Rankine, 1857) with failure planes oriented at an angle $(45 + \bar{\phi}/2)$ degrees from the horizontal. The relationship between the horizontal and vertical intergranular pressures for this state is expressed by Eq. 199-1.

$$\bar{\sigma}_a = K_a \bar{\sigma}_v \qquad (199\text{-}1)$$

where

$$K_a = \tan^2\left(45 - \frac{\bar{\phi}}{2}\right)$$

The coefficient K_a is referred to as the coefficient of active earth pressure and is derived in the example of Fig. 201-1.

Figure 199-2   Mohr's circles for cohesionless soils.

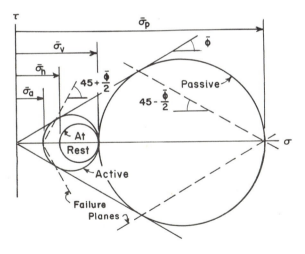

Figure 200-1   Mohr's circles for soils with a cohesion intercept.

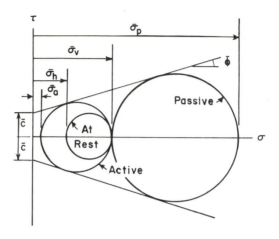

Equation 199-1 and Fig. 199-2 are based on effective stress; they are both valid for a cohesionless material such as sand and gravel and for normally consolidated clay. A more general expression of the active state that applies for soils with a cohesion intercept is given by Eq. 200-2 and is shown by Mohr's circle on Fig. 200-1.

$$\bar{\sigma}_a = K_a \bar{\sigma}_v - 2\bar{c}\sqrt{K_a} \qquad (200\text{-}2)$$

Equation 200-2 applies to overconsolidated and compacted clays and is also based on the effective stress envelope. For practical purposes the cohesion intercept and K_a should be based on the total stress envelope. For a more detailed discussion, refer to p. 207.

Passive State

The passive Rankine state of plastic equilibrium can be illustrated by again considering the element of sand below a level soil surface. For this case the horizontal pressure is increased under drained conditions while the vertical pressure is held constant. The horizontal intergranular pressure will then increase from its at-rest value until it reaches a limiting value greater than the vertical intergranular pressure, as illustrated on Mohr's circle of Fig. 199-2.

Figure 201-1

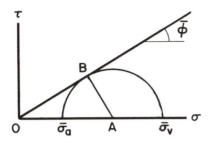

Example Problem Derive the relationship between the horizontal and vertical intergranular pressures for the active state as given by Eq. 199-1.

Solution

$$\bar{\sigma}_a = OA - AB$$
$$AB = OA \sin\bar{\phi}$$
$$\bar{\sigma}_a = OA(1 - \sin\bar{\phi})$$
$$\bar{\sigma}_v = OA + AB$$
$$\bar{\sigma}_v = OA(1 + \sin\bar{\phi})$$
$$\frac{\bar{\sigma}_a}{\bar{\sigma}_v} = \frac{1 - \sin\bar{\phi}}{1 + \sin\bar{\phi}} = \tan^2(45 - \phi/2)$$

Therefore

$$\bar{\sigma}_a = \tan^2(45 - \bar{\phi}/2)\bar{\sigma}_v$$

and

$$K_a = \tan^2(45 - \bar{\phi}/2)$$

Using a procedure similar to that in Fig. 201-1, an expression for the passive Rankine pressure can be developed. This expression for sand is given by Eq. 201-2 and for soils with a cohesion intercept by Eq. 202-2.

$$\bar{\sigma}_p = K_p\bar{\sigma}_v \qquad\qquad (201\text{-}2)$$

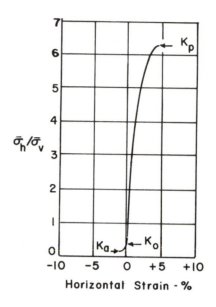

Figure 202-1 Horizontal strain required to reach active and passive states in a dense sand as measured from triaxial shear tests. (After Lambe and Whitman, 1969)

where

$$K_p = \tan^2(45 + \bar{\phi}/2)$$

$$\bar{\sigma}_p = K_p\bar{\sigma}_v + 2\,\bar{c}\sqrt{K_p} \qquad (202\text{-}2)$$

The passive case for soils with a cohesion intercept is shown by Mohr's circle in Fig. 200-1. The orientation of the failure planes for the passive case is $(45 - \bar{\phi}/2)$ degrees from the horizontal, and can be developed from Mohr's circle.

Yield Conditions

The yield conditions (horizontal strain) required to attain the active Rankine state are much less than those required to obtain the passive Rankine state. On the basis of an analysis of data from triaxial shear tests on dense sand,

Lambe and Whitman (1969) suggest that the relationship shown in Fig. 202-1 represents the relative horizontal strain required to reach each state of plastic equilibrium. They draw the following conclusions from Fig. 202-1.

1. Very little horizontal strain, less than –0.5%, is required to reach the full active pressure.
2. Little horizontal strain (compressive), about 0.5%, is required to reach one-half the maximum passive resistance.
3. Much more horizontal strain is required to reach the full passive resistance.

For loose sands, Lambe and Whitman suggest that the first two conclusions remain valid and that the horizontal strain required to reach full passive resistance may be as high as 15%. Figure 202-1 is based on triaxial shear test data, and the magnitudes for the field conditions are not likely to be the same. However, Fig. 202-1 should provide insight into the relative magnitude of the yield required to attain each state. The fact that much more strain is required to attain a state of passive plastic equilibrium is reasonable, because the stress change required to move from the at-rest state to the passive state is much greater than the stress change to reach the active state.

RANKINE'S METHOD APPLIED TO RETAINING STRUCTURES

The concept of Rankine's state of plastic equilibrium can be used to evaluate the lateral pressures that act against various retaining structures. Rankine's method assumes that there is no friction or adhesion between the soil (backfill) and the retaining structure. This assumption is, of course, not correct and leads to some inaccuracies in the method. The inaccuracies are generally small for the active case and are on the conservative side for many, although not all, practical conditions. The error from neglecting the friction between the wall and the soil for the passive case can be quite large and is not always conservative. The effect of wall friction is discussed in detail in a later section of this chapter (Lateral Pressures against Rough Walls). Before using Rankine's method to compute lateral pressures, the effect of neglecting wall friction should be investigated.

Active Pressure—Cohesionless Soil

The intergranular active pressure produced by a horizontal backfill of cohesionless soil acting at any depth on a vertical wall can be evaluated by Eq. 203-1.

$$\bar{\sigma}_a = K_a \bar{\sigma}_v \qquad (203\text{-}1)$$

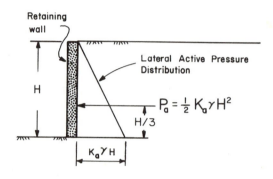

For a homogeneous dry cohesionless soil, this expression may be written as

$$\bar{\sigma}_a = K_a \gamma z \tag{204-2}$$

where

γ = unit weight of the material

z = depth below ground surface

Equation 204-2 states that the horizontal active pressure varies linearly with depth, as shown in Fig. 204-1. The total lateral force per unit length of the retaining wall of Fig. 204-1 is

$$P_a = \frac{1}{2} K_a \gamma H^2 \tag{204-3}$$

and acts at a point $H/3$ above the base of the wall.

If the soil deposit behind the wall is stratified so that K_a and γ are not constant with depth, the lateral pressure as determined by Rankine's method will not increase linearly, but will change abruptly at the soil strata interfaces. For these cases the appropriate value of K_a must be used for each strata, and the magnitude and position of the resultant lateral force must be determined from the pressure diagram. A typical Rankine lateral pressure diagram for a cohesionless stratified soil deposit is shown in Fig. 205-1.

The coefficient of active earth pressure K_a must be applied only to the intergranular vertical stress when computing the active pressure against a

Figure 205-1 Rankine's lateral pressure diagram for a stratified soil deposit.

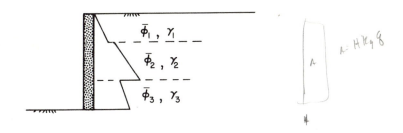

Figure 205-2 Rankine's lateral pressure diagram with saturated soil.

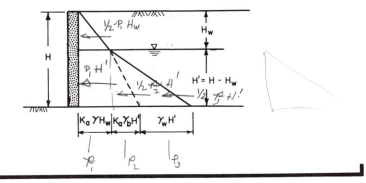

retaining structure with a submerged backfill. The intergranular active pressure is, therefore, computed on the basis of total unit weight, γ, above the water table and of buoyant unit weight, γ_b, below the water table. The pore pressure below the water table must be added to the active pressure to obtain the total horizontal pressure acting against the retaining structure. Figure 205-2 illustrates the shape of Rankine's lateral pressure diagram for a retaining structure in which the backfill is saturated below a depth of H_w.

A uniform surcharge applied to the backfill material may be easily accounted for when using Rankine's method to compute the active case. Equation 203-1 is still used to compute the active Rankine pressure, but the

Figure 206-1 Rankine's lateral pressure diagram with a uniform surcharge.

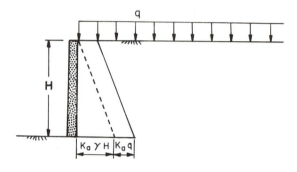

vertical intergranular pressure will be increased by an amount equal to the uniform surcharge. Figure 206-1 illustrates Rankine's lateral pressure diagram for the case of a dry homogeneous cohesionless backfill with a uniform surcharge q. The total force per unit length of wall can be computed from Eq. 206-2, and the line of action can be determined from the pressure diagram of Fig. 206-1 and the principle of moments.

$$P_a = \frac{1}{2}K_a\gamma H^2 + qK_aH \qquad (206\text{-}2)$$

A retaining structure is of finite height; therefore, when the wall yields to develop the active case, the entire semi-infinite soil mass will not be in a state of plastic equilibrium. Figure 207-1 shows the active zone that would develop if the retaining wall a-b yields away from the soil. The required yield at the top of the wall can be specified in terms of the width of the active zone at the top of the wall and can be written as Δ/ℓ. Since ℓ decreases linearly to zero at the bottom of the wall, the required displacement Δ at the bottom of the wall would also be zero. Hence, the active case will develop if the wall yields by tipping to position a'-b. On the basis of results from triaxial shear tests, the yield strain Δ/ℓ required to establish the active case is approximately 0.005, as shown in Fig. 202-1. For a friction angle, $\bar{\phi}$, of 37°, this would correspond to a movement Δ at the top of the wall of approximately $0.0025H$. If the wall yields by rotating about the top, a nonlinear pressure diagram similar to that shown in Fig. 207-2 will develop.

Figure 207-1 Active zone for a wall of finite height.

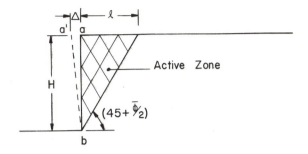

Figure 207-2 Approximate active pressure distribution for a wall that yields by rotation about the top.

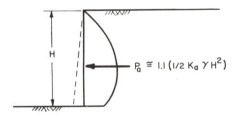

Active Pressure—Cohesive Soil

In this discussion cohesive soil refers to soil with a cohesion intercept on the strength envelope. The strength envelope for a normally consolidated clay does not have a cohesion intercept and should be treated as a cohesionless soil; however, lateral pressure problems involving normally consolidated clays are rarely encountered. In practical applications the strength envelope for clay soil will nearly always have a cohesion intercept for the following reasons:

1. Soils near the surface may be desiccated and show overconsolidation characteristics.

Rankine's Method Applied to Retaining Structures   **207**

2. If clay soils are compacted in the backfill, they will appear to be overconsolidated.
3. Above the water table, capillary pressures in clay cause apparent cohesion.
4. For short-term loading conditions, the unconsolidated, undrained strength envelope (total stress envelope) is generally applicable, and this envelope has a cohesion intercept.

The limiting plastic states (active and passive) should not be used for long-term loading conditions of clay backfill, because these soils tend to creep and, under an active or passive state of stress, will tend to seek the at-rest condition.

In this section total stress parameters c and ϕ are used because pore pressures in the cohesive backfills are usually unknown.

The pressure discussed for the active case can be evaluated using Eq. 200-2, and is shown in Fig. 208-1. At a depth z_o, the active pressure is zero; at depths less than z_o, the active pressure is negative. A negative pressure indicates that the soil is in tension, and this may cause cracks in the soil. Since soil is not capable of applying a negative pressure to any type of retaining wall, the common practice in computing the total horizontal active force against a retaining structure is to include only the shaded portion of the pres-

Figure 208-1   Rankine's active pressure distribution for a cohesive soil.

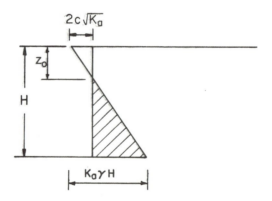

6   Lateral Pressures and Retaining Structures

sure distribution diagram shown in Fig. 208-1. This total force is given by Eq. 209-1.

$$P_a = \frac{1}{2} K_a \gamma H^2 - 2cH\sqrt{K_a} + \frac{2c^2}{\gamma} \tag{209-1}$$

The concept of active horizontal pressures can be used to estimate the height to which an open cut can be made in a cohesive soil. In Fig. 208-1, if no retaining wall were present to take the horizontal force, Rankine analysis would predict a local shear failure at every point on the vertical soil face below z_o. The critical height for open cuts for brittle clay soils should, therefore, be equal to z_o. In plastic soils, however, the excess of stress over strength near the vertical face will be distributed over the entire failure surface before a general shear failure occurs, and observation on actual cuts indicate that in these soils the critical height may approach $2z_o$. From Eq. 200-2 the depth z_o is

$$z_o = \frac{2c}{\gamma\sqrt{K_a}} \tag{209-2}$$

It is common practice for short-term field loading conditions (no time for drainage) to determine the strength of clay soil by the unconfined compression test, which is a special case of the UU triaxial shear test. Under these conditions the failure envelope is nearly horizontal, and, therefore, the friction angle ϕ is zero. If the average value of the unconfined compressive strength is used to characterize the strength of a clay soil deposit, then Eq. 209-2 reduces to Eq. 209-3.

$$z_o = \frac{2c}{\gamma} \tag{209-3}$$

where

$$c = \frac{q_u}{2}$$

Passive Pressures

The passive pressure distribution can be evaluated by Rankine's method with Eq. 202-2. This distribution for a soil above the water table is shown in Fig. 210-1. The effects of a water table, a uniform surcharge or a stratified soil deposit, are handled in the same manner as for the active case. On the basis of the pressure distribution diagram of Fig. 210-1, the total passive force

Figure 210-1 Rankine's passive lateral pressure diagram for a cohesive soil (yield by tipping about the base).

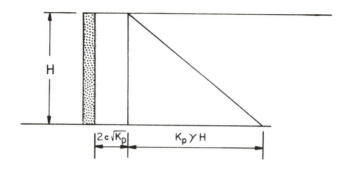

Figure 210-2 Approximate passive pressure distribution for a wall that yields by rotation about the top (cohesionless soil).

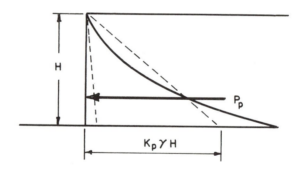

can be expressed by Eq. 210-3.

$$P_p = \frac{1}{2}\ K_p \gamma H^2 + 2c\sqrt{K_p}H \tag{210-3}$$

The location of P_p is determined by the principle of moments. For cohesionless soils use $c = 0$ and $\phi = \bar{\phi}$.

6 Lateral Pressures and Retaining Structures

The passive state develops as the retaining structure moves toward the backfill. The yield requirement to develop the passive case is much greater than that required to develop the active case (Fig. 202-1). For a wall of finite height, the passive pressure develops by the wall tipping about its base toward the soil. Because the explanation for the fact that tipping can cause the passive case to develop is the same as that presented for the active case, it will not be repeated here. If the wall rotates about the top of the wall, a nonlinear pressure distribution similar to that shown in Fig. 210-2 will develop.

LATERAL PRESSURES AGAINST ROUGH WALLS

Coulomb (1776) developed a method of determining lateral pressures that includes the effect of friction between the soil and the wall. A plane failure surface is assumed, and the lateral force required to maintain equilibrium is calculated. The procedure is repeated for several trial failure surfaces, and the one producing the critical force (largest for the active case and smallest for the passive case) is selected. This general analysis procedure is a type of limiting equilibrium method. The method readily accommodates a geometrically irregular backfill and sloping wall.

Active Case—Cohesionless Soil

Figure 212-1 shows a retaining wall with a sloping face and a sloping dry granular backfill. Coulomb's method can be used to determine the active lateral force acting against the wall by using the following procedure.

1. Assume a trial plane failure surface (such as A-B of Fig. 212-1a).
2. Draw a free-body diagram of the assumed failure mass (Fig. 212-1b).
3. Draw a force polygon of the forces acting on the free-body diagram (Fig. 212-1c). The magnitude and direction of the weight, W, are known; the directions of the resultant force, R, along the failure surface and the active force, P_a, are also known.
4. The magnitude of P_a required to close the force polygon is determined and represents the lateral force that would be required to prevent failure along the assumed failure plane.
5. Steps 1 to 4 are repeated for other failure surfaces; the surface that yields the maximum lateral force is the critical failure surface, and this force is the active lateral force.

By plotting the results of the analysis directly above the intersection of the failure surface with the ground surface, the maximum active force can easily

Figure 212-1 Coulomb's method for the active force against a retaining wall.

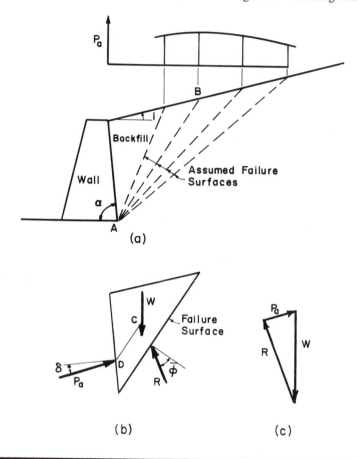

be selected. As pointed out earlier, Coulomb's method can accommodate irregular backfills such as a stepped backfill. However, there may be several local maximum values of P_a as the geometry changes.

Since each trial analysis assumes that failure is impending (a limiting equilibrium condition), the direction of the resultant force, R, along the assumed failure surface is inclined at an angle $\bar{\phi}$ to the normal to the failure surface, as shown in Fig. 212-1b. The angle $\bar{\phi}$ is the friction angle of the soil and is, therefore, the maximum angle of obliquity between the resultant and the normal to the failure surface (see Fig. 167-1).

212   **6** Lateral Pressures and Retaining Structures

As the retaining wall yields (by an outward movement or by tipping about the base) to develop the active case, the failure wedge moves downward along the failure surface and also downward along the soil-wall interface. Since there is movement between the soil and the wall, the active force, P_a, will be inclined at an angle δ to the normal to the retaining wall, as shown in Fig. 212-1b. The angle δ is the friction angle between the soil and the wall. For concrete walls, δ is generally about 2/3 ϕ. For most practical conditions the backfill moves down relative to the wall, and the active force P_a is inclined at the angle δ below the normal to the wall, as shown in Fig. 212-1b. If the wall is supported on a soft foundation, however, the wall may settle to an extent that the movement of the wall will be downward relative to the backfill. Under this condition, the active force, P_a, would be inclined at an angle δ above the normal to the wall. By drawing a force polygon with P_a above the normal to the wall, this case can be shown to produce a higher value of P_a. If the friction angle, δ, between the wall and the soil is zero, Coulomb's method will yield the same results as Rankine's method for a vertical wall and level backfill. The example in Fig. 214-1 illustrates Coulomb's method.

The point of application of the active force, P_a, is approximately at the point of intersection, D, of the back of the retaining wall, with a line C-D drawn from the centroid of the failure wedge, C, and parallel to the failure surface, as shown in Fig. 212-1b (Terzaghi and Peck, 1967).

Coulomb's method uses a plane sliding surface. The actual surface is somewhat curved, but for the active case the error is small and Coulomb's method provides reasonable results. Coulomb's method can be used to derive a general equation for the active pressure for the retaining wall shown in Fig. 215-1. This wall has a constant sloping backfill and the wall is sloped. The equation is derived by expressing P_a in terms of γ, H, α, θ, i, δ, and $\bar{\phi}$ (all the variables are defined in Fig. 215-1), and setting the derivative of P_a with respect to θ equal to zero and solving for θ. This angle is the orientation of the critical failure surface, and substituting it into the expression for P_a results in Eq. 213-1.

$$P_a = \frac{\gamma H^2}{2} \left\{ \frac{\sin^2 (\alpha + \bar{\phi})}{\sin^2\alpha \, \sin (\alpha - \delta) \left[1 + \sqrt{\dfrac{\sin (\bar{\phi} + \delta) \sin (\bar{\phi} - i)}{\sin (\alpha - \delta) \sin (\alpha + i)}} \right]^2} \right\} \quad (213\text{-}1)$$

Surcharge Loads and Submerged Backfill

Coulomb's method can also accommodate surcharge loads and the effect of a fully or partially submerged backfill. The general procedure is the same for

Figure 214-1

Example Problem   Use Coulomb's method to determine the active force against a concrete retaining wall 3.5 m high. The backfill is dry, loose sand with a unit weight of 15.6 kN/m³ and a friction angle $\bar\phi$ of 32°. Assume the friction angle, δ, between the soil and the wall is 20°.

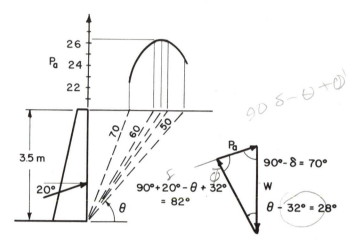

Force polygon for $\theta = 60°$

For a level backfill,

$\phi = 35$
$\delta = 23.3$
$\theta = 50$

$$W = \frac{1}{2}\, \gamma H^2 \cot \theta$$

$$W = 95.6 \cot \theta$$

From the law of sines,

$$P_a = W \left[\frac{\sin(\theta - \bar\phi)}{\sin(90 + \delta - \theta + \bar\phi)} \right]$$

For $\theta = 60°$

$$P_a = 95.6 \cot \theta \left[\frac{\sin(\theta - 32)}{\sin(142 - \theta)} \right]$$

$$P_a = 26.2 \text{ kN/m}$$

6   Lateral Pressures and Retaining Structures

(*Note:* For a more complex geometry P_a would be computed
graphically from the force polygon.)

θ	P_a (kN/m)	
50	24.8	
60	26.2 —	60
70	22.5	
55	26.2	
57	26.3 ←	

Figure 215-1 Retaining wall with a sloping face and sloping backfill.

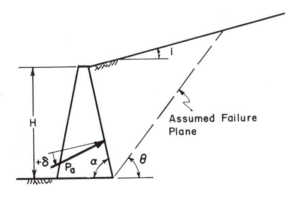

these cases as previously outlined except that additional forces are added to
the freebody diagram and to the force polygon.

For the case of a surcharge load the magnitude and direction of the additional load is known, and it can be added to the known weight vector in constructing the force polygon.

If the backfill is fully or partially submerged, there will be a neutral pressure acting normal to the failure surface, and the lateral force will consist of a
neutral force component normal to the wall and the active force from the soil.
In this case the magnitude and direction of the weight and both neutral forces

Lateral Pressures Against Rough Walls

215

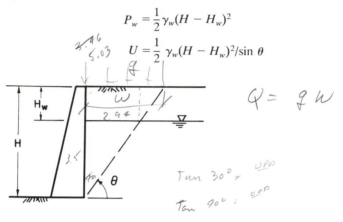

$$P_w = \frac{1}{2}\gamma_w(H - H_w)^2$$

$$U = \frac{1}{2}\gamma_w(H - H_w)^2/\sin\theta$$

$$Q = \mathcal{g}\, w$$

$$Tan\ 30^\circ = \frac{opp}{?}$$

$$Tan\ 90^\circ = \frac{opp}{?}$$

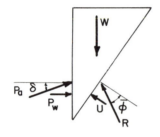

Free-Body Diagram   *Force Polygon*

W = total weight of failure wedge
W_d = weight above the water table
W_b = buoyant weight below the water table
W_w = equivalent weight of water below the water table

are known, and the direction of the resultant intergranular force along the failure plane and the active lateral force are known. The free-body diagram and force polygon from one trial failure surface are illustrated in Fig. 216-1.

The force polygon shows that P_a could have been obtained without computing U or P_w by using the buoyant unit weight below the water table to compute the weight, W, of the failure wedge. The total force against the retaining wall is the vector sum of P_a and P_w, and its line of action must be obtained by the principle of moments.

Culmann's Graphical Construction

Culmann (1866) developed a convenient graphical procedure for using Coulomb's method. The basic procedure is given below and is illustrated in Fig. 218-1.

1. Draw the retaining wall and associated features to some convenient scale (include positions of external loads, ground features, wall dimensions, etc.).
2. From point A (refer to Fig. 218-1) draw line AC at an angle $\bar{\phi}$ to the horizontal.
3. Lay off the line AD at an angle β with the line AC.

$$\beta = \alpha - \delta$$

4. Draw an assumed failure wedge (try to obtain a geometric shape).
5. Compute the weight, W, of the wedge.
6. Plot the magnitude of W (weight of the failure wedge) to a convenient scale, starting at the base of the wall and measured along the line AC.
7. Draw a line through the point along AC, established in step 6 and parallel to AD, and locate the point of intersection with the assumed failure wedge. The length of this line represents the magnitude of P_a required to maintain equilibrium for the assumed failure plane.
8. Repeat steps 4 to 7 for several assumed failure planes.
9. Through the locus of points established in 7, draw a smooth curve (the Culmann line).
10. Draw a line parallel to AC and tangent to the Culmann line at the position of maximum force.
11. Find the magnitude of the largest value of P_a, which is measured from the tangent point to AC and parallel to AD.

The method is demonstrated in Fig. 218-1 for the case of a level backfill, but it can easily be applied to an irregular or stepped backfill as well. Note that Culmann's construction produces Coulomb's force polygon rotated such that the weight is plotted along the line AC (Fig. 218-1) instead of being vertical. The line of action of P_a (resulting from the backfill only) is determined in the same manner as that for Coulomb's method.

Figure 218-1

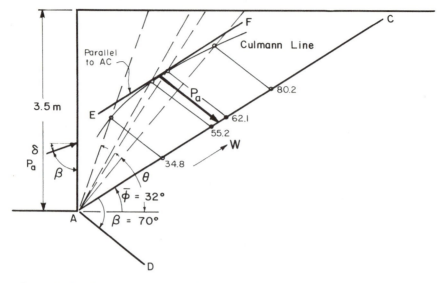

Example Problem Use Culmann's graphical construction to solve the problem of Fig. 214-1.

$$\gamma = \ 15.6 \ \text{kN/m}^3$$
$$\bar{\phi} = \ 32°$$
$$\delta = \ 20°$$
$$\alpha = \ 90°$$
$$\beta = \ 90° - 20° = 70°$$

θ	W (kN/m)
50°	80.2
57°	62.1
60°	55.2
70°	34.8

From point of tangency of EF with the Culmann line,

$$P_a = 26 \ \text{kN/m}$$

Figure 219-1

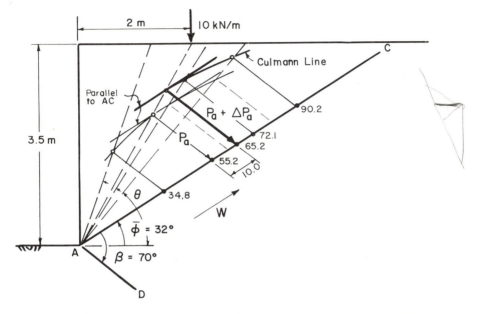

Example Problem Solve Fig. 218-1 with a line load of 10 kN/m applied 2 m behind the wall.

$$\gamma = 15.6 \text{ kN/m}^3$$
$$\overline{\phi} = 32°$$
$$\delta = 20°$$
$$\alpha = 90°$$
$$\beta = 90° - 20° = 70°$$

θ	W (kN/m)	$W + q$ (kN/m)
50°	80.2	90.2
57°	62.1	72.1
60°	55.2	65.2
70°	34.8	34.8

$P_a + \Delta P_a = 31$ kN/m (with line load)

P_a on same wedge as $(P_a + \Delta P_a)_{max}$.

Figure 220-1 Point of application of lateral load resulting from a line load.

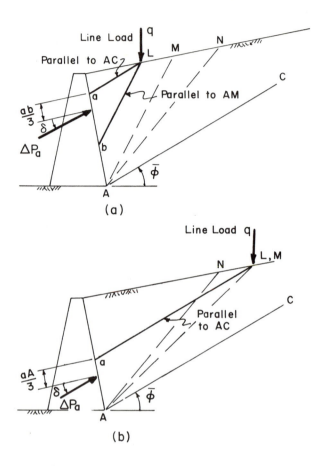

(a)

(b)

Culmann's construction can easily be used to include the effects of a uniform surcharge or line load applied to the backfill. Each additional load that falls within the assumed failure wedge is included in the force polygon by adding it to the weight of the failure wedge. Figure 219-1 modifies the problem solved in Fig. 218-1 to include a line load of 10 kN/m. Several line loads can be handled in the same manner. When a line load is added to the problem, the Culmann line becomes discontinuous as shown in Fig. 219-1.

The value $P_a + \Delta P_a$ is scaled from AC to the maximum point on the Culmann line that includes the line load. P_a is then found by using the same wedge and scaling the distance from AC to the Culmann line that does not include the line load. The difference between these two values is ΔP_a. An approximate method of determining the point of application of the additional load ΔP_a is illustrated in Fig. 220-1 (Terzaghi and Peck, 1967). Points L, M, and N in Fig. 220-1 are, respectively the point of application of the line load, the intersection of the failure plane with the backfill slope, and the intersection of the failure plane on the backfill slope when the line load is not included. If the line load is applied between the top of the wall and M, the point of application of ΔP_a is determined, as shown in Fig. 220-1a. If the line load acts beyond point N, the point of application of ΔP_a is determined, as shown in Fig. 220-1b.

Passive Pressure—Rough Walls

Coulomb's method can also be used to evaluate the passive pressure against a structure. In this case the failure wedge moves in the opposite direction (see Fig. 222-1) from the active case. The direction of the boundary forces opposes movement and is shown on Fig. 222-1. The procedure for Coulomb's method for the passive case is the same as that described for the active case except that the critical failure surface is that which produces the minimum value of the passive force P_p. The direction and magnitude of W are known, and the direction of R and P_p are known. If the structure of Fig. 222-1b moves up relative to the soil, the friction angle, δ, between the soil and the structure would be measured below the normal, and δ is said to be negative. Negative wall friction produces a value of passive resistance lower than that for positive wall friction. It is, therefore, important to determine the relative movement of the structure and backfill when calculating the passive pressure.

Culmann's graphical construction can also be used to determine passive forces. The procedure for cohesionless soils is the same as for the active case except that the line AC is laid off at an angle $\bar{\phi}$ below the horizontal instead of above, and the minimum value of P_p is selected from construction of the Culmann line as the critical force.

Coulomb's method assumes a plane failure surface. For granular soils the assumption of a plane failure surface provides reasonable results only if the wall friction δ is near zero. With increasing wall friction the failure surface becomes curved as in Fig. 222-1, and analysis with a plane failure surface yields a passive wall force that is too high and is, therefore, unconservative. The curved failure surface may be assumed to be either a logarithmic spiral or a circular arc. The logarithmic spiral method has been described in detail by Terzaghi and Peck (1967). The ϕ-circle method is described in a following section.

Figure 222-1 Effect of wall friction on the shape of the failure surface (passive pressure).

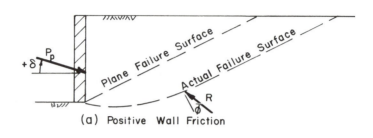

(a) Positive Wall Friction

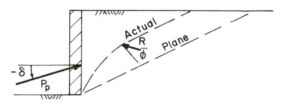

(b) Negative Wall Friction

The direction of the planes of failure or slip planes behind a vertical wall with zero wall friction and a horizontal backfill is shown to be at an angle of $45° - \overline{\phi}/2$ with the horizontal in Fig. 199-2. When wall friction is considered, the direction of the slip lines must be determined from the Mohr stress circle. Figure 223-1 shows the direction of the slip lines for the case where $\overline{\phi} = 40°$ and $\delta = 25°$. The failure sector may then be considered to consist of a region of curved slip lines below a passive Rankine zone of triangular shape.

Critical Failure Arc

The traditional procedure by both the logarithmic spiral and the ϕ-circle methods has been to analyze several trial failure arcs to find the minimum wall force P_p. It now appears that the critical arc for a level backfill is closely

6 Lateral Pressures and Retaining Structures

Figure 223-1 Orientation of failure plane.

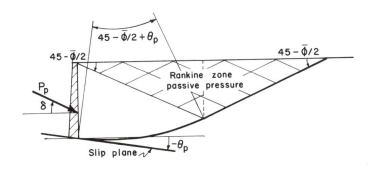

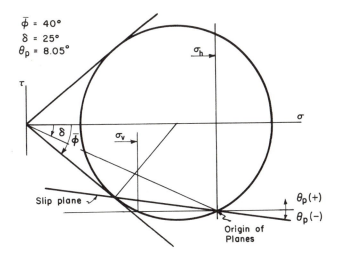

approximated by a slope parallel to the slip lines defined by the given wall friction δ and the soil friction $\bar{\phi}$ (Abdul-Baki, 1966; Parcher and Means, 1968). This approximation becomes less accurate with increasing slope of the backfill. The procedure has been found to closely define the critical surface with both the logarithmic spiral and the ϕ-circle methods for walls with level backfill, thus eliminating the need for trial solutions.

Lateral Pressures Against Rough Walls

The slope of the slip lines θ_p is defined from Eq. 224-1. The angle θ_p is measured from horizontal, counterclockwise being considered positive in Fig. 223-1.

$$\theta_p = \frac{1}{2}\left[90 - \arcsin\left(\frac{\sin \delta}{\sin \bar{\phi}}\right) - \bar{\phi} - \delta\right] \tag{224-1}$$

Passive Wall Force by ϕ-Circle Method

A Rankine passive zone abf is defined by drawing a line at an angle of $45° - \bar{\phi}/2$ from a point a in Fig. 224-2. The circular failure arc is then drawn with slope θ_p at the base of the wall and tangent to line $\overline{fb}$ at the intersection with line $\overline{fa}$.

Now consider the equilibrium of the freebody $adfe$ in Fig. 224-2 and Fig. 225-2. For a noncohesive soil the analysis is shown in Fig. 224-2. If the soil has a cohesive component of strength, an independent analysis shown in Fig. 225-2 is added. In both figures the soil in wedge ebf is considered to be in the passive Rankine state of plastic equilibrium so that the force E or E'

Figure 224-2   Passive wall force by ϕ-circle method.

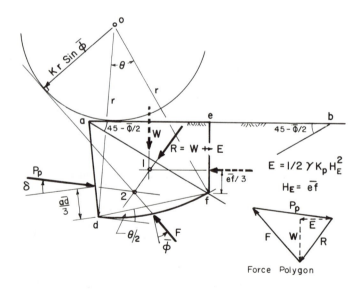

Figure 225-1   Friction circle correction factors.

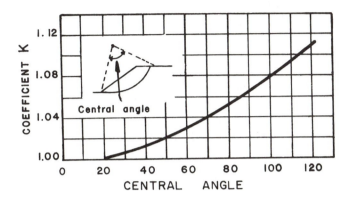

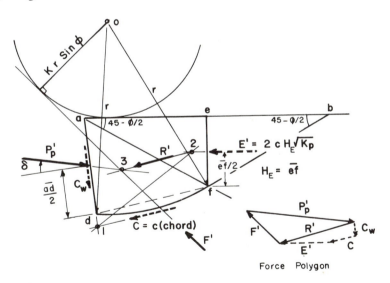

acting on face $\overline{ef}$ is known. The free body *adfe* is then in equilibrium with three forces: (1) the resultant of the Rankine passive force E and the weight W in Fig. 224-2, or the resultant of E' and the cohesive forces C and C_w in Fig. 225-2; (2) the reaction F or F' along the curved surface; (3) the passive wall force P_p or P'_p, respectively, in Fig. 224-2 and Fig. 225-2.

The ϕ-circle method is based on the concept that since all forces dF acting along the failure arc act at angle $\overline{\phi}$ with the normal to the arc, they also act tangent to a circle of radius $r \sin \overline{\phi}$ drawn at the center of the failure arc, where r is the radius of the failure arc. The resultant F of these forces will then pass nearly tangent to this ϕ circle, but will actually be tangent to a circle of radius $Kr \sin \overline{\phi}$, where the correction factor K is a number slightly greater than 1 and may be estimated from Fig. 225-1.

The procedure for the graphical solution is as follows (see Fig. 224-2).

1. On a scale drawing of the wall, draw a tangent line at the slope θ_p at the base of the wall.
2. From the tangent, measure the deflection angle $1/2 \, (45° - \overline{\phi}/2 - \theta_p)$ and extend the chord of the arc to intersect line *af*, and locate the center of the circle by erecting perpendiculars to the tangents at each end of the curve.
3. Draw the ϕ-circle radius of $Kr \sin \overline{\phi}$.
4. Determine the weight W of the wedge *adfe* and force $E = 1/2 \, K_p \gamma H_E^2$, and determine their resultant. Through the intersection of W and E (point 1) extend the resultant R to intersect with the line of action of P_p (point 2).
5. The forces P_p, R, and F intersect at point 2 if they are in equilibrium. The line of action of F is thus directed through intersection 2 and tangent to the ϕ circle.
6. Complete the force polygon, using the known direction of F and P_p. Determine the value of P_p.

If the backfill is a cohesive soil, an additional analysis is required as shown in Fig. 225-2. For the reasons explained on pages 207–208 total stress parameters c and ϕ are used to evaluate P_p and P'_p. The minimum value of the sum $P_p + P'_p$ represents the wall force causing failure.

1. On a drawing of the free body *adfe*, locate the force $E' = 2c\sqrt{K_p} H_E$ at the midpoint of face $\overline{ef}$. The force P'_p acts at the midpoint of face $\overline{ad}$ at the angle of wall friction δ.
2. The force C is equal to the sum of the unit cohesive forces parallel to the chord $\overline{df}$, as shown by the following equation.

$$C = \Sigma \, c(\cos \theta) \, dl = cL_{\text{chord}}$$

where θ is the angle between the chord and the arc. The force C is located a distance s from the center O such that the moment of the cohesive force C equals the moment of the sum of the unit cohesive forces on the arc about point O as follows.

$$sC = scL_{\text{chord}} = rcL_{\text{arc}}$$

and

$$s = r\frac{L_{\text{arc}}}{L_{\text{chord}}}$$

The wall adhesive force

$$C_w = c_w(\overline{ad})$$

acts along the face of the wall. The unit wall cohesion c_w may be conservatively estimated to be less than c.

3. On a force polygon determine the resultant of the force C and wall cohesion C_w.
4. From the intersection of C and C_w (point 1) extend their resultant to intersect the line of action of force E' (intersection 2).
5. Find the resultant of the cohesive forces and force E' from the force polygon, and through point 2 extend the resultant to intersect the line of action of P_p' (intersection 3).
6. The direction of the force F' is through intersection 3 and tangent to the ϕ circle.
7. With the known direction of F' and P_p' close the polygon and measure P_p'.
8. Add P_p' to P_p.

The ϕ-circle and the logarithmic spiral methods yield comparable results for P_p, but these values of P_p may still be larger than the real values, particularly with large values of wall friction. Shields and Tolunay (1973) have proposed a technique based on the method of slices (see Chapter 7) that gives results for P_p that are less than those given by the circle and spiral methods and closer to reported measurements of P_p.

Figure 228-1 compares values of the passive pressure coefficient, K_p, for a level backfill and vertical wall, as determined by Rankine's method, Coulomb's method, the ϕ-circle method, the log spiral method, and the method of slices, where K_p is defined from the equation

$$P_p = \frac{1}{2}K_p\gamma H^2$$

The comparisons are made for various values of $\overline{\phi}$ and δ. These comparisons clearly illustrate the effect of wall friction on the magnitude of passive pressures, as well as the effect of the shape of the assumed failure surface in the various methods of analysis.

Figure 228-1 Comparison of values of K_p as determined by various methods.

$\bar{\phi}$ (degrees)	30			35			40		
δ degrees	0	10	20	0	10	20	0	10	20
Rankine	3.00	—	—	3.69	—	—	4.60	—	—
Coulomb	3.00	4.14	6.10	3.69	5.31	8.32	4.60	6.95	11.77
ϕ-Circle	3.00	4.01	5.25	3.69	5.12	7.00	4.60	6.65	9.64
Log Spiral	3.00	4.02	5.26	3.69	5.13	7.04	4.60	6.69	9.67
Method of Slices	3.00	3.80	4.40	3.69	4.84	5.80	4.60	6.26	7.80

DESIGN CONSIDERATIONS

The design of many structures requires the computation of lateral pressures using the principles that have been introduced in the previous sections. Lateral pressures are required to proportion the structure so that it will be both internally and externally stable. Conventional retaining walls, for example, must be externally stable against sliding and overturning, and soil pressure should not exceed the allowable bearing pressure of the soil under the toe of the wall. Moreover, the elements of the retaining wall must be adequately designed for moment and shear at all points. The design requirements of other retaining structures may be stated differently, but in any case the magnitude and distribution of lateral pressures will be required.

As previously discussed, the magnitude and distribution of lateral pressures are functions of many variables and boundary conditions, including the yield of the structure, the type and properties of the backfill material, the friction at the soil structure interface, the presence of groundwater, the method of placement of the backfill material, and the foundation conditions for the structure. This section is a general summary of some of the important considerations that must be made in designing structures that are subject to lateral pressures.

Yield Conditions

Yield conditions are very important in evaluating the magnitude and distribution of lateral pressures (see Figs. 202-1, 207-2, and 210-2). It is generally not difficult to distinguish between cases that tend toward the active case

6 Lateral Pressures and Retaining Structures

Figure 229-1   Reinforced concrete retaining wall on a nonyielding rock foundation.

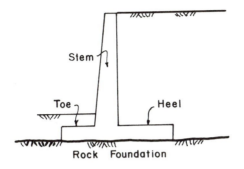

and those that tend toward the passive case. The difficulty in the analysis is to determine whether there can be sufficient yield to develop one of the limiting cases without compromising the integrity of the structure. Most conventional retaining walls yield enough to develop the active case. An exception might be a cantilever retaining wall that is supported on a rock foundation, as shown in Fig. 229-1. The reinforced concrete stem of the wall cannot deflect enough to develop the active case; and since the foundation is essentially nonyielding, very little tilting of the wall will occur. If the wall is supported on a soil foundation, however, the soil under the toe would compress enough to allow the wall to tip and thus, develop the active case. A bridge abutment that is restrained at the top by the bridge deck before the backfill is placed is another example of a retaining structure that probably will not yield enough to develop the active case. Structures that are restrained from yielding should be designed to resist at-rest lateral pressures, which can be more than twice the magnitude of active pressures.

Some structures are allowed to yield, but not in the manner required to develop active pressures (see Fig. 207-2). The sequence of construction of braced excavations, for example, is such that the top of the excavation is totally restrained, with yield developing and increasing with depth. Current design practice generally uses an apparent pressure diagram (Fig. 230-1) to design the bracing system (struts) for a braced excavation. The apparent pressure diagrams that are currently used (Terzaghi and Peck, 1967) were developed from field measurements and represent an envelope of the most critical condition. The diagrams are, of course, functions of the soil type.

Design Considerations

Figure 230-1 Suggested apparent pressure diagram for design of struts in open cuts in sand. (After Terzaghi and Peck, 1967.)

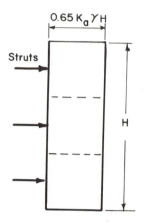

$$0.65\,K_a\,\gamma\,H$$

Struts

H

Considerations in Designing the Backfill System

Backfill systems for retaining structures should be designed to minimize the lateral pressure that the structure must support. A good backfill material has two important attributes.

- High long-term strength.
- Free draining.

In general, granular materials make the best type of backfill because they maintain an active state of stress indefinitely and are usually free draining. Clay soils, on the other hand, tend to creep and have a very low permeability. The tendency to creep causes clays to seek the at-rest case, which causes an increase in the lateral pressure with time. Thus, if a retaining structure has a clay backfill but was designed for active pressures, it will either fail structurally or deform to an extent that it becomes unusable.

Control of the water table in the backfill is also an important consideration. The total lateral force against a retaining wall for a fully submerged backfill will be two and one-half to three times that for a dry backfill. Control of

6 Lateral Pressures and Retaining Structures

groundwater can be most easily accomplished with a free-draining backfill material. Some cases may also require an underdrain system in the backfill as well as control of surface drainage.

The extent of the backfill should be to the boundaries of the active wedge if the lateral pressures are to be computed on the basis of the properties of the backfill material.

The strength properties of soils usually improve with increased density; therefore, it may seem desirable to compact the backfill material as much as possible, although this is generally not the case. Overcompaction may increase lateral pressures. For nonyielding walls excessive compaction can tend to increase the at-rest lateral pressure coefficient and, thus, the lateral pressure against the structure. Additionally, if heavy equipment is used to compact the backfill, the compaction equipment may induce loads on the structure that are much greater than the design loads. Therefore, it is important to consider carefully the backfill compaction requirements and to be specific in the construction specifications about the required compaction and the type of equipment that can be used.

PROBLEMS

6-1 Estimate the total at-rest lateral pressure at a depth of 7 m in a loose sand deposit with a water table at a depth of 3.0 m. Make reasonable estimates for the unit weight of the sand (see Fig. 18-1).

6-2 Derive Eq. 200-2 and determine the orientation of the failure planes.

6-3 Derive Eq. 201-2 and determine the orientation of the failure planes.

6-4 A smooth vertical wall 3.5 m high retains a mass of dry loose sand. The dry unit weight of the sand is 15.6 kN/m^3 and the friction angle, $\bar{\phi}$, is 32°. Estimate the total force per meter acting against the wall if the wall is prevented from yielding. Estimate the force if the wall is allowed to yield sufficiently to develop the active Rankine case.

6-5 Use Rankine's method to determine the magnitude and location of the resultant active force against a retaining wall 6 m high. There is a water table 2.5 m below the top of the wall and the backfill supports a uniform surcharge of 30 kN/m^2. The backfill material is sand with a friction angle, $\bar{\phi}$, of 35°, a unit weight above the water table of 18.7 kN/m^3, and a saturated unit weight of 21.2 kN/m^3.

6-6 A cave-in occurred in a clay soil during excavation of a vertical trench when the trench was 5.1 m deep. The unit weight of the soil was 18.7 kN/m^3. Estimate the average unconfined compressive strength of the clay.

Problems

6-7 A smooth retaining wall 4 m high supports a cohesive backfill with a unit weight of 17 kN/m^3. The soil has a cohesion intercept of 10 kPa and a friction angle, ϕ, of 10°. Estimate the total force acting against the wall and the depth to the point of zero lateral pressure.

6-8 Estimate the movement at the top of a retaining wall 5 m high that would be required to establish the active case. Assume that the backfill is sand with a friction angle, $\bar{\phi}$, of 35°.

6-9 A smooth vertical retaining structure 6 m high supports a backfill of compacted saturated clay. The results of a CU triaxial shear test on the backfill material are presented in problem 5-22. Estimate the total thrust that can be applied against the retaining structure (pushing toward the backfill). Use total stress parameters. Assume $\gamma = 17$ kN/m^3.

6-10 Make a sketch similar to Fig. 207-1 showing the yield conditions for the passive case and the family of lines that represent failure planes.

6-11 Neglecting wall friction in computing the active lateral pressure against a retaining wall is conservative if the backfill settles relative to the wall. However, if the wall settles relative to the backfill, neglecting wall friction is unconservative. Show that this is true.

6-12 Use Coulomb's method to determine the magnitude and direction of the total lateral force against the retaining wall of problem 6-5. Assume the friction angle between the wall and the soil to be $2/3\bar{\phi}$. Make one trial only, using an assumed failure surface at an angle specified by your instructor.

6-13 Compare the total force against the retaining wall of Fig. 218-1 with that of problem 6.4. Explain the difference.

6-14 Use Culmann's graphical construction to determine the active force against a vertical retaining wall 5.0 m high. The backfill is level and is dry sand with a unit weight of 16.2 kN/m^3 and a friction angle, $\bar{\phi}$, of 36°. Assume the friction angle, δ, between the soil and the wall is $2/3\bar{\phi}$. Also use Eq. 213-1 to compute the active force and compare the results with Culmann's method.

6-15 Solve problem 6-14 with a line load of 12 kN/m applied 1.5 m behind the wall. Estimate the point at which the failure surface intersects the surface of the backfill. Determine the point of application on the wall of the additional lateral force from the line load.

6-16 Use Eq. 224-1 and the ϕ-circle method to compute the total passive force that can be applied against a vertical wall in contact with a sand backfill having a level surface. The wall is 5.0 m high. The unit weight of the sand is

16.2 kN/m³ and the friction angle is 36°. Assume the friction angle between the wall and the backfill to be 20°. Compare the force with that computed using Rankine's method and explain why there is a difference.

6-17 Describe three situations in which passive pressures are developed.

6-18 Why is granular material preferable over clay for the backfill for a retaining wall?

6-19 What are the possible consequences of overcompacting the backfill behind a retaining structure? Explain.

6-20 Solve problem 6-9 using the ϕ-circle method. The adhesion and friction angle between the wall and the soil are $2/3c$ and $2/3\phi$, respectively. Use total stress parameters.

7

SLOPE STABILITY ANALYSIS

The varied topographic features of the earth's surface are possible only because the shear strength of the soil or rock exceeds shearing stresses imposed by gravity or other loading. Normally one expects the steepest slopes to be the least stable, but there are examples of failure in relatively flat slopes as well. Factors leading to instability can generally be classified as (1) those causing increased stress and (2) those causing a reduction in strength. Factors causing increased stress include increased unit weight of soil by wetting, added external loads such as buildings, steepened slopes either by natural erosion or by excavation, and applied shock loads. Loss of strength may occur by adsorption of water, increased pore pressures, shock or cyclic loads, freezing and thawing action, loss of cementing materials, weathering processes, and strength loss with excessive strain in sensitive clays. The presence of water is a factor in most slope failures, since it causes both increased stresses and reduced strength.

The rate of slide movement in a slope failure may vary from a few millimeters per hour to very rapid slides in which large movements take place in a few seconds. Slow slides occur in soils having a plastic stress-strain characteristic where there is no loss of strength with increasing strain. Rapid slides occur in situations where there is an abrupt loss of strength, as in liquefaction of fine sand or a sensitive clay. Many slopes exhibit creep movements (a few millimeters per year) on a more or less continuous basis as a result of seasonal changes in moisture and temperature. Such movements are not to be confused with a shear failure.

Some slopes cannot be readily analyzed. Examples include slopes of complex geology or badly weathered slopes where the varied materials and their strength cannot be readily identified. Slopes involving heavily overconsolidated clays and shales or stiff fissured clays are difficult to analyze, i.e., Waco Dam Slide (Beene, 1967).

Slopes that may be analyzed include natural slopes, slopes formed by excavation through natural materials, and artificial embankments.

The most common methods of slope-stability analysis are based on limit equilibrium. In this type of analysis the factor of safety with regard to the slope's stability is estimated by examining the conditions of equilibrium when incipient failure is postulated along a predefined failure plane, and then comparing the strength necessary to maintain equilibium with the available strength of the soil. All limit equilibrium problems are statically indeterminate and, since the stress-strain relationship along the assumed failure surface is not known, it is necessary to make enough assumptions so that a solution using only the equations of equilibrium is possible. The number and type of assumptions that are made leads to the major difference in the various limit equilibrium methods of analysis.

A second method of slope analysis is based on use of the theory of elasticity or plasticity to determine the shearing stresses at critical places within a slope for comparison with the shearing strength. Recently developed finite element computer techniques are an example of this type of analysis. An introduction to the finite element method is presented in Chapter 10.

HISTORICAL BACKGROUND

The development of limit equilbrium methods based on the plastic equilibrium of trial failure surfaces began in Sweden in 1916, following the failure of a number of quay walls at Gothenburg Harbor.[1] Petterson (1955) and Hultin (1916) in separate publications reported that the failure surfaces in the soft clays of Gothenburg Harbor closely resembled arcs of circles (see Fig. 237-1). Over the next few years the friction circle method of analysis was devised, results from simple undrained shear tests were used with reasonable success in predicting stability (S_u or $\phi = 0$ analysis), and the method of slices

---

[1]A. Collin (1846) reported circular sliding surfaces in clay during canal construction in France. A stability analysis based on shear tests was attempted. The work was forgotten and was not rediscovered until about 1940.

Figure 237-1 The Stigberg quay sliding surface 1916, Gothenburg Harbor. (After Petterson, 1955.)

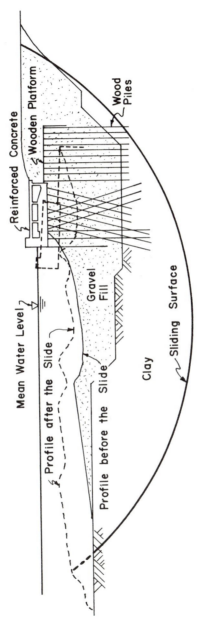

was introduced (Fellenius, 1927, 1936). The concept of pore water pressure and the effective stress method of analysis was introduced by Terzaghi (1936b). Improved soil strength measurements resulted from better sampling techniques, the development of the triaxial shear test, and the measurement of pore pressures. Improved methods of analysis that included the side forces between slices were developed, beginning with Fellenius (1936) and Bishop (1955). More rigorous analytical methods usually involving the use of digital computers are available (Morgenstern and Price, 1965; Janbu, 1973; Bailey and Christian, 1969). However, despite the use of more rigorous methods of analysis and improved soil-testing techniques, many uncertainties remain in predicting the stability of slopes. These uncertainties are primarily associated with the measurement of soil strength (Johnson, 1975) and the prediction of pore pressures.

EFFECTIVE STRESS METHOD AND TOTAL STRESS METHOD

Two approaches are available for specifying strength parameters and for evaluating strength requirements. They are referred to as the effective stress method and the total stress method. In the effective stress method, the pore pressures along the assumed failure surface are estimated for use in the analysis, and the shear strength is then based on effective strength parameters. Effective stress parameters are obtained from a consolidated drained shear test or from a consolidated undrained shear test with pore pressure measurements. In the total stress method, laboratory tests are performed in a manner designed to simulate the condition in the embankment, and the shear strength is determined in terms of total stress. It is assumed that the pore pressures that develop in the sample during the laboratory test would be equal to those that develop in the embankment at failure. There is no basic difference in the reliability of the two methods. The uncertainties in both methods stem from an inability to estimate the pore pressure behavior within the slope. In the construction of an earth embankment the effective stress method may be advantageous in that pore pressures measured during construction can be used to check the stability.

INFINITE SLOPES

As an introduction to the limiting equilibrium methods, an infinite slope will be considered. From a practical standpoint any slope of great extent with soil conditions essentially the same along every vertical line can be considered an infinite slope. The soil strata may be varied but for the purpose of this discussion a homogeneous soil is assumed.

Infinite Slopes in Dry Sand

Consider equilibrium of the wedge of dry cohesionless soil one unit thick in Fig. 239-1. The vertical distance h is the depth of soil down to the assumed failure plane. In an infinite slope the side forces will be the same on every vertical section and, therefore, will cancel out. The weight of the wedge can be expressed as

$$W_t = bh(1)\gamma$$

The weight of the wedge is balanced by the normal force N and shearing force T on the bottom of the wedge. The normal and shearing forces are

$$N = W \cos \beta$$

$$T = W \sin \beta$$

Dividing these forces by the base dimension of the wedge gives the normal stress, $\bar{\sigma}$, and the shear stress, τ, required for equilibrium.

$$\bar{\sigma} = \frac{N}{b/\cos \beta} = h\gamma \cos^2 \beta$$

$$\tau = \frac{T}{b/\cos \beta} = h\gamma \sin \beta \cos \beta$$

The safety factor of an infinite slope is defined as the ratio of soil strength to the required shear stress for equilibrium. Since the strength of a

Figure 239-1 A typical free-body and force polygon from an infinite slope analysis.

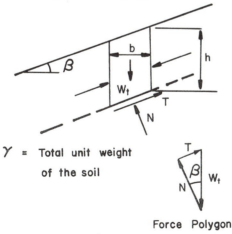

γ = Total unit weight of the soil

Force Polygon

cohesionless soil can be expressed as

$$S = \bar{\sigma} \tan \bar{\phi}$$

The safety factor becomes

$$F = \frac{S}{\tau} = \frac{\bar{\sigma} \tan \bar{\phi}}{h\gamma \sin \beta \cos \beta}$$

or

$$F = \frac{\tan \bar{\phi}}{\tan \beta} \tag{240-1}$$

With a safety factor of 1 in Eq. 240-1, the maximum slope β is equal to $\bar{\phi}$. This provides an estimate of the maximum slope for dry cohesionless materials.

Infinite Submerged Slopes in Sand

The analysis of underwater slopes (submerged slopes) is essentially the same as the analysis presented in the preceding section for dry slopes except that the normal intergranular stress and the shearing stress are calculated on the basis of submerged unit weight rather than total unit weight. However, since the unit weights cancel, the safety factor for a submerged slope can also be expressed by Eq. 240-1.

Infinite Slope in a c-ϕ Soil with Seepage Parallel to the Slope

A slope with seepage parallel to and at the surface is shown in Fig. 241-1a. The strength of this homogeneous isotropic soil can be represented by $\bar{c} + \bar{\sigma} \tan \bar{\phi}$ (the soil is called a c-ϕ soil). A freebody diagram of a wedge of soil is shown in Fig. 241-1b. The side forces on the soil slice due to water and soil are equal and opposite. The weight of the slice is calculated with the total or saturated unit weight (water plus soil). Pore pressure along the potential failure surface can be evaluated from the partial flow net of Fig. 241-1a and is

$$u = \gamma_w h \cos^2 \beta$$

Thus, the uplifting water force is

$$U = \frac{ub}{\cos \beta}$$

or

$$U = \gamma_w bh \cos \beta$$

Figure 241-1 Infinite slope in a c-ϕ soil with seepage parallel to the surface.

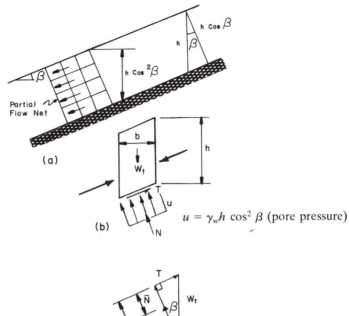

$u = \gamma_w h \cos^2 \beta$ (pore pressure)

(b)

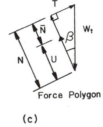

Force Polygon

(c)

The remaining forces on the free-body diagram are

$$W_t = bh \,(1)\, \gamma$$
$$N = W \cos \beta$$
$$\overline{N} = N - U$$
$$T = W \sin \beta$$
$$\overline{N} = bh \cos \beta (\gamma - \gamma_w)$$
$$\overline{N} = bh \cos \beta \, \gamma_b$$

The effective normal stress and the shearing stress along the bottom of the slice can now be obtained from $\bar{N}$ and T by dividing by the length along the bottom of the wedge.

$$\bar{\sigma} = \frac{\bar{N}}{b/\cos\beta} = \gamma_b h \cos^2\beta.$$

$$\tau = \frac{T}{b/\cos\beta} = \gamma h \cos\beta \sin\beta$$

The shear strength for a c-ϕ soil is expressed as

$$S = \bar{c} + \bar{\sigma} \tan\bar{\phi},$$

which becomes

$$S = \bar{c} + \gamma_b h \cos^2\beta \tan\bar{\phi}$$

The safety factor for an infinite slope in a c-ϕ soil with seepage parallel to and at the surface can now be expressed as

$$F = \frac{S}{\tau}$$

or

$$F = \frac{\bar{c} + \gamma_b h \cos^2\beta \tan\bar{\phi}}{\gamma h \cos\beta \sin\beta} \qquad (242\text{-}1)$$

If $\bar{c} = 0$, Eq. 242-1 reduces to

$$F = \frac{\gamma_b}{\gamma}\frac{\tan\bar{\phi}}{\tan\beta}$$

Since $\gamma_b/\gamma \approx 1/2$, the safety factor for a cohesionless infinite slope with full seepage is about one-half that for the slope without seepage.

Assuming a factor of safety of unity and solving for h leads to Eq. 242-2, which indicates that for cohesive soils there is a limiting thickness for stability.

$$h = \frac{\bar{c}\,\sec^2\beta}{\gamma[\tan\beta - (\gamma_b/\gamma)\tan\bar{\phi}]} \qquad (242\text{-}2)$$

This value of h applies only for the case of full seepage parallel to the surface, but a similar relationship could be established for other cases. A dimensionless parameter called a stability number is often useful in plotting slope stability data. The stability number for the case of c-ϕ soil with full seepage can be defined by rewriting Eq. 242-2 as

$$\frac{\bar{c}}{h\gamma} = \frac{\tan\beta - \dfrac{\gamma_b}{\gamma}\tan\bar{\phi}}{\sec^2\beta}$$

Stability numbers are used later in this chapter to develop stability charts for analyzing slopes.

In the preceding analysis pore pressures were calculated and included, and parameters from Mohr's effective strength failure envelope were used. Therefore, this is referred to as an effective stress analysis.

SLOPES OF FINITE HEIGHT

Plane Failure Surfaces

Plane failure surfaces often occur when a soil deposit or embankment has a specific plane of weakness. Excavations into stratified deposits where the strata are dipping toward the excavation may fail along a plane parallel to the strata. Many earth dams have sloping cores of relatively weak material compared to the shell of the dam, and a possible failure along a plane through the core should be investigated. Methods of analysis that consider blocks or wedges sliding along plane surfaces have been developed to analyze cases where there is a specific plane of weakness (Seed and Sultan, 1967).

Although a plane failure surface is not a practical assumption to make for slopes of homogeneous soil, it is a simple failure mechanism and is shown in this section for purposes of illustration and for later comparison with results using circular failure surfaces. Such a plane failure surface was analyzed by Culmann (1866).

Consider the equilibrium of the triangular wedge formed by the assumed failure surface in Fig. 244-1. The three forces considered are the weight of the wedge W, a cohesive force C_r, parallel to the potential sliding surface, and the resultant P, of the normal and frictional forces. The relationship of these forces is shown on the force polygon. The cohesive force C_r equals the required unit cohesive strength c_r times the length of the potential failure surface L. The unit cohesion and friction angle are the values required for equilibrium and will be equal to or less than the available cohesion and friction. A safety factor with respect to cohesion, F_c, is defined as the ratio of available cohesion to required cohesion, and a factor of safety with respect to friction, F_ϕ, is defined as the ratio $\tan \phi$ (available) to $\tan \phi_r$ (required). If either safety factor is assumed to be 1, the other will be greater than 1 if the slope is stable. The factor of safety with respect to strength, F_s, or correct factor of safety for the assumed failure mechanism occurs when $F_c = F_\phi = F_s$. For a c-ϕ soil, F_s is determined by trial and error. A value of F_ϕ is assumed, and this establishes a value of ϕ_r. The value of F_c is then computed from the force polygon of Fig. 244-1b and the definition of F_c. The procedure is then repeated until $F_c = F_\phi$. This safety factor represents the safety factor with respect to strength, F_s, for the assumed failure plane.

When there is a well-defined plane of weakness, the critical failure plane may be obvious. See the example problem in Fig. 244-2. However, for a homogeneous

Figure 244-1 Slope analysis assuming a plane failure surface.

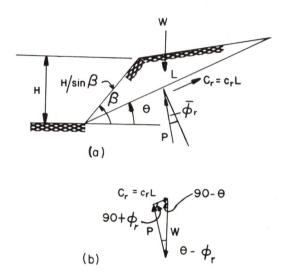

(a)

(b)

Figure 244-2

Example Problem A 60° cut slope is to be excavated to a depth of 6 m, as shown. A subsurface investigation has revealed bedding planes that dip toward the excavation at a slope of approximately 45°.

Determine the safety factor of the 60° slope with 10° backslope, as shown. Direct shear tests were run on samples to evaluate the strength parallel to the bedding planes. (CU tests).

Given:

$$c = 14 \text{ kN/m}^2 \qquad \phi = 30°$$

$$\gamma = 18 \text{ kN/m}^3 \qquad H = 6 \text{ m}$$

Solution

$$L = 9.25 \text{ m}$$

$$W = \frac{\gamma}{2}(a \sin \alpha)b = (18/2)9.25(\sin 15°)6.93 = 149.3 \text{ kN}$$

where $\alpha = \beta - \theta$

244

7 Slope Stability Analysis

Use trial and error. Estimate $F_\phi = 2.0$.

$$\phi_r = \tan^{-1}\left(\frac{\tan 30°}{2}\right) = 16.1°$$

$$\frac{C_r}{\sin(\theta - \phi_r)} = \frac{W}{\sin(90 + \phi_r)}$$

$$C_r = \frac{149.3 \sin(28.9)}{\sin 106.1} = 75.1 \text{ kN}$$

$$c_r = \frac{C_r}{L} = \frac{75.1}{9.25} = 8.11$$

$$F_c = \frac{c}{c_r} = \frac{14}{8.11} = 1.72$$

Plot F_ϕ versus F_c with repeated trials as shown below.

$$F_s = F_c = F_\phi = 1.80$$

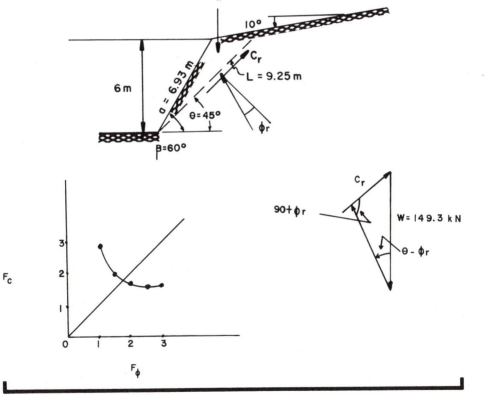

soil slope there are an infinite number of possible failure planes. The critical plane can be established using the following procedure. From the force polygon of Fig. 244-1b,

$$\frac{C_r}{W} = \frac{\sin (\theta - \phi_r)}{\sin (90 + \phi_r)} = \frac{\sin (\theta - \phi_r)}{\cos \phi_r}$$

and, from Fig. 244-1a,

$$\frac{C_r}{W} = \frac{c_r L}{\frac{1}{2}\gamma L (H/\sin \beta) \sin (\beta - \theta)}$$

Combining these expressions and solving for $c_r/\gamma H$ yields

$$\frac{c_r}{\gamma H} = \frac{\sin (\theta - \phi_r) \sin (\beta - \theta)}{2 \cos \phi_r \sin \beta} \tag{246-1}$$

The critical failure plane (the one that will yield the lowest factor of safety) is defined by the value of θ that yields the maximum value of $c_r/\gamma H$. Differentiate Eq. 246-1 with respect to θ and set equal to zero to find θ_c.

$$\theta_c = \frac{1}{2} (\beta + \phi_r)$$

Substitution of this value into Eq. 246-1 yields the maximum stability number for the slope.

$$\left(\frac{c_r}{\gamma H}\right)_{max} = \frac{1 - \cos (\beta - \phi_r)}{4 \sin \beta \cos \phi_r} \tag{246-2}$$

Equation 246-2 and the iterative procedure just explained can be used to find the safety factor with respect to strength F_s.

Equation 246-2 is the stability number for plane failure surfaces for c-ϕ soil. If $\phi = 0$, Eq. 246-2 reduces to

$$\left(\frac{c_r}{\gamma H}\right)_{max} = \frac{1 - \cos \beta}{4 \sin \beta} = \frac{1}{4} \tan \frac{\alpha}{2}$$

A plot of this function in Fig. 248-1 shows that a plane failure surface is not the critical failure surface but approaches the results for curved failure surfaces at very steep slopes.

As pointed out earlier, the critical failure surface for a homogeneous soil slope will not be a plane surface, and so Eq. 246-2 is of little practical value. It was developed to demonstrate the general procedure for slope stability problems using a simple mechanism.

This method is most suited for a total stress analysis. Therefore, pore pressures are not included in the analysis and the safety factor should be

computed on the basis of total unit weights and the Mohr-Coulomb total
strength envelope.

Friction Circle Method

A circular failure arc is drawn from a trial center in Fig. 247-1. At the
center a friction circle is drawn at a radius $r \sin \phi_r$ such that all lines tangent
to the friction circle and cutting the circular failure arc form the angle ϕ_r with
the normal. These lines represent the direction of the combined normal and
mobilized frictional forces distributed around the failure arc. The resultant
normal and frictional force is assumed also to be tangent to the ϕ_r circle.
Actually, the resultant is tangent to a slightly larger ϕ_r circle or radius $Kr \sin \phi_r$,
where K is a factor greater than one. Values of K can be estimated from
Fig. 225-1. Note that the two forces dP shown in Fig. 247-1 intersect slightly
outside the friction circle.

    The equilibrium of the circular wedge is analyzed by considering three
vectors; the weight, W, a resultant cohesive force, C_r, and the resultant
normal and friction force, P.

    The weight vector is equal to the area of the wedge times the unit weight
of the soil and acts through the centroid of the wedge.

Figure 247-1   Friction circle method.

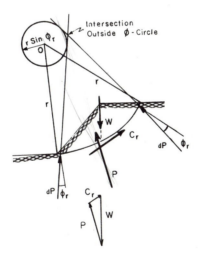

Figure 248-1 Comparison of plane and circular failures.

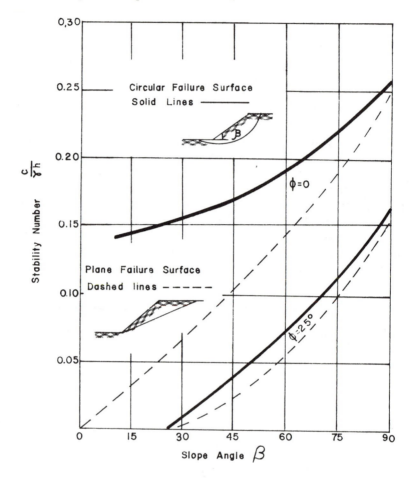

The cohesive force C_r acts parallel to the chord of the failure arc and is equal to $c_r L_{chord}$. As explained in Chapter 6, the force C_r is located a distance s from the center O, where

$$s = r \frac{L_{arc}}{L_{chord}}$$

or a distance slightly greater than the radius r.

7 Slope Stability Analysis

The intersection of the forces W and C_r establishes a point through which the third force, P, must act. The direction of P is established by drawing a line tangent to the adjusted ϕ circle from the intersection of W and C_r.

The weight of the wedge is known, but the magnitudes of P and C_r must be established by trial and error similar to the method illustrated in the preceding section for plane failure surfaces. If F_ϕ is assumed, the friction circle for ϕ_r equal to $\tan^{-1}(\tan \phi/F_\phi)$ can be drawn and the equilibrium triangle completed for the required cohesive force, C_r. The factor of safety with respect to cohesion $F_c = C/C_r$ is then compared with the assumed F_ϕ.

By plotting F_ϕ versus F_c, a new estimate for F_ϕ can be made and the process repeated until $F_c = F_\phi = F_s$.

Additional centers and arcs must be analyzed to determine the minimum factor of safety. The iterative procedure can be shortened somewhat by finding the critical circle as that which yields the minimum safety factor with respect to cohesion. Iteration to find F_s would then be required on the critical circle.

The friction circle method is primarily limited to homogeneous soils and a total stress analysis. The influence of pore pressure or boundary water forces can be included; however, the method of slices is more adaptable for the more complex slope situations.

Stability numbers based on friction circle analyses are plotted in Fig. 248-1 and compared with numbers based on a plane failure surface.

Slope Stability Charts

Slopes that approximate simple sections of relatively uniform soil may be analyzed by the slope stability charts of Figs. 250-1 and 251-1. Three of the parameters, c_r, γ, and H, are combined into a dimensionless stability number, which is plotted as a function of the slope β for various values of ϕ_r. When $\phi_r = 0$, a sixth parameter, D, the depth of the soil layer, becomes important and appears on the left side of Fig. 250-1. For steeper slopes the failure arc goes through the toe, and with flatter slopes the failure arc extends below the toe. In $\phi = 0$ soils with slopes less than 53°, the failure extends as deep as possible, hence, the need to define the depth D of the soil in relationship to height of the slope (see Fig. 251-1).

The factor of safety for a $\phi = 0$ soil may be obtained directly from Fig. 251-1, but a trial and error procedure is required for a c-ϕ soil. All points in the charts represent a factor of safety of unity. As in the example problem with a plane failure surface (Fig. 244-2), a factor of safety with respect to friction, F_ϕ, is initially assumed and is compared with the resulting factor of safety with respect to cohesion, F_c. The assumed F_ϕ is adjusted until $F_\phi = F_c = F_s$.

Slopes of Finite Height

Figure 250-1 Stability numbers for homogeneous simple slopes. (After Taylor, 1948.)

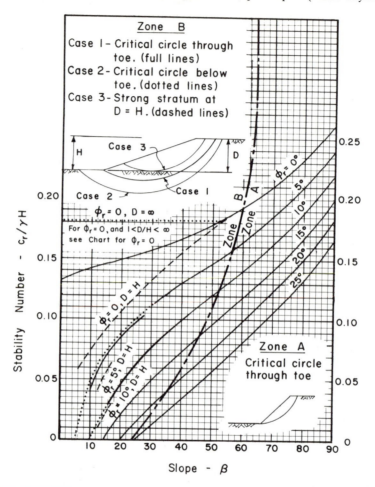

7 Slope Stability Analysis

Figure 251-1 Stability numbers for homogeneous simple slopes for $\phi = 0$. (After Taylor, 1948.)

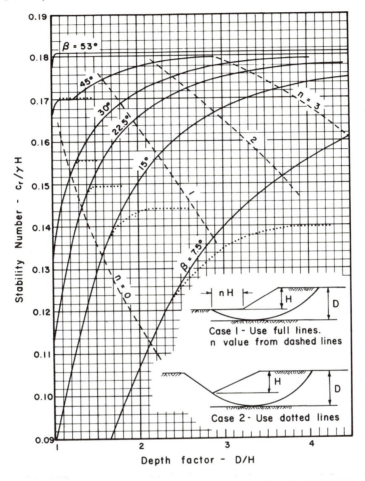

Figure 252-1

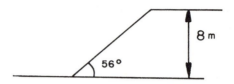

Example Problem Determine the safety factor F_s for the slope shown.

$$\gamma = 17 \text{ kN/m}^3$$

$$c = 22 \text{ kPa}$$

$$\phi = 24°$$

Assume $F_\phi = 1.8$.

$$\phi_{required} = \text{arc tan}\left(\frac{\tan 24°}{1.8}\right) = 13.9°$$

From Fig. 250-1, at $\beta = 56°$, $\phi_r = 13.9°$,

$$c_r/\gamma H = 0.115$$

$$c_r = 0.115(17)(8) = 15.6 \text{ kPa}$$

$$F_c = \frac{22}{15.6} = 1.41 \neq 1.8$$

Try $F_\phi = 1.56$.

$$c_r = (0.104)(17)(8) = 14.14 \text{ kPa}$$

$$F_c = \frac{22}{14.14} = 1.56 = F_\phi = F_s$$

The example problem of Fig. 252-1 uses the stability chart to find the factor of safety for a simple slope.

Method of Slices

With this method the trial failure arc is divided into a reasonable number of slices, as shown in Fig. 255-1. The overturning moment is determined by summing the moment of the weight of each slice about the trial center O.

7   Slope Stability Analysis

Note that slices to the left of O have a negative moment ($\sin \alpha$ is negative).

The overturning moment is

$$\text{OM} = \Sigma W_n a_n = r \Sigma W_n \sin \alpha_n$$

The side forces on each slice are not included in the moment equations, since, when all slices are considered, the net moment of the side forces will be zero. The moment required for equilibrium is due to the tangential force $T_n = S'_n/F$ on the base of each slice. The force S'_n is the sum of the cohesive and frictional strength at the base of each slice. For stability

$$r \Sigma W_n \sin \alpha_n = r \Sigma T_n = r \Sigma \frac{S'_n}{F} = \frac{r \Sigma (c_n \ell_n + P_n \tan \phi_n)}{F}$$

and the safety factor F is

$$F = \frac{\text{RM}}{\text{OM}} = \Sigma \frac{(c_n \ell_n + P_n \tan \phi_n)}{\Sigma W_n \sin \alpha_n} \qquad (253\text{-}1)$$

The safety factor is defined as the ratio of resisting moment to overturning moment. When the analysis is based on total stress parameters c and ϕ, the equation for F is Eq. 253-1.

If effective stresses $\bar{c}$ and $\bar{\phi}$ are used, the normal force is reduced by the water force $U = u_n \ell_n$, where u_n is the average pore pressure on the bottom of the slice. The factor of safety based on effective stress parameters is Eq. 253-2.

$$F = \frac{\text{RM}}{\text{OM}} = \frac{\Sigma [\bar{c}_n \ell_n + (P_n - u_n \ell_n) \tan \bar{\phi}_n]}{\Sigma W_n \sin \alpha_n} \qquad (253\text{-}2)$$

Although the side forces cancel out of the overall moment equation, they do influence the magnitude of the normal reaction P_n on the base of the slice and thus the frictional shear strength at the base of the slice.

The side forces are actually indeterminate but can be approximated in various ways. Johnson (1975) has presented a summary of methods that consider side forces. Two commonly used methods of analysis, the Ordinary method of slices and Bishop's simplified method, are described in the following sections.

Ordinary Method of Slices

The side forces on the individual slices were neglected in the initial development of the method of slices (Fellenius, 1936). Each slice was considered to be in equilibrium under three forces, the weight W_n, the normal reaction $P_n = W_n \cos \alpha_n$, and the tangential force $T_n = W_n \sin \alpha_n$. By comparing the shear strength with the force T_n for each slice, it is apparent that the safety factor for each slice will vary. Upper slices have low safety factors and lower slices have high safety factors. The overall safety factor from Eq. 253-2 thus represents an average safety factor of the individual slices.

With this method, referred to as the Ordinary method of slices, the value of P_n in Eq. 253-2 equals $W_n \cos \alpha_n$. The resulting factor of safety is conservative for soils where ϕ is greater than zero. This method has been widely used for many years. Despite the development of more refined methods of analysis, the use of the method is still justified, since the accuracy of most slope analyses is limited more by the inability to accurately evaluate soil strength parameters than by the method of analysis used.

Bishop's Simplified Method of Slices

The Ordinary method of slices satisfies only the overall moment equilibrium, neglects the moment equilibrium of the individual slice, and only approximates the force equilibrium of each slice. Methods of analysis that satisfy all three equilibrium conditions are necessarily more complicated and difficult to apply. Bishop (1955) found that by including horizontal side forces to compute P_n and also satisfying the overall moment equilibrium, the resulting safety factor was only slightly less than that found by more rigorous methods. (0–6%; Wright et al, 1973).

Figure 255-1b shows a typical slice including side forces represented by horizontal components E and vertical components X. The force P_n and in turn the strength on the bottom of the slice will differ from the case where side forces are neglected. Each slice is assumed to have the same factor of safety, F, and a required strength T_n equal to the available strength at the bottom of the slice divided by F.

$$T_n = \frac{S'_n}{F} = \frac{\bar{c}_n \ell_n}{F} + \frac{P_n - u_n \ell_n}{F} \tan \bar{\phi}_n$$

As a simplification, Bishop assumed that the sum of the vertical side forces on each slice $(X_n + X_{n+1})$ equaled zero. Then from a summation of vertical forces,

$$\Sigma F_v = P_n \cos \alpha_n + T_n \sin \alpha_n - W_n = 0$$

Substituting the expression for T_n into the above equation, solving for P_n, and subtracting $u\ell$ from each side of the equation leads to the following equation.

$$P_n - u_n \ell_n = \left(W_n - \frac{\bar{c}_n \ell_n}{F} \sin \alpha_n - u_n \ell_n \cos \alpha_n \right) \left(\frac{\sec \alpha_n}{1 + \left(\frac{\tan \bar{\phi}_n \tan \alpha_n}{F} \right)} \right)$$

Figure 255-1 Method of slices.

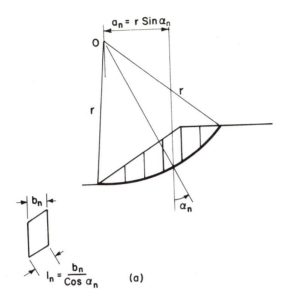

$$a_n = r \sin \alpha_n$$

O

r

r

$$l_n = \frac{b_n}{\cos \alpha_n}$$

(a)

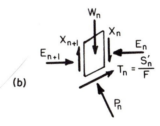

(b)

Substituting this expression for $P_n - u_n \ell_n$ into Eq. 253-2 and replacing ℓ_n by $b_n \sec \alpha_n$ leads to Eq. 257-1 for the safety factor F. This equation still offers some difficulty, because F appears on both sides of the equation. The safety factor F is calculated from successive trials starting with an initial estimate for F. If the calculated value of F is used in succeeding trials, convergence is quite rapid. A tabular form convenient for carrying out the solution is shown in Fig. 258-1 for the example problem of Fig. 256-1.

Slopes of Finite Height

Figure 256-1

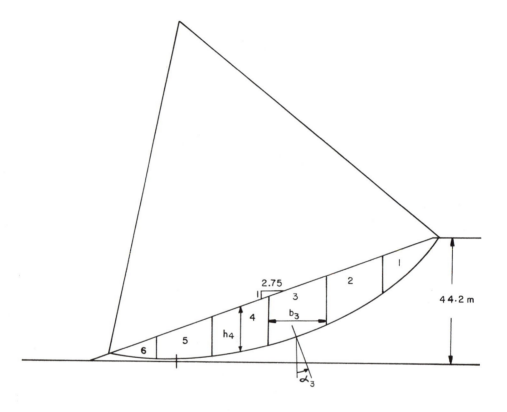

Example Problem Determine the safety factor for the given trial center and slope conditions. Use Bishop's modified method.

The pore pressure is estimated by using a pore pressure coefficient r_u. The pore pressure is then proportional to the total pressure given by the equation $u = r_u \gamma h$

$$\overline{\phi} = 37.5°$$

$$\overline{c} = 16.8 \text{ kN/m}^2$$

$$\gamma = 22.8 \text{ kN/m}^3$$

$$r_u = 0.4$$

$$u = r_u \gamma h \text{ (estimated pore pressure)}$$

From the scaled drawing of the slope and trial failure arc the width and height of each wedge are measured and tabulated in Fig. 258-1. The weight of each wedge is calculated.

$$W = bh\gamma$$

The value of α is measured and tabulated. Other columns in the table are calculated from given equations. The symbol m represents the numerator of Eq. 257-1 when written as

$$F = \frac{\Sigma m_n}{\Sigma W_n \sin \alpha_n}$$

The failure arc predicted by Bishop's simplified method has been found to compare well with the actual failure surface (Sevaldson, 1956). The equation for Bishop's simplified method follows.

$$F = \frac{\Sigma \left\{ [\bar{c}_n b_n + (W_n - u_n b_n) \tan \overline{\phi}_n] \dfrac{\sec \alpha_n}{1 + (\tan \overline{\phi}_n \tan \alpha_n / F)} \right\}}{\Sigma W_n \sin \alpha_n} \qquad (257\text{-}1)$$

Either effective or total stress parameters may be used in these equations, although the equations show only the effective stress parameters. In the case of $\phi = 0$ soils, Eq. 257–1 may be solved directly for F, and the equation reduces to that of the conventional slices method.

DESIGN CONSIDERATIONS

Critical Design States

Earth dams must be safe against slope and foundation failure for all operating conditions. There are three generally recognized critical stages based on pore pressure conditions for which the stability of the embankment should be ascertained. These three situations are:

- End of construction.
- Steady-state seepage.
- Rapid draw down.

Design Considerations

Figure 258-1

Example Problem

$$\gamma = 22.8 \ \text{kN/m}^3 \qquad \bar{c} = 16.8 \ \text{kN/m}^2 \qquad r_u = 0.4 \qquad \bar{\phi} = 37.5° \qquad u = 0.4\gamma h$$

Slice number	b (m)	h (m)	W	α (degrees)	W sin α	u	ub	c̄b	(W − ub) × tan φ̄	(W − ub) × tan φ̄ + c̄b	First Trial F = 1.8 $\dfrac{\sec \alpha}{1 + \frac{\tan \bar{\phi} \tan \alpha}{F}}$	m (F = 1.57)	Second Trial F = 1.57 factor	m (F = 1.54)	Third trial F = 1.54 factor	m
1	20	8.8	4,010	41.1	2,636	80	1,605	336	1,845	2,181	0.967	2,109	0.930	2,029	0.925	2017
2	20	16.2	7,390	31.34	3,843	148	2,955	336	3,403	3,739	0.930	3,477	0.902	3,374	0.898	3359
3	20	18.8	8,570	20.74	3,035	171	3,429	336	3,944	4,281	0.921	3,943	0.902	3,863	0.900	3851
4	20	17.0	7,750	10.95	1,472	155	3,101	336	3,567	3,903	0.941	3,673	0.931	3,632	0.929	3631
5	20	12.0	5,470	1.41	135	109	2,189	336	2,518	2,854	0.990	2,825	0.988	2,821	0.988	2820
6	17	4.5	1,740	-5.87	-178	41	698	286	800	1,085	1.051	1,141	1.058	1,148	1.060	1150
Σ					10,940					18,043		17,167		16,867		16827
					$\Sigma \ W_n \sin \alpha$							F = 1.57		F = 1.54		F = 1.54 *Ans.*

$$\Sigma m = 17{,}167 \qquad F = 1.57$$
$$\Sigma m = 16{,}867 \qquad F = 1.54$$
$$\Sigma m = 16827 \qquad F = 1.54 \quad Ans.$$

The first trial value of F used is 1.8 and yields F = 1.57. On the third trial the value has converged to 1.54.

Construction pore pressures usually reach their maximum values when the embankment reaches maximum height. After the reservoir has been filled for a long time, pore pressures are determined by steady-state seepage conditions and may be estimated from a flow net where gravitational flow conditions govern. Rapid lowering of the reservoir produces the third critical situation, particularly for slow-draining soils.

The upstream slope stability may be critical for the construction or rapid draw-down condition; the downstream slope should be checked for the construction and steady-state seepage condition.

Minimum Factor of Safety

The minimum factor of safety allowed will depend on the hazard involved with a failure as well as on the method of analysis, the reliability of the measured strength parameters, and the estimated pore pressures. A list of minimum safety factors for an earth dam, assuming the use of Bishop's modified analysis, is suggested below.

- End of construction, 1.3.
- Steady-state seepage, 1.5.
- Rapid draw down, 1.3.

PROBLEMS

7-1 Derive an equation for the safety factor of an infinite slope, assuming that seepage is emerging from the slope at an angle α, which is less than the slope angle β.

154292.1

(a) For a cohesionless soil.
(b) For a c-ϕ soil.

7-2 The soil of an 18° infinite slope is subject to full depth seepage. The soil properties are as follows: saturated unit weight, 17 kN/m³, effective cohesion, 10 kN/m², effective friction, 14°.
Determine the limiting depth, h, of this soil, measured vertically.

7-3 An infinite slope of granular soil has a slope of 23°. The saturated unit weight of the sand is 21.5 kN/m³ and the effective friction angle is 35°. The sand has a depth of 5 m over ledge rock and is subject to partial seepage parallel to the slope to a depth of 2.5 m. The depth of sand is measured vertically. Assume that the sand above the seepage surface is 70% saturated. Find F.

Problems

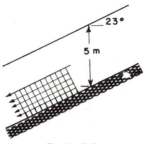

Prob. 7-3

7-4 For an infinite slope of cohesive soil without seepage, construct a plot of stability number versus slope for values of $\phi = 0°$, $\phi = 10°$, and $\phi = 20°$.

7-5 Determine the safety factor for the trial arc shown, using the friction circle method. This is not necessarily the most critical circle.

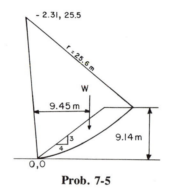

$W = 769$ kN
$\phi = 20°$
$c = 14.4$ kN/m^2
$\gamma = 15.7$ kN/m^3

Prob. 7-5

7-6 Determine the safety factor for the data of problem 7-5, using Taylor's slope stability chart.

7-7 What is the safety factor based on a plane failure surface in the example problem in Fig. 244-2 if the back slope is horizontal rather than at 10°?

7-8 Determine the safety factor of an 8 m deep excavation with side slopes of 30° in a soft clay that weighs 18 kN/m^3 and has a cohesion of 30 kN/m^2. A firm material lies below the soft clay. $\phi = 15°$, $D/H = 1$.

7-9 What slope angle must be used in problem 7-8 if a safety factor of 1.4 is desired?

7-10 Using the stability charts determine the factor of safety of a 50° slope 10 m high. The soil weighs 20 kN/m^3, has a friction angle of 25°, and cohesion of 25.5 kN/m^2.

7-11 A 52 m high homogeneous embankment with a 2.25-to-1 slope has an estimated pore pressure coefficient $r_u = 0.25$ at the end of construction. Choose a trial failure arc and determine the safety factor, using Bishop's modified method. Soil properties: $\bar{\phi} = 36°$, $\bar{c} = 16$ kN/m², $\gamma = 21$ kN/m³. Assume a firm base under the embankment.
Answer = (minimum factor of safety, 1.4)

7-12 Use Bishop's modified method to determine the safety factor for the trial arc shown. Determine the pore pressure from the flow net. Soil properties: $\bar{\phi} = 30°, \bar{c} = 20$ kN/m², γ above top flow line $= 17$ kN/m³, $\gamma_{sat} = 20$ kN/m³.

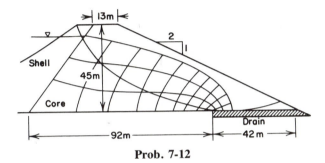

Prob. 7-12

7-13 Derive an equation for the stability number of a submerged infinite slope of soil.

Problems

8

BEARING CAPACITY AND FOUNDATIONS

All structures located on land are supported on foundation systems at or below the ground surface. There are many types of foundation systems, and the selection of the appropriate type for a given structure and set of subsurface conditions is the responsibility of the geotechnical engineer. Selection of the foundation system to be used is essentially an economic study of alternatives. The considerations include not only the cost of materials and labor, but also the costs involved for other items such as groundwater control, measures to minimize damage to adjacent facilities, and time required for construction.

In general, foundations can be grouped as shallow foundations or deep foundations. The most common types of shallow foundations are spread footings and continuous footings. The footings are used to "spread" highly concentrated column or bearing wall loads over soil strata near the ground surface. Deep foundations include pile and pier foundations in which piles or piers transfer structural loads to deep load-bearing strata.

The general theories for evaluating the supporting capability of shallow and deep foundations are developed in this chapter. Specific foundation design methods are discussed in detail in textbooks on foundation design.

The distinction between a shallow and a deep foundation is a relative matter. A foundation is usually considered to be shallow if a rotational bearing failure is possible. This is generally possible when the depth to the base of the footing is less than one to two times the footing width. Footings should be placed at least deep enough to avoid soil volume changes due to moisture change or freezing and thawing. The depth of future adjacent excavations should also be considered.

In designing shallow foundations, two possible failure mechanisms must be considered.

- A shear failure in the soil.
- Excessive settlement leading to differential settlement in excess of that tolerable for the supported structures.

Under a sufficiently heavy load the footing may sink into the soil as a result of shear failure. The pressure that causes this condition is called the ultimate bearing capacity, q'_{ult}. A failure of this type would be characterized by the soil heaving on one or both sides of the footing. Analysis of this failure mechanism is a type of limit equilibrium analysis. In some cases the footing may settle enough to cause structural distress without producing a shear failure in the soil. The settlement is caused by the pressure the footing exerts on the soil. Methods discussed in Chapters 3 and 4 are used to evaluate the stress increase with depth and the resulting settlement.

A complete design includes finding the size of footing that will produce a bearing pressure less than the allowable bearing capacity determined by one of the two failure criteria above. It must be less than the ultimate bearing capacity divided by a safety factor and low enough to limit the settlement to less than some prescribed allowable settlement. The allowable settlement is dependent upon the characteristics of the structure being supported.

ULTIMATE BEARING CAPACITY
OF SHALLOW FOUNDATIONS

Factors that determine the ultimate bearing capacity of a horizontal footing with a centric load include (1) the unit weight, shear strength, and deformation characteristics of the soil, (2) the size, shape, depth, and roughness of the footing, and (3) the water table conditions and initial stresses in the foundation soil. Most methods of analysis are based on the assumption of zones of plastic equilibrium in the soil supporting the footing. Well-defined slip lines are assumed to extend from the edge of the footing to the adjacent ground surface (Fig. 265-1a). The ultimate load will be well defined on a load settlement diagram, as in Fig. 265-1a. Failure of an actual footing where stress conditions control would be sudden and catastrophic.

Figure 265-1 Shear failures for footings.

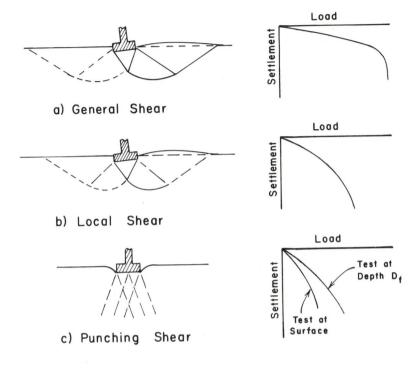

a) General Shear

b) Local Shear

c) Punching Shear

This type of failure is described as a general shear failure and is characteristic of narrow footings of shallow depth resting on stronger denser soils that are relatively incompressible.

For weaker, more compressible soils and wider or deeper footings, two other failure modes have been defined (Vesic, 1973). An intermediate failure mode or local shear failure is characterized by well-defined slip lines immediately below the footing but extending only a short distance into the soil mass (Fig. 265-1b). Under the circumstances of high compressibility, a punching shear mode of failure results; this is characterized by lack of a well-defined slip line below the footing (Fig. 265-1c). Settlement with increasing load is primarily due to the compression of the soil immediately under the footing, with soil to the side being uninvolved.

In the case of local shear and punching shear, the ultimate load is not well defined, as seen in Figs. 265-1b and 265-1c. In these cases the ultimate

Ultimate Bearing Capacity of Shallow Foundations

265

Figure 266-1 Assumed slip lines for rough footing.

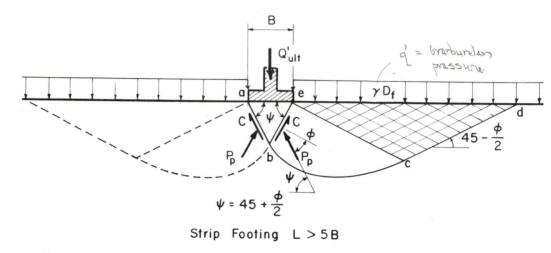

Strip Footing L > 5B

load may be chosen as the load at a given percentage settlement in proportion to the footing width.

Theoretical methods for predicting ultimate footing loads are generally based only on the general shear failure case. For the other two failure modes a reduction in the ultimate load due to compressibility effects is applied to the solution for the general shear case.

The general form of the bearing-capacity equation for a c-ϕ soil can be developed as follows. Consider the ultimate load of a rough footing of width B in Fig. 266-1. The length of the footing L is large with respect to B, and plane strain conditions prevail ($L > 5B$). The wedge of soil directly beneath the footing acts as part of the footing. The sloping edge of the soil wedge $\overline{eb}$ bears against a radial shear zone defined by a special failure surface $\overline{bc}$. A Rankine passive pressure zone edc exists to the side, and an overburden pressure γD_f acts down on the Rankine zone. This situation can be compared with the retaining wall passive pressure case (Fig. 222-1). The footing and the soil wedge below it are supported by the vertical components of the passive force P_p and cohesive force C acting on the faces $\overline{ab}$ and $\overline{eb}$. Thus, neglecting the weight of the soil wedge abe,

$$Q'_{ult} = 2P_p \sin\left(45 + \frac{\phi}{2}\right) + 2C \sin\left(45 + \frac{\phi}{2}\right) \qquad (266\text{-}2)$$

8   Bearing Capacity and Foundations

A rough approximation for P_p can be obtained by referring to Eq. 210-3, which defines the passive pressure for a vertical retaining wall with zero wall friction and a level backfill. To this equation is added a third term, which represents the increase in P_p due to a uniform surcharge γD_f.

$$P_p = \frac{1}{2}\gamma H^2 K_p + 2\,cH\sqrt{K_p} + \gamma D_f H K_p$$

Assume that P_p in Fig. 266-1 can be obtained by substituting the distance $\overline{eb}$ for H in the preceding equation. The resulting magnitude for P_p is not correct, but the equation is dimensionally correct.

The length $\overline{eb} = B/(2\cos(45 + \phi/2))$ is analogous to H in the above equation. The force C is

$$C = \frac{cB}{2\cos(45 + \phi/2)}$$

If these relationships are substituted into Eq. 266-2 and the trigonometric terms are combined into functions of ϕ, $f(\phi)$, $g(\phi)$, and $h(\phi)$, the result is the form of the general bearing-capacity equation. The functions of ϕ are bearing-capacity factors N_c, N_q, and N_γ, which must be independently evaluated.

$$Q'_{ult} = Bc\,f(\phi) + \gamma D_f\,Bg(\phi) + \frac{\gamma B^2}{2}\,h(\phi)$$

$$q'_{ult} = \frac{Q'_{ult}}{B} = cN_c + \gamma D_f N_q + \frac{\gamma B}{2}\,N_\gamma \qquad (267\text{-}1)$$

This is the Terzaghi-Buisman bearing-capacity equation. The factors N_c, N_q, and N_γ have been separately evaluated, as explained below.

Prandtl (1921) first developed a solution for N_c, assuming a weightless cohesive material with a shear strength c. He assumed that the curved failure surface $\overline{bc}$ of Fig. 266-1 is a logarithmic spiral that reduces to a circle for $\phi = 0$.

Reissner (1924) first provided the solution for N_q, which defines the component of bearing capacity due to friction of a weightless material on which a surcharge γD_f is acting.

The equations for N_q and N_c as found by Prandtl and Reissner are as follows.

$$N_q = e^{\pi\tan\phi}\left[\tan^2\left(45 + \frac{\phi}{2}\right)\right]$$

$$N_c = (N_q - 1)\cot\phi$$

Results of investigators using other techniques have shown general agreement with these equations.

The application of plastic theory to soils with weight involves additional complications. Values of N_γ presented in the literature vary over a wide range depending primarily on the shape of the assumed failure surface used. Terzaghi (1943) and later Meyerhof (1955) presented values of N_γ that have been widely used. Vesic (1973) suggested the equation below, which gives values somewhat higher than the Meyerhof values. However, the following equation is believed to be the best representation for N_γ.

$$N_\gamma = 2(N_q + 1) \tan \phi$$

Tabular values of N_c, N_q, and N_γ are given in Fig. 268-1. Plotted values are shown in Fig. 269-1.

For footings on soft clay or sand of low relative density (local shear or punching shear), reduced values equal to two-thirds of c and $\tan \phi$ have been

Figure 268-1   Bearing capacity factors.

ϕ	N_c	N_q	N_γ
0	5.14	1.0	0.0
5	6.5	1.6	0.5
10	8.3	2.5	1.2
15	11.0	3.9	2.6
20	14.8	6.4	5.4
25	20.7	10.7	10.8
30	30.1	18.4	22.4
32	35.5	23.2	30.2
34	42.2	29.4	41.1
36	50.6	37.7	56.3
38	61.4	48.9	78.0
40	75.3	64.2	109.4
42	93.7	85.4	155.6
44	118.4	115.3	224.6
46	152.1	158.5	330.4
48	199.3	222.3	496.0
50	266.9	319.1	762.9

Figure 269-1

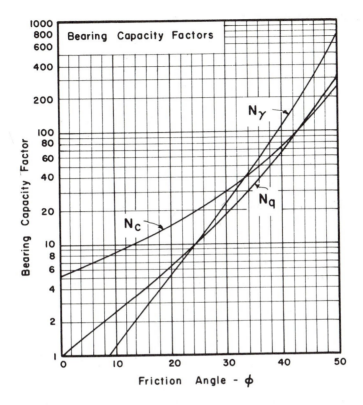

suggested (Terzaghi and Peck, 1967). Vesic (1973) related the correction of tan ϕ for granular soils to the relative density by the equation

$$\text{correction factor} = 0.67 + D_r - 0.75D_r^2 \quad \text{for} \quad 0 \le D_r \le 0.67$$

Equation 267-1 assumes that results for weighted soils can be superposed with the results for N_c and N_q using weightless soil. This superposition is believed to lead to errors on the safe side (up to 20% at $\phi = 40°$, Vesic 1973).

There is also evidence, not shown by Eq. 267-1, of a scaling effect with increasing footing size. The actual bearing capacity of wider footings is probably less than given by Eq. 267-1. Pope (1975) has developed a nondimensional chart for bearing pressure that takes into account this scaling

Ultimate Bearing Capacity of Shallow Foundations

269

effect. Though not included in current design practice, scaling effects will probably receive more attention in the future.

Effect of Footing Shape

The Terzaghi bearing capacity equation (Eq. 267-1) applies to long strip footings. For circular, square, or rectangular footings, the shape of the failure surface is three dimensional rather than two dimensional, as in the strip footing. Thus the bearing capacity is altered. Because of the complexity of the problem, the approach is largely empirical. The terms of Eq. 267-1 are multiplied by shape factors given in Fig. 271-1. The factors in the table are representative of suggestions found in the literature. Tests show that the shape factor for N_c (cohesive soils) is reasonably accurate. Shape and depth factors for granular soil are often neglected.

Correction for Depth of Footing and Load Inclination

$$q'_{ult} = \left[\frac{1}{2} \gamma B N_\gamma S_\gamma + q' N_q S_q \wedge_h \right]$$

Correction factors for the depth of footing and for inclined loads are also shown in Fig. 271-1. The general bearing-capacity equation as modified by shape, depth, and load inclination factors is

$$q'_{ult} = cN_c(s_c)(d_c)(i_c) + \gamma D_f N_q (s_q)(i_q) + \frac{1}{2} \gamma B N_\gamma(s_\gamma)(i_\gamma) \qquad (270\text{-}1)$$

Net Bearing Pressure

Equation 267-1 and other bearing-capacity equations to this point have represented the gross ultimate bearing capacity. The net bearing capacity is defined as the gross bearing capacity less the surcharge pressure γD_f adjacent to the base of the footing. The net bearing pressure represents the increase in pressure over the pressure existing as a result of the adjacent overburden weight.

$$q_{ult} = q'_{ult} - \gamma D_f$$

Equation 267-1 then becomes

$$q_{ult} = cN_c + \gamma D_f (N_q - 1) + \frac{\gamma}{2} B N_\gamma \qquad (270\text{-}2)$$

A similar adjustment can be made to Eq. 270-1.

The allowable bearing pressure is the net bearing pressure divided by an appropriate safety factor F.

$$q_a = \frac{q_{ult}}{F}$$

$$q_{ult} = \frac{1}{2} \gamma_o B (N_\gamma)$$
$$+ q' N_q + c N_c$$

Figure 271-1 Correction factors for shape of footing, depth of footing and load inclination for shallow footings. (See Vesic, 1973.)

Shape factors

Footing Shape	s_c	s_q	s_γ
Strip footing	1.0	1.0	1.0
Rectangle	$1 + \dfrac{B}{L}\left(\dfrac{N_q}{N_c}\right)$	$1 + \dfrac{B}{L}\tan\phi$	$1 - 0.4\dfrac{B}{L}$
Circle and square	$1 + \dfrac{N_q}{N_c}$	$1 + \tan\phi$	0.60

Limitation $B \leq L$

Depth Factors

d_c	d_q	d_γ
$1 + \dfrac{0.2D_f}{B}$	1.0	1.0

Inclination Factors*

i_c	i_q	i_γ
$\left(i_q - \dfrac{1 - i_q}{N_c \tan\phi}\right)$	$\left(1 - \dfrac{H}{V + BLc \cot\phi}\right)^2$	$(i_q)^{3/2}$

*For $\phi = 0$, $i_c = 1 - \dfrac{2H}{BLcN_c}$;

H = horizontal load, V = vertical load.

Water Table Effects

The presence of a water table around a footing reduces the effective shear strength of a granular soil and, hence, its bearing capacity. For a fully submerged footing the unit weight used with the N_q and N_γ terms of the

Ultimate Bearing Capacity of Shallow Foundations

for $\phi \leq 28°$

$\tan\phi_{L-s} = \frac{2}{3}\tan\phi$

bearing-capacity equation is the buoyant unit weight. Since the buoyant unit weight of soil is about one-half the moist unit weight, the bearing capacity of a submerged footing is about one-half that of the footing well above the water table. If the water table is a distance B below the footing, it is assumed to have no effect. When the water table is at the base of the footing, the buoyant unit weight is used only with the N_γ term.

For water table positions intermediate between the ground surface and base of the footing or between the base of the footing and a distance B below the footing, adjust the respective unit weight by linear interpolation.

Eccentric Loads

Footings frequently must resist an overturning moment, which results in an eccentrically applied load. Meyerhof (1953) has proposed that the dimensions B' and L' of an equivalent concentric footing be found and used to determine the allowable bearing pressure. The equations for B' and L' are

$$B' = B - 2e_b \quad \text{and} \quad L' = L - 2e_\ell$$

where e_b and e_ℓ are the eccentricity in each respective direction. The total allowable load is then $q_a(B'L')$.

Other more conservative approaches can be found in foundation textbooks.

SETTLEMENT CONSIDERATIONS

Structural damage can result when there is a differential settlement between various parts of a structure. Differential settlement is difficult to predict accurately, but it is generally related to the maximum settlement at any point within a structure. The allowable maximum settlement that is specified for a given structure depends on the type of foundation and on the structural framing system. For example, there is likely to be less differential settlement for a structure supported on a mat foundation than for the same structure supported on individual spread footings, even though the maximum settlement is the same. The allowable total settlement for the mat foundation can, therefore, be larger than for the spread footing foundation.

Settlement of Footings on Sand

Settlement criteria rather than ultimate bearing capacity usually governs the allowable bearing capacity for footings on sand when the footing width exceeds 1 m. Thus methods for predicting settlement are needed for economical design.

8 Bearing Capacity and Foundations

Figure 273-1

Example Problem

1. Find the allowable net bearing load per meter of length of a long wall footing 2 m wide on a stiff, saturated clay. The depth of the footing is 0.5 m. The unit weight of the clay is 17 kN/m³, and its shear strength c is 120 kN/m². Assume the load to be applied rapidly such that undrained ($\phi = 0$) conditions prevail. Use of a safety factor of 3. Neglect settlement.

From Eq. 270-2,

$$\phi = 0, \qquad N_\gamma = 0$$

$$q_{ult} = cN_c(s_cd_c) + \gamma D_f(N_qs_qd_q - 1)$$

$$N_c = 5.14 \qquad N_q = 1 \qquad N_qs_qd_q - 1 \approx 0$$

$$s_c = 1 \qquad d_c = 1 + 0.2D_f/B = 1 + 0.2\left(\frac{0.5}{2}\right)$$

$$d_c = 1.05$$

$$q_{ult} = 120 \text{ kN/m}^2(5.14)(1)(1.05) = 648 \text{ kN/m}^2$$

$$q_a = \frac{q_{ult}}{F} = \frac{648}{3} = 216 \text{ kN/m}^2$$

Allowable load per meter of length $= 2(216) = 432$ kN/m

2. Determine the allowable load for a rectangular footing 2×3 m for the same soil conditions as in part 1.

$$q_{ult} = cN_c(s_cd_c)$$

$$N_c = 5.14 \qquad s_c = 1 + \frac{B}{L}\left(\frac{N_q}{N_c}\right) = 1 + (2/3)\frac{1}{5.14} = 1.13$$

$$d_c = 1 + 0.2\frac{D_f}{B} = 1 + 0.2\left(\frac{0.5}{2}\right) = 1.05$$

$$q_{ult} = 120(5.14)(1.13)(1.05) = 732 \text{ kN/m}^2$$

$$Q_a = \frac{q_{ult}}{F}BL = \frac{732}{3}(2)3 = 1464 \text{ kN}$$

3. Determine the allowable load for the rectangular footing in part 2 if the eccentricity is 1/6 m in each direction.

Adjust the footing dimensions,

$$B' = B - 2e = 2 - 2(\tfrac{1}{6}) = 1.67 \text{ m.}$$

$$L' = B - 2e = 3 - 2(\tfrac{1}{6}) = 2.67 \text{ m}$$

Settlement Considerations

$$\frac{q_{ult}}{F_s} \cdot \quad \frac{1}{2}\gamma B d\gamma \quad S\gamma 1\gamma$$

$$B+2B \qquad = \frac{1}{2}\gamma B d\gamma$$

273

$$s_c = 1 + \frac{N_q}{N_c} \frac{B'}{L} = 1 + \frac{1}{5.14} \frac{1.67}{2.67} = 1.12$$

$$d_c = 1 + 0.2\left(\frac{0.5}{1.67}\right) = 1.06$$

$$q_{ult} = 120(5.14)(1.12)(1.06) = 732 \text{ kN/m}^2$$

$$Q_a = \frac{q_{ult}}{F} B'L' = \frac{732}{3}(1.67)(2.67) = 1088 \text{ kN}$$

Laboratory triaxial tests on disturbed sand samples that duplicate the field stress path provide vertical strain data from which footing settlement can be predicted (Lambe, 1967). The plate-bearing test, the cone penetrometer test, and the standard penetration test are all field tests used to predict settlement of footings on sand. These methods are described in the following section.

Since settlement of sand occurs rapidly, the appropriate loads to use in the analysis are dead load plus maximum live load.

Settlement by Plate-Bearing Test (Granular Soils)

A plate-bearing test involves applying a load to a given area in increments and measuring the settlement as the incremental loads are applied. The test area is generally much smaller than the actual footing area, and, therefore, some adjustment must be made for the size of the footing. Some insight into the effect of footing size on settlement can be obtained by considering settlement predicted by the theory of elasticity. Lambe and Whitman (1969) indicate that the settlement of a rigid circular footing on an elastic half-space can be calculated by Eq. 274-1.

$$\Delta H = q \frac{R}{E} \frac{\pi}{2} (1 - \nu^2) \tag{274-1}$$

where

q = average uniform load
R = radius of footing
ν = Poisson's ratio
E = modulus of elasticity

Equation 274-1 indicates that the settlement of a rigid circular footing for a given average uniform load is directly proportional to the radius of the area.

8 Bearing Capacity and Foundations

Figure 275-1

Example Problem

1.   Determine the allowable load for a square footing 3 × 3 m at a depth of 2 m in a medium dense sand. The sand has a friction angle of 36° and has moist unit weight of 16 kN/m³. Use a safety factor of 2.5.

From Eq. 270-2, with shape and depth factors

$$q_{ult} = \gamma D_f(N_q s_q - 1) + \tfrac{1}{2}\gamma B N_\gamma(s_\gamma)$$

$$s_q = 1 + \tan \phi = 1.73 \qquad s_\gamma = 0.6$$

$$(N_q s_q - 1) = 64.22, \; N_\gamma = 56.3$$

$$q_{ult} = (16)(2)(64.22) + \frac{16}{2}(3)(56.3)(0.6)$$

$$q_{ult} = 2055 + 811 = 2866 \text{ kN/m}^2$$

$$Q_a = \frac{2866}{2.5}(3)^2 = 10318 \text{ kN}$$

2.   Repeat part 1 assuming a maximum water table position 1 m below the base of the footing. The saturated unit weight of the sand is 18 kN/m³. The buoyant unit weight of sand $\gamma_{sat} - \gamma_w = 18 - 9.81 = 8.2$ kN/m³. Calculate a weighted unit weight for the sand below the footing to a depth equal to $B = 3$ m.

$$\gamma = \frac{1(16) + 2(8.2)}{3} = 10.8 \text{ kN}$$

$$q_{ult} = 2055 + \frac{10.8}{2}(3)(56.3)(0.6) = 2055 + 547 = 2602 \text{ kN}$$

$$Q_a = \frac{2602}{2.5}(3)^2 = 9367 \text{ kN}$$

The settlement in both cases is probably greater than would normally be acceptable. Allowable bearing pressure based on settlement is discussed in following sections.

Therefore, the settlement of a circular footing could theoretically be predicted for an average uniform load on the basis of a plate-bearing test by the following equation.

$$\Delta H = \Delta H_o \left(\frac{R}{R_o} \right) \tag{275-2}$$

Settlement Considerations

where

ΔH_o = settlement of test plate under the given uniform load q
R_o = radius of the test plate
R = radius of the footing

A similar procedure can be applied to square and rectangular footings by using the least plan dimension of the footing in place of the radius. Experience has shown that even for very deep uniform soil deposits there is some error in Eq. 275-2. This can be partly explained in that the modulus of elasticity increases with depth as the intergranular stress increases. Modifications of Eq. 275-2 have, therefore, been proposed. Terzaghi and Peck (1967) suggest the following equation to predict the settlement of a footing on the basis of a plate-bearing test with a 0.305 m (1 ft) square plate.

$$\Delta H = \Delta H_o \left(\frac{6.56B}{3.28B + 1} \right)^2 \qquad (276\text{-}1)$$

where

ΔH = settlement of a footing with a bearing pressure q
ΔH_o = settlement of a 0.305 m square test plate with a bearing pressure q
B = width of footing in meters

Equation 276-1 can be applied to circular, square, rectangular, or continuous footings. The equation can be used to account for a larger test plate or to compare the behavior of two different-size footings on cohesionless soil by writing it in the following form.

$$\Delta H = \Delta H_o \left(\frac{B}{B_o} \right)^2 \left(\frac{3.28\,B_o + 1}{3.28\,B + 1} \right)^2 \qquad (276\text{-}2)$$

where B_o is the width of the test plate in meters.

The results of a plate-bearing test are presented as a curve of settlement versus bearing pressure for the test plate, as shown in Fig. 277-1. The settlement-versus-bearing-pressure curve and Eq. 276-2 can then be used by a trial-and-error procedure to proportion a footing on the basis of a minimum allowable settlement. This procedure is illustrated in the example of Fig. 277-3.

In using the results of a plate-bearing test to predict settlements, there are two very important considerations. First, the surface of the soil where the test plate is placed must be in an undisturbed condition so that it is representative of the soil deposit. Second, the soil deposit must be homogeneous to a depth that is great compared to the width of the footing. This second consideration is important, because the depth of stress influence for the test plate will be much less than for the footing. This idea is presented in Fig. 104-1 and

Figure 277-1 Load settlement curve from plate-bearing test.

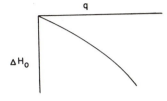

Figure 277-2 Stress influence of test plate and footing.

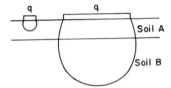

Figure 277-3

Example Problem A plate-bearing test was performed with a 0.305-m square plate on a dense sand deposit, and the following curve was obtained. Use Eq. 276-2 to proportion a square footing to support a column load, Q, of 4000 kN. The maximum allowable settlement is 25 mm.

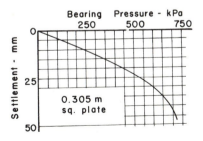

Settlement Considerations

The solution process is trial and error, using Eq. 276-2 and the plate-bearing test curve.

The trial-and-error procedure can be summarized in tabular form as:

q_a (kPa)	B (m)	ΔH_o (mm)	ΔH (mm)
250	4	11	38.0 > 25
150	5.16	6	21.4 < 25
175	4.78	7.5	26.5 > 25
170	4.85	7	24.8 ≈ 25

Use $B = 4.85$ m

In carrying out the trial-and-error procedure to develop the preceding table, an allowable bearing pressure, q_a, was first assumed, and then the required footing width, B, was calculated from

$$B^2 = \frac{Q}{q_a}$$

ΔH_o was then obtained from the settlement versus bearing pressure curve from the plate-bearing test and ΔH was obtained from Eq. 276-2. The trial-and-error procedure continues until ΔH = allowable settlement.

is illustrated in Fig. 277-2, where the major stress influence for the test plate is entirely within soil A whereas the same stress influence for the footing extends well into soil B.

Settlement by Cone Penetrometer Test

The cone penetrometer is a 35.6 mm (1.4 in.) diameter probe that is forced into the sand. The measured point resistance q_c is a measure of the soil strength, which can be used to predict settlement of footings over sand. A more detailed explanation of the cone penetrometer test is presented in Chapter 11. To estimate settlement, European engineers have long used a method based on the data from the Dutch cone penetrometer (De Beer, 1967). A simpler method was proposed by Schmertmann (1970).

From the theory of elasticity the vertical strain beneath uniformly loaded circular or rectangular footings is given by Eq. 278-1.

$$\epsilon_y = \frac{q_a}{E} I_y \tag{278-1}$$

8 Bearing Capacity and Foundations

where q_a is the net pressure increase, E is the elastic modulus of the soil, and I_y the vertical strain influence factor. The maximum vertical strain occurs at a depth of about 0.3 to 0.6 of B below the footing. The vertical strain varies with Poisson's ratio, footing shape, and the location under the footing. The variation of I_y below a circular footing is shown by the curves in Fig. 279-1.

Schmertmann (1970) suggested that for practical purposes a single approximate distribution of the strain influence factor could be used to compute the static settlement of isolated, rigid, shallow footings. The I_y distribution suggested by Schmertmann is shown in Fig. 279-1, where the maximum I_y is 0.6 at depth $B/2$. Settlement below a depth of $2B$ is assumed negligible.

For footings below grade, a correction factor C_1 reduces I_y to account for strain relief due to embedment.

$$C_1 = 1 - 0.5 \frac{\overline{\sigma}_o}{q' - \overline{\sigma}_o}$$

Figure 279-1 Vertical Strain Influence Factor. (After Schmertmann, 1970.)

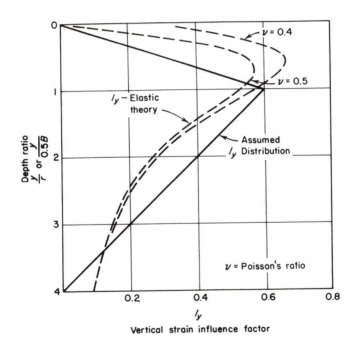

Vertical strain influence factor

where $\bar{\sigma}_o$ is the effective overburden pressure at the foundation level and $q' - \bar{\sigma}_o$ is the net pressure increase q_a. A second correction factor C_2 adjusts for creep, an effect similar to secondary compression in clay.

$$C_2 = 1 + 0.2 \log \left(\frac{t \ (\text{yrs})}{0.1} \right)$$

where t is the time in years after loading.

The soil modulus E is determined from cone penetrometer data. From screw-plate tests in several sand deposits Schmertmann (1970) found that the equation

$$E = 2q_c$$

gave a good correlation with measured settlement.

The equation for settlement is then

$$\Delta H = \int_0^\infty \epsilon_y \, d_y \approx C_1 C_2 q_a \sum_0^{2B} \left(\frac{I_y}{E} \right) \Delta y$$

Note that q_a is the net bearing pressure at the base of the footing.

Application of this method to a number of sites with considerable variation in footing geometry, loading, and soil variables demonstrated that the predicted settlement was conservative (greater than measured) in most cases. The example problem of Fig. 281-1 illustrates the use of the cone penetrometer to predict the settlement of a sand deposit.

Settlement by Standard Penetration Test

As illustrated on Fig. 120-1, the settlement of a footing on sand depends on the initial relative density of the sand deposit. It has, therefore, become common practice to correlate various penetration resistance methods with relative density of sand deposits and, hence, settlement of footings on sand deposits. The most commonly used field penetration test is the Standard penetration test. The test involves determining the number of blows, N, required to drive a split spoon sampler 0.305 m (1 ft) with a 623 N (140 lb) weight dropping through a distance of 0.763 m (30 in.). The split spoon sampler has an outside diameter of 50.8 mm (2.0 in.) and an inside diameter of 34.9 mm ($1\frac{3}{8}$ in.). The Standard penetration test is performed at regular intervals as a soil boring is advanced.

Terzaghi and Peck (1948) presented a family of curves that related the Standard penetration test blow count, N, to the bearing pressure that would cause a settlement of 25 mm (1 in.). These curves also included the effect of footing width. The curves presented by Terzaghi and Peck (1948) were developed from settlement observations of many footings on sand deposits of known standard penetration resistance.

Figure 281-1

Example Problem Given the log of cone penetrometer bearing capacity, estimate the settlement under the structure shown in the figure. The gross bearing pressure is 48 kN/m² and the least width B of the structure is 12.2 m. (Ramage and Stewart, 1978.)

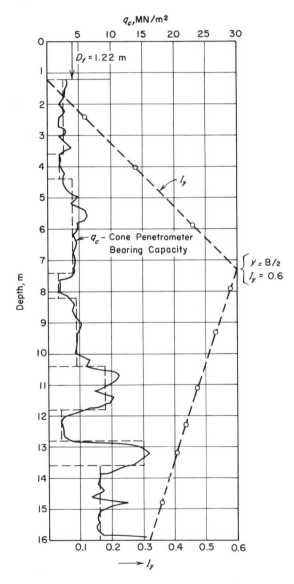

Settlement Considerations

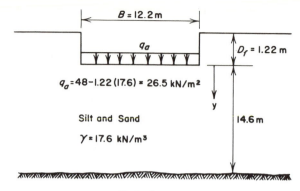

$B = 12.2\,\text{m}$

q_a

$D_f = 1.22\,\text{m}$

$q_a = 48 - 1.22\,(17.6) = 26.5\ \text{kN/m}^2$

y

Silt and Sand

$14.6\,\text{m}$

$\gamma = 17.6\ \text{kN/m}^3$

Bedrock

Solution   The soil below the structure is divided into nine layers of approximately uniform values of q_c and the results tabulated. Maximum $I_y = 0.6$ is plotted at $B/2$ below the footing and lines extended to $I_y = 0$ at $y = 0$ and $y = 2B$. Record values of I_y at the mid point of each of the nine layers.

$$\Delta H = C_1 C_2 \cdot q_a \sum_0^{2B} \left(\frac{I_y}{E_s}\right) \Delta y$$

$$\gamma = 17.62\ \text{kN/m}^3$$

$$q_a = 48.0 - 21.5 = 26.5\ \text{kN/m}^2$$

	$\bar{y}$ (m)	q_c (MN/m²)	E_s (MN/m²)	Δy (m)	I_y	$\dfrac{I_y}{E_s}\Delta y$
1	1.2	2.5	5.0	2.4	0.12	5.8×10^{-8}
2	2.8	2.0	4.0	0.8	0.28	5.6×10^{-8}
3	4.7	4.0	8.0	3.0	0.46	17.2×10^{-8}
4	6.6	1.5	3.0	0.8	0.58	15.5×10^{-8}
5	8.1	4.5	9.0	2.2	0.53	12.9×10^{-8}
6	9.9	9.0	18.0	1.4	0.47	3.6×10^{-8}
7	11.1	2.0	4.0	1.0	0.44	11.0×10^{-8}
8	12.0	15.0	30.0	0.8	0.41	1.1×10^{-8}
9	13.5	8.0	16.0	2.4	0.36	5.4×10^{-8}
						78.8×10^{-8}

$$C_1 = 1 - 0.5\,\frac{21.5}{26.5} = 0.59$$

$$C_2 = 1 + 0.2 \log\left(\frac{t}{0.1}\right) = 1 + 0.2 \log\left(\frac{1}{0.1}\right) = 1.2$$

$$\Delta H = (0.59)(1.2)(26.5 \times 10^3)(78.8 \times 10^{-8}) = 14.7 \times 10^{-3}\ \text{m} = 14.7\ \text{mm}$$

Peck, Hanson, and Thornburn (1974) have somewhat revised the Terzaghi and Peck curves to account for more recent research and observational data. Their new relationship relating standard penetration test resistance, allowable net bearing pressure, and settlement can be expressed as

$$q_a = C_w (0.41)N \, \Delta H \tag{283-1}$$

where

q_a = allowable net bearing pressure in kPa that will cause a settlement of ΔH in millimeters

N = Average corrected standard penetration test blow count

ΔH = settlement in millimeters

C_w = water table correction factor (explained below)

When the location of the water table is near the base of the footing, it will have some effect on the settlement of a footing on sand. As explained in the section on ultimate bearing capacity, the water table reduces the effective unit weight of the material by about one-half. For settlement consideration, the assumption is made that the allowable bearing pressure should be reduced by one-half if the water table is at or above the ground surface, and that it should receive no reduction if it is deeper than B below the base of the footing. A linear interpolation is made for intermediate positions of the water table. A correction factor for Eq. 283-1 can be stated as follows.

$$C_w = 0.5 + 0.5 \frac{D_w}{D_f + B}$$

where

C_w = water table correction factor $(0.5 < C_w < 1.0)$
D_w = depth to water table from ground surface
D_f = depth to base of footing from ground surface
B = width of footing

Gibbs and Holtz (1957) showed that for a constant relative density the standard penetration blow count increased with increasing effective overburden pressure. To account for the effect of overburden pressure on N values, Peck, Hanson, and Thornburn (1974) suggested an effective overburden stress correction factor, C_N, for the N values to be used in Eq. 283-1. This correction is shown on Fig. 284-1. The recorded field value of N should be multiplied by C_N for use in Eq. 283-1.

Since soil deposits are generally somewhat erratic, the N values used in Eq. 283-1 should be based on the results of several test borings. For the design of footings, the general practice is to determine N values at 0.76 m (2.5 ft) intervals as the test boring is advanced. Each N value should then be

Figure 284-1 Blow count correction factor for overburden pressure. (After Peck, Hanson, and Thornburn, 1974.)

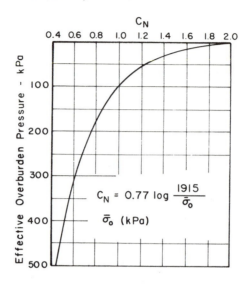

Figure 284-2

Example Problem

1. Determine the allowable load for a 3 × 3 m footing at a depth of 2 m in a medium dense sand if the blow count N is 28 blows per foot. The total settlement is to be limited to 25 mm. Assume the moist unit weight of the sand is 16 kN/m³. The water table is more than 3 m below the footing.

Solution The overburden pressure at $B/2$ below the footing is

$$\bar{\sigma}_o = 16(2 + 3/2) = 56 \text{ kPa}$$

The correction C_N from Fig. 284-1 is

$$C_N = 1.18$$

The corrected blow count is

$$N_{corr} = 1.18(28) = 33$$

From Eq. 283-1,

$$q_a = 0.41(33)(25) = 338 \text{ kPa}$$

284

$$Q_a = (3)^2 338 = 3044 \text{ kN}$$

as limited by total settlement of 25 mm. Compare this answer with the example problem of Fig. 275-1, part 1.

2. Determine the allowable load for the footing in part 1 if there is a water table 1 m below the footing. Assume the buoyant unit weight of the sand is 8.2 kN/m³. The blow count measures 26 blows per foot. Total settlement is limited to 25 mm.

Solution   The effective overburden pressure at $B/2$ below the footing is

$$\bar{\sigma}_o = [16(3) + 8.2(\tfrac{1}{2})] = 52 \text{ kPa}$$

From Fig. 284-1,

$$C_N = 1.2$$

The corrected blow count is

$$N_{\text{corr}} = 1.2(26) = 31$$

The water table correction is

$$C_w = 0.5 + (0.5)\frac{3}{2 + 3} = 0.8$$

From Eq. 283-1,

$$q_a = 0.8(0.41)(31)(25) = 254 \text{ kPa}$$

$$Q_a = 254(3)^2 = 2288 \text{ kN}$$

Compare this answer with example problem of Fig. 275-1, part 2.

corrected for overburden pressure. The average corrected value of N over a distance from the base of the footing to a depth B below the base of the footing should then be used in Eq. 283-1. When several borings are made, the lowest average value should be used.

The example problem of Fig. 284-2 illustrates the procedure used to proportion a footing on the basis of the Standard penetration test.

Settlement of Footings on Saturated Clay

The settlement of footings on saturated clay can be estimated using Eq. 285-1, which is developed from Eq. 128-4.

$$\Delta H = \Sigma \left[HC \log\left(\frac{\bar{\sigma}_o + \Delta\sigma}{\bar{\sigma}_o}\right) \right] \qquad (285\text{-}1)$$

Settlement Considerations

Figure 286-1 Plot of initial intergranular stress $\bar{\sigma}_o$, stress increase $\Delta\sigma$, and maximum past pressure.

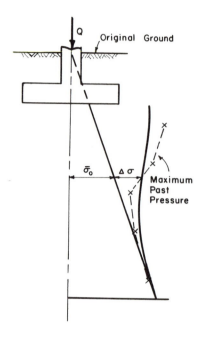

This equation shows that settlement is a function of the existing intergranular pressure $\bar{\sigma}_o$, the pressure increase $\Delta\sigma$, the compressibility of the soil C, and the thickness of the compressible layer H. The general settlement analysis procedure involves dividing the compressible clay layers below the footing into sublayers and computing the settlement of each sublayer. The total settlement is the sum of the sublayer settlements. The stress increase in each sublayer can be evaluated using the stress analysis methods of Chapter 3.

Most natural soil deposits are not uniform with depth. Either the soil type changes or the compressible properties of the soil change with depth. This would be the case for a sand stratum overlying a soft compressible clay or for a clay deposit in which the upper layers are very stiff from overconsolidation by desiccation. Since conditions can change with depth, it is desirable to plot the initial intergranular pressure, the stress increase, and the preconsolidation pressure as a function of depth below the footing as shown

Figure 287-1

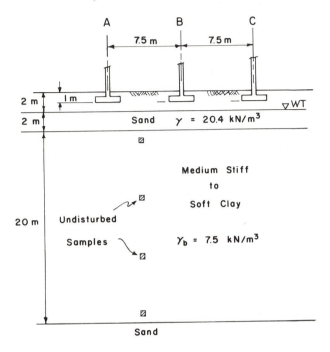

Example Problem A six-story building of masonry block construction is proposed for a given site. Loads will be transferred to the foundation system through bearing walls spaced on 7.5-m centers. Anticipated bearing wall loads at foundation level include 275 kN/m dead load and 47 kN/m sustained live load.

One possible foundation system for the proposed building would utilize continuous footings to support the bearing walls. For this system the footings would be 2.25 m wide and would be founded 1.0 m below ground surface, as shown in the figure. In order to evaluate whether this foundation system will be acceptable, the settlement of the footings will have to be determined.

A subsurface investigation at the site revealed the idealized soil profile shown above.

Predict the settlement of the continuous footing B that would result from compression of the 20-m medium stiff to soft clay stratum.

The compression of the clay can be predicted using Eq. 285-1.

$$\Delta H = \Sigma \left(HC \log \frac{\overline{\sigma}_o + \Delta\sigma}{\overline{\sigma}_o} \right)$$

The increase in stress $\Delta\sigma$ will be a function of depth and can be evaluated for the continuous footings of this problem from Fig. 104-1. Some judgment will generally be required to determine whether a wall footing, such as those in this problem, should be treated as a continuous footing or as several uniformly loaded areas. For this problem it is assumed that the wall footings are long enough to be treated as continuous footings.

Footing pressure Use dead load plus sustained live load.

$$q = \frac{Q}{B} = \frac{(275 + 47)\text{kN/m}}{2.25 \text{ m}}$$

$$q = 143.1 \text{ kPa}$$

Both A and C will contribute equally to the stress increase under footing B. The stress increase below footing B is given in the following table.

Stress increase below footing B

Depth (m)	y (m)	$\dfrac{y}{B}$	Footing B (kPa)	Footings A and C (kPa)	Total (kPa)
				$\Delta\sigma$	
4	3	1.33	62.6	2.7	65.4
9	8	3.56	25.5	14.7	40.2
14	13	5.78	15.7	17.8	33.5
19	18	8.00	11.4	16.5	27.9
24	23	10.22	8.9	14.5	23.4

The initial vertical intergranular pressure, $\bar{\sigma}_o$, and the stress increase, $\Delta\sigma$, are plotted below as a function of depth.

Consolidation tests were performed on undisturbed samples from depths of 5, 11, 17 and 23 m. The maximum past pressure was determined by Casagrande's method. (Fig. 130-1.)

The maximum past pressure is shown on the same plot with the initial intergranular pressure and the pressure increase resulting from the continuous footing loads. This plot indicates that the upper part of the soil profile is overconsolidated with a transition to a normally consolidated clay near the bottom of the clay stratum. Based on the plot of initial vertical intergranular pressure, vertical stress increase, and maximum past pressure, the clay stratum has been divided into three regions, as shown on the plot. An evaluation of the ϵ versus $\log \bar{\sigma}$ curves (not shown in this example) indicated that a value of 0.02 can be used for the slope of C of the ϵ-$\log \bar{\sigma}$ curve in the overconsolidated region and 0.18 in the normally consolidated region.

8   Bearing Capacity and Foundations

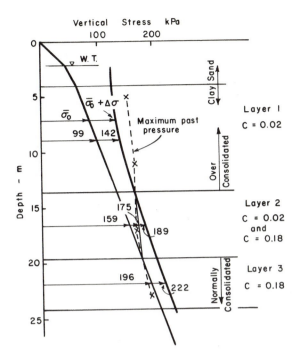

The settlement in layer 2 (13.5 to 19.5 m) will occur partly in the over-consolidated range and partly in the normally consolidated range. This is illustrated in the settlement computation table where $C = 0.02$ until the vertical stress exceeds the maximum past pressure; then a value of $C = 0.18$ is used. Hence, two additive sets of computations are shown for layer 2.

Settlement Computation Table

Layer	C	H	$\bar{\sigma}_o$	$\bar{\sigma}_o + \Delta\sigma$	ΔH
1	0.02	9.5	99	142	0.0298
2	0.02	6.0	159	175	0.0050
2	0.18	6.0	175	189	0.0361
3	0.18	5.5	196	222	0.0536
				$\Sigma\Delta H$	0.1244 m

The total settlement below footing B resulting from compression of the clay stratum will be approximately 0.12 m. In addition to this settlement there will be some compression of the overlying sand that will contribute to the total settlement.

A total settlement of this magnitude would generally be considered excessive, and alternative foundation systems may be considered.

in Fig. 286-1. This diagram can aid in dividing the compressible layers into substrata for the settlement analysis. Each substratum should then be assigned a value of C for use in Eq. 285-1.

Settlement of clay soil generally occurs slowly. Therefore the appropriate loads to use for a settlement analysis of a footing on clay are the dead load plus the sustained live load. The sustained live load is that portion of the live load that is applied for a significant time.

In many cases a footing can contribute to an increase in stress below an adjacent footing and, hence, can increase the settlement of the adjacent footing. Stress overlap must always be considered in predicting the stress increase below a footing.

The example problem of Fig. 287-1 illustrates a settlement analysis for a foundation system that consists of several parallel continuous footings.

ALLOWABLE BEARING PRESSURE

As indicated earlier, the allowable bearing pressure for shallow foundations is limited by either the ultimate bearing capacity divided by a reasonable safety factor or the pressure that causes the tolerable settlement. Generally, the settlement criteria will control the design for both sand and clay foundations.

Footings on Sand

The allowable bearing capacity for footings on sand can be determined from consideration of Eq. 270-2 for the net ultimate bearing capacity and of Eq. 283-1 for settlement. The ultimate bearing capacity requires determination of the friction angle $\bar{\phi}$ for the sand. Drained conditions normally apply for granular soils (except for shock or dynamic loads), and effective strength parameters would apply. These parameters can be measured by a *CD* triaxial shear test or direct shear test.

Because it is difficult to duplicate field conditions in the laboratory, a common way of evaluating $\bar{\phi}$ is from the results of a Standard penetration test. Figure 291-1 shows a relationship between N and $\bar{\phi}$. The value of N used in determining $\bar{\phi}$ from Fig. 291-1 should be the average N over a distance from the base of the footing to a depth of approximately B below the base of the footing. Equation 270-2 shows that the ultimate bearing capacity for a footing on sand is a function of the width and depth of the footing and the friction angle $\bar{\phi}$. An appropriate safety factor should be used with the ultimate bearing-capacity equation. Generally, a safety factor of 3 is used for dead load plus the normal or sustained live load, and 2 for dead load plus maximum live load. Equation 283-1 indicates that the allowable bearing pressure based on settlement is a function of the corrected standard

Figure 291-1 Relationship between standard penetration resistance N (corrected for overburden) and friction angle $\bar{\phi}$. (After Peck, Hanson, and Thornburn, 1974.)

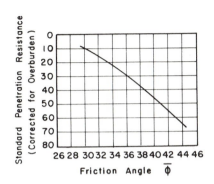

penetration resistance N. These criteria can be combined to produce an allowable bearing-pressure curve for a given footing depth, soil deposit, and allowable settlement. The example problem of Fig. 292-1 shows how to develop an allowable bearing-pressure curve for a footing on sand.

Footings on Clay

The design of shallow foundations on clay is frequently controlled by settlement criteria. For footings on relatively homogeneous clay deposits, the footing width generally has little effect on the magnitude of settlement. Therefore, if settlement controls, it is often necessary to select an alternative foundation system rather than simply increase the footing width. This might involve overexcavation with replacement fill, using a compensating foundation (one or more basement levels), a mat foundation, preloaded foundation, or perhaps a pile foundation.

Since the settlement computations for footings on clay are rather involved, a trial footing size is generally based on the ultimate bearing capacity with an appropriate factor of safety, and then settlement is checked.

Soft clay layers often occur below an overlying stratum of sand. For these cases it is generally necessary to check both the settlement of the clay layer and the ultimate bearing capacity of the clay.

Allowable Bearing Pressure

Figure 292-1

Example Problem   Develop an allowable bearing pressure chart for continuous footings on sand. The average corrected N value to a depth in excess of 5 m is 20. The water table is at a depth of 7 m, the footing is to be founded at a depth of 0.5 m below ground surface, and the unit weight of the sand is 18 kN/m³. Use a safety factor of 2 with respect to an ultimate bearing capacity failure and limit the settlement to 25 mm. Neglect the depth factor. Consider ultimate bearing capacity.

$$q_{ult} = D_f\gamma(N_q - 1) + \frac{\gamma}{2}BN_\gamma$$

From Fig. 291-1 for $N = 20$,

$$\bar{\phi} = 33°$$

$$N_\gamma = 35$$

$$N_q = 26$$

$$q_a = \frac{1}{2}[0.5(18)(25) + \frac{18}{2}(35)B]$$

$$q_a = 112 + 158B \text{ kPa} \tag{a}$$

Consider settlement.

$$q_a = 0.41N(\Delta H)$$

$$q_a = 0.41(20)(25)$$

$$q_a = 205 \text{ kPa} \tag{b}$$

The allowable bearing pressure chart can now be developed by plotting Eqs. a and b as shown in the figure.

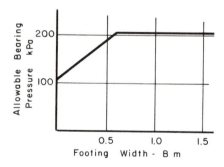

As shown on the allowable bearing pressure curve for this case, the ultimate bearing capacity would only control for a footing less than 0.25 m wide. Even for very lightly loaded structures, a minimum footing width of about 0.5 m is used.

In evaluating the ultimate bearing capacity of a clay deposit, the question arises as to the appropriate shear strength parameter and the type of shear test to use. On saturated clays the time of loading is usually rapid compared with drainage rate, and the undrained condition applies. Thus, use the $\phi = 0$ analysis with the cohesion c determined by unconfined compression, UU triaxial shear, or vane shear tests.

The approximate safety factors to use with ultimate bearing capacity of footings on clay are generally the same as for footings on sand. For dead load plus maximum live load use 2 and for dead load plus normal load use 3.

Footings on Silt

Silty soils fall into two categories.

1. Nonplastic silt.
2. Plastic silt.

In general, a nonplastic silt can be treated as a sand in evaluating the allowable bearing pressure for footing design. Plastic silt should be treated as a clay soil.

Allowable Settlements

Large settlements may affect the appearance of buildings. Even when no structural damage occurs, the vertical or horizontal alignment of the building may not fit the surroundings because of excess settlements. Large settlements may also interfere with the function of the building in unusual cases—for example, buildings that contain sensitive machinery or that are used for technical functions such as radar. Structural failure of building members is another problem caused by excessive settlement.

Settlements that allow damage to buildings have been correlated with maximum settlement, differential settlement, and slope of the settlement curve $d(\Delta H)/dx$. Data compiled by Grant et al (1974) for a relatively large number of different types of buildings on varied foundations seem to indicate that there is a correlation between maximum settlement (ΔH_{max}) and maximum slope of the deflection curve and also between differential settlement (δ_{max}) and maximum slope of the deflection curve. These relations may be stated as follows.

Clay	**Sand**
$\Delta H_{max} = 1200 \dfrac{d(\Delta H)}{dx}$	$\Delta H_{max} = 600 \dfrac{d(\Delta H)}{dx}$
$\delta_{max} = 650 \dfrac{d(\Delta H)}{dx}$	$\delta_{max} = 350 \dfrac{d(\Delta H)}{dx}$

Allowable Bearing Pressure

Figure 294-1

Type of Building or Situation	Allowable Slope of the Deflection Curve $d(\Delta H)/dx$
High brick walls	0.0008
One-story brick buildings	0.0015
Reinforced concrete buildings	0.003
Continuous steel frames	0.002
Simple steel frames	0.005
Cracking in panel walls	0.006
Buildings with sensitive machinery	0.001
Smoke stacks and water towers	0.004

Since these correlations exist, it is possible to use any one of the three variables as the criterion for specific allowable settlement. The most common and easy criterion is maximum settlement. The maximum settlement is a satisfactory criterion for small jobs or jobs with limited foundation investigation. For larger jobs, it may be more appropriate to use a thorough investigation of the character of the soil and its variation and to specify allowable settlements in terms of differential settlement or slope of the deflection curve. Some limiting values of the slope of the deflection curve for different types of buildings and situations that have been extracted from the current literature are listed in Fig. 294-1.

Rate of settlement is another important factor that must be considered in possibilities of damage due to excessive settlement. The tolerable slopes of the deflection curve listed in Fig. 294-1 hold primarily for soils that contain clay and have moderately fast settlements. Buildings on thick clay with very slow settlement appear to be able to tolerate greater values of $d(\Delta H)/dx$, and buildings on sand where the settlement may occur during the construction period can also tolerate greater values of the slope of the deflection curve.

Presumptive Bearing Capacity

Some building codes specify allowable bearing pressures that can be used for proportioning footings. These allowable bearing pressures are referred to as presumptive bearing capacities and in some instances are based on precedent from many years of experience. Allowable bearing pressures are specified in the Uniform Building Code (UBC), which has been adopted in many cities. Selection of an allowable bearing pressure from the Uniform Building Code would not be based on local precedent. The code does make provision, however, for the local building official to require a more extensive subsurface investigation.

Allowable bearing pressures as specified in codes are generally based on a visual classification of the near-surface soils. These allowable pressures, therefore, neglect important considerations such as the shape, width, and depth of the footing; the location of the water table; settlement and the stress history of the soil deposit. In most cases presumptive bearing-capacity values are conservative; in some cases they can be too high (unconservative).

Presumptive bearing capacities should generally be used only for preliminary purposes or for minor structures where there is a knowledge of the general subsurface conditions in the area.

DEEP FOUNDATIONS

Piles and drilled piers are the most common types of deep foundations. Both pile and drilled pier foundations are used to transfer surface loads to deep load-bearing strata when the surface soils are not capable of supporting the applied loads. The mechanism of deriving support is essentially the same for both systems. The main difference in pile and pier foundations is in the method of installation. Piles are installed by driving the pile or pile casing into place with an impact or vibratory pile-driving hammer. A number of pile-driving hammers are available, and selection of the appropriate type and size depends on the piles that are being used and the subsurface conditions. Drilled piers are installed by drilling or excavating a cylindrical hole to the desired depth and then backfilling the hole with concrete. The hole may or may not be excavated with a casing. If a casing is used, it is sometimes withdrawn as the concrete is placed. In some cases the bottoms of drilled piers are enlarged or belled to increase the end-bearing capacity of the pier.

Since piles and piers both derive support by essentially the same means, the remaining discussion is devoted to pile foundations. However, much of the discussion also applies to pier foundations.

Piles generally obtain support from a combination of friction along the surface of the pile shaft and from end bearing at the bottom of the shaft as shown in Fig. 296-1. In many cases, however, either the friction component or the end-bearing component will be predominant and the pile will be referred to as a friction pile or as an end-bearing pile.

Applied loads are generally transmitted to piles through reinforced concrete pile caps. A group of piles are usually embedded into each pile cap

Figure 296-1 Support components for piles and piers.

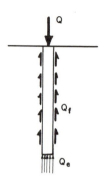

and the group supports the applied loads. In evaluating the capacity of the foundation system, it will, therefore, be necessary to consider both individual and group action of the pile system. Whether individual action or group action controls will depend on the subsurface conditions, the spacing of the piles, and the structural capacity of the piles. Settlement of the pile group must also be considered in evaluating group capacity.

Estimating Individual Pile Capacity

The capacity of the individual piles can be estimated by static formulas, by dynamic formulas, or by pile load tests. Static formulas are used to estimate the capacity of an individual pile on the basis of the strength properties of the soil. Dynamic formulas are generally used to establish pile-driving criteria for pile installation. By far the most reliable method of estimating individual pile capacity is through the use of pile load tests. A test pile is driven and loaded to failure. The results of the test are then used to establish both pile capacity and pile-driving criteria.

When designing a pile foundation, consideration must also be given to the structural design of the pile. The pile must be able to resist not only the stresses imposed by the structural loads, but also the handling and driving stresses to which the pile will be subjected during construction.

Static Formulas

The ultimate load capacity of piles can be estimated by calculating the resistance derived from the end-bearing and friction components of the total

                                8 Bearing Capacity and Foundations

pile capacity. The ultimate pile capacity can, therefore, be expressed as

$$Q_{ult} = Q_e + Q_f \qquad (297\text{-}1)$$

where

Q_{ult} = ultimate pile capacity
Q_e = end-bearing capacity
Q_f = side friction capacity

As with shallow foundations, a safety factor must be applied to the ultimate pile capacity in arriving at an allowable pile load.

The end-bearing component of pile capacity can be estimated by a method similar to that for estimating the ultimate bearing capacity of shallow foundations. Although the end-bearing failure surface is somewhat different for deep foundations than for shallow foundations, the form of the equation for end-bearing capacity is the same as Eq. 270-2.

$$Q_e = q_{ult}\,(\text{Area}) = \left[cN_c + \gamma D_f(N_q - 1) + \frac{\gamma}{2}BN_\gamma\right](\text{Area}) \qquad (297\text{-}2)$$

Since the depth of a pile foundation is much greater than the diameter of the individual piles, the N_γ term of Eq. 297-2 is small compared to the N_q term. Therefore, the ultimate end-bearing capacity of an individual pile is generally expressed as

$$Q_e = [cN_c + \gamma D_f(N_q - 1)]\,(\text{Area}) \qquad (297\text{-}3)$$

The bearing-capacity factors N_c and N_q must be adjusted somewhat to account for the shape of the pile (usually circular or square) and the high depth to width ratio. When shape and depth are taken into account, N_c is approximately 6. It is generally agreed that for granular soils the value of N_q should be much higher for pile foundations than for shallow foundations, but there is a wide range in the recommended values for N_q. This wide range in N_q values is illustrated in Fig. 298-1 (Vesic, 1967).

For piles that are end bearing in clay, the undrained strength of the clay is usually appropriate for analysis ($\phi = 0$ concept). In granular soils the cohesion, c, is zero. Since granular soils are generally free draining, pore pressures can dissipate rapidly and, therefore, the drained strength is appropriate.

The friction component of pile capacity can be estimated by considering the adhesion and friction between the soil and the pile.

For clay soils the friction component is treated as an adhesion between the pile shaft and the soil. It is generally calculated as the undrained shear strength times the surface area of the pile. Since shear strength properties generally vary with depth, it is usually necessary to divide the pile into

Deep Foundations

Figure 298-1 Bearing capacity factor for circular deep foundations. (After Vesic, 1967.)

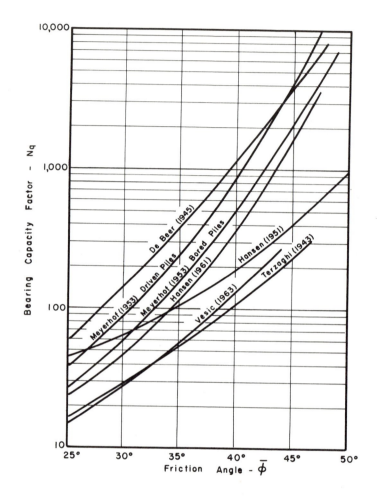

vertical increments and to express the frictional component of resistance as the sum of the resistance of each increment. This can be expressed as

$$Q_f = \Sigma \pi \, d_i \ell_i S_{ui} \qquad (298\text{-}2)$$

8 Bearing Capacity and Foundations

where

d_i = pile diameter of increment i
ℓ_i = length of increment i
S_{ui} = undrained shear strength of clay in increment i

For sandy soils the friction component of pile resistance depends on the coefficient of friction between the soil and the pile and the horizontal intergranular stress adjacent to the pile surface. The intergranular stress will increase with depth, and it is necessary to use the average stress. The horizontal stress at any depth is equal to the vertical intergranular stress times a lateral earth pressure coefficient, K. The earth pressure coefficient depends on the density of the soil deposit and the method of driving the pile. For piles that are driven into place (displacement piles), the lateral earth pressure coefficient will generally be between 1.0 and 3.0. The friction component of the pile resistance for sand soils can be expressed as

$$Q_f = \Sigma\pi\, d_i\ell_i K\, \bar{\sigma}_{vi} \tan \delta_i \qquad (299\text{-}1)$$

where

K = lateral pressure coefficient
$\bar{\sigma}_{vi}$ = average vertical intergranular stress over increment i
$\tan \delta_i$ = coefficient of friction between the soil and the pile over increment i

The example problem of Fig. 306-1 illustrates how to compute the ultimate resistance of a pile.

Dynamic Formulas

The load capacity of a pile is often estimated from the resistance of the pile to penetration during driving. Several dynamic pile formulas have been developed on the basis of the following energy consideration: Energy input from the pile hammer is equal to the energy used to drive the pile plus energy losses. This can be stated as

$$E = Ps + \Delta E \qquad (299\text{-}2)$$

where

E = energy delivered to the pile by the pile-driving hammer
P = penetration resistance of the pile
s = movement of the pile tip
ΔE = energy losses

The major difference in most of the dynamic pile formulas is in the energy loss term. The energy loss is basically the result of heat and elastic

Deep Foundations

deformation of the pile. However, the dynamic formulas neglect the time-dependent behavior of the elastic deformations. In other words, the elastic deformations are assumed to occur at the time of impact of the hammer on the pile rather than in the form of elastic compression and tension waves that propagate through the pile and soil.

A further problem with dynamic formulas is that the friction resistance of piles driven into clay is much less during driving than after the pile has been driven and allowed to set for several days. This phenomenon for piles driven in clay stems from the fact that the remolded strength of clay is much less than the undisturbed strength. Driving a pile into clay remolds the clay adjacent to the pile. Several days after driving, the clay regains some strength as·a result of the thixotropic properties of clay, and the static capacity is much higher than the dynamic capacity that was observed during driving.

In spite of the many shortcomings of pile-driving formulas, they are widely used. If used with an understanding of their limitation, they can provide a useful guide for establishing pile-driving criteria.

Most foundation engineering textbooks present several dynamic pile-driving formulas based on Eq. 299-2.

The Wave Equation

The major problem with dynamic pile-driving formulas is that they do not simulate soil-pile interaction or the time-dependent nature of the problem. The impact of a pile-driving hammer produces elastic deformations in a pile in the form of elastic compression and tension waves that propagate through the pile. These pile deformations interact with the soil both along the surface of the pile and at the pile tip. In addition to the characteristics of the pile and soil, the pile-driving accessories have an influence on pile-driving behavior. Pile-driving accessories include the hammer, anvil, capblock, pile cap, and cushion, and must be selected for each pile-driving job so that a pile can be driven to the required depth without excessive damage to the pile.

Smith (1960) presented a method of modeling the complete pile-driving operation, including pile-driving accessories, soil-pile interaction, and the time-dependent nature of the elastic pile deformations. The model presented by Smith was based on the propagation of an elastic wave through a long rod. The model can be described by a partial differential equation that is solved numerically. The idealized model of the physical system used to obtain the numerical solution is shown in Fig. 301-1.

With the development of computers, a numerical solution of the wave equation became realistically possible, and it is now generally accepted and used in geotechnical engineering practice. A wave equation analysis of a pile system can be used to select pile-driving equipment, to predict pile-bearing capacity, and to establish pile-driving criteria.

Figure 301-1 Pile representation for wave equation analysis.

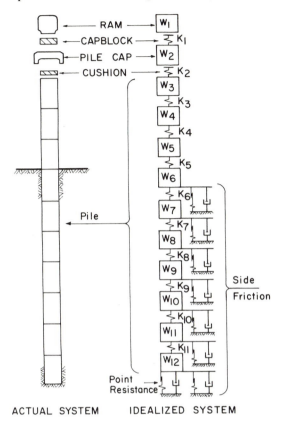

ACTUAL SYSTEM   IDEALIZED SYSTEM

Pile Load Tests

The most reliable method of determining the capacity of a pile or drilled pier is by a load test. A test pile or pier is installed using the procedure proposed for construction, and it is then loaded to a near-failure condition. Accurate records are maintained during installation and during the load test. Although a load test is rather expensive, on large projects substantial economic benefits can be derived from the load tests. Savings can be achieved by avoiding an overconservative design and by determining whether or not the design pile lengths are correct. Even on some smaller projects, pile load tests can be justified by performing them at the beginning of construction.

Deep Foundations

Figure 302-1 Load testing for pile capacity.

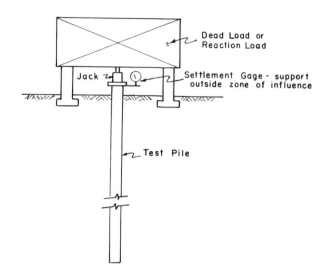

Dead Load or
Reaction Load

Jack

Settlement Gage - support
outside zone of influence

Test Pile

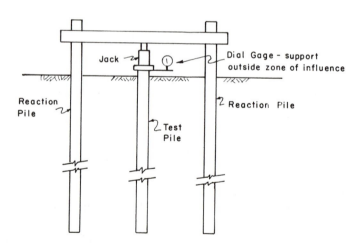

Jack

Dial Gage - support
outside zone of influence

Reaction
Pile

Reaction Pile

Test
Pile

8 Bearing Capacity and Foundations

In addition to verifying the support capacity of the pile at the design elevation, the construction driving criteria can be established by correlating the driving records for the test pile with the measured load capacity of the pile. This correlation can then be used with one of the dynamic pile-driving formulas or with the wave equation to establish the driving specification. The friction and end-bearing components of the pile support capacity can be separated during a pile load test by running a tension test on the pile. Only the friction component is capable of resisting a tension force.

After the pile has been installed, the load test is performed by applying an axial load to the top of the pile in increments. This is usually accomplished by jacking the top of the pile against a dead load that has been positioned over the pile. The jacking reaction can also be provided by tension piles that have been driven adjacent to the test pile. Two possible methods of applying the test load are illustrated in Fig. 302-1.

Friction piles driven into saturated clay should not be tested until some time has elapsed to allow the remolded clay to regain strength.

Capacity of Pile Groups

Individual structural loads are generally supported by several piles acting as a group. The structural load is applied to a pile cap that distributes the load to the piles, as shown in Fig. 304-1. If the piles are spaced a sufficient distance apart, then the capacity of the group is the sum of the individual capacities. However, for closely spaced friction piles (where the end-bearing component is small compared to the friction component), the capacity of the piles may be limited on the basis of group action, and the capacity may be much less than the sum of the individual pile capacities. Group action is evaluated by considering the piles to fail as a unit around the perimeter of the group. Both end bearing and friction are considered in evaluating the group capacity. End bearing is evaluated by considering the area enclosed by the perimeter of the pile group as the area of a footing located at a depth corresponding to the elevation of the pile tips. The friction component of pile support is evaluated by considering the skin friction that can be generated around the perimeter of the group over the length of the piles. These two components are illustrated in Fig. 304-1.

Settlement of Pile Groups

The settlement of pile groups must also be considered. Methods have been developed to evaluate the settlement of pile groups based on the theory of elasticity (Poulos, 1968). Meyerhof (1976) discusses methods for predicting the settlement of pile groups in both sand and clay soils. The same general

Deep Foundations   **303**

Figure 304-1 Load capacity of pile group.

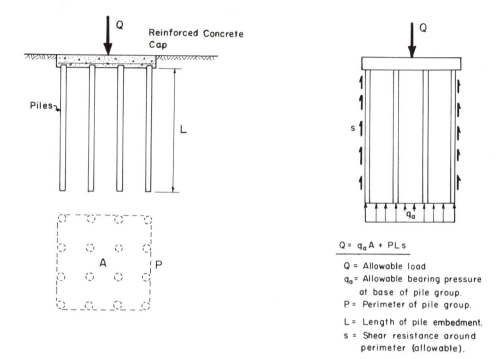

$$Q = q_a A + PLs$$

Q = Allowable load
q_a = Allowable bearing pressure
at base of pile group.
P = Perimeter of pile group.

L = Length of pile embedment.
s = Shear resistance around
perimeter (allowable).

methods presented for shallow foundations are used to predict the settlement of deep foundations. Settlement in clay soils is generally more troublesome than in sandy soils and, therefore, is discussed below.

The most common way of performing a settlement analysis on clay soil is based on the principles presented in Chapter 4. For pile groups that are essentially end bearing on a very stiff clay or for pile groups on a dense sand stratum that is underlain by a much softer clay, the load is assumed to be applied at the pile tips as a uniform load over the area of the group. The stress increase below the tips can be evaluated using the methods of Chapter 3. A simplified approach is illustrated in Fig. 305-1 and in Fig. 96-1, and the settlement of the softer clay is then computed on the basis of this stress increase. For friction piles in clay, the load is assumed to be applied as a uniform load at a depth of about two-thirds the length of the piles, as shown in Fig. 305-2.

8 Bearing Capacity and Foundations

Figure 305-1 Stress distribution for settlement analysis.

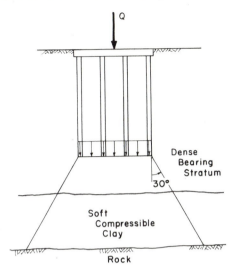

Figure 305-2 Stress distribution for settlement analysis.

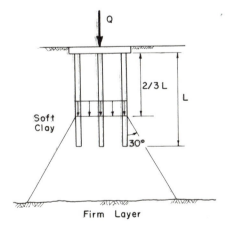

Figure 306-1

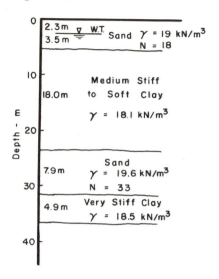

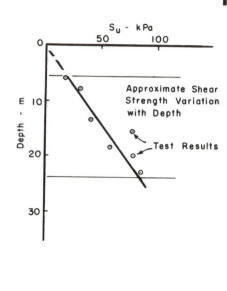

Example Problem A 10-story building of steel frame construction is proposed for a given site. Loads will be transferred to the foundation system through steel columns. The column loads will be approximately 5000 kN per column.

A subsurface investigation at the site revealed the above idealized soil profile shown. The medium stiff to soft clay soil is slightly over-consolidated. The undrained shear strength was determined at various depths and the results are plotted as shown. Standard penetration tests were performed in the various sand strata, and the average corrected value for each stratum is shown on the soil profile. Assume the sand above the water table is saturated also.

A pile foundation is being considered as one possible foundation scheme. Assume that the bottom of the pile cap would be at a depth of 1.2 m below ground surface.

Determine the single *ultimate* pile capacity for a 0.3-m diameter pipe pile driven to a depth of 13 m.

Solution First consider the pile driven to a depth of 13 m. The top of the pile will be at a depth of 1.2 m and will develop skin friction in the sand to a depth of 5.8 m. The clay will provide skin friction over the remaining 7.2 m of the pile length as well as some end-bearing capacity. The ultimate pile capacity for this problem can be stated from Eq. 297-1 as

$$Q_{ult} = Q_{f(sand)} + Q_{f(clay)} + Q_{e(clay)}$$

Equation 299-1 can be used to express the support provided by skin friction in the sand.

$$Q_f = \Sigma \pi d_i \ell_i K \bar{\sigma}_{vi} \tan \delta_i$$

Use

$$K = 1.0 \qquad \text{(assume the minimum value)}$$

and

$$\delta_i = 2/3\bar{\phi}$$

From the average corrected value of N and Fig. 291-1,

$$\phi = 33°$$
$$\delta = 22°$$
$$\ell_i = 5.8 \text{ m} - 1.2 \text{ m} = 4.6 \text{ m}$$

The average intergranular stress over the portion of the pile embedded in the sand is

$$\bar{\sigma}_{vi} = 2.3(19.0) + 1.2(19.0 - 9.81)$$
$$\bar{\sigma}_{vi} = 54.7 \text{ kPa}$$
$$Q_{f(\text{sand})} = \pi(0.3)(4.6)(1.0)(54.7) \tan 22°$$
$$Q_{f(\text{sand})} = 95.8 \text{ kN}$$

Equation 298-2 can be used to express the support provided by skin friction in the clay.

$$Q_f = \Sigma \pi d_i \ell_i S_{ui}$$
$$\ell_i = 13.0 - 5.8 = 7.2 \text{ m}$$

Evaluate the undrained shear strength at a depth corresponding to the mid-point of ℓ_i.

$$\text{Midpoint depth} = 5.8 + \frac{7.2}{2} = 9.4 \text{ m}$$
$$S_u = 33 \text{ kPa}$$
$$Q_{f(\text{clay})} = \pi(0.3)(7.2)(33)$$
$$Q_{f(\text{clay})} = 223.9 \text{ kN}$$

The end-bearing capacity of the pile can be evaluated using Eq. 297-3.

$$Q_e = [cN_c + \gamma D_f(N_q - 1)] \text{ (Area)}$$

for $\phi = 0$

$$N_q = 1$$
$$N_c = 6$$
$$c = S_u$$
$$\text{Area} = \pi(0.15)^2 = 0.0707 \text{ m}^2$$

Deep Foundations

Evaluate S_u at a depth corresponding to the depth of the pile tips.

$$\text{Depth to pile tips} = 13 \text{ m}$$
$$S_u = 47 \text{ kPa}$$
$$Q_e = 47(6)(0.0707) = 19.9 \text{ kN}$$
$$Q_{ult} = 95.8 + 223.9 + 19.9 = 340 \text{ kN}$$

Determine the *ultimate* capacity of a pile group consisting of 16 piles driven to the depth indicated above. Assume a pile spacing of 0.75 m.

$$b = 3(0.75) + 0.3 = 2.55 \text{ m}$$
$$\text{Perimeter} = 4(2.55) = 10.2 \text{ m}$$
$$\text{Area} = 2.55^2 = 6.5 \text{ m}^2$$
$$Q_{ult} = Q_{f(sand)} + Q_{f(clay)} + Q_e$$
$$Q_{f(sand)} = 10.2(4.6)(1.0)(54.7) \tan 22°$$
$$Q_{f(sand)} = 1037 \text{ kN}$$
$$Q_{f(clay)} = 10.2(7.2)(33) = 2424 \text{ kN}$$
$$Q_e = 47(6)6.5 = 1833 \text{ kN}$$
$$Q_{ult} = 1037 + 2424 + 1833 = 5294 \text{ kN}$$

Compare this value with the *ultimate* capacity of the group based on individual pile capacity.

$$Q_{ult} = 16(340) = 5440 \text{ kN} > 5294 \text{ kN}$$

Therefore, group capacity governs.

PROBLEMS

8-1 Show that Prandtl's equation for N_c (page 267) reduces to 5.14 when ϕ equals zero. (Use L'Hospital's rule.)

8-2 Plot a curve showing the ultimate bearing capacity as a function of footing width for a strip footing at the surface for:

(a) A granular soil—$c = 0$, $\bar{\phi} = 30°$, $\gamma = 17 \text{ kN/m}^3$.
(b) A cohesive soil—$\phi = 0$, $c = 60 \text{ kN/m}^2$.

8-3 What is the net ultimate bearing capacity for a square footing at a depth of 1 m for:

(a) A granular soil? $c = 0$, $\bar{\phi} = 30°$, $B = 2 \text{ m}$, $\gamma = 18 \text{ kN/m}^3$.
(b) A saturated cohesive soil? $\phi = 0$, $c = 60 \text{ kN/m}^2$, $B = 2 \text{ m}$.

8-4 Determine the net ultimate bearing capacity for the two soils given in problem 8-3 if the water table will be at:

(a) A depth of 0.5 m.
(b) A depth of 1 m.

$$\gamma_{sat} = 20 \text{ kN/m}^3$$

8-5 Plot the net ultimate bearing capacity as a function of ϕ for a strip footing 1 m wide and 1 m deep in a c-ϕ soil, where $c = 50 \text{ kN/m}^2$. Use ϕ from zero to 36°. $\gamma = 17.5 \text{ kN/m}^3$.

8-6 Plot the ultimate bearing capacity as a function of c for a strip footing 1 m wide and 1 m deep and $\phi = 15°$. Use c from 0 to 100 kN/m². $\gamma = 17.5 \text{ kN/m}^3$.

8-7 Use Eq. 270-2 with correction factors to determine the size of a square footing at a depth of 1 m in a cohesive soil to carry a vertical load of 600 kN with a safety factor of 3. $q_u = 120 \text{ kN/m}^2$ and $\phi = 0$.

8-8 What is the safety factor for a 1.5 m² footing at a depth of 1 m if the vertical load is 1000 kN and the horizontal load is 200 kN applied at the base of the footing? The soil is granular with $\bar{\phi} = 34°$, and $\gamma = 19 \text{ kN/m}^3$.

8-9 Using Eq. 270-2, determine the allowable load per meter of length of a continuous footing at a depth of 2 m in a sand. Use a factor of safety of 3.

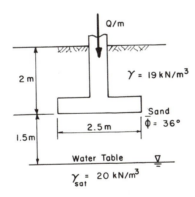

8-10 The unconfined compressive strength of a relatively stiff clay is 150 kN/m². Determine the size of a square footing to the nearest 0.1 m to carry a 300 kN concentric load. Use a safety factor of 3. The footing is 1.5 m deep. Assume the load to be rapidly applied so that undrained conditions prevail ($\phi = 0$).

Problems

8-11 What size should the footing in problem 8-10 be if the load has an eccentricity of 1/6 m?

8-12 Determine the size of a square footing on a granular soil to carry a 1300 kN concentric load as determined by an allowable total settlement of 25 mm. Assume the corrected blow count of the sand is 25 blows per foot. The depth of the footing is 1 m.

8-13 The standard penetration test shows a blow count of 30 blows per foot at a depth of 8 m in a granular deposit. The site will be excavated and a footing located near this depth. What is the corrected blow count to use for design based on total settlement? $\gamma = 17.5$ kN/m^3.

8-14 The corrected blow count for the sand is 20 blows per foot. Plot the allowable bearing capacity versus the width of a strip footing. The allowable bearing capacity is to be determined with a safety factor of 2.5 or by an allowable total settlement of 25 mm, whichever governs. The footing is 0.5 m deep and the site will be completely submerged. $\gamma = 19.5$ kN/m^3.

8-15 For the soil profile given in the example problem of Fig. 306-1, evaluate the ultimate load-carrying capacity of a 0.3 m diameter pipe pile driven 0.5 m into the deep sand stratum.

8-16 For the 10-story structure and the soil profile of the example problem of Fig. 306-1, determine how many piles driven to a depth of 17 m will be required to support the 5000 kN column load. What would be the minimum pile spacing so that group action does not control? Use a factor of safety of 2.

8-17 Estimate the settlement of the pile group in problem 8-16 if the value of C is 0.15 for the upper clay layer.

9

IMPROVING
SOIL CONDITIONS
AND PROPERTIES

Because of the high cost of importing large quantities of suitable material, it is generally economical to use local soil for fills and foundations. Yet as more engineering structures are built, it becomes increasingly difficult to find a site with suitable soil properties. The properties at many sites must be improved by the use of some form of static or dynamic loading, grading, drainage or by the use of admixtures. It is increasingly important for the engineer to know the degree to which soil properties may be improved and the cost involved in this improvement. The following sections present a brief introductory coverage of the current methods used to improve soil conditions and properties. Space limitations prevent a complete discussion of this topic. See the Bibliography reference to American Society of Civil Engineers, Specialty Conference (1968), for a comprehensive reading list.

COMPACTION

Soil compaction is the most obvious and simple way of increasing the stability and supporting capacity of soil. Compaction is defined as the process of increasing soil unit weight by forcing soil solids into a tighter state and reducing the air voids. This is accomplished by applying static or dynamic loads to the soil. The purpose of compaction is to produce a soil having physical properties appropriate for a particular project.

Compaction Theory

Standard compaction tests were developed in the 1930s by Proctor (1933). Two basic compaction tests, the standard Proctor compaction test and the modified Proctor compaction test, are in use today. Some details for these tests are shown in Fig. 313-1. ASTM or AASHTO standards should be consulted for other test details. The tests are conducted by compacting the moistened soil sample (at controlled water contents) into a mold in a specified number of layers. Each layer is compacted by a given number of blows of a hammer of specified mass and drop height. The compactive effort measured in terms of energy per unit volume of compacted soil is given in Fig. 313-1. The dry unit weight and water content of the soil in the mold are measured and plotted as in Fig. 314-1.

For a given compactive effort the dry unit weight of compacted soil will vary with the water content at the time of compaction. The curves shown in Fig. 314-1 are for a sandy silt. The water content at which the dry unit weight is a maximum is termed the optimum water content. The upper curve in Fig. 314-1 represents a greater compactive effort than was used for the lower curve. Compactive effort is measured in terms of energy per unit volume of compacted soil. For less compactive effort the compaction curve for the same soil is lower and is displaced to the right, indicating a higher optimum water content. It is useful to plot on the same graph the dry unit weight as a function of water content, assuming 100% saturation or zero air voids. This curve represents an upper bound for dry unit weight. Other curves for 80 or 90% saturation will plot below the zero air-voids curve, as shown in Fig. 314-1. The position of the compaction curve below the zero air-voids curve indicates that this particular soil has a saturation of about 90% when compacted to an above-optimum water content. This is typical of most soils. The unit weight curve for any given saturation can be plotted using Eq. 17-1e.

$$\gamma_d = \frac{G}{1+e}\, \gamma_w = \frac{G}{1+(wG/S_r)}\, \gamma_w$$

The compaction tests are designed to simulate the unit weight of soils compacted by field methods. For control of field compaction, construction specifications require that the dry unit weight of field soils be equal to or

Figure 313-1 Proctor Compaction Tests

Test	ASTM number	AASHTO number	Mold Volume		Hammer		Drop Height		Compactive Effort (Energy)
			(m³)	(ft³)	Mass (kg)	Weight (lb)	(m)	(ft)	
Standard Proctor	D 698-70	T-99	945×10^{-6}	1/30	2.5	5.5	0.3	1.0	25 blows/layer 3 layers (590 kJ/m³ or $12{,}375$ ft lb/ft³)
Modified Proctor	D 1557-70	T-180	945×10^{-6}	1/30	4.5	10.0	0.46	1.5	25 blows/layer 5 layers ($2{,}700$ kJ/m³ or $56{,}250$ ft lb/ft³)

*See standards for test details using a larger mold (213×10^{-3} m³, 1/13.33 ft³).

Figure 314-1 Compaction curve for a sandy silt.

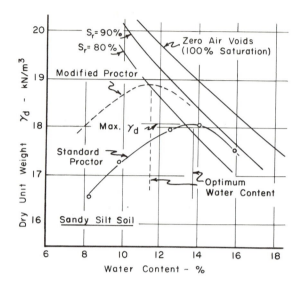

greater than a given percentage of the maximum dry unit weight obtained in one of the standard compaction tests (typically 90 to 100%). This dry unit weight can be obtained by controlling the field compaction effort and water content of the soil during compaction. The standard Proctor test is adequate compaction for most applications such as retaining wall backfill, highway fills, and earth dams. The modified Proctor test is used for heavier load applications such as airport and highway base courses.

Properties of Compacted Granular Soils

Compacting granular soils to a high unit weight increases the shear strength and reduces the compressibility, thus making it desirable to specify as high a unit weight as can reasonably be attained. Field control of the unit weight of granular soils with less than 12% fines is based either on the maximum dry unit weight from one of the standard compaction tests or on relative density. Relative density is defined by Eq. 22-1b.

$$D_r = \frac{\gamma_{d\ max}}{\gamma_d} \left(\frac{\gamma_d - \gamma_{d\ min}}{\gamma_{d\ max} - \gamma_{d\ min}} \right) 100$$

9 Improving Soil Conditions and Properties

Figure 315-1 Friction angle as a function of the relative density of sand.

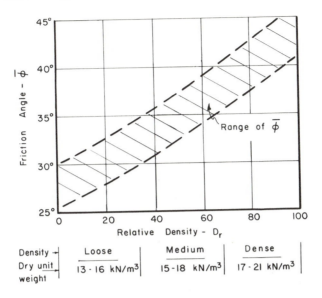

An approximate relationship between relative density and friction angle is shown in Fig. 315-1. This relationship was developed from the work of many researchers and includes results from shear tests on many different cohesionless soils. For example, the strength gain obtained from compacting cohesionless soils can be shown by observing the lower bound of the band in Fig. 315-1. Along the lower bound the friction angle increases from approximately 25° at zero relative density (soil in its loosest state) to approximately 41° at 100% relative density (very dense state).

Properties of Compacted Silts and Clays

Because the physical properties of granular soils are improved by compaction to the maximum dry unit weight, there is a tendency to assume that this applies to all soils. However, in the case of fine-grained soils the shear strength, compressibility, swelling potential, and permeability are not necessarily improved by compaction to a maximum unit weight. Establishing the optimum compaction conditions for a given soil may involve extensive testing. However, research has provided some guidelines. Many (but not all) fine-grained soils exhibit engineering properties related to particle orientation

Compaction

caused by the compaction conditions. The usual random soil structure is said to be "flocculated" (Fig. 31-1). Under certain conditions some soils can be forced toward greater particle orientation or a "dispersed" structure. Soils compacted at water contents dry of optimum tend to remain flocculated regardless of the method of compaction used. Soil compacted wet of optimum by methods that cause appreciable shear strain may have its structure dispersed or altered toward greater particle orientation. Compaction methods that do not produce appreciable shear strain will still leave the soil with a generally flocculated structure.

Consideration of the electrical environment surrounding fine-grained soil particles offers some explanation for these structural changes (Fig. 14-1). Soils compacted on the dry side of optimum do not have enough water to develop the double layer and the attractive forces (edge to face of particles) are relatively strong. This makes it difficult to change the flocculated structure to a dispersed structure by compactive effort. The relatively high capillary tension in the water films also restricts particle rearrangement, making it difficult to reach high dry unit weights at low water contents. With increasing moisture content the electric double layer becomes more fully developed. Repulsive electrical forces increase, permitting greater particle orientation, reduction in void size, and higher unit weight for the same compactive effort.

Higher unit weights are achieved by forcing air from the soil voids. With increasing water content, discontinuities develop in the air voids, making it more difficult to force air out and further reduce the air voids. This second effect becomes predominant at water contents over optimum, and the maximum unit weight that can be attained with higher water content decreases.

The question then is, how do these effects influence the engineering properties of compacted clays and silts?

Shearing Strength

First, consider the strength characteristics of compacted clays and silts. Figure 317-1 shows the strength of a silty clay as a function of the molding water content. The strength is measured at 5% strain. The five samples were compacted with the same compactive effort. The sample with maximum unit weight does not have the highest strength, and a large loss of strength is noted as the water content increases from just below to just above optimum. Some of the effects of compaction on shear strength are summarized here.

1. For a constant compaction water content, the shear strength will increase with increasing compactive effort until a critical degree of saturation is reached and then decrease rapidly with further increase in dry unit weight. The attempt to compact a soil on the wet side of optimum to a high degree of saturation results in a large strength loss.

9 Improving Soil Conditions and Properties

Figure 317-1 Strength of compacted samples of silty clay. (After Seed et al., 1959.)

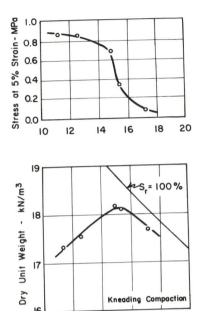

2. Comparing two samples of the same dry unit weight, the sample compacted dry of optimum (flocculated structure) will exhibit a higher strength at low strain than soils compacted wet of optimum (dispersed structure). At higher strain the initial soil structure is destroyed and the ultimate strength is about the same regardless of the molding water content.

3. If two samples are compacted to the same dry unit weight, one dry of optimum and the other wet of optimum, and then soaked without volume change to the same water content near saturation, the sample compacted dry of optimum (flocculated structure) will show greater strength.

4. The combination of conditions that will produce the greatest strength varies widely with the soil type and must be determined by a series of tests.

Compaction

5. On the wet side of optimum, kneading compaction methods produce a greater tendency toward a dispersed structure than do static compaction methods and would, therefore, tend to cause smaller values of shear strength.

Compressibility and Volume Change
Some of the effects of compaction on the compressibility of silts and clays are summarized here.

1. Soils compacted on the dry side of optimum tend to shrink less on drying and swell more when soaked than those compacted on the wet side of optimum.
2. The compressibility of a compacted cohesive soil increases, not only with an increase in liquid limit but also with an increase in the degree of saturation and with a decrease in dry density.
3. Soils subjected to a pressure that causes greater internal stresses than the compaction process caused will have a high degree of compressibility, probably due to additional breakdown in the structure under the higher pressure.
4. The compressibility characteristics for soils compacted on the wet side of optimum are influenced heavily by the type of compaction used. This is apparently not a factor on soils compacted dry of optimum. For soils compacted wet of optimum, the kneading and impact types of compaction that tend to create a dispersed structure also affect most greatly the compressibility property, and soils compacted wet of optimum using kneading methods have higher compressibility than those which have been compacted using other methods.

Permeability
A soil compacted wet of optimum (dispersed structure) exhibits a much lower permeability than a similar sample compacted dry of optimum (flocculated structure) to the same dry unit weight. The ratio of permeabilities can exceed 100.

Compaction of Expansive Soils
It is sometimes necessary to use expansive soils as subgrade or foundation material. Expansive soils may be identified by their high PI and fine particle size. If a soil has these properties and also has an expanding crystalline structure (montmorillonite) and an availability of water, swelling is a distinct problem.

Swelling in expansive soils can be partially controlled in the following ways.

- Raise the water content of the soil to a high value before construction and keep it constant. This is sometimes difficult to accomplish, especially under structures where evaporation is restricted and water flow is uncontrolled.
- Design structures so that the soil stresses are greater than the swelling pressure in the soil.
- Stabilize the soil by chemical injection, grouting, or compaction. If compaction is used, care must be taken not to overcompact. Overcompaction may occur if the optimum moisture content is used and if the compactive effort is too great. Overcompacted expansive soils may swell greatly if water is added later. Proper compaction would include using a moisture content considerably greater than optimum and relatively small compactive energy. Under these conditions the soil will have a lower bearing capacity, but this may be more tolerable than a high swelling potential.

Summary
With an understanding of the properties of compacted silts and clays, an engineer can select the appropriate compaction conditions to produce a material of the desired engineering properties. Some examples of problems where judgment is necessary follow.

1. Where a compacted soil is used under a pavement, the problem of selecting the appropriate compaction condition becomes complex. Water will tend to accumulate in the soil and approach a saturated condition. The soil may tend to swell if the surcharge is not high enough to prevent it. Both the increase in water content and reduction in unit weight by swelling lead to a reduction in strength.
2. For compacted soil under the floor of a building, high bearing capacity may not be as important but expansion would be detrimental. Expansion would be minimized by compacting at a water content and unit weight to produce a relatively high degree of saturation yet not so high as to reduce the strength below an adequate level.
3. For the case of an earth dam, compaction on the wet side would produce a more impermeable material with plastic stress-strain characteristics. A plastic core is desirable to prevent cracking, particularly in situations where large foundation settlements are anticipated. Here again a balance must be made with the resultant lower strength and potential pore pressure problems caused by using a higher initial water content.
4. The shell of an earth dam should be compacted on the dry side of optimum for the dual advantage of greater strength and greater permeability.

Field Equipment for Compaction

Several types of equipment for compacting soils are commercially available. The choice of the proper equipment will depend primarily on the type of soil and economic considerations. Generally, some type of vibratory equipment will work most efficiently in granular noncohesive soils, whereas cohesive soils respond best to equipment that penetrates the layer to be compacted. Some of the most common types of field equipment are discussed below.

Tampers

Hand-held tampers operated by compressed air or gasoline power are commonly used to compact all types of soil adjacent to structures or confined areas where use of larger compaction equipment is impractical. The tamper is repeatedly thrown into the air about 0.3 m and falls to the soil surface to compact the soil on impact.

Vibrating Plate

Hand-held vibrating plates operate efficiently on clean granular soils, but also achieve satisfactory compaction on some fine-grained soils. Vibrating plates are also gang-mounted on machines for use in less restricted areas.

Figure 320-1 Smooth-drum roller.

9 Improving Soil Conditions and Properties

Figure 321-1 Pneumatic-tired roller.

Smooth Drum Roller
Smooth drum rollers (Fig. 320-1) are used most often for finishing operations, on fills, or in highway construction. They are not suitable for producing uniform compaction in deep layers and, when used on cohesive soils, they may produce a density stratification with high densities at the surface of each layer.

Pneumatic-tired Roller
The action produced by the pneumatic-tired roller (Fig. 321-1) is somewhat better than the smooth drum roller in that it produces a combination of

Compaction

Figure 322-1   Sheepsfoot roller

pressure and kneading action on the soil. Rubber-tired rollers are effective for a wide range of soils from clean sand to silty clay. Rollers of this type range in mass up to 200 metric tons (t) and are available in a wide variety of wheel configurations, tire pressures, and wheel loads.

Sheepsfoot Roller

The sheepsfoot roller (Fig. 322-1) consists of a drum with a large number of projections or feet that penetrate into the soil layer during the rolling operations. With the first passes these projections penetrate through the layer to compact the lower portion. In successive passes compaction is obtained in the middle and top sections of the layer. The high concentration of stress under the projections makes this roller particularly suitable for compacting cohesive soils when bonding of lifts is important. The lift thickness should be small enough that the roller feet will penetrate to the lower section of the layer on the initial passes. The sheepsfoot roller is not suitable for compacting clean granular soils.

Vibratory Roller

Rollers with vibratory action (Fig. 323-1) are efficient in compacting granular soils. Accessory equipment to provide the vibration may be attached to

Figure 323-1 Vibratory smooth drum roller.

smooth-drum, pneumatic-tired, or sheepsfoot rollers. For clean granular soils the maximum compaction occurs at depths of 0.3 to 0.4 m, therefore, thick lifts may be used (D'Appolonia et al, 1969).

**Comparison of Laboratory
Compaction Methods with Field
Compaction Methods**

Laboratory compaction methods listed in order of increasing shear action are static, vibratory, impact, and kneading compaction. The standard Proctor compaction tests use the impact method, which simulates to some degree the action of the sheepsfoot roller. There is no field equivalent of static compaction that consists of pressing the soil into a mold with a given uniform pressure over the entire surface. The kneading compactor is a special device that repeatedly forces a small pressure foot into the soil sample at the controlled pressure to simulate the action of the sheepsfoot roller.

The rubber-tired roller and sheepsfoot roller cause large shear strain and would be effective in changing a wet cohesive soil from flocculated to a dispersed state. A smooth drum or smooth drum vibratory roller would produce less shearing action.

Compaction

STABILIZATION BY ADMIXTURES

The physical properties of soils can often be economically improved by the use of admixtures. Some of the more widely used admixtures include lime, portland cement, and asphalt and are discussed in the following section.

Lime Stabilization

In general, lime stabilization improves the strength, stiffness, and durability of fine-grained soils. In addition, lime is sometimes used to improve the properties of the fine fraction of granular soils. Lime has also been used as a stabilizer for soils in the base courses of pavement systems, under concrete foundations, on embankment slopes and canal linings.

Adding lime to soils produces a lower maximum density and a higher optimum moisture content than in the untreated soil. Moreover, lime produces a decrease in plasticity index.

Lime stabilization has been extensively used to decrease swelling potential and swelling pressures in clays. The addition of lime produces a high concentration of calcium ions in the double layer around the clay particles, hence decreasing the attraction for water.

Ordinarily the strength of wet clay is improved when the proper amount of lime is added. This improvement in strength is due partly to the decrease in plastic properties of the clay and partly to the pozzolanic reaction of lime with soil, which produces a cemented material that increases in strength with time. Lime-treated soils, in general, have greater strength and a higher modulus of elasticity than untreated soils.

Cyclic freezing and thawing cause loss of strength of lime-treated soils, but there is evidence of an independent healing action that may offset this, so there is no cumulative strength loss.

Care must be taken in the design of lime-treated soils. Strength and fatigue properties may be adversely affected if the incorrect amount of lime is used or the procedure for construction is improper.

Cement Stabilization

Portland cement and soil mixed at the proper moisture content has been used increasingly in recent years to stabilize soils in special situations. Probably the main use has been to build stabilized bases under concrete pavements for highways and airfields. Soil-cement mixtures are also used to provide wave protection on earth dams and, more extensively, as cores for entire sections of some smaller earth dams. Mitchell and Freitag (1959) have described three categories of soil cement.

1. Normal soil-cement usually contains 5 to 14% cement by volume and is used generally for stabilizing low-plasticity soils and sandy soils.

Figure 325-1 Strength properties of soil cement mixtures.

Property	Granular soils	Fine-grained soils
Unconfined compressive strength (kN/m^2)	(500 to 1000) × (cement content in percent)	(300 to 600) × (cement content in percent)
Cohesion	$c = 50 + 0.255 \times$ (unconfined compression strength), (kN/m^2)	
Friction angle	40–45 degrees	30–40 degrees
Flexural strength	(1/5 to 1/3) × (compressive strength)	
Modulus (compression)	7×10^3–35×10^3 MN/m^2	7×10^5–7×10^6 (kN/m^2)
Poisson's ratio	0.1–0.2	0.15–0.35

2. Plastic soil-cement has enough water to produce a wet consistency similar to mortar. This material is suitable for use as waterproof canal linings and for erosion protection on steep slopes where road-building equipment may not be used.

3. Cement-modified soil is a mix that generally contains less than 5% cement by volume. This forms a less rigid system than either of the other types, but improves the engineering properties of the soil and reduces the ability of the soil to expand by drawing in water.

Several different criteria have been used to design soil-cement mixes. Among these are compressive strength, reaction to freeze-thaw tests, and durability tests. Soil-cement designs may, in some cases, be made more efficient by use of admixtures such as sodium chloride or calcium chloride. These admixtures are especially effective in sandy soils or in fine-grained soils containing organic material in which proper setting of the soil-cement is retarded. Strength properties in soil-cement mixtures as a function of cement content are given in Fig. 325-1. It is apparent that with high percents of cement, compressive strength approaching that of a lean mix concrete can be produced in the soil cement.

Asphalt Stabilization

Bituminous materials (asphalts, tars, and pitches) are used in various consistencies to improve the engineering properties of soils. Mixed with cohesive soils, bituminous materials improve bearing capacity and soil

Stabilization by Admixtures

325

strength by waterproofing the soil and preventing high moisture content. Bituminous materials added to sand act as a cementing agent and produce a stronger, more coherent mass. The amount of bitumen added ranges from 4 to 7% for cohesive material to 4 to 10% for sandy material (Winterkorn and Fang, 1975). The primary use of bituminous materials is in road construction, where it may be the primary ingredient for the surface course or be used in the subsurface and base courses for stabilizing soils.

INJECTION AND GROUTING

Grouting has been extensively used primarily to control groundwater flow. Since the process fills soil voids with some type of stabilizing material, grouting is also used to increase soil strength and prevent excessive settlement.

Many different materials have been injected into soils to produce changes in the engineering properties of the soil. Injection may be accomplished in several ways. In one method a casing is driven and the injection is made under pressure to the soil at the bottom of the hole as the casing is withdrawn. In another method a grouting hole is drilled and at each level in which injection is desired, the drill is withdrawn and a collar is placed at the top of the area to be grouted and the grout is forced into the soil under pressure. Another method is to perforate the casing in the area to be grouted and leave the casing permanently in the soil. This allows for multiple injections over a period of time in the same soil mass (Caron et al, 1975).

Grout may be designed (1) to penetrate through the void system of the soil mass, or (2) to displace soil and improve the properties by densifying low-density soils or by restricting the flow of water by the formation of lenses of grout material in the soil.

Penetration grouting may involve portland cement or fine-grained soils such as bentonite or other materials of particulate nature. These materials penetrate only a short distance through most soils and are primarily useful in very coarse sands or gravels. Viscous fluids, such as a solution of sodium silicate, may be used to penetrate fine-grained soils. Some of these solutions form gels that restrict permeability and improve compressibility and strength properties.

Displacement grouting usually consists of using a groutlike portland cement and sand, which when forced into the soil displaces and compacts the surrounding material about a central core of grout. Injection of lime is sometimes used to produce lenses in the soil that will block the flow of water and reduce compressibility and expansion properties of the soil. The lenses are produced by hydraulic fracturing of the soil mass.

The injection and grouting methods are generally expensive compared with other stabilization techniques and are used primarily in special situations.

DYNAMIC STABILIZATION

Vibration or other dynamic stabilization techniques are often efficient in stabilizing cohesionless soils. Only a brief description of these methods is included in this chapter. For a more complete description of the methods and principles of design, consult ASCE (1978), Brown (1977), Menard and Broise (1975).

Vibroflotation

The vibroflot (Fig. 327-1) is a cylindrical tube containing water jets at top and bottom and equipped with a rotating eccentric weight, which develops a horizontal vibratory motion. The vibroflot is sunk into the soil using the lower jets and is then raised in successive small increments, during which the surrounding material is compacted by the vibration process. The enlarged hole around the vibroflot is backfilled with suitable granular material. This method is very effective for increasing the density of a sand deposit for depths up to 30 m. Figure 328-1 shows the recommended particle size for best results. Probe spacings of compaction holes should be on a grid pattern of about 2 m to produce relative densities greater than 70% over the entire area. If the sand is coarse, the spacing may be somewhat larger.

Figure 327-1 Vibroflotation compaction. (After Brown, 1977.)

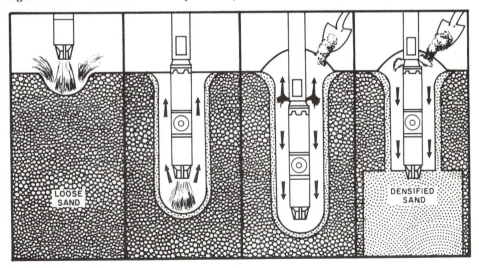

Figure 328-1 Soil gradation for effective compaction by vibroflotation.

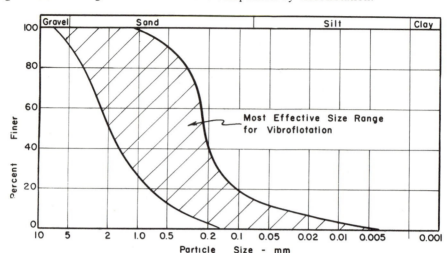

In soft cohesive soil and organic soils the vibroflotation technique has been used with gravel as the backfill material. The resulting densified stone column effectively reinforces softer soils and acts as a bearing pile for foundations. The stone columns also furnish drainage channels for moisture in the soil and accelerate consolidation.

Terra-probe

A method of densifying granular soils, which is somewhat similar to vibroflotation, uses an open-ended pipe pile (terra-probe) and a vibratory pile driver. The pile is driven to the desired depth up to 20 m and then extracted. This process is repeated over a grid pattern on the area to be densified. Saturated soil conditions are best for maximum effect.

Blasting

Buried explosive charges are used to densify cohesionless material. The process is one of causing liquefaction and is generally effective below 1 m. The upper soil is displaced in a random manner, which does not produce great density. Blasting is a useful method for deep densification of cohesionless soils in cases where surrounding structures are at a great enough distance to prevent damage.

9 Improving Soil Conditions and Properties

Compaction Piles

Soils that are free draining may be densified by capping a pipe pile, driving the pile and, as it is extracted, backfilling with granular material.

Both the process of driving the pile and the displacement of soil due to the pile produce densification of surrounding sandy material. Resulting densities can be computed by using the displacement volume of the pile.

Schroeder and Byington (1972) presented a comparison between vibroflotation, vibrating probe, and compaction piles on several projects (Fig. 330-1). The method chosen in each case depends on economical considerations.

Heavy Tamping

Heavy tamping utilizes a heavy mass (5 to 40 t) dropped from a height of 7 to 35 m and was originally used to compress cohesionless soils or fill materials. The method produces enough energy to compact the soil to a great depth.

Heavy tamping has recently been used to compact fine-grained soils. The principles are not well understood, but Menard and Broise (1975) suggest that cohesive soils are liquefied and the increased pore pressure is dissipated rapidly through radial fissures produced by the impact of the falling weight. The method has been primarily used by Menard's consulting firm, with headquarters in France. However, Menard has used the method in several other countries around the world, and it seems to have promise for compaction of both cohesive and granular materials at a reasonable cost if the necessary equipment is available without too great a shipping charge.

PRECOMPRESSION

Large-scale construction sites composed of weak silts and clays or organic materials, sanitary land fills, and other compressible soils may often be stabilized effectively and economically by precompression.

Precompression involves preloading the soil to produce settlement before construction begins. The preload is generally in the form of an earth fill, and it must be left in place long enough to induce the required settlement. A settlement monitoring system, consisting of settlement plates and possibly piezometers, should be used to determine when sufficient settlement has occurred. After the required settlement has occurred, part or all of the preload may be removed prior to construction of buildings or other structures at the site.

Occasionally an additional load, called a surcharge, is used to decrease the time required to achieve the required settlement. The surcharge load

Figure 330-1 Densities attainable by deep compaction methods. (After Schroeder and Byington, 1972.)

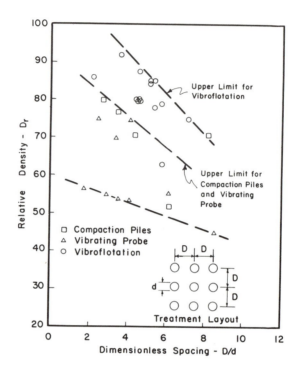

causes total soil pressures that are higher than the soil pressures under the final design load and, hence, the ultimate settlement under the design load will be only a percentage of the ultimate settlement under the full surcharge load. Since only a portion of the ultimate settlement under the full surcharge load will be required, the time to achieve that settlement will be less than without the surcharge. This idea can be demonstrated by considering Eqs. 128-4, 137-1, and 137-3 and Fig. 139-1.

Preloads may also involve a pattern of vertical sand drains to decrease the time of primary consolidation. The sand drains may be driven, augered, or jetted. When in place, they decrease the length of the consolidation drainage path and, hence, decrease the time to achieve the required settlement.

330   **9** Improving Soil Conditions and Properties

Soft compressible soils that may require a preload to improve foundation conditions generally have low shear strength. Therefore, the stability of the foundation soils under the weight of the preload is a necessary consideration in evaluating the feasibility of a preload.

Some uses of preloads are to reduce settlement under earth abutments and piers, to improve foundation conditions under warehouses and other large buildings, and to reduce settlement of highways and airfields.

Careful consideration must be given to the design of a preload in using precompression to improve foundation conditions. Sufficient soil data must be developed so that the magnitude and rate of settlement can be predicted, using the principles introduced in Chapter 4, and so that deep stability can be evaluated, using the principles of Chapter 7 and 8.

DRAINAGE

Engineering properties of soils may be greatly improved by reducing the water content. This is especially important in the construction of foundations in granular soils with high permeability, but may also be necessary on a permanent basis in some tighter soils. (See Chapter 9, "Drainage and Stabilization," in *Foundation Engineering*, by Peck Hanson, Thornburn, 1974.) Drainage may be accomplished to a fairly shallow depth using ditches and sumps, or to a deeper depth using well points and deep wells.

Drains projected horizontally into slopes are often used to permanently stabilize slopes that are susceptible to surface erosion and instability caused by seepage toward the surface of the slope.

Care should be taken that lowering the position of the water table will not cause excessive settlement under or around structures.

REINFORCED EARTH

The idea of reinforcing fill behind retaining walls by the use of layers of flat metal strips is a recent innovation in soil stabilization. Vidal (1969) introduced the idea, which has received considerable attention in research and application since that time.

An application of reinforced earth for a retaining wall is shown in Fig. 332-1. This method of stabilizing soils may also be used to improve the bearing capacity of granular soils overlying weaker cohesive soils (Binquet and Lee, 1975). The stability of walls and footings designed using reinforced earth becomes a function of the strength and spacing of the metal strips and friction between the strip and surrounding soil. Some design procedures are available (Lee et al, 1973; Lee, 1976).

The problem of corrosion of the metal strips is a limiting factor and, in many instances, seriously reduces the cost advantage that could otherwise be obtained.

Fabrics have been used to provide horizontal reinforcement of soils. These materials have a reasonably high tensile strength and allow only small horizontal strain. Fabrics have also been used to encapsulate soils to prevent their migration, but at the same time to permit water movement. Soils may be encased in a waterproof membrane to keep water out and preserve soil strength.

SUMMARY

A summary of soil stabilization methods used for structural foundations is given in Fig. 333-1.

Figure 332-1 Reinforced earth retaining wall.

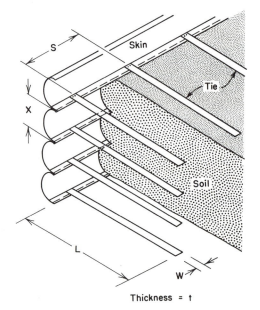

9 Improving Soil Conditions and Properties

Figure 333-1 Stabilization of Soils for Foundations of Structures. (After Mitchell, 1977.)

Method	Principle	Most Suitable Soil Conditions/types	Maximum Effective Treatment Depth	Economical Size of Treated Area
Compaction (with or without admixtures)	Densification by rolling vibration or kneading	Inorganic soils	A few meters	Any size
Grouting	Addition of chemicals, cement, asphalt, or clay to soil pores	Sand or coarse soils	Unlimited	Small
Vibroflotation	Densification by vibration and compaction of back-fill material	Cohesionless soils with less than 20 percent fines	30 m	> 1500 m²
Vibro replacement stone columns	Hole jetted into soft, fine-grained soil and back-filled with densely compacted gravel	Soft clays and alluvial deposits	20 m	> 1500 m²
Terraprobe	Densification by vibration; liquefaction-induced settlement under overburden	Saturated or dry, clean sand	20 m (ineffective above 4 m depth)	> 1500 m²
Blasting	Shock waves and vibrations cause liquefaction, displacement and remolding	Saturated, clean sands; partly saturated sands and silts after flooding	20 m	Small areas can be treated economically
Compaction piles	Densification by displacement of pile volume and by vibration during driving	Loose, sandy soils; partly saturated clayey soils; loess	20 m	Small to moderate
Heavy tamping (dynamic consolidation)	Repeated application of high-intensity impacts at surface	Cohesionless soils best, other types can also be improved	15–20 m	> 5000 m²
Preloading and surcharge (with or without sand drains)	Load is applied sufficiently in advance of construction so that compression of soft soils is completed prior to development of the site	Normally consolidated soft clays, silts, organic deposits, completed sanitary landfills	—	> 1000 m²
Strips and membranes	Horizontal tensile strips or membranes buried in soil under footings	All soil types	A few meters	Small

Summary

333

10

NUMERICAL METHODS AND COMPUTER APPLICATIONS

The development of the digital computer and the use of numerical methods to solve complex mathematical problems have had a significant impact on engineering problem solving. Computer-controlled systems are now routinely used for many industrial processes and for laboratory and field testing. Pocket calculators and desk-top computers are used to solve routine engineering problems. In short, the "computer impact" has reached nearly all levels of the engineering profession. Geotechnical engineering is no exception.

As a word of caution, it should be recognized that results obtained from computer programs can be misused or misleading. The engineer must exercise common sense and sound judgment when interpreting computer results. Misuse can result from use of incorrect input data, unfamiliarity with assumptions or limitations of a computer program, or such deep involvement with the intricacies of the mathematical analysis as to lose sight of the real problem or basic principles. Good engineering judgment must be used in selecting analysis parameters, and the output must be interpreted with a knowledge of precedents. The role of the computer is to provide increased computational power; it does not eliminate the need for engineering judgment and experience.

The application of computers as an analytical tool in geotechnical engineering can be divided into two categories.

1. The computer can be programmed to solve problems by following established manual procedures. An example of this type of application would be slope stability analysis by Bishop's simplified method.

2. More importantly, the computer can be used to obtain numerical solutions to problems described by partial differential equations for which analytical solutions are not available and for which the numerical solution is not practical by manual methods. Complex stress analysis and seepage analysis problems fall into this category.

The second category of problems involves the use of either the finite difference method or the finite element method. Since an understanding of these methods is essential for application to engineering problems, an introduction of each method is presented, followed by a discussion of computer applications in general.

NUMERICAL SOLUTIONS OF PARTIAL DIFFERENTIAL EQUATIONS

Several types of geotechnial engineering problems are well suited for solution by the finite difference method. Examples include seepage analysis, consolidation of saturated clay, and pile analysis by the wave equation. Although other problems can be solved by finite difference methods, these problems represent a cross section of applications.

Finite Difference Approximations

To represent ordinary or partial differential equations in finite difference form, the derivatives are defined by finite difference approximations. An approximation of the first derivative of a function

$$y = f(x) \text{ at } x_i$$

can be stated as

$$y_i' = \frac{y_{i+1} - y_{i-1}}{2(\Delta x)} \qquad \text{central difference} \qquad (336\text{-}1)$$

Figure 337-1 Central difference approximation of the first derivative at x_i.

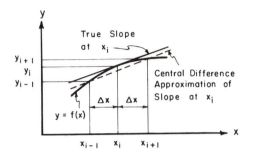

The increment Δx represents a finite distance along the x-axis between specific points x_{i-1}, x_i, x_{i+1}, etc. The subscripts indicate corresponding values of y_i for specific values of x_i. The subscript increases in the positive x direction. The finite difference approximation of the first derivative as given by Eq. 336-1 is illustrated in Fig. 337-1. From the figure, the derivative of the function at x_i (slope of the curve at x_i) is approximated by considering the slope of a line between two points on the curve, a finite distance on each side of x_i. Equation 336-1 represents the central difference approximation of the first derivative. The first derivative at x_i can also be approximated by considering only the point ahead of or the point in back of the point at x_i. These approximations are respectively the forward and backward approximations of the first derivative. Expressed algebraically

$$y_i' = \frac{y_{i+1} - y_i}{\Delta x} \qquad \text{forward difference} \qquad (337\text{-}2)$$

$$y_i' = \frac{y_i - y_{i-1}}{\Delta x} \qquad \text{backward difference} \qquad (337\text{-}3)$$

Equations 337-2 and 337-3 are less exact approximations of the derivative at x_i than Eq. 336-1. This is readily visualized by sketching these approximations in Fig. 337-1.

The example problem of Fig. 338-1 illustrates the use of the central difference approximation and forward difference approximation for evaluating the first derivative of a function.

The central difference approximation of the second derivative can be obtained by writing expressions for the first derivative at points $(\Delta x/2)$ on

Numerical Solutions of Partial Differential Equations

Figure 338-1

Example Problem Find the numerical value of the first derivative at $x = 1.0$ of the following function. Use the central difference approximation and compare the solution with the exact value and with the forward difference approximation.

$$y = x^3 - 36x + 10$$

Solution Obtain the central difference approximation of y' at $x = 1$ by using Eqs. 336-1. Use $\Delta x = 0.2$.

$$y_i' = \frac{y_{i+1} - y_{i-1}}{2(\Delta x)}$$

$$x_i = 1.0, \qquad x_{i+1} = 1.2, \qquad x_{i-1} = 0.8$$

$$y_{i+1} = (1.2)^3 - 36(1.2) + 10 = -31.47$$

$$y_{i-1} = (0.8)^3 - 36(0.8) + 10 = -18.29$$

$$y' = \frac{-31.47 - (-18.29)}{2(0.2)} = -32.95$$

The derivative may be evaluated analytically as

$$\frac{dy}{dx} = 3x^2 - 36$$

At

$$x = 1$$
$$\frac{dy}{dx} = 3(1)^2 - 36 = -33.00$$

For Δx of 0.1 the central difference approximation is 32.99.

The forward difference approximation of the derivative at $x = 1.0$ is -32.35 for $\Delta x = 0.2$ and -32.70 for $\Delta x = 0.1$.

each side of x_i ($x_{i-1/2}$ and $x_{i+1/2}$) and then considering the second derivative as the rate of change in the first derivative between the two points.

$$y_i'' = \frac{y'_{i+1/2} - y'_{i-1/2}}{\Delta x} \tag{338-2}$$

where

$$y'_{i+1/2} = \frac{y_{i+1} - y_i}{\Delta x} \tag{338-3}$$

10   Numerical Methods and Computer Applications

and

$$y'_{i-1/2} = \frac{y_i - y_{i-1}}{\Delta x}$$ (339-1)

Substitution of Eqs. 338-3 and 339-1 into Eq. 338-2 yields the central difference approximation of the second derivative

$$y''_i = \frac{y_{i+1} - 2y_i + y_{i-1}}{(\Delta x)^2}$$ (339-2)

Higher order finite difference approximations are available by making use of Taylor's series expansion (James et al, 1967).

Partial derivatives can be approximated in the same manner as was illustrated for ordinary derivatives. For a function with two independent variables such as

$$\phi = f(x, y)$$

it is convenient to use a grid with a mesh of Δx by Δy, as shown on Fig. 339-4, to represent specific points of the independent variables.

Two subscripts are now required to describe the x, y location of the point where the function and its derivatives are to be evaluated. The central difference approximations for the first and second derivatives are as follows.

$$\left(\frac{\partial \phi}{\partial x}\right)_{i,j} = \frac{\phi_{i+1,j} - \phi_{i-1,j}}{2(\Delta x)}$$ (339-3)

Figure 339-4 Grid for finite difference approximation of partial derivatives.

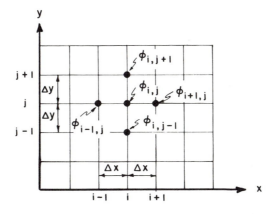

$$\left(\frac{\partial \phi}{\partial y}\right)_{i,j} = \frac{\phi_{i,j+1} - \phi_{i,j-1}}{2(\Delta y)} \tag{340-1}$$

$$\left(\frac{\partial^2 \phi}{\partial x^2}\right)_{i,j} = \frac{\phi_{i+1,j} - 2\phi_{i,j} + \phi_{i-1,j}}{(\Delta x)^2} \tag{340-2}$$

$$\left(\frac{\partial^2 \phi}{\partial y^2}\right)_{i,j} = \frac{\phi_{i,j+1} - 2\phi_{i,j} + \phi_{i,j-1}}{(\Delta y)^2} \tag{340-3}$$

Forward and backward difference approximations of partial derivatives may also be defined in a manner similar to those shown for ordinary derivatives.

Form of Partial Differential Equations

Many partial differential equations used in geotechnical engineering have the general form

$$a\frac{\partial^2 \phi}{\partial x^2} + b\frac{\partial^2 \phi}{\partial x\, \partial y} + c\frac{\partial^2 \phi}{\partial y^2} = f\left(x, y, \phi, \frac{\partial \phi}{\partial x}, \frac{\partial \phi}{\partial y}\right) \tag{340-4}$$

where the coefficients a, b, and c are functions of x and y. Partial differential equations of the form of Eq. 340-4 may take one of three classifications or types, depending on the value of $b^2 - 4ac$. These classifications are referred to as elliptic, parabolic, and hyperbolic (James et al, 1967).

$$b^2 - 4ac < 0 \qquad \text{elliptic} \tag{340-5}$$

$$b^2 - 4ac = 0 \qquad \text{parabolic} \tag{340-6}$$

$$b^2 - 4ac > 0 \qquad \text{hyperbolic} \tag{340-7}$$

The general method of solution is the same for equations within the same classification.

Steady-State Seepage

The distribution of head in a two-dimensional porous media flow situation is described by the LaPlace equation.

$$\frac{\partial^2 h}{\partial x^2} + \frac{\partial^2 h}{\partial y^2} = 0$$

This equation was derived in Chapter 2 and was first presented as Eq. 64-1*b*. Several methods of solving the LaPlace equation were discussed. The most widely used method in geotechnical engineering is the graphical solution by sketching flow nets. Seepage problems can also be readily solved by numerical methods.

In light of Eq. 340-5, the LaPlace equation is an elliptic partial differential equation. Elliptic partial differential equations are characterized by a closed domain in which boundary conditions are prescribed on the entire boundary. Problems involving this type of equation are referred to as boundary value problems.

The two-dimensional flow problem of Fig. 63-2 is used to illustrate numerical solutions of seepage problems. Since this problem is symmetrical, a solution of either the left or the right half provides the complete solution. Only the right side of Fig. 63-2 as shown in Fig. 341-3 is used. The necessary boundary conditions are shown and a grid has been superimposed on the flow field of Fig. 341-3. The LaPlace equation must be satisfied at all points within the flow field. This can be assured in the numerical solution by writing the LaPlace equation in finite difference form for all grid points within the flow field. At a general point i, j,

$$\frac{h_{i,j+1} - 2h_{i,j} + h_{i,j-1}}{(\Delta y)^2} + \frac{h_{i+1,j} - 2h_{i,j} + h_{i-1,j}}{(\Delta x)^2} = 0 \qquad (341\text{-}1)$$

By using a square grid ($\Delta x = \Delta y$), Eq. 341-1 (the difference equation) can be written as

$$h_{i,j+1} + h_{i,j-1} - 4h_{i,j} + h_{i+1,j} + h_{i-1,j} = 0 \qquad (341\text{-}2)$$

Figure 341-3 Two-dimensional flow field with boundary conditions and grid for finite difference solution (from Fig. 63-2). Assuming a head loss of 10 m across the wall, the head at the line of symmetry will be 5 m.

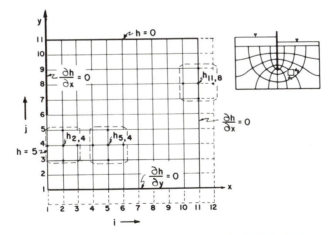

When the difference equation is written for all points in the flow field where the head is unknown, a system of linear algebraic equations is generated. This system of equations can then be solved for the unknown heads. The heads at all of the grid points obtained in this manner will, of course, satisfy the difference equations and, hence, the LaPlace equation. The accuracy of the solution will depend on the size of the grid used.

For the interior grid point $i = 5$ and $j = 4$ of Fig. 341-3, the difference equation (Eq. 341-2) becomes

$$h_{5,5} + h_{5,3} - 4h_{5,4} + h_{6,4} + h_{4,4} = 0$$

The heads $h_{i,j}$ at the five concerned grid points are unknown.

For the grid point $i = 2$ and $j = 4$, adjacent to the boundary where $i = 1$, the difference equation (Eq. 341-2) becomes

$$h_{2,5} + h_{2,3} - 4h_{2,4} + h_{3,4} + h_{1,4} = 0$$

For this case, however, $h_{1,4} = 5$, and the difference equation is written as

$$h_{2,5} - 4h_{2,4} + h_{3,4} + h_{1,4} = -5$$

For all grid points adjacent to a boundary with a known head, the value of the head of at least one grid point will be known, as for the preceding case.

At a grid point located on a flow line boundary, the head will be an unknown quantity and, therefore, the difference equation must be written for that point. The boundary condition for a flow line boundary is that the head gradient normal to the boundary must be zero, because the flow is parallel to the boundary. For flow boundaries along the x and y directions, this condition can be stated as

$$\frac{\partial h}{\partial y} = 0 \quad \text{and} \quad \frac{\partial h}{\partial x} = 0$$

respectively. Flow boundaries that are not along the x and y directions can also be accommodated but are not discussed here (refer to James et al, 1967). In the finite difference solution, the flow line boundary condition can be satisfied by assuming imaginary grid points adjacent to the boundary, as shown in Fig. 341-3. In order for the derivative across the boundary to be zero, the value of the head at the imaginary grid point must be equal to the head at the mirror-image grid point in the flow field. This is illustrated by setting the central difference approximation of the first derivative as given by Eq. 339-3 equal to zero.

$$\left(\frac{\partial h}{\partial x}\right)_{i,j} = \frac{h_{i+1,j} - h_{i-1,j}}{2(\Delta x)} = 0$$

Application of the flow line boundary condition can be illustrated by considering the grid point at $i = 11$ and $j = 8$. The difference equation for this point would be

$$h_{11,9} + h_{11,7} - 4h_{11,8} + h_{12,8} + h_{10,8} = 0$$

but

$$h_{12,8} = h_{10,8}$$

therefore, the difference equation becomes

$$2h_{10,8} - 4h_{11,8} + h_{11,9} + h_{11,7} = 0$$

Equations can be similarly written for all grid points that fall on a flow line boundary.

For the flow field of Fig. 341-3, there are 105 grid points at which the head is unknown and 17 grid points at which the head is known. This will require 105 difference equations that can be solved simultaneously for the unknown heads. Several methods can be used to solve the resulting system of equations.

The Gauss-Seidel method is frequently used to solve large systems of equations. This method uses an iterative procedure in which initial estimates are made for the values of the unknown heads at all grid points, and then each iteration improves the values of the initial estimates. To insure convergence of this method, there is a dominant coefficient requirement. The system of equations that are generated for a finite difference solution of seepage problems meet the dominant coefficient requirement. For a detailed explanation, refer to texts on linear algebra or numerical methods.

The steps in the Gauss-Seidel method can be summarized as follows.

1. Assume initial values for all unknown heads appearing in the grid.
2. Solve for new values of head at each grid point by sequentially solving the difference equation (Eq. 341-2) for the unknown head $h_{i,j}$. Always use the most recent value of head for each grid point. After a new value of head is found at every grid point, one iteration has been completed.
3. Iterate through step 2 until each new value of head differs from its previous value by an amount less than some prescribed epsilon.

The required iterative procedure is conveniently carried out for seepage problems by rewriting the difference equation (Eq. 341-2) in the following form.

$$h_{i,j} = \frac{1}{4} \left(h_{i,j+1} + h_{i,j-1} + h_{i+1,j} + h_{i-1,j} \right) \qquad (343\text{-}1)$$

Figure 344-1

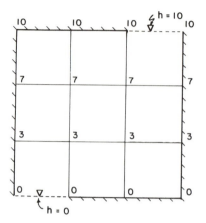

Initial values

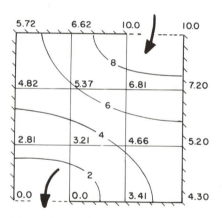

After 12 iterations

Example Problem   Given the above two-dimensional flow field, use the finite difference method to solve for the heads within the flow field.

Solution   Divide the flow field into a grid system, as shown on the figure, and assume initial values of head at all grid points. Apply the difference equation to all unknown head points. The new values of head after one iteration are calculated below. After about 12 iterations, the head values are shown on the figure.

Known values of head:

$$h_{1,1} = h_{2,1} = 0$$
$$h_{3,4} = h_{4,4} = 10$$

Heads for first iteration:

$$h_{3,1} = \tfrac{1}{4}[0 + 0 + 2(3)] = 1.50$$
$$h_{4,1} = \tfrac{1}{4}[2(1.5) + 2(3)] = 2.25$$
$$h_{1,2} = \tfrac{1}{4}[2(3) + 7 + 0] = 3.25$$
$$h_{2,2} = \tfrac{1}{4}(3 + 3.25 + 7 + 0) = 3.31$$
$$h_{3,2} = \tfrac{1}{4}(3 + 3.31 + 7 + 1.5) = 3.70$$
$$h_{4,2} = \tfrac{1}{4}[2(3.7) + 7 + 2.25] = 4.16$$
$$h_{1,3} = \tfrac{1}{4}[2(7) + 10 + 3.25] = 6.81$$
$$h_{2,3} = \tfrac{1}{4}(7 + 6.81 + 10 + 3.31) = 6.78$$
$$h_{3,3} = \tfrac{1}{4}(7 + 6.78 + 10 + 3.70) = 6.87$$
$$h_{4,3} = \tfrac{1}{4}[2(6.87) + 10 + 4.16] = 6.98$$
$$h_{1,4} = \tfrac{1}{4}[2(10) + 2(6.81)] = 8.40$$
$$h_{2,4} = \tfrac{1}{4}[10 + 8.40 + 2(6.78)] = 7.98$$

This form of the equation would be modified slightly when applied to a flow line boundary. For example, along the flow line boundary where $i = 11$ in Fig. 341-3, the difference equation would be

$$h_{11,j} = \frac{1}{4} \left(2h_{10,j} + h_{11,j+1} + h_{11,j-1}\right)$$

Similar adjustments would be made for the other flow line boundaries. Using Eq. 343-1 or its modified form for flow line boundaries, the value of $h_{i,j}$ can be solved for specific values of i and j. By systematically changing the values of i and j, the iterative procedure can be carried out until the desired accuracy is achieved. The rate of convergence can be improved by using an overrelaxation factor, but this topic is beyond the scope of this text. For discussions on overrelaxation factors, refer to other texts on numerical methods (James et al, 1967; Carnahan et al, 1969).

The example problem of Fig. 344-1 illustrates the finite difference method for use in solving seepage problems. This simple example was easily solved by hand. The only practical way to solve more complex problems with many more grid points is by use of a computer. A FORTRAN computer program to solve the seepage problem of Fig. 341-3 is given in Appendix B. The profile converged to the solution shown in Fig. 346-1 after 36 iterations. Use of an overrelaxation factor speeded the convergence, so only 15 iterations were required.

Numerical Solutions of Partial Differential Equations

Figure 346-1

Example Problem Solve for the distribution of head over the grid shown in the example problem of Fig. 341-3. Use the computer program in Appendix B.

$D = 10.0$ DEL $= 1.0$ EPS $= 0.0005$ RF $= 1.66$ MAX $= 150$

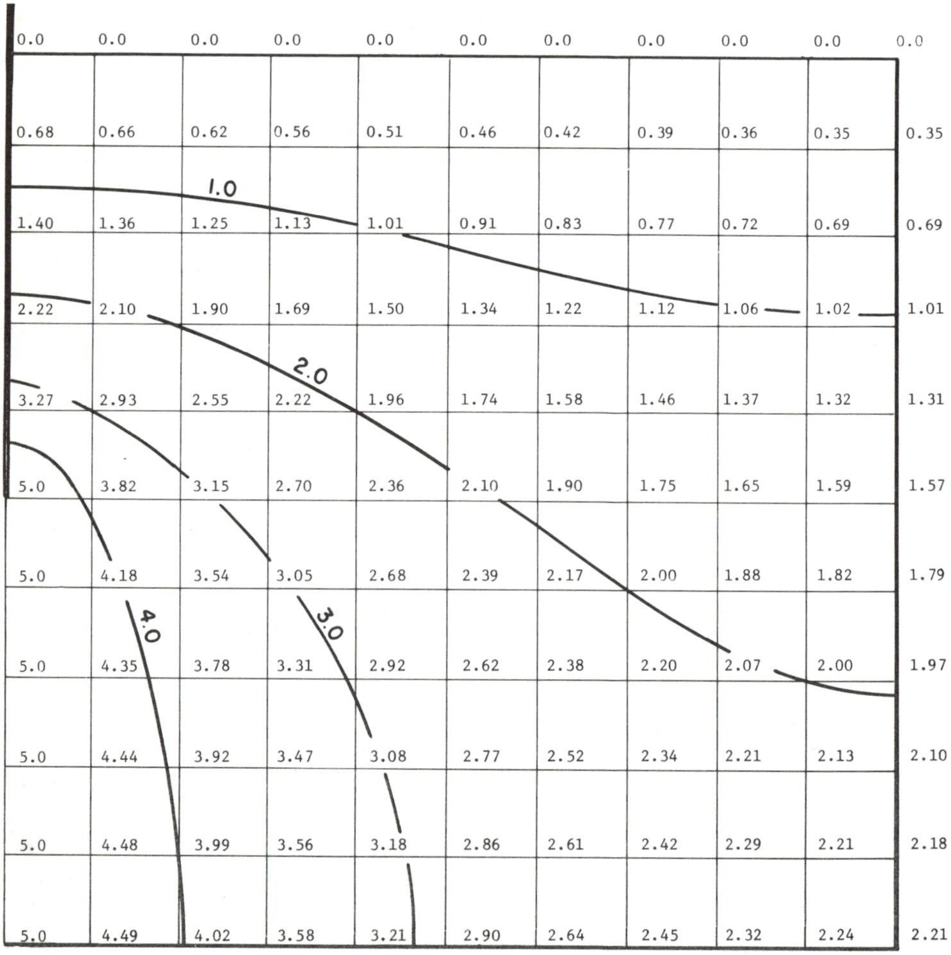

0.0	0.0	0.0	0.0	0.0	0.0	0.0	0.0	0.0	0.0	0.0
0.68	0.66	0.62	0.56	0.51	0.46	0.42	0.39	0.36	0.35	0.35
1.40	1.36	1.25	1.13	1.01	0.91	0.83	0.77	0.72	0.69	0.69
2.22	2.10	1.90	1.69	1.50	1.34	1.22	1.12	1.06	1.02	1.01
3.27	2.93	2.55	2.22	1.96	1.74	1.58	1.46	1.37	1.32	1.31
5.0	3.82	3.15	2.70	2.36	2.10	1.90	1.75	1.65	1.59	1.57
5.0	4.18	3.54	3.05	2.68	2.39	2.17	2.00	1.88	1.82	1.79
5.0	4.35	3.78	3.31	2.92	2.62	2.38	2.20	2.07	2.00	1.97
5.0	4.44	3.92	3.47	3.08	2.77	2.52	2.34	2.21	2.13	2.10
5.0	4.48	3.99	3.56	3.18	2.86	2.61	2.42	2.29	2.21	2.18
5.0	4.49	4.02	3.58	3.21	2.90	2.64	2.45	2.32	2.24	2.21

(Contour lines labeled 1.0, 2.0, 3.0, and 4.0 are drawn across the grid.)

More advanced applications are found in the literature (Jeppson, 1968). Jeppson (1970) has also applied finite difference methods to the solution of unsaturated flow problems, seepage from canals, and three-dimensional flow problems.

Consolidation of Clay Soil

The consolidation of saturated clay soil was discussed in Chapter 4. The differential equation of consolidation defined by Terzaghi (Eq. 135-3) is repeated here as Eq. 347-1.

$$c_v \frac{\partial^2 \epsilon}{\partial y^2} = \frac{\partial \epsilon}{\partial t} \qquad (347\text{-}1)$$

The solution of this equation yields the strain in the soil as a function of depth and time. The shape of the solution surface for a double-drained condition is shown in Fig. 136-2. For the case of drainage at one surface only the general shape of the solution surface would be similar to that shown in Fig. 347-2.

An analytical solution to the differential equation of consolidation in terms of dimensionless factors was presented in Chapter 4 (Eq. 137-2). The differential equation of consolidation can also be solved numerically by a finite difference technique. An advantage of the numerical solution is that it can conveniently accommodate stratified soil deposits in which c_v varies from stratum to stratum.

Figure 347-2 Solution surface and boundary condition for one-dimensional consolidation with single drainage.

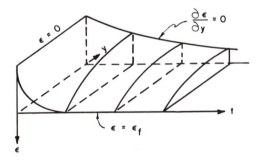

Figure 348-1   Open ended consolidation grid.

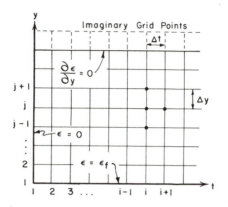

When the coefficients of the consolidation equation (Eq. 347-1) are compared with the general form for partial differential equations (Eq. 340-4), it is seen to be a parabolic type. This type of partial differential equation is characterized by an open-ended solution domain as illustrated in Fig. 348-1. The solution starts with known initial conditions and advances outward (in this case with respect to time) from the initial values, while always satisfying boundary conditions.

A numerical solution can be obtained by considering the grid and boundary conditions shown in Fig. 348-1 and then by writing Eq. 347-1 in finite difference form. The grid size is Δt by Δy. The position of a grid point along the y-axis is given by the subscript j and along the t-axis by the subscript i. The forward difference approximation will be used for the first derivative of ϵ with respect to time

$$\left(\frac{\partial \epsilon}{\partial t}\right)_{i,j} = \frac{\epsilon_{i+1,j} - \epsilon_{i,j}}{\Delta t}, \qquad (348\text{-}2)$$

and the central difference approximation will be used for the second derivative of ϵ with respect to y.

$$\left(\frac{\partial^2 \epsilon}{\partial y^2}\right)_{i,j} = \frac{\epsilon_{i,j+1} - 2\epsilon_{i,j} + \epsilon_{i,j-1}}{(\Delta y)^2} \qquad (348\text{-}3)$$

Substitution of these approximations into Eq. 347-1 yields the finite difference approximation of the differential equation of consolidation and is given as Eq. 349-1.

$$\epsilon_{i+1,j} = \epsilon_{i,j} + c_v \frac{\Delta t}{(\Delta y)^2} \ (\epsilon_{i,j+1} - 2\epsilon_{i,j} + \epsilon_{i,j-1}) \qquad (349\text{-}1)$$

It can be seen from Eq. 349-1 that the strain at a grid point $(i + 1, j)$ can be obtained in terms of the strain at the grid points (i, j), $(i, j + 1)$ and $(i, j - 1)$ of the previous time increment. After all the strains have been determined along the $(i + 1)$ column, the strains for the next time increment can be solved in terms of those just obtained. Thus, the solution always advances forward in time by obtaining strains at a new time increment in terms of those from the previous increment.

To start the solution, the strains along the initial condition boundary at $t = 0$ must be known. Boundary conditions must also be satisfied along the boundaries at $y = 0$ and $y = H$. For the one-dimensional consolidation case with drainage at $y = 0$ only, the boundary conditions are shown in Figs. 347-2 and 348-1 and can be stated as

$$\epsilon = 0 \quad \text{at } t = 0$$

$$\epsilon = \epsilon_f \quad \text{at } y = 0$$

$$\frac{\partial \epsilon}{\partial y} = 0 \quad \text{at } y = H$$

The boundary condition at $y = 0$ is satisfied by setting all values for strain equal to ϵ_f along this boundary. The boundary condition at $y = H$ is satisfied by using an imaginary line of grid points at $y = H + \Delta y$ that have the same value of strain as at $y = H - \Delta y$. These imaginary grid points will assure that

$$\frac{\partial \epsilon}{\partial y} = 0$$

along the boundary $y = H$.

This procedure is called an explicit method. For stable solution for which the truncation error does not grow exponentially, the following condition must be satisfied (James et al, 1967).

$$c_v \frac{\Delta t}{(\Delta y)^2} \leqslant 0.5$$

For a given value of c_v, this criterion can be satisfied by adjusting the size of the grid Δt and Δy.

A FORTRAN computer program is presented in Appendix B for the solution of one-dimensional consolidation of homogeneous single-drained systems.

Other finite difference procedures such as the Crank-Nickolson method and the implicit method are available and have some numerical advantages over the explicit method. The explicit method was developed here because it is more straightforward and still illustrates numerical solutions for this type of problem. For discussions of these other procedures, refer to other texts on numerical methods (Carnahan et al, 1969; Forsythe and Wasow, 1960).

The finite difference method for the solution of consolidation problems can easily be programmed for solution on the digital computer. Jordan and Schiffman (1967) have applied the finite difference method to determine the rate of consolidation in the SEPOL subsystem of ICES (Integrated Civil Engineering System). SEPOL is a general-purpose SEttlement Problem Oriented Language and can handle very complex soil profile and loading conditions, including gradually applied loads.

Barden (1965) used a finite difference procedure to solve a consolidation problem involving a nonlinear partial differential equation that included secondary consolidation.

Wave Equation

The wave equation was discussed in Chapter 8 as a method of analyzing the dynamic characteristics of pile driving. The equation was not developed nor is a numerical solution presented here. However, it should be pointed out that the only practical solution to the equation uses a finite difference technique. The wave equation is a hyperbolic partial differential equation and has an open-ended solution domain similar to that for the parabolic partial differential equations. Bowles (1974) presents a computer program for solution of the wave equation as applied to a pile-driving problem. Computer solutions for the pile-driving problem are also available elsewhere (such as in Hirsh et al, 1976).

FINITE ELEMENT METHOD

The finite element was first introduced in engineering by Turner et al (1956) for the solution of stress analysis problems as they related primarily to the aircraft industry. Since that time, this method has become a useful and accepted tool in geotechnical engineering and in other areas of civil engineering. Its use in geotechnical engineering has been related primarily to static and dynamic stress analysis (Kulhawy and Duncan, 1972; Clough and Woodward, 1967; Idriss et al, 1974; Duncan, 1977), although it has also been used to solve seepage and consolidation problems (Desai, 1977b; Schiffman and Arya, 1977; Christian and Boehmer, 1970; Desai, 1972).

The main distinction between the finite element method and the finite difference method is in where the approximation comes in the solution process. In the finite difference method, exact differential equations are used to describe a physical system, and an approximate (numerical) solution of the exact equations is then obtained. In the finite element method, the approximation is made to the physical system, and an exact analysis of the approximate system is then made.

The use of the finite element method is becoming very common in geotechnical engineering practice. The literature contains many references to its use; furthermore, many "canned" computer programs are available for finite element stress analysis. Because of its impact on the engineering profession, a fundamental understanding of the finite element method is becoming essential for practicing engineers. The purpose of this section is to provide the reader with a physical understanding of the finite element method as it relates to solving stress analysis problems. Accordingly, the method is developed for basic elastic plane-strain stress analysis. Additional study of the finite element method is required to assess the state of the art, to apply the method to complex problems, and to understand other applications, such as porous media flow. The basic and brief development presented here will help to eliminate the mysticism that, for many engineers, is currently associated with the finite element method.

Matrix Notation

Because the finite element method involves the manipulation of systems of algebraic equations, it is convenient to present the development of the method in matrix notation. It is beyond the scope of this text to present a discussion on matrix algebra, and the reader is referred to the many references on the subject.

To understand the development of the finite element method, one should be familiar with the concepts of what a matrix represents, matrix addition, matrix multiplication, the transpose of a matrix, and the partitioning of matrices. In the following discussion a two-dimensional matrix is shown as $[A]$ and a one-dimensional matrix (vector) is shown as $\{A\}$.

Stress Analysis of Linearly Elastic Systems

The most common use of the finite element method in geotechnical engineering has to do with stress analysis problems. Stress analysis problems

Figure 352-1 Possible finite element assemblage for an earth dam and its foundation.

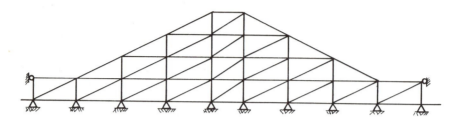

include such applications as static analysis of stresses and movements in embankments (Kulhawy and Duncan, 1972; Duncan, 1972); earthquake stress analysis of embankments (Clough and Chopra, 1966; Idriss et al, 1974); soil-structure interaction problems (Clough, 1972; Duncan, 1977); and the analysis of stresses induced by foundations.

The basic idea behind the finite element method for stress analysis is that a continuum is represented by a number of elements connected only at the element nodal points (joints), as shown for an earth dam in Fig. 352-1. An analysis of this substitute system (finite element assemblage) is performed to solve for the unknown nodal displacements. Once the nodal displacements are known, the stresses and strains within each element can be obtained.

The elements do not need to be triangles, as used in Fig. 352-1. Rectangles and trapezoids are commonly used in finite element problems. Regardless of the shape, the element is a basic structural unit of the assemblage, just as beams and columns are basic structural units of frames. Each element is continuous, and stresses and strains can be evaluated at any point within an element.

Element Stiffness

In order to perform an analysis of the finite element assemblage, it is first necessary to establish the stiffness of each element. The stiffness is the relationship between nodal displacements and the corresponding nodal forces induced by the displacements. Establishing the stiffness of an element requires making assumptions about the variation of strains within the element combined with the stress-strain properties of the material.

Consider a typical triangular element from a finite element assemblage as shown in Fig. 353-1. The three nodal points are labeled, counterclockwise, i,

Figure 353-1 Typical triangular element showing nodal displacement components.

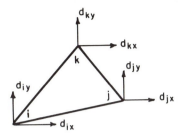

j, and k. The nodal displacement vector for the element will be represented by

$$\{\delta\} = \begin{Bmatrix} d_{ix} \\ d_{iy} \\ d_{jx} \\ d_{jy} \\ d_{kx} \\ d_{ky} \end{Bmatrix} \tag{353-2}$$

where the subscripts represent the node and direction of the displacement. Displacements d_x and d_y at any point within the element are related to nodal coordinates and displacements by a displacement function. Turner et al (1956) used a linear displacement function. The Turner shape function can be stated as

$$d_x = C_1 + C_2x + C_3y \tag{353-3}$$

$$d_y = C_4 + C_5x + C_6y \tag{353-4}$$

This linear displacement function guarantees continuity of displacements between elements. Since the displacements vary linearly along the side of the triangles, identical displacements at the common nodes of adjacent elements will produce the same displacements at all points along the interface of the two adjacent elements. Other finite element formulations of this same problem are possible by using other displacement functions. A number of other possibilities exist and are discussed in textbooks on the finite element method (see Huebner, 1975, and Zienkiewicz, 1977).

Finite Element Method

353

The constants C_1 to C_6 in Eqs. 353-3 and 353-4 can be evaluated in terms of the six nodal displacements and the respective coordinates of the nodes. For example, C_1, C_2, and C_3 can be determined from the equations

$$d_{ix} = C_1 + C_2 x_i + C_3 y_i \tag{354-1}$$

$$d_{jx} = C_1 + C_2 x_j + C_3 y_j \tag{354-2}$$

$$d_{kx} = C_1 + C_2 x_k + C_3 y_k \tag{354-3}$$

Solving for C_1, C_2, and C_3 from Eqs. 354-1, 354-2, and 354-3 yields

$$C_1 = \frac{1}{2\Delta} \{ (x_j y_k - x_k y_j) d_{ix} - (x_i y_k - x_k y_i) d_{jx} + (x_i y_j - x_j y_i) d_{kx} \}$$

$$C_2 = \frac{1}{2\Delta} \{ (y_j - y_k) d_{ix} + (y_k - y_i) d_{jx} + (y_i - y_j) d_{kx} \}$$

$$C_3 = \frac{1}{2\Delta} \{ (x_k - x_j) d_{ix} + (x_i - x_k) d_{jx} + (x_j - x_i) d_{kx} \}$$

where

$$\Delta = \frac{1}{2} \{ (x_j y_k - x_k y_j) - (x_i y_k - x_k y_i) + (x_i y_j - x_j y_i) \}$$

The constants C_1, C_2, and C_3 are now known in terms of the nodal coordinates and nodal displacements. In a similar manner, the constants C_4, C_5, and C_6 can be obtained by writing expressions for the vertical displacements at each node i, j, and k. The solutions are identical to C_1, C_2, and C_3 if d_{ix}, d_{jx}, and d_{kx} are replaced with d_{iy}, d_{jy}, and d_{ky}.

The next major step in formulating the element stiffness is to express the strain-displacement and stress-strain relationships of the problem. The element strains may be expressed in terms of the displacements by

$$\{\epsilon\} = \begin{Bmatrix} \epsilon_x \\ \epsilon_y \\ \gamma_{xy} \end{Bmatrix} = \begin{Bmatrix} \dfrac{\partial d_x}{\partial x} \\[2ex] \dfrac{\partial d_y}{\partial y} \\[2ex] \dfrac{\partial d_x}{\partial y} + \dfrac{\partial d_y}{\partial x} \end{Bmatrix} \tag{354-4}$$

When the appropriate partial derivatives of Eqs. 353-3 and 353-4 are substituted into Eq. 354-4, the strains may be expressed as

Figure 355-1 Typical element geometry.

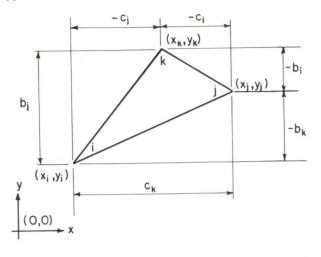

$$\begin{Bmatrix} \epsilon_x \\ \epsilon_y \\ \gamma_{xy} \end{Bmatrix} = \frac{1}{2\Delta} \begin{bmatrix} b_i & 0 & b_j & 0 & b_k & 0 \\ 0 & c_i & 0 & c_j & 0 & c_k \\ c_i & b_i & c_j & b_j & c_k & b_k \end{bmatrix} \begin{bmatrix} d_{ix} \\ d_{iy} \\ d_{jx} \\ d_{jy} \\ d_{kx} \\ d_{ky} \end{bmatrix} \qquad (355\text{-}2)$$

where

$$b_i = (y_j - y_k)$$
$$c_i = (x_k - x_j)$$
$$b_j = (y_k - y_i)$$
$$c_j = (x_i - x_k)$$
$$b_k = (y_i - y_j)$$
$$c_k = (x_j - x_i)$$

as illustrated in Fig. 355-1, or, in shorthand matrix notation,

$$\{\epsilon\} = [B]\{\delta\} \qquad (355\text{-}3)$$

Note that all the elements of the $[B]$ matrix are known in terms of the element's nodal coordinates.

Finite Element Method

355

The stress-strain relationship for the material is given by Hooke's law. In matrix notation, Hooke's law for plane strain conditions may be stated as

$$\left\{\begin{matrix} \sigma_x \\ \sigma_y \\ \tau_{xy} \end{matrix}\right\} = \frac{E(1-\nu)}{(1+\nu)(1-2\nu)} \begin{bmatrix} 1 & \nu/(1-\nu) & 0 \\ \nu/(1-\nu) & 1 & 0 \\ 0 & 0 & (1-2\nu)/2(1-\nu) \end{bmatrix} \left\{\begin{matrix} \epsilon_x \\ \epsilon_y \\ \gamma_{xy} \end{matrix}\right\}$$

(356-1)

where

$$E = \text{modulus of elasticity}$$
$$\nu = \text{Poisson's ratio}$$

In shorthand, matrix notation Eq. 356-1 becomes

$$\{\sigma\} = [D]\{\epsilon\},$$

but, since

$$\{\epsilon\} = [B]\{\delta\},$$

$$\{\sigma\} = [D][B]\{\delta\}$$

(356-2)

Equation 356-2 relates the stresses at any point within an element to the six nodal displacements of the element.

The next step in the development of the element stiffness matrix is to represent the stresses at the edges of the elements by equivalent nodal forces. The stresses along the edges of the elements are really internal stresses in the continuum. When the continuum is idealized as an assemblage of finite elements, the stresses must be placed along the edges of the element similar to a free-body diagram. They are then treated in the analysis as equivalent nodal forces. Figure 357-1 shows these equivalent nodal forces for a triangular element of *unit thickness*. These equivalent forces may be defined in matrix notation as

$$\{F\} = \left\{\begin{matrix} F_{ix} \\ F_{iy} \\ F_{jx} \\ F_{jy} \\ F_{kx} \\ F_{ky} \end{matrix}\right\} = \frac{1}{2} \begin{bmatrix} b_i & 0 & c_i \\ 0 & c_i & b_i \\ b_j & 0 & c_j \\ 0 & c_j & b_j \\ b_k & 0 & c_k \\ 0 & c_k & b_k \end{bmatrix} \left\{\begin{matrix} \sigma_x \\ \sigma_y \\ \tau_{xy} \end{matrix}\right\}$$

(356-3)

Figure 357-1 Representation of the stresses along the element boundaries by equivalent nodal forces.

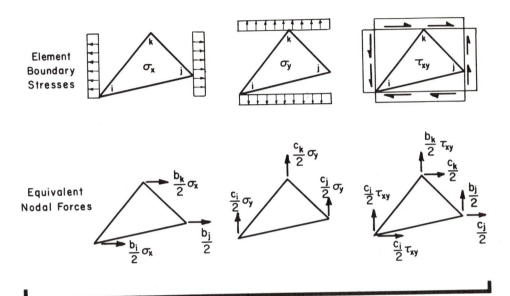

Equation 356-3 may be written in shorthand form as

$$\{F\} = \Delta[B]^T\{\sigma\} \qquad (357\text{-}2)$$

It should be noted that $[B]^T$ is the transpose of $[B]$ as used in Eq. 355-2. Substitution of Eq. 356-2 into Eq. 357-2 yields

$$\{F\} = \Delta[B]^T[D][B]\{\delta\}$$

Let

$$[k] = \Delta[B]^T[D][B] \qquad (357\text{-}3)$$

then

$$\{F\} = [k]\{\delta\} \qquad (357\text{-}4)$$

Finite Element Method

In expanded form, Eq. 357-4 becomes

$$
\left\{
\begin{array}{c}
F_{ix} \\
F_{iy} \\
F_{jx} \\
F_{jy} \\
F_{kx} \\
F_{ky}
\end{array}
\right\}^e
=
\begin{bmatrix}
k^e_{ix,ix} & k^e_{ix,iy} & k^e_{ix,jx} & k^e_{ix,jy} & k^e_{ix,kx} & k^e_{ix,ky} \\
k^e_{iy,ix} & k^e_{iy,iy} & k^e_{iy,jx} & k^e_{iy,jy} & k^e_{iy,kx} & k^e_{iy,ky} \\
k^e_{jx,ix} & k^e_{jx,iy} & k^e_{jx,jx} & k^e_{jx,jy} & k^e_{jx,kx} & k^e_{jx,ky} \\
k^e_{jy,ix} & k^e_{jy,iy} & k^e_{jy,jx} & k^e_{jy,jy} & k^e_{jy,kx} & k^e_{jy,ky} \\
k^e_{kx,ix} & k^e_{kx,iy} & k^e_{kx,jx} & k^e_{kx,jy} & k^e_{kx,kx} & k^e_{kx,ky} \\
k^e_{ky,ix} & k^e_{ky,iy} & k^e_{ky,jx} & k^e_{ky,jy} & k^e_{ky,kx} & k^e_{ky,ky}
\end{bmatrix}
\left\{
\begin{array}{c}
d_{ix} \\
d_{iy} \\
d_{jx} \\
d_{jy} \\
d_{kx} \\
d_{ky}
\end{array}
\right\}^e
\qquad (358\text{-}1)
$$

The superscript e refers to the element number. It can be seen that the elements of the 6-by-6 matrix $[k]$ represent nodal forces due to unit nodal displacements. The matrix $[k]$ is the element stiffness matrix for a plane strain triangular element of *unit thickness*. For a given element geometry and given material properties (E and ν), all the elements of the element stiffness matrix $[k]$ are known.

The element stiffness matrix could have been developed from a more general approach by using energy considerations.

Stiffness Matrix for Complete Assemblage

The final step in using the finite element method to solve stress analysis problems is to perform a structural analysis of the finite element assemblage. This is accomplished by analyzing equilibrium at all joints.

In order to relate the stiffness matrices of the individual elements to the complete assemblage, it is convenient to establish a system of numbering the lines of possible nodal displacements. The displacement at each node can be specified in terms of displacement components along the x-axis and the y-axis. Each possible displacement component is referred to as a line of action. Since each node has two lines of action, the total number of lines of action in the finite element assemblage equals two times the number of nodes.

To illustrate the structural analysis of the finite element assemblage, the very simple two-element assemblage of Fig. 359-1 is used. The use of only two elements to approximate the continuum of Fig. 359-1 is not a realistic idealization, but it provides a simple illustration of how to form a stiffness matrix for a complete assemblage. Each node is assigned a specific node number and each element a specific element number, as shown in Fig. 359-1. The lines of action at each node are numbered as follows.

$$x \text{ direction} = 2n - 1 \qquad (358\text{-}2)$$

$$y \text{ direction} = 2n \qquad (358\text{-}3)$$

where n is the node number.

Figure 359-1 Simple two-element assemblage.

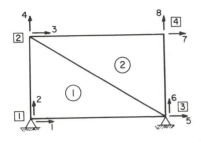

◯ Element number

↑ Lines of action

③ Node number

△ Node restraint in the x and y direction

Each element has an $i, j,$ and k node, as shown in Fig. 353-1. So that the element stiffness matrices are consistent, the $i, j,$ and k nodes must be assigned in a counterclockwise direction. For the problem shown in Fig. 359-1, the $i, j,$ and k nodes for each element are assigned as follows.

	Node Number		
Element Number	i	j	k
1	1	3	2
2	2	3	4

Numerical values for the subscripts of Eqs. 358-1 can now be assigned on the basis of the line of action numbers computed by Eqs. 358-2 and 358-3.

	Line of Action Number					
Element Number	ix	iy	jx	jy	kx	ky
1	1	2	5	6	3	4
2	3	4	5	6	7	8

In terms of these line of action numbers, Eqs. 358-1 for each element can be written as shown in Fig. 360-1.

Equilibrium equations can now be developed for each node. This will generate a system of eight equations (two at each node) for the example of

Finite Element Method

359

Figure 360-1 Force equations for the elements in Fig. 359-1.

$$
\begin{Bmatrix} F_1^1 \\ F_2^1 \\ F_5^1 \\ F_6^1 \\ F_3^1 \\ F_4^1 \end{Bmatrix} =
\begin{bmatrix}
k_{1,1}^1 & k_{1,2}^1 & k_{1,5}^1 & k_{1,6}^1 & k_{1,3}^1 & k_{1,4}^1 \\
k_{2,1}^1 & k_{2,2}^1 & k_{2,5}^1 & k_{2,6}^1 & k_{2,3}^1 & k_{2,4}^1 \\
k_{5,1}^1 & k_{5,2}^1 & k_{5,5}^1 & k_{5,6}^1 & k_{5,3}^1 & k_{5,4}^1 \\
k_{6,1}^1 & k_{6,2}^1 & k_{6,5}^1 & k_{6,6}^1 & k_{6,3}^1 & k_{6,4}^1 \\
k_{3,1}^1 & k_{3,2}^1 & k_{3,5}^1 & k_{3,6}^1 & k_{3,3}^1 & k_{3,4}^1 \\
k_{4,1}^1 & k_{4,2}^1 & k_{4,5}^1 & k_{4,6}^1 & k_{4,3}^1 & k_{4,4}^1
\end{bmatrix}
\begin{Bmatrix} d_1 \\ d_2 \\ d_5 \\ d_6 \\ d_3 \\ d_4 \end{Bmatrix}
\qquad (360\text{-}1a)
$$

$$
\begin{Bmatrix} F_3^2 \\ F_4^2 \\ F_5^2 \\ F_6^2 \\ F_7^2 \\ F_8^2 \end{Bmatrix} =
\begin{bmatrix}
k_{3,3}^2 & k_{3,4}^2 & k_{3,5}^2 & k_{3,6}^2 & k_{3,7}^2 & k_{3,8}^2 \\
k_{4,3}^2 & k_{4,4}^2 & k_{4,5}^2 & k_{4,6}^2 & k_{4,7}^2 & k_{4,8}^2 \\
k_{5,3}^2 & k_{5,4}^2 & k_{5,5}^2 & k_{5,6}^2 & k_{5,7}^2 & k_{5,8}^2 \\
K_{6,3}^2 & k_{6,4}^2 & k_{6,5}^2 & k_{6,6}^2 & k_{6,7}^2 & k_{6,8}^2 \\
k_{7,3}^2 & k_{7,4}^2 & k_{7,5}^2 & k_{7,6}^2 & k_{7,7}^2 & k_{7,8}^2 \\
k_{8,3}^2 & k_{8,4}^2 & k_{8,5}^2 & k_{8,6}^2 & k_{8,7}^2 & k_{8,8}^2
\end{bmatrix}
\begin{Bmatrix} d_3 \\ d_4 \\ d_5 \\ d_6 \\ d_7 \\ d_8 \end{Bmatrix}
\qquad (360\text{-}1b)
$$

Fig. 359-1. For example, from the free-body diagram of node 2 as shown on Fig. 361-1, the force equilibrium equation for the y direction becomes

$$[\Sigma F_4 = 0]$$

$$R_4 - F_4^1 - F_4^2 = 0 \qquad (360\text{-}2)$$

The forces F_4^1 and F_4^2 are the internal equivalent nodal forces from elements 1 and 2 as obtained from Eqs. 360-1a and 360-1b. The equilibrium equation (Eq. 360-2) can thus be written as

$$
k_{4,1}^1 d_1 + k_{4,2}^1 d_2 + (k_{4,3}^1 + k_{4,3}^2) d_3 + (k_{4,4}^1 + k_{4,4}^2) d_4 + (k_{4,5}^1 + k_{4,5}^2) d_5
$$
$$
+ (k_{4,6}^1 + k_{4,6}^2) d_6 + k_{4,7}^2 d_7 + k_{4,8}^2 d_8 = R_4 \qquad (360\text{-}3)
$$

The equilibrium equations along all lines of action will generate a system of equations of the following form.

$$[K]\{\delta\} = \{R\} \qquad (360\text{-}4)$$

where

$[K]$ = stiffness matrix for the complete assemblage
$\{\delta\}$ = nodal displacement vector
$\{R\}$ = vector of applied nodal loads

Figure 361-1   Free-body diagram of joint.

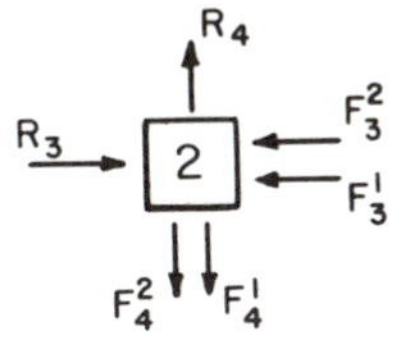

The elements of the complete stiffness matrix will be the coefficients of the equilibrium equation. For example, the elements of the fourth row are obtained from Eq. 360-3, which is the equilibrium equation along line of action 4 (vertical direction at joint 2). These coefficients are

$$K_{4,1} = k_{4,1}^1$$
$$K_{4,2} = k_{4,2}^1$$
$$K_{4,3} = (k_{4,3}^1 + k_{4,3}^2)$$
$$K_{4,4} = (k_{4,4}^1 + k_{4,4}^2)$$
$$K_{4,5} = (k_{4,5}^1 + k_{4,5}^2)$$
$$K_{4,6} = (k_{4,6}^1 + k_{4,6}^2)$$
$$K_{4,7} = k_{4,7}^2$$
$$K_{4,8} = k_{4,8}^2$$

Application of Boundary Conditions

Equation 360-4 in its present form cannot be directly solved for the unknowns, because some unknowns are displacements (at the unrestrained nodes) and some unknowns are external nodal forces (reactions at the restrained nodes). It is therefore necessary to reorganize the equations so that the unknown displacements are grouped. After reorganizing the equilibrium equations, Eq. 360-4 can be written in a partitioned form as

$$\begin{bmatrix} K_{DD} & \vdots & K_{DR} \\ \cdots & \vdots & \cdots \\ K_{RD} & \vdots & K_{RR} \end{bmatrix} \begin{Bmatrix} \delta_D \\ \delta_R \end{Bmatrix} = \begin{Bmatrix} R_D \\ R_R \end{Bmatrix} \qquad (361\text{-}2)$$

The subscript D of the submatrices indicates the lines of action that are free to displace, and the subscript R indicates the restrained lines of action.

From Eq. 361-2,

$$[K_{DD}]\{\delta_D\} + [K_{DR}]\{\delta_R\} = \{R_D\}$$

or

$$[K_{DD}]\{\delta_D\} = \{R_D\} - [K_{DR}]\{\delta_R\}$$

(362-1)

For many problems the displacements of the restrained joints are zero, and Eq. 362-1 reduces to

$$[K_{DD}]\{\delta_D\} = \{R_D\}$$

(362-2)

In some cases there may be specified displacements at the restrained joints. When some of the elements of $\{\delta_R\}$ are nonzero, the product of $[K_{DR}]\{\delta_R\}$ in Eq. 362-1 can simply be subtracted from $\{R_D\}$ and the resulting equation will have the form of Eq. 362-2. These equations can be solved for the unknowns $\{\delta_D\}$.

Expanding Eq. 362-2 for the problem of Fig. 359-1 yields

$$\begin{bmatrix} K_{3,3} & K_{3,4} & K_{3,7} & K_{3,8} \\ K_{4,3} & K_{4,4} & K_{4,7} & K_{4,8} \\ K_{7,3} & K_{7,4} & K_{7,7} & K_{7,8} \\ K_{8,3} & K_{8,4} & K_{8,7} & K_{8,8} \end{bmatrix} \begin{Bmatrix} d_3 \\ d_4 \\ d_7 \\ d_8 \end{Bmatrix} = \begin{Bmatrix} R_3 \\ R_4 \\ R_7 \\ R_8 \end{Bmatrix}$$

(362-3)

The system of equations represented in matrix form by Eq. 362-3 can now be solved for the unknown nodal displacements d_3, d_4, d_7, and d_8 resulting from the external nodal forces R_3, R_4, R_7, and R_8. The external nodal forces could represent equivalent nodal forces from loads applied to the boundaries of the system; they could represent equivalent nodal forces from the weight of each element; or they could represent other types of applied nodal loads.

Obtaining Element Stresses

After obtaining all the nodal displacements, the element strains and stresses can be obtained from Eqs. 355-3 and 356-2.

Geotechnical Engineering Stress Analysis

Since soil has nonlinear stress-strain properties, a linear elastic stress analysis is generally not appropriate for geotechnical engineering problems. However, an understanding of the finite element method for linear elastic stress analysis problems provides a basis for understanding the details of more advanced problems and is sufficient for an understanding of the basic principles behind the finite element method. The nonlinear property of soils is generally handled by an equivalent linear procedure. This involves making an assumption for the modulus of elasticity E and obtaining an elastic analysis. On the basis of the computed strain at all points in the finite element system, new values of E are selected that are compatible with the computed strains, and another elastic analysis is performed. This procedure is repeated several times until E is compatible with the strains.

COMPUTER APPLICATIONS

The tremendous computational power of the computer is a valuable asset in solving engineering problems. For example, slope stability analysis is more accurate if a large number of trial surfaces are examined.

The use of computer programs does not reduce the quality or quantity of "engineering" required to evaluate slope stability problems. If anything, it increases the engineer's responsibility. In performing a computer-aided analysis, an engineer is responsible to:

- Idealize a model of the problem that is to be analyzed.
- Determine the appropriate critical design state.
- Select for each design state the appropriate strength parameters for the various soils.
- Select, thoroughly understand, and prepare the data for the computer program that is to be used for the analysis.
- Interpret the results of the analysis and examine them for reasonableness.

The only task missing from an analysis that is performed without the aid of a computer is to carry out the time-consuming computations.

In interpreting the results of an analysis, an engineer must always examine the solution to determine if it is reasonable. Even the most thoroughly tested and documented computer programs often have "bugs" that go undetected and do not show up until a certain configuration and set of parameters are encountered. An engineer is responsible for the solution even when the computations were performed by a computer (or a technician).

General professional acceptance of computer utilization for solving geotechnical engineering problems began in the early 1960s. Since that time, there have been many advances in numerical solution techniques, computer program (software) development, and computer hardware development. Furthermore, computer utilization has become a basic part of nearly all undergraduate and graduate civil engineering curricula.

Because the development of computer software and hardware has been so extensive, an engineer should ask and seek answers to the following questions before undertaking the development of a computer program to solve a particular problem.

1. Does the problem warrant a computer solution?
   (a) Will an approximate manual solution be just as appropriate and fast in light of the known physical parameters?
   (b) Will the program have future applications?
2. Are there already available computer programs that will solve the problem?
3. Does use of the computer allow use of a more effective technique for the problem than just computerization of a manual technique?

In general, the computer is a fine tool for engineering analysis and leaves the engineer free to concentrate on decisions between solution alternatives instead of spending time on computations.

11

SUBSURFACE EXPLORATION

It is important that an adequate subsurface investigation be made before acquiring a building site or making other investments dependent upon a particular site. Failure to recognize this requirement can result in the need for making costly relocation plans or for spending additional money to utilize a poor site. Adequate ground support of structures, while absolutely essential, may seem unimportant and is often taken for granted until an actual failure occurs. As a result, engineers frequently must point out to clients the need for an adequate subsurface investigation as an essential step for construction projects.

This chapter describes methods of exploring and classifying subsurface soils for engineering uses. Techniques for evaluation of the subsurface site conditions range from general reconnaissance to detailed soil testing. The specific procedures to be used in a given instance depend on the geographical and geological conditions at the site, the nature of the proposed construction, and the experience of the engineer. Where possible, standard procedures for investigation, sampling, and testing should be used. Where applicable, reference has been made here to standards described by the American Association of State Highway and Transportation Officials (AASHTO) and by the American Society for Testing and Materials (ASTM).

Subsurface soil and rock investigations can provide the following kinds of information for making design decisions in a variety of project situations.

- Establish suitable horizontal and vertical location of a proposed structure.
- Locate and evaluate borrow material for construction of earth embankments for highways or earth dams.
- Locate and evaluate sands and gravels suitable for highway aggregate, concrete aggregate, filter material, or slope protection.
- Determine need for subgrade or foundation treatment to support loads or to control water movement.
- Estimate foundation settlement or evaluate the stability of slopes or foundations.

A subsurface site investigation usually includes the following phases: site reconnaissance, exploration planning, subsurface investigation (field testing, laboratory testing, and analysis of data), and preparation of a final report. Refer to the introduction of Chapter 1 for a more general and philosophical discussion of project analysis.

SITE RECONNAISSANCE

Reconnaissance of the project area is the obvious first step in a site investigation. Excavations or cuts in the area reveal information on subsurface conditions. For most areas, topographic maps, aerial photos, and geologic or mineral resource maps are available. County soil survey reports prepared by the U.S. Department of Agriculture Soil Conservation Service, when available, show soil characteristics from depths of 3 to 15 ft. The more recently prepared soil survey maps include engineering soil classification information. Reports on subsurface conditions may be available for adjacent sites. State agencies (State Engineer or Water Resource Department) may have on file drilling logs for water wells in the area. Information from these sources can be obtained at relatively little cost and serve as useful guides in planning the extent of exploratory testing.

EXPLORATION PLANNING

A profile of subsurface conditions can be prepared from logs of soil and rock exposed in cuts and excavations and from logs of previous test borings. Soil profile data are only definite at the boring site and care must be exercised in extrapolating information between borings.

The spacing of borings is dependent upon the complexity of the geology and the bearing that continuity of the soil and rock formations have on the project design. Initial borings may be widely spaced. In such cases additional borings can be spaced at closer intervals in those areas where initial findings indicate the need for more detailed information of subsurface conditions. With more uniform geologic conditions, wider-spaced borings are acceptable.

The depth of boring will depend on the type of construction and depth of influence of surface loads as discussed in Chapter 3. For a rectangular loaded area (building) the recommended minimum boring depth varies from one to two times the minimum dimension of the area. For highways the depth of boring may be as little as 2 m below the subgrade.

Supplemental information between borings can be obtained by use of geophysical methods. The seismic refraction and electrical resistivity methods permit tracing the depth of soil formations between widely spaced borings. These techniques are based respectively on changes in density and electrical resistivity of subsurface layers.

SUBSURFACE INVESTIGATION

A variety of techniques are available for sampling and exploring subsurface soils (Hvorslev, 1949). Engineering judgment must be used in selecting the appropriate method for each circumstance. Methods of investigation can be categorized as (1) methods of sample retrieval and (2) in-situ testing (evaluation of soil properties in place). The choice of the procedure used depends upon the nature and depth of the soils and the tolerable degree of sample disturbance permitted. Methods of underground exploration and sampling are summarized in Fig. 368-1.

Soil Sampling

Methods of retrieving soil samples range from simple hand-dug test pits to sophisticated drilling and sampling devices.

Test Pits and Hand Augers

For shallow exploration a test pit dug by hand shovel may suffice. Excavation with a tractor-mounted back hoe is a rapid means of digging test pits 3 to 4 m deep. Shoring may be needed for safety in test pits deeper than 2 m. Normally, large disturbed samples are taken from a test pit, but hand-carved specimens of undisturbed soil can be readily obtained.

Hand augers from 50 mm to 150 mm in diameter are frequently used for shallow exploration. They operate best in cohesive soils above the water table. The holes are normally not cased. Retention of dry sand in the sampler or penetration of rocky soils may be impossible. Below the water table, retention of cohesive samples may also prove difficult. Operating a hand

Figure 368-1 Methods of underground exploration and sampling.

Common name of method	Materials in which used	Method of advancing the hole
Wash borings	All soils. Cannot penetrate boulders.	Washing inside a driven casing.
Dry-sample boring	All soils. Cannot penetrate boulders or large obstructions.	Washing inside a driven casing.
Undisturbed sampling	Samples obtained only from cohesive soils.	Usually washing inside a 6-in. casing. Augers may be used.
Auger boring	Cohesive soils and cohesionless soils above groundwater elevation.	Augers rotated until filled with soil and then removed to surface.
Well drilling	All soils, boulders, and rock.	Churn drilling with power machine.
Rotary drilling	All soils, boulders, and rock.	Rotating bits operating in a heavy circulating liquid.
Core borings	Boulders, sound rock, and frozen soils.	Rotating coring tools: diamond shot, or steel-tooth cutters.
Test pits	All soils. In pervious soils below groundwater level pneumatic caisson or lowering of groundwater is necessary.	Hand digging in sheeted or lagged pit. Power excavation occasionally used.

Note: Test rods, sounding rods, jet probings, geophysical methods, and so forth are not included in this table, because no samples are obtained. This table is taken from "Exploration of Soil Conditions and Sampling Operations" by H. A. Mohr, Soil Mechanics, Series 21, Third Revised Edition, Publication 376 of the Graduate School of Engineering, Harvard University, November 1943.

11 Subsurface Exploration

Method of sampling	Value for foundation purposes
Samples recovered from the wash water.	Almost valueless and dangerous because results are deceptive.
Open-end pipe or spoon driven into soil at bottom of hole.	Most reliable of inexpensive methods. Data on compaction of soil obtained by measuring penetration resistance of spoon.
Special sampling spoon designed to recover large samples.	Used primarily to obtain samples of compressible soils for laboratory study.
Samples recovered from material brought up on augers.	Satisfactory for highway exploration at shallow depths.
Bailed sample of churned material or samples from clay socket.	Clay socket samples are dry samples. Bailed samples are valueless.
Samples recovered from circulating liquid.	Samples are of no value.
Cores cut and recovered by tools.	Best method of determining character and condition of rock.
Samples taken by hand from original position in ground.	Materials can be inspected in natural condition and place.

auger to depths greater than about 2 m is possible but becomes increasingly tedious.

Power Drills
Various types of power augers are available in diameters up to 1 m and can operate up to 30 m in depth.

Augers provide only disturbed soil samples. Accurate identification of the level from which a particular soil comes depends primarily on how the auger is operated. Flight augers that lift the soil continuously from the bottom of the hole would produce a mixed sample. The auger may be used to advance the hole a short distance. After withdrawing the auger, a suitable sampling device is driven or pushed into the soil at the bottom of the hole to obtain a representative sample from that level.

Hollow-stem augers may be used to advance the hole and may be left in place while the soil at the bottom of the boring is sampled through the hollow stem and plug. Inside diameters of 63 mm ($2\frac{1}{2}$ in.) to 86 mm ($3\frac{3}{8}$ in.) are commonly used (Fig. 371-1).

Drilling through cohesionless soil may require that the boring be cased to prevent the hole from collapsing, particularly below the water table. Wash boring equipment is frequently used to advance the hole. Drilling fluid (usually a bentonite slurry) is circulated through the drill bit and carries cuttings to the surface. The drilling fluid may also serve to keep the hole from collapsing, making it unnecessary to case the hole. The soil cuttings lifted by the drilling fluid are thoroughly mixed and are not of much use in identifying the soil at the bottom of the hole. At intervals, the drill is withdrawn and a sampler device is driven or pushed into the soil at the bottom of the hole.

For drilling through harder materials or rock, special rotary-drill bits are used with either compressed air or drilling fluid to lift the cuttings to the surface. Diamond core drills are used where a core of the rock material is to be obtained. Pressure drills are not often used for exploratory drilling.

Thick-Walled Samplers
Thick-walled samplers range from 50 mm (2 in.) to 115 mm ($4\frac{1}{2}$ in.) outside diameter and are driven into soil to obtain samples suitable for identification and classification tests. Split-barrel samplers (Fig. 372-1a) make removal of the sample easier than do solid-tube samplers. Recovery of the sample may be difficult in dry coarse soils.

Thin-walled Samplers
When undisturbed samples of cohesive or semicohesive soils for testing shear strength or consolidation characteristics are required, thin-walled samplers[1]

[1]AASHTO T-207

Figure 371-1 Hollow-stem Auger

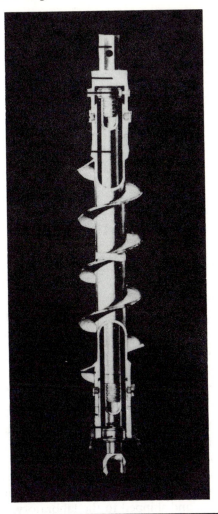

are used. Hvorslev (1949) has defined a thin-walled sampler as having a wall area less than 10% of the area of the sample. Hard-drawn seamless steel tubing 50 to 140 mm (2 in. to $5\frac{1}{2}$ in.) in diameter, known as Shelby tubing, is frequently used (Fig. 372-1*b*). The tubing is typically 1 m long of 12 to 18 Ga metal. A simple adapter is used to attach the tubing to the end of a hollow drill rod. The sample is taken by forcing the sampler into the soil in smooth

Subsurface Investigation

Figure 372-1

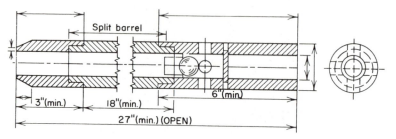

(*a*) Thick-walled tube sampler, standard penetrometer.

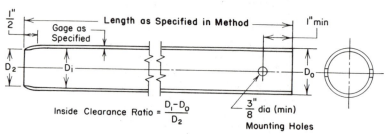

(*b*) Thin-walled tube sampler.

continuous strokes. The cutting edge of the sampler is rolled in and sharpened to cut a sample slightly smaller than the inside diameter of the sampler tube to minimize the drag on the sample as it moves into the tube. Fitting a ball check valve in the adapter to provide a vacuum as the sampler is withdrawn helps to recover the sample. The filled sampler tube is removed from the adapter, sealed, and shipped to the laboratory.

For dense soils containing hard materials that would bend the thin-walled tube sampler, samplers such as the Denison type are used. The Denison sampler has a rotating outer shell with teeth that cut through the soil, allowing the sample to slide into the nonrotating inner barrel of the sampler.

For soft, sensitive clays, the Swedish foil sampler is used to take long continuous samples (up to 20 m) with minimum disturbance to the sample. The sampler has a piston attached to 16 thin steel foil ribbons housed in the outer barrel of the sampler. The piston remains stationary as the sample barrel is pushed into the clay. The steel foils are threaded through a slot behind the cutting edge of the sampler and line the inside of the 67.8 mm diameter sampler tube so the sample does not drag on the wall of the tube. The filled tube is taken apart in about 2 m long sections and sent to the laboratory, where the sample is extracted with minimum disturbance by pulling the foils.

There are many other types of soil-sampling devices. Hvorslev (1949) has described many such devices and techniques.

In-situ Testing

In situ tests of subsurface soils often provide better information at lower cost than taking and testing samples. Several types of in-situ tests are described in the following paragraphs.

Vane Shear Test

The shear strength of in-place soft-to-medium stiff clays can be measured by pushing a four-bladed vane (Fig. 163-1) into the clay and measuring the torque required for rotation. Suitable vanes are relatively easy to make. Recommended dimensions for field vanes are given by the AASHTO specification.[2] The vane is attached to a stiff rod of sufficient length to reach the bottom of the boring. This test is suitable only for soft-to-medium stiff clays (Fig. 30-1) that do not contain sand lenses, gravel, or other solid particles.

Field Permeability Tests

The permeability of pervious layers below the water table can be found by measuring the draw-down in piezometers located at distances from a pumped well (Fig. 58-1). The well should be constructed and developed according to good well practice. Tests should be made at several pumping rates and pumping continued until reasonably steady-state conditions are achieved. Computation of the permeability is made from the appropriate equations depending upon whether the well is artesian or of the open aquifer type.

Various types of borehole permeability tests made by pumping water either into or out of the boring are used (Fig. 56-1). The bore hole may or may not extend below the water table. The Earth Manual of the U.S. Bureau of Reclamation gives procedures and calculations.

Plate-bearing Test

Soil-bearing capacity for the design of spread footings and highway or airport pavements is often measured by plate-bearing tests. Circular steel plates at

[2]AASHTO T-223.

least 25 mm thick varying in diameter from 300 to 760 mm are prescribed for standard tests.[3] Load settlement data permits determination of the coefficient of subgrade reaction.

Care should be exercised in predicting the bearing capacity of large footings based on test results from small test areas, since the zone of significant stress influence extends only to a depth below the footing equal to little more than the footing width. The plate-bearing test is discussed in Chapter 8 as a means of predicting the settlement of footings on sand.

Penetration or Sounding Tests

The force or energy required to push or drive a probe into the earth is a measure of soil strength or bearing capacity. Over the years engineers have developed a great variety of equipment and techniques based on this simple idea. The two principal methods in use are the Standard penetration test and the cone penetration test. The Standard penetration test is used primarily in the United States. The cone penetration test was developed in Holland and is widely used in Europe and other parts of the world. Recently there has been an increasing use of the cone penetrometer test in the United States.

Standard Penetration Test[4]

The number of blows required to drive a standard split barrel sampler (see Fig. 372-1) 0.3 m (12 in.) into the soil is the standard penetration resistance N. The standard split barrel sampler has a 50.8 mm OD (2 in.) and a 34.9 mm ID (inside diameter) ($1\frac{3}{8}$ in.). It is approximately 0.76 m long and is connected to an A-size (or larger) drill rod. The sampler is driven by a 64 kg mass (140 lb) falling 0.762 m (30 in.). The sampler is lowered to the bottom of a bore hole and set by driving it an initial 0.15 m (6 in.). It is then driven an additional 0.3 m (12 in.), and the number of blows required to advance the sampler the additional 0.3 m is recorded as the standard penetration resistance N. A disturbed but representative soil sample is normally recovered from the barrel of the sampler for a visual identification.

The Standard penetration test is most frequently used to measure the relative density of granular soils. Although the test is sometimes used as a measure of the shear strength of cohesive soils, the correlation is not as reliable for cohesive soils. Several factors can influence the blow count obtained, and care should be exercised when evaluating the test results. With increasing depth in granular soils and increasing overburden weight, the number of blows needed to drive the sampler increases for sand of the same relative density. Figure 375-1 shows an estimated relationship between blow count and overburden pressure for coarse sand of constant relative density.

[3]AASHTO T-222; ASTM D-1196.
[4]ASTM D-1586; AASHTO T-206.

Figure 375-1 Relationships among standard penetration resistance, relative density, and effective overburden pressure for coarse sand.

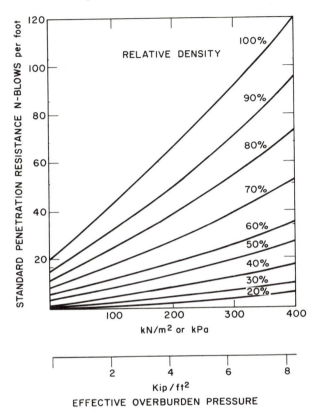

In testing below the water table, it is important that the water level in the bore hole be kept above the groundwater level. Quick withdrawal of the drill tools could drop the water level in the bore hole below the groundwater level. Groundwater flowing in at the bottom of the boring will loosen granular soils, causing the blow count for the next penetration test to be low.

Standard penetration test results are affected by many other factors, such as presence of gravel in the soil, use of a drill rod heavier than type *A*, dull sampler shoe, carelessness in dropping the ram the required distance, and many other factors. Despite the somewhat crude character of the test, it is widely used.

Subsurface Investigation

The use of the standard penetration test to predict settlement of footings on sand is presented in Chapter 8 as Eq. 283-1. Relationships between the standard penetration resistance, N, friction angle, ϕ, and relative density, D_r, are presented in Chapters 8 and 9 as Figs. 291-1 and 315-1.

Cone Penetration Test[5]

The cone penetrometer consists of a 60° cone with a circular base area of 1000 mm^2 (see Fig. 376-1). The cone is pushed into the soil at a rate of 10 to

[5]ASTM D-3441-75T.

Figure 376-1   Mechanical friction-cone penetrometer.

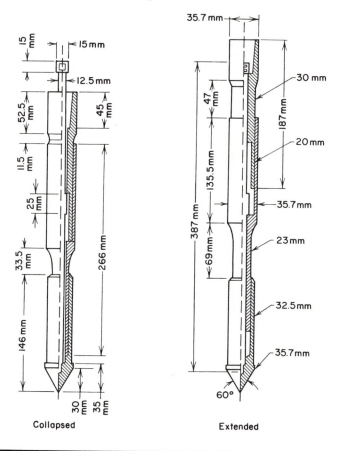

Collapsed   Extended

20 mm/sec by hydraulic pressure applied to drill rods extending from the cone to the ground surface. The penetration resistance q_c is found by dividing the measured force by the 1000 mm² cone area.

The Begeman friction cone (Begeman, 1965) has a friction sleeve mounted on the rods behind the cone. A system of inner and outer rods permits advancing either the cone alone or the cone and sleeve together. Thus both a cone-bearing capacity q_c and a soil-sleeve friction f_s are measured.

The penetrometer is pushed to the initial sampling depth using the outer push rods. The inner rods are used to force the cone ahead 40 mm. At this point a shoulder on the shaft engages the sleeve, and again, using the inner rods, both the cone and the sleeve are advanced an additional 40 mm. The measured force in each case is recorded and used to calculate q_c and f_s. The outer push rods are then advanced 200 mm, closing up the spacing between the cone, sleeve, and outer push rods, leaving the unit in a position 200 mm below the initial position. Note that the bearing and friction measured on one push are not at the same level. The bearing pressure q_c measured on one push must be correlated with the friction f_s measured on the following push.

In addition to q_c and f_s, the friction ratio R_f equal to f_s/q_c (%) is also calculated. Plots are then made showing the variation of q_c, f_s, and R_f with depth.

The parameters q_c, f_s, and R_f have been correlated with soil characteristics in several ways. The values of q_c and R_f can be used in Fig. 377-1 for an approximate identification of the soil type.

Figure 377-1 Cone penetrometer bearing capacity.

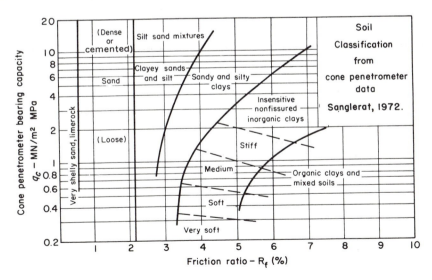

Figure 378-1   Limiting cone resistance as a function of friction angle. (After Meyerhof, 1974.)

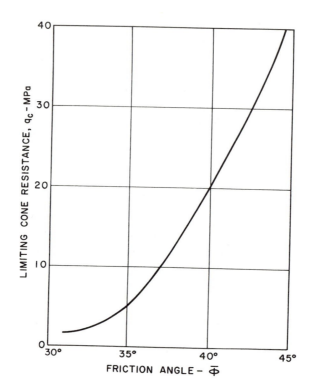

FRICTION ANGLE – $\overline{\phi}$

Several methods are available for evaluating the drained shear strength, $\overline{\phi}$, of sands from cone penetration resistance. For a review of such methods, see Mitchell (1978). One such relationship suggested by Meyerhof (1974) is shown in Fig. 378-1. The value of q_c used is the limiting or maximum cone resistance measured at a critical depth below which the penetration pressure shows no increase with further penetration. Difficulty in identifying the limiting q_c in soils where q_c continues to increase with depth may limit the usefulness of this method.

The undrained shear strength of clay can also be estimated from cone penetration data. Schmertmann (1975) suggests that $S_u = f_s$ is a lower limit

11   Subsurface Exploration

for the undrained strength. From the bearing capacity equation (Eq. 267-1) the undrained shear strength becomes

$$S_u = \frac{q_c - \gamma z}{N_c} \qquad (379\text{-}1)$$

where γz is the total overburden pressure at the depth at which q_c is measured. The factor N_c has been shown to vary from 5 to 70; however, for recent clay deposits of low sensitivity, a plasticity index less than 10% and an overconsolidation ratio of less than 2, an N_c value of 16 is recommended (Schmertmann, 1975). As a better procedure, a few undrained shear tests to measure values of S_u should be used in Eq. 379-1 to better ascertain the correct value for N_c.

The cone penetrometer data is also useful for identifying compressible layers that will settle excessively when loaded. Suggested criterion for identifying compressible layers is

$$q_c < \frac{1MN}{\text{m}^2} \text{ (Sanglerat, 1972)}$$

Pile capacity can be estimated from q_c and f_s using the following formula based on Eq. 297-1.

$$Q_{\text{allowable}} = \frac{1}{F} \left(q_c A_p + \sum_o^L f_s \times \Delta A_s \right)$$

where

q_c = the average cone resistance at the pile tip
f_s = the average sleeve friction for ΔA_s
ΔA_s = pile perimeter $\times \Delta L$ = incremental pile surface area
F = safety factor
A_p = Area of pile tip.

Some advantages of using the cone penetrometer over conventional bore procedures with standard penetration tests and tube sampling include:

1. More accurate and directly useful information about soil properties is obtained.
2. The rate of probing is faster and more economical.
3. The shorter sampling interval permits more accurate identification of subsurface layers.

Some disadvantages of the cone penetrometer are:

1. No samples of subsurface material are obtained.
2. Groundwater conditions cannot be readily evaluated.

3. The procedure is limited to soils that do not contain large rock or hard layers, which prevent penetration of the cone.
4. In some soils the interpretation of cone resistance data may be difficult.

It is not recommended that the cone penetrometer be used without some supplementary information from conventional soil borings. Data from conventional borings, that is, Standard penetration blow count, unconfined shear strength from tube samplers, visual identification of soils, and water table data, are vital to accurate interpretation of cone penetrometer data.

The use of the cone penetrometer for predicting settlement of footings on sand is presented in Chapter 8.

Recording Field Data

Careful records of exploratory work should be kept to realize maximum benefit from the effort. The following information should be routinely recorded in field notes.

1. Date of the work and names of personnel involved.
2. Horizontal location of borings or test pits with respect to an established coordinate system.
3. Elevation of the ground surface with respect to an established bench mark.
4. Elevation of the water table or explanation of other special circumstances associated with the groundwater conditions.
5. A field classification of the soils encountered.
6. Elevation of the upper level of each soil type encountered.
7. Elevation at which soil samples were taken for laboratory analysis. Such samples should be carefully tagged as to boring number, elevation and soil description.
8. Record of Standard penetration blow count, cone penetration resistance, or vane shear data where these tests are used.
9. Description of the drilling equipment and techniques used. Was it necessary to case the hole? Any difficulties encountered or changes in drilling technique can provide valuable information and should be recorded.

After tests are completed on laboratory samples, information from the field and laboratory should be assembled and summarized. Soil profiles or boring logs with complete soil information can be plotted.

Typical soil boring plots are shown in Fig. 381-1 and 382-1. The soil in Fig. 381-1 is primarily cohesive and was sampled with a Swedish foil sampler. The boring at the granular site of Fig. 382-1 was made with a wash boring rig. Standard penetration tests were made at intervals.

11 Subsurface Exploration

Figure 381-1

LOG OF FOUNDATION INVESTIGATION – UNDISTURBED SAMPLES

PROJECT <u>Dike Investigation</u> HOLE NO. <u>PSII</u> LOCATION <u>Sta 180 + 90</u> Log Sheet <u>3</u> of <u>5</u>

Ground Surface El. <u>4244.0</u> Ground Water or Water Surface El. <u>4244.0</u> Top of Rock El. _____ Top of Casing El. _____

Sample Size <u>66 mm</u> Ave. Sampling Prog. ft/hr _____ Investigation Begun <u>8-3-57</u> Investigation Fin. _____

Jar No. or Tube	Sample No.	Elevation	Depth	Stratification	DESCRIPTION OF MATERIALS		PROCEDURE Description of Operation, General Remarks on Investigation	Logged by
		4224.0	20					
FSII – B	8				Medium stiff, grey tailings of low to medium plasticity.			
	9				Firm, brown, fine sand; fairly clean.			
	10		21		Medium stiff, grey tailings of low to medium plasticity.			
	11				Firm, brown, fine sand; fairly clean.			
	12				Medium stiff, grey tailings of low to medium plasticity.			
	13				Firm, brown, fine sand.			
	14		22		Medium stiff, grey, interbedded tailings of low to medium plasticity.			
	15		23					
	16							
	17		24					
	18				Firm, grey, fine sand; fairly clean.			
FSII – C	1		25		Medium stiff, grey, interbedded tailings of low to medium plasticity; predominately low plasticity tailings.			
	2				Low plasticity tailings.			
	3		26		Low plasticity tailings.		Foil stress – 900 kg. Strain – 5.3 mm Clevis attached to piston chain broke at 27.4' Stopped sampling.	
	4		27					
					Not sampled			
FSII – D	1		28		Firm, grey, silty, fine sand.		8/3/57 Lowered sampler to 27.6' and began sampling. Initial foil stress – 100 kg.	
	2				Medium stiff, grey, interbedded tailings of low to medium plasticity; predominately low plasticity tailings.			
	3		29					
	4							
	5	4214.0	30					

Driller:_____ Checked by _____ Resident Engineer:_____

Equipment used:_____ Remarks: _____

Figure 382-1 Typical Soil Boring Log.

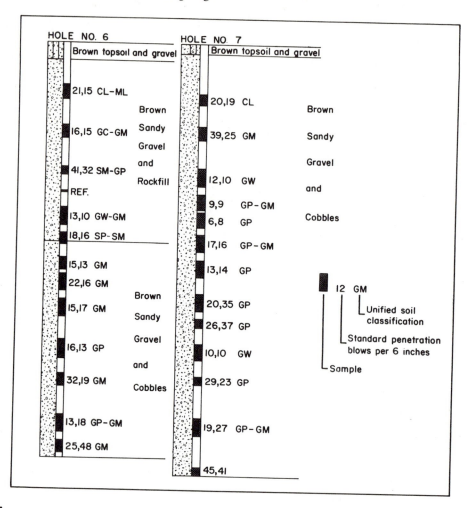

Field Soil Classification[6]

As soils are sampled, they should be classified and described for at least two reasons. First, the immediate information on soil types encountered may be useful

[6]ASTM D-2488.

in adjusting plans for further exploratory work, and second, a comparison between the laboratory and the field classifications serves to verify the accuracy of the sample identification.

Laboratory analysis is necessary to definitely classify a soil according to the AASHTO (Fig. 35-1) or Unified soil classification system (Fig. 37-1), although an experienced technician can field-classify a soil with reasonable accuracy.

The most obvious classification is based on the particle size of the soil. Size definitions adequate for field classifications are shown in Fig. 33-1.

The coarser soils can be distinguished by eye. The presence of a fine sand can be detected by a gritty feeling when the sand is rubbed between the fingers. Particles passing the No. 200 sieve are about as small as can be seen with the unaided eye; therefore, silts and clays are distinguished by pasticity, dilatancy, or dry strength tests.

The plasticity of a soil is measured in terms of the range of water content over which the soil behaves as a plastic material (Fig. 28-1). A rough comparison of soil plasticity can be made by kneading a soil between the fingers and noting the time it takes the wet soil to dry from near the liquid limit to a crumbly state near the plastic limit. A silty soil will reach the crumbly state sooner than a clay.

The dilatancy test is made by vibrating a pat of wet soil held in the open palm of the hand. If the soil reacts to this test, a shiny film of water forms on the soil surface as the vibration increases the density of the soil and forces water to the surface. When the soil is squeezed between the fingers, the surface becomes dull as the squeezing causes the soil to increase in volume or dilate, thus sucking the surface water back into the soil pores. A fine sand reacts readily to this test, a silt somewhat slower, and a clay not at all.

The strength of an air-dried sample of soil also enables one to distinguish between a silt and clay. The more clay present in a soil the more difficult it is to crush between the fingers.

Other distinguishing features of the soil should be noted, such as color, odor, and presence of organic material, rock fragments, or other materials. Soils containing organic matter usually have a characteristic odor. The color of a soil often varies with the water content and time of exposure to air, so this should be noted. For example, some black organic soils turn white on their surface after a few hours of exposure to air.

Laboratory Testing

Soil samples arriving at the laboratory must be in a condition still representative of the field soil if the laboratory tests results are to be of greatest value.

Sample Handling

Careful handling of samples should be observed so that changes that are likely to influence the laboratory test results do not occur. Large disturbed samples may be handled in bags or cans; bags should be of closely woven material to prevent loss of soil fines. Small samples are often placed in pint or quart jars or in plastic bags.

Undisturbed samples require special treatment to prevent loss of moisture or sample distortion. The ends of sample tubes may be sealed with wax or capped by other means. The time between sampling and testing should be kept as short as possible.

Disturbed-Sample Preparation

Initial preparation of disturbed-soil samples may be made in two ways: the dry method or the wet method. Because less time and effort are involved in the dry method, it is used more frequently. Test results on samples prepared by the wet method have been shown to differ from samples prepared by the dry method (Gillette, 1956). Higher liquid and plastic limits obtained for samples prepared by the wet method are believed to be due to greater retention of fine colloidal material. These two procedures are described by standards as follows.

- Dry method — ASTM D421-58.
    AASHTO T87-72.

- Wet method — ASTM D2217-63T.
    AASHTO T146-49.

REPORT PREPARATION

The report of a subsurface investigation might range from a one- or two-page letter to a lengthy formal document. The procedures of good engineering report writing should be observed. Since the conclusions or recommendations are the most important sections of a report, they should have a prominent place in the report. They should not, however, be placed first, because the reader would not have read any introductory information. Thus, a long formal report should include a brief summary or background. The remainder of the report should include a more detailed introduction, a description of field investigations and laboratory investigations, and analysis of data. The body of the report might include summarized data in the form of graphs or charts. Normally, more detailed information on test results is contained in an appendix.

As a general guide in preparation of a report, the topics or elements common to many engineering reports are presented here.

Cover and Title Page

Letter of Transmittal

Table of Contents

List of Figures and Tables

    Abstract (For a more formal report an abstract may be used, with the Summary and Conclusions placed at the end of the report)

Introduction
  Purpose or Scope of Investigation
  Background Information
  Outline of Investigative Procedure

Field Investigation
  Phases of Investigation
  Typical Results

Laboratory Tests
  Description of Test Procedures
  Typical Results
  Interpretation of Results

Analysis of Problems
  Analytical Procedures

Summary and Conclusions
  Authorization
  Need for Investigation
  Summarized Investigation
  Conclusions and Recommendations

Appendix
  Detailed results of tests and analyses

Appendix A

INTERNATIONAL SYSTEM OF UNITS

The SI system (Système International d'Unités) is used in this text except in a few instances where English units have been given for purpose of comparison. In the SI system the basic units for time, length, and mass are, respectively, the second, the meter, and the kilogram. The unit of force is the Newton, which is defined from Newton's second law as the product of mass times acceleration.

$$F = Ma, \text{Newton} = (\text{kilogram}) \left(\frac{\text{meter}}{\text{second}^2} \right)$$

In geotechnical engineering, force units are used extensively; however, laboratory equipment is calibrated in terms of mass units (kilogram or gram). Data recorded in mass units is converted to force units by multiplying the measured mass in kilograms by the acceleration due to gravity. The average value of acceleration at the earth's surface is 9.81 m/sec². The units for the SI system are summarized in the following table. In using the SI system it is recommended that numerical values be kept between 0.1 and 1000. For larger or smaller values the appropriate prefix should be applied to the units to keep the numerical result within the recommended range. Prefixes and symbols are shown in the following table.

387

SI Base Units

Dimension	Symbol	SI Unit
Mass	M	kilogram (kg)
Length	L	meter (m)
Time	T	second (sec)
Force	F	Newton (N)

SI Prefixes and Symbols

Multiplier	Prefix	Symbol
10^9	giga	G
10^6	mega	M
10^3	kilo	k
10^{-3}	milli	m
10^{-6}	micro	μ
10^{-9}	nano	n

Thus if the mass of soil in a 10.2 cm diameter mold 11.7 cm high measures 1.91 kg, its unit weight is found as follows.

$$\frac{\text{mass} \times \text{acceleration of gravity}}{\text{volume}} = \frac{(1.91 \text{ kg}) \, 9.81 \text{ m/sec}^2}{\frac{\pi}{4} (10.2)^2 (11.7) \text{ cm}^3} \left[\frac{(100 \text{ cm})^3}{(\text{m})^3} \right]$$

$$= 19599 \, \frac{N}{m^3} = 19.6 \text{ kN/m}^3$$

A convenient unit of pressure is the Pascal, which is equal to 1 N/m^2. The Pascal is abbreviated as Pa.

A number of useful engineering quantities are now given in both the English and the SI systems.

Weights
Man, 175 lb—778 N or 0.778 kN
Automobile, 3000 lb—13.34 kN
Pile capacity, 100 tons—890 kN

Unit Weights
Water, 62.5 lb/ft³—9.81 kN/m³
Soft clay, 96 lb/ft³—15.1 kN/m³
Dense sand, 130 lb/ft³—20.4 kN/m³

Appendix A   International System of Units

Pressure or Stress
 1 atm. 14.7 lb/in.^{2}—101 kN/m^2 or kPa

Bearing Capacity of Granular Soil
 2000 lb/ft^2—96 kN/m^2 or kPa
 4000 lb/ft^2—192 kN/m^2 or kPa

Allowable stress for Mild Steel
 20,000 lb/in.^{2}—138,000 kN/m^2 or kPa
       —138 MN/m^2 or MPa

Compressive Strength of Concrete
 3500 lb/in.^{2}—24 MN/m^2 or MPa

Modulus of Elasticity of Steel
 30 × 10^6 lb/in.^{2}—207 GN/m^2 or GPa

Appendix B

COMPUTER PROGRAMS

Two FORTRAN computer programs that utilize the finite difference method are presented here. The purpose of presenting these programs is to illustrate how finite difference methods can be programmed for solution on a digital computer. Both programs solve only specific and very simple problems. This was purposely done so that the logic required for a more general program would not detract from a clear illustration of the finite difference logic. Both programs include comment statements to indicate the purpose of various parts of the program.

Seepage Analysis

The program of Fig. 392-1 utilizes a finite difference solution of the LaPlace equation to solve the problem of seepage under a sheet pile wall as presented in Figs. 63-2, 341-3, and 346-1. The solution is obtained in terms of total head at the grid points of the flow field.

Consolidation Analysis

The program of Figs. 394-1, and 396-1 utilizes a finite difference solution of the Terzaghi differential equation of consolidation (Eqs. 135-4 and 347-1). The results are obtained in terms of the strain as a function of depth and time and the average strain as a function of time. The example illustrates the consolidation of a 10-m single-drained clay layer.

Figure 392-1 Seepage analysis program.

```
      C
            DIMENSION H(15,30)
      C
      C     THIS PROGRAM SOLVES FOR THE DISTRIBUTION OF HEAD FOR A TWO-
      C     DIMENSIONAL POROUS MEDIA FLOW SITUATION DESCRIBED BY THE
      C     LAPLACE EQUATION.
      C
      C     THIS PROGRAM ENCORPORATES THE USE OF AN OVERRELAXATION FACTOR
      C     TO SPEED CONVERGENCE (FOR RF = 1, THIS METHOD IS IDENTICAL
      C     TO THE GAUSS-SEIDEL METHOD)
      C
      C
      C     D = DEPTH OF SOIL
      C     DEL = DISTANCE BETWEEN NODES
      C     EPS = ERROR CRITERIA
      C     HT = TOTAL HEAD DIFFERENCE
      C     RF = RELAXATION FACTOR
      C     MAX = MAXIMUM NUMBER OF ITERATIONS
      C
         1  READ 100,D,DEL,EPS,HT,RF,MAX
       100  FORMAT(5F10.0,I5)
      C
      C     DETERMINE CONTROL DATA
      C
         2  NC=1.*D/DEL+1.
            NR=(NC-1)/1+1
            NJ=(NR-1)/2+1
            PRINT 8000,D,DEL,EPS,HT,RF,MAX
      8000  FORMAT(4X,5F10.3,I5)
      C
      C     INITIALIZE ALL INTERIOR NODAL VALUES TO 0.25 TIMES
      C     THE TOTAL HEAD DIFFERENCE
      C
            DO 10 I=1,NR
            DO 10 J=1,NC
        10  H(I,J)=0.25*HT
      C
      C     SET BOUNDARY CONDITIONS
      C
            DO 11 I=NJ,NR
        11  H(I,1)=HT*0.5
            DO 12 J=1,NC
        12  H(1,J)=0.0
      C
      C     PRINT INITIAL AND BOUNDARY CONDITIONS
      C
            DO 61 I=1,NR
        61  PRINT 1002,(H(I,J),J=1,NC)
            N=0
        50  K=0
            DO 13 I=2,NR
            DO 13 J=1,NC
      C
      C     SET UP LOGIC TO DETERMINE THE LOCATION OF THE PROPER FINITE
      C     DIFFERENCE EQUATION
      C
            IF(J-1)14,14,16
```

Seepage analysis program continued.

```
      14 IF(I-NJ)15,13,13
      16 IF(J-NC)17,18,18
      17 IF(I-NR)19,20,20
      18 IF(I-NR)21,22,22
C
C         EQUATION FOR NODE AT CORNER BOUNDARY
      22 HEAD=RF*.25*(2.*H(I,J-1)+2.*H(I-1,J))+(1.-RF)*H(I,J)
         GO TO 30
C
C         EQUATION FOR NODES AT SHEET PILE
      15 HEAD=RF*.25*(H(I+1,J)+H(I-1,J)+2.*H(I,J+1))+(1.-RF)*H(I,J)
         GO TO 30
C
C         EQUATION FOR NODES AT RIGHT BOUNDARY
      21 HEAD=RF*.25*(H(I+1,J)+H(I-1,J)+2.*H(I,J-1))+(1.-RF)*H(I,J)
         GO TO 30
C
C         EQUATION FOR NODES IN INTERIOR
      19 HEAD=RF*.25*(H(I+1,J)+H(I-1,J)+H(I,J+1)+H(I,J-1))+(1.-RF)*H(I,J)
         GO TO 30
C
C         EQUATION FOR NODES AT BOTTOM BOUNDARY
      20 HEAD=RF*.25*(2.*H(I-1,J)+H(I,J+1)+H(I,J-1))+(1.-RF)*H(I,J)
         GO TO 30
C
C         DETERMINE THE DIFFERENCE IN HEAD FROM PREVIOUS
C         ITERATION AND IF LARGER THAN EPS PERFORM ANOTHER ITERATION
C
      30 DIF=HEAD-H(I,J)
         IF(ABS(DIF)-EPS)31,31,32
      32 K=K+1
      31 H(I,J)=HEAD
      13 CONTINUE
         N=N+1
         IF(K)70,70,71
C
C         STOP IF PROGRAM EXCEEDS MAXIMUM NUMBER OF ITERATIONS
C
      71 IF(N-MAX)50,3,3
       3 PRINT 1001,K,N
    1001 FORMAT(1H1,9X,34HTHE SOLUTION FAILED TO CONVERGE AT,I5,7H  NODES//
         $10X,53HTHE RESULTING HEADS AT EACH NODE ARE AS FOLLOWS AFTER,I5,
         *11H ITERATIONS////)
         GO TO 99
      70 PRINT 2000,N
    2000 FORMAT(1H1,9X,32HTHE SOLUTION HAS CONVERGED AFTER, I5,11H ITERATIO
         $NS////)
         DO 60 I=1,NR
      60 PRINT 1002,(H(I,J),J=1,NC)
    1002 FORMAT(1X//15F7.3)
      99 STOP
         END
```

Figure 394-1 Consolidation analysis program.

```
C
C
C       THIS PROGRAM SOLVES FOR THE DISTRIBUTION OF STRAIN FOR THE
C       ONE-DIMENSIONAL CONSOLIDATION OF A  HOMOGENEOUS SINGLE DRAINED
C       SYSTEM (DIFFERENTIAL EQUATION DEFINED BY TERZAGHI).
C
C
C       ND = NUMBER OF DIVISIONS
C       H = HEIGHT OF SOIL LAYER (SINGLE DRAINED)
C       DELZ = WIDTH OF INCREMENT OF DEPTH (M)
C       CV = COEFFICIENT OF CONSOLIDATION (M** /DAY)
C       TIME = MAXIMUM TIME OF INTEREST (DAYS)
C       DELT = INCREMENT OF TIME (DAYS)
C       TM1 = TIME INTERVAL FOR WHICH OUTPUT IS WANTED IN THE
C             FIRST ONE-FOURTH OF THE MAXIMUM TIME
C       TM2 = TIME INTERVAL FOR WHICH OUTPUT IS WANTED IN THE
C             LAST THREE-FOURTHS OF THE MAXIMUM TIME
C
        DIMENSION STR1(22), STR(22)
        DIMENSION STRAV(1000), TX(1000)
        IN=5
        IOUT=6
C
C       READ INPUT DATA
C
        READ(IN,100)DELZ,DELT,H,CV
   100 FORMAT (4F10.0)
        READ(IN,101)TIME,TM1,TM2
   101 FORMAT (3F10.0)
        WRITE(IOUT,901)
   901 FORMAT ( 1H1)
C
C       WRITE INPUT DATA
C
        WRITE (IOUT,902)DELZ,H,CV
   902 FORMAT (7H DELZ= F8.2,5X,7H      H= F8.2,5X,7H    CV= E12.4)
        WRITE (IOUT,903)TIME,DELT,TM1,TM2
   903 FORMAT (7H TIME= F10.2,5X,7H DELT= F10.2,5X,7H   TM1= F10.4,5X,7H
       1TM2= F10.4///)
C
C       DETERMINE CONTROL DATA
C
        IT=0
        KL = 0
        ND = H/DELZ + 1.001
        XND=ND
        NDP = ND+1
        T=0.
        TCHECK=0.
        ONFRTH = 0.25*TIME
C
C       DETERMINE THE CONSTANT USED IN THE FINITE DIFFERENCE EQUATION
C
        CONST = CV*DELT/(DELZ*DELZ)
C
C       INITIALIZE VALUES OF STRAIN
C
        STR(1) = 100.0
```

Consolidation analysis program continued.

```
        STR1(1) = 100.0
        DO 10 I=2,NDP
    10 STR1(I)=0.0
C
C      PRINT INITIAL VALUES
C
        WRITE(IOUT,1000)
  1000 FORMAT (7X,4HTIME,30X,23HPERCENT OF TOTAL STRAIN /5X,6H(DAYS)/)
        WRITE (IOUT,899)T,(STR1(I),I=1,ND)
    899 FORMAT(5X,F6.0,5X,11F10.3/16X,11F10.3)
        T=T+DELT
    20 STR1(NDP)=STR1(ND-1)
    25 DO 11 I=2,ND
C
C      FINITE DIFFERENCE EQUATION
C
    11 STR(I)=STR1(I) + CONST*(STR1(I-1) - 2.0*STR1(I) + STR1(I+1))
        STR(NDP) = STR(ND-1)
        TCHECK=TCHECK+DELT
C
C      LOGIC TO LIMIT OUTPUT
C
        IF (T .GT. ONFRTH) TM1=TM2
        IF (TCHECK .LT. TM1) GO TO 71
        WRITE(IOUT,899)T,(STR(I),I=1,ND)
        IT=IT+1
        TX(IT)=T
C
C      DETERMINE THE AVERAGE STRAIN FOR THIS ITERATION
C
        SUM=0.
        DO 13 I=1,ND
    13 SUM = SUM + STR(I)
        STRAV(IT) = SUM/XND
        TCHECK=0.
C
C      STOP IF TIME EXCEEDS THE MAXIMUM TIME ALLOTTED
C
        IF(T.GE.TIME)GO TO 99
C
C      CHANGE VALUES OF STRAIN TO THOSE VALUES JUST CALCULATED
C
    71 DO 12 I=2,ND
    12 STR1(I) = STR(I)
    70 T = T + DELT
        GO TO 20
C
C      PRINT VALUES OF AVERAGE STRAIN AND THE TIME
C      AT WHICH THEY OCCUR
C
    99 WRITE(IOUT,1001)
  1001 FORMAT (1H1,12X,4HTIME,13X,14HAVERAGE STRAIN/13X,4HDAYS,12X,16HPER
       1CENT OF TOTAL/)
        WRITE (IOUT,1002) (TX(I), STRAV(I),I=1,IT)
  1002 FORMAT(1X,F16.1,F22.2)
        STOP
        END
```

PARTIAL OUTPUT FROM CONSOLIDATION ANALYSIS PROGRAM

DELZ= 1.00   H= 10.00   CV= .2800-01   TM2= 100.0000
TIME= 1600.00   DELT= 1.00   TM1= 10.0000

PERCENT OF TOTAL STRAIN

DEPTH →

TIME (DAYS)	0	1	2	3	4	5	6	7	8	9	10
0.	100.000	.000	.000	.000	.000	.000	.000	.000	.000	.000	.000
10.	100.000	22.097	2.636	.197	.010	.000	.000	.000	.000	.000	.000
20.	100.000	35.464	8.031	1.269	.148	.013	.001	.000	.000	.000	.000
30.	100.000	44.219	13.769	3.173	.565	.080	.009	.001	.000	.000	.000
40.	100.000	50.356	19.078	5.586	1.302	.248	.040	.005	.001	.000	.000
50.	100.000	54.901	23.777	8.233	2.326	.547	.109	.019	.003	.000	.000
60.	100.000	58.419	27.882	10.933	3.569	.985	.233	.048	.009	.001	.000
70.	100.000	61.237	31.466	13.581	4.966	1.555	.422	.100	.021	.004	.001
...	...	...	...	...	...	...	...	...	...	...	...
1400.	100.000	92.428	85.043	78.027	71.555	65.784	60.856	56.893	53.991	52.221	51.627
1500.	100.000	92.934	86.042	79.495	73.453	68.066	63.466	59.765	57.056	55.403	54.847
1600.	100.000	93.406	86.974	80.863	75.224	70.195	65.901	62.447	59.917	58.373	57.855

TIME DAYS	AVERAGE STRAIN PERCENT OF TOTAL
10.0	11.36
20.0	13.18
30.0	14.71
40.0	16.06
50.0	17.26
60.0	18.37
70.0	19.40
...	...
1400.0	69.86
1500.0	71.87
1600.0	73.74

Figure 396-1

Bibliography

Abdul-Baki, A. F. (1966). "A Direct Method for Determination of Earth Pressures on Retaining Walls." Ph.D. Dissertation, Oklahoma State University, Stillwater.

Alpan, I. (January 1967). "The Empirical Evaluation of the Coefficient K_o and K_{or}," *Soils and Foundations*, Japanese Society of Soil Mechanics and Foundation Engineering, *VII* (I), Tokyo, p. 31.

American Association of State Highway and Transportation Officials (July 1978). "Standard Specifications for Transportation Materials and Methods of Sampling and Testing," Part II.

American Society of Civil Engineers (August 1968). "Placement and Improvement of Soil to Support Structures," Specialty Conference, Cambridge, Mass.

American Society of Civil Engineers (May to August 1972). "Subsurface Investigation for Design and Construction of Foundations of Buildings. Parts I, II, III, and IV, *Journal of the Soil Mechanics and Foundations Division* 98 (SM5, SM6, SM7, SM8). Also published as *ASCE Manuals and Reports on Engineering Practice* (56), 1976.

American Society of Civil Engineers (February 1978). *Soil Improvement History, Capabilities, and Outlook*, Committee on Placement and Improvement of Soils, Geotechnical Engineering Division.

Anderson, L. R., and J. Ramage (April 1976). "Underwater Placement of Sand Fill at Port of Portland Terminal #6," ASCE National Water Resources and Ocean Engineering Convention, San Diego, Calif., Preprint 2718.

Atterberg, A. (1911). "Über die physikalishe Bodenuntersuchung und über die Plastizität der Tone" (On the investigation of the physical properties of soils and on the plasticity of clays) Int. Mitt. für Bodenkunde, *1*, pp. 10-43.

Bailey, A., and J. T. Christian (April 1969). "Slope Stability Analysis—User's Manual," ICES LEASE. 1, *Soil Mechanics Publication No. 235*, Massachusetts Institute of Technology, Cambridge.

Barden, L. (December 1965). "Consolidation of Clay With Non-Linear Viscosity," *Geotechnique, 15* (4), pp. 1-31.

Barden, L. and P. L. Berry (September 1965). "Consolidation of Normally Consolidated Clay," *Journal of the Soil Mechanics and Foundations Division ASCE, 91* (SM5), pp. 15-36.

Beene, R. W. (July 1967). "Waco Dam Slide," *Journal of Soil Mechanics and Foundation Division, ASCE, 93* (SM4), pp. 35-44.

Begeman, H. K. (1965). "The Friction Cone as an Aid in Determining the Soil Profile," *Proceedings, 6th International Conference on Soil Mechanics and Foundation Engineering, I*, pp. 17-20.

Binquet, J., and K. L. Lee (December 1975). "Bearing Capacity Tests on Reinforced Earth Slabs," *Journal of Geotechnical Engineering Division, ASCE*, 101 (GT12), pp. 1241-1256.

Bishop, A. W. (March 1955). "The Use of the Slip Circle in the Stability Analysis of Slopes," *Geotechnique, 5* (1), pp. 7-17.

Bjerrum, L. (June 1967). "Engineering Geology of Norwegian Normally-Consolidated Marine Clays as Related to Settlements of Buildings," *Geotechnique, 17* (2), pp. 83-117.

Boussinesq, J. (1885). "Application des Potentiels a L'etude de L'equilibre et duo mouvement des Solids Elastiques," Gauthier-Villars, Paris.

Bowles, J. E. (1974). *Analytical and Computer Methods in Foundation Engineering*, McGraw-Hill Book Company, New York.

Brooker, E. W., and H. O. Ireland (1965). "Earth Pressure at Rest Related to Stress History," *Canadian Geotechnical Journal, 2*, (1), pp. 1-15.

Brown, E. (December 1977). "Vibroflotation Compaction of Cohesionless Soils," *Journal of the Geotechnical Engineering Division, ASCE, 103* (GT12), pp. 1437-1451.

Carnahan, B., H. A. Luther, and J. O. Wilkes (1969). *Applied Numerical Methods*, John Wiley & Sons, New York.

Caron, C., P. Cattin, and T. F. Herbst (1975). "Injections," *Foundation Engineering Handbook*, edited by H. F. Winterkorn, and H. Y. Fang, pp. 337-353. Van Nostrand Reinhold Company, New York.

Casagrande, A. (1936). "The Determination of the Pre-Consolidation Load and Its Practical Significance," *Proceedings, First International Conference on Soil Mechanics*, Cambridge, Mass., *3*, pp. 60-64.

Casagrande, A. (1937). "Seepage Through Dams," *Journal New England Water Works Association, 51* (2), pp. 131-172. Reprinted in *Contributions to Soil Mechanics, 1925-1940*, Boston Society of Civil Engineers, 1940, and as Harvard University Soil Mechanics Series No. 5.

Casagrande, A. (1948). "Classification and Identification of Soils," *Transactions, ASCE, 113*, pp. 901-992.

Christian, J. T., and J. W. Boehmer (July 1970). "Plane Strain Consolidation by Finite Elements," *Journal of the Soil Mechanics and Foundations Division, ASCE, 96* (SM4), pp. 1435-1457.

Clough, G. W. (1972). "Application of the Finite Element Method to Earth-Structure Interaction," *Proceedings of Symposium, Applications of the Finite Element Method in Geotechnical Engineering*, III, U.S. Army Engineer Waterways Experiment Station, Vicksburg, Mississippi, pp. 1057-1116.

Clough, R. W., and A. K. Chopra (April 1966). "Earthquake Stress Analysis in Earth Dams," *Journal of the Engineering Mechanics Division, ASCE, 92* (EM2), pp. 197-211.

Clough, R. W., and R. J. Woodward, III (July 1967). "Analysis of Embankment Stress and Deformations," *Journal of the Soil Mechanics and Foundations Division, ASCE, 93* (SM4), July 1967, pp. 529-549.

Collin, A. (1846). *Landslides in Clays*, University of Toronto Press, 1956. (Translated by W. R. Schreiver from the French, 1846.)

Coulomb, C. A. (1776). "Essai sur une Application des R'egles des Maximis et Minimis à quelques Problèmes de Statique Relatifs à l'Architecture," *Mem. Acad. Roy, pres. divers Savants, 7*, Paris, p. 38.

Culmann, K. (1866). *Die Graphische Statik*. Zurich.

D'Appolonia, D. J., R. V. Whitman, and E. D'Appolonia (January 1969). "Sand Compaction with Vibratory Rollers," *Journal of the Soil Mechanics and Foundations Division, ASCE, 95* (SM1), pp. 263-284.

Darcy, H. (1856). "Les fontaines publiques de la ville de Dijon," (The water supply of the city of Dijon). Dalmont, Paris, 674 pp.

De Beer, E. E. (1945). "Etude des Fondations sur Pilotis et des Fondations Directes," *Annales des Travaux Public de Belgique, 46*, pp. 1-78.

De Beer, E. E. (1967). "Bearing Capacity and Settlement of Shallow Foundations on Sand," *Proceedings of a Symposium, Bearing Capacity and Settlement of Foundations*, edited by A. S. Vesic, Duke University, Durham, N. Car., pp. 15-33.

Desai, C. S. (September 1972). "Finite Element Procedures for Seepage

Analysis Using an Isoparametric Element," *Proceedings of Symposium, Applications of the Finite Element Method in Geotechnical Engineering,* II, U.S. Army Engineer Waterways Experiment Station, Vicksburg, Miss., pp. 799-824.

Desai, C. S. (1977b). "Flow Through Porous Media," *Numerical Methods in Geotechnical Engineering,* edited by C. S. Desai and J. T. Christian, McGraw-Hill Book Company, New York.

Desai, C. S., and J. T. Christian (eds.) (1977a). *Numerical Methods in Geotechnical Engineering,* McGraw-Hill Book Company, New York.

Duncan, J. M. (1972). "Finite Element Analyses of Stresses and Movements in Dams, Excavations and Slopes, State of the Art," Proceedings of Symposium, Applications of the Finite Element Method in Geotechnical Engineering, I, U.S. Army Engineering Waterways Experiment Station, Vicksburg, Miss., pp. 267-326.

Duncan, J. M. (September 1977). "Behavior and Design of Long-Span Metal Culvert Structures." Paper prepared for presentation at the ASCE Convention in San Francisco.

Ernst, L. F. (1950). "Een nieuwe formule voor de berekening van de doorlaatfactor met de boorgatenmethode," Rap. Landbouwproefsta. en Bodemkundig Inst. T.N.O., Groningen, Netherlands (Mimeographed).

Evans, H. E., and R. W. Lewis (March 1970). "Effective Stress Principle in Saturated Clay," *Journal of the Soil Mechanics and Foundation Division, ASCE, 96* (SM2), pp. 671-684.

Fellenius, W. (1927). *Erdstatische Berechnungen* (Calculation of stability of slopes), rev. Ed., Berlin, 1939, 48 pp.

Fellenius, W. (1936). "Calculation of the Stability of Earth Dams," *Transactions 2nd Congress on Large Dams, 4,* Washington, D.C., pp. 445.

Forsythe, G. E., and W. Wasow (1960). *Finite-Difference Methods for Partial Differential Equations,* John Wiley & Sons, New York.

Gibbs, H. J. and W. G. Holtz (1957). "Research on Determining the Density of Sands by Spoon Penetration Testing," *Proceedings Fourth International Conference on Soil Mechanics and Foundation Engineering,* London, I, pp. 35-39.

Gillette, H. S. (1956). "Preparing Base-Course Materials for Disturbed Soil Indicator Test," *Highway Research Board Bulletin 122,* pp. 1-12.

Grant, R., J. T. Christian, and E. H. Vanmarcke (September 1974). "Differential Settlement of Buildings," *Journal of the Geotechnical Engineering Division, 100* (GT9), pp. 973-991.

Grim, R. E. (1953). *Clay Mineralogy,* McGraw-Hill Book Company, New York.

Hansen, J. B. (1961). "A General Formula for Bearing Capacity," Bulletin

No. 11, Danish Technical Institute, Copenhagen.

Hansen, J. B. (1951). "Simple Statical Computation of Permissible Pile Loads," pp. 14-17. Christiani and Nielsen Post, Copenhagen, Denmark.

Hendron, A. J., Jr. (1963). "The Behavior of Sand in One-Dimensional Compression," Ph.D. Dissertation, Department of Civil Engineering, University of Illinois, Urbana.

Hirsch, T. J., L. Carr, and L. L. Lowery, Jr. (1976). "Pile Driving Analysis —Wave Equation Users Manual TTI Program," I-IV., U.S. Department of Transportation, Federal Highway Administration, Washington, D.C.

Hooghoudt, S. B. (1936). "Bijdragen tot de kennis van eenige natuurkundige grootheden van den grond, 4." Versl. Lamdb., Ond. *42* (13) B:449-541. Algemeene Landsdrukkerij, The Hague.

Huebner, K. H. (1975). *The Finite Element Method for Engineers*, John Wiley & Sons, New York.

Hultin, S. (1916). "Grufyllander for Kejbyggnader," *Teknish Tidsskeift*, V.U., 31, Stockholm, p. 275.

Hvorslev, M. J. (1949). "Subsurface Exploration and Sampling of Soils for Civil Engineering Purposes, U.S. Army Engineer Waterways Experiment Station, Vicksburg, Miss., 521 pp.

Idriss, I. M., H. B. Seed, and N. Serff (January 1974). "Seismic Response by Variable Damping Finite Elements," *Journal of the Geotechnical Engineering Division, ASCE, 100* (GT1), pp. 1-13.

James, M. L., G. M. Smith, and J. C. Wolford (1967). *Applied Numerical Methods for Digital Computation With FORTRAN*, International Textbook Company, Scranton, Penn.

Janbu, N. (1973). "Slope Stability Computations," *Embankment-Dam Engineering, Casagrande Volume*, edited by R. C. Hirschfield and S. J. Poulos, pp. 47-86. John Wiley & Sons, New York.

Jeppson, R. W. (January 1968). "Seepage From Canals—Solutions by Finite Differences," *Journal of the Hydraulics Division, ASCE, 94* (HY1), pp. 259-283.

Jeppson, R. W., and R. W. Nelson (1970). "Inverse Formulation and Finite Difference Solution to Partially Saturated Seepage from Canal," Soil Science Society of America Proceedings, *34* (1), pp. 9-14.

Johnson, S. J. (January 1970). "Foundation Precompression With Vertical Sand Drains," *Journal of the Soil Mechanics and Foundations Division, ASCE, 96* (SM1), pp. 145–175.

Johnson, S. J. (1975). "Analysis and Design Relating to Embankments," *Proceedings of Conference on Analysis and Design in Geotechnical Engineering, II, ASCE*, Austin, Texas.

Jordan, J. C., and R. L. Schiffman (December 1967). "A Settlement Problem Oriented Language—User's Manual," *Soil Mechanics Publica-*

tion No. 204 ICES SEPOL-1, Massachusetts Institute of Technology, Cambridge.

Kulhawy, F. H., and J. M. Duncan (July 1972). "Stress and Movement in Oroville Dam," *Journal of the Soil Mechanics and Foundations Division, ASCE, 98* (SM7), pp. 653-665.

Lambe, T. W. (May 1958). "The Engineering Behavior of Compacted Clay," *Journal of Soil Mechanics and Foundation Engineering, ASCE, 84* (SM2). [Also in *Transactions ASCE, 125*, Part 1, p. 718 (1960).]

Lambe, T. W. (November 1967). "Stress Path Method," *Journal of Soil Mechanics and Foundation Division, ASCE, 93* (SM6), pp. 309-331.

Lambe, T. W., and R. V. Whitman (1969). *Soil Mechanics*, John Wiley & Sons, New York, 553 pp.

Lane, E. W. (1935). "Security From Under-Seepage, Masonry Dams on Earth Foundations," *Transactions ASCE, 100*, pp. 1235-1351.

Lee, K. L. (1976). "Reinforced Earth—An Old Idea in a New Setting," *New Horizons in Construction Materials*, I, Edited by Hsai-Yang Fang, Envo Publishing Co.

Lee, K. L., B. D. Adams, and J. J. Vagneron (October 1973). "Reinforced Earth Retaining Walls," *Journal of the Soil Mechanics and Foundations Division, ASCE, 99* (SM 10), pp. 745-764.

Lowe, J., III (1960). "Current Practice in Soil Sampling in the United States Highway Research Board, *Special Report No. 60*, pp. 142-154.

Lowe, J., III (July 1967). "Stability Analysis of Embankments," *Journal of the Soil Mechanics and Foundations Division, 93* (SM4), pp. 1-33.

Menard, L., and Y. Broise (1975). "Theoretical and Practical Aspects of Dynamic Consolidation," The Institution of Civil Engineers, London.

Meyerhof, G. G. (1953). "The Bearing Capacity of Foundations Under Eccentric and Inclined Loads," *Proceedings, 3rd International Conference Soil Mechanics and Foundation Engineering*, Zurich, *1*, pp. 400-445.

Meyerhof, G. G. (1955). "Factors Influencing Bearing Capacity of Foundations," *Geotechnique*, London, *5*, pp. 227-242.

Meyerhof, G. G. (January 1956). "Penetration Tests and Bearing Capacity of Cohesionless Soils," *Journal of the Soil Mechanics and Foundations Division, ASCE, 82* (SM1).

Meyerhof, G. G. (1974). "Penetration Testing Outside Europe," *General Report, Proceedings, European Symposium on Penetration Testing*, 2.1 Stockholm.

Meyerhof, G. G. (March 1976). "Bearing Capacity and Settlement of Pile Foundations," *Journal of the Geotechnical Engineering Division, ASCE, 102* (GT3), pp. 195-228.

Mitchell, J. K. (1977). "Soil Improvement Methods and Applications," Prepared for Utah State University Special Course for U.S. Soil

Conservation Engineers, Logan, Utah.

Mitchell, J. K. and D. R. Freitag (December 1959). "A Review and Evaluation of Soil-Cement Pavements," *Journal of the Soil Mechanics and Foundations Division, ASCE, 85* (SM6), pp. 49-73.

Mitchell, J. K., and T. A. Lunne (July 1978). "Cone Resistance as Measure of Sand Strength," *Journal of the Geotechnical Engineering Division, ASCE, 104* (GT7), pp. 995-1012.

Mohr, H. A. (1943). "Exploration of Soil Conditions and Sampling Operations," *Soil Mechanics*, Series 21, Third Rev. Ed., Publication 376, Graduate School of Engineering, Harvard University, Cambridge, Mass.

Moore, C. A. (September 1971). "Effect of Mica on K_o Compressibility of Two Soils," *Journal of the Soil Mechanics and Foundations Division, ASCE, 97* (SM9), pp. 1275-1291.

Morgenstern, N. R., and V. E. Price (1965). "The Analysis of the Stability of General Slip Surfaces," *Geotechnique, 15*, pp. 79-93.

Newmark, N. M. (1942). "Influence Charts for Computation of Stresses in Elastic Foundations," *University of Illinois Engineering Experiment Stations Bulletin, Urbana, Ill. 338*, p. 28.

Parcher, J. V., and R. E. Means (1968). *Soil Mechanics and Foundations*, Charles E. Merrill Publishing Company, Columbus, Ohio.

Peck, R. B. (1962). "Art and Science in Subsurface Engineering," *Geotechnique, 12* (1), p. 60.

Peck, R. B., W. E. Hanson, and T. H. Thornburn (1974). *Foundation Engineering*, Sec. Ed., John Wiley & Sons, New York.

Pettersson, K. E. (1955). "The Early History of Circular Sliding Surfaces," *Geotechnique, 5*, p. 275.

Pope, R. G. (September, 1975). "Non-dimensional Chart for the Ultimate Bearing Capacity of Surface and Shallow Foundations," *Geotechnique, 25* (3), pp. 593-604.

Poulos, H. G. (1968). "Analysis of the Settlement of Pile Groups," *Geotechnique, 18*, pp. 449-471.

Prandtl, L. (1921). "Uber die Eindringungsfestigkeit (Harte) Plastischer Baustoffe und die Festigkeit von Schneiden" [On the penetrating strengths (hardness) of plastic construction materials and the strength of cutting edges], *Zeit. angew. Math. Mech., 1* (1), pp. 15-20.

Proctor, R. R. (1933). "Four articles on the design and construction of rolled-earth dams," *Engineering News-Record, III*, pp. 245-248, 286-289, 348-351, 372-376.

Ramage, J., and W. S. Stewart, Jr. (1978). "Evaluation of Sensing Systems for Measuring Properties of Ground Masses: Static Cone Type Penetrometer Tests," U.S. Department of Transportation, Federal

Highway Administration, Washington, D.C., 100 pp.

Rankine, W. J. M. (1857). "On the Stability of Loose Earth," *Phil. Trans., Royal Soc.*, London.

Reissner, H. (1924). "Zum Erddrackproblem" (Concerning the earth pressure problem), *Proceedings, 1st International Congress of Applied Mechanics* Delft, Netherlands, pp. 295-311.

Reynolds, O. (1883). "An Experimental Investigation of the Circumstances Which Determine Whether the Motion of Water Shall be Direct or Sinuous and of the Law of Resistance in Parallel Channels," *Phil. Trans. Roy. Soc., 174* (3), p. 935.

Richart, Jr., F. E., J. R. Hall, Jr., and R. D. Woods (1970). *Vibrations of Soils and Foundations*, Prentice-Hall, Englewood Cliffs, New Jersey.

Sanglerat, G. (1972). *The Penetrometer and Soil Exploration*, Elsevier Publishing Company, Amsterdam, 464 pp.

Schiffman, R. L., and S. K. Arya (1977). "One-Dimensional Consolidation," *Numerical Methods in Geotechnical Engineering*, edited by C. S. Desai and J. T. Christian, pp. 364-398, McGraw-Hill Book Company, New York.

Schmertmann, J. H. (1955). "The Undisturbed Consolidation Behavior of Clay," *Transactions, ASCE, 120*, pp. 1201-1227.

Schmertmann, J. H. (May 1970). "Static Cone to Compute Static Settlement Over Sand," *Journal of the Soil Mechanics and Foundation Division, ASCE, 96* (SM3), pp. 1011-1043.

Schmertmann, J. H. (1975). "The Measurement of In-Situ Shear Strength," *Proceedings, ASCE Specialty Conference on In-Situ Measurement of Soil Properties, 2.* Raleigh, N. Car., pp. 57-138.

Schroeder, W. L., and M. L. Byington (1972). "Experiences With Compaction of Hydraulic Fills," *Proceedings, 10th Annual Engineering Geology and Soils Engineering Symposium*, Moscow, Idaho.

Seed, H. B. (September 1968). "Landslides During Earthquakes Due to Liquefaction," *Journal of the Soil Mechanics and Foundations Division, ASCE, 94* (SM5), September 1968, pp. 1053-1122.

Seed, H. B., and C. K. Chan (1959). "Structure and Strength Characteristics of Compacted Clays," *Journal of the Soil Mechanics and Foundations Division, ASCE, 85* (SM5), October 1959, pp. 87-128.

Seed, H. B., and K. L. Lee (November 1966). "Liquefaction of Saturated Sands During Cyclic Loading," *Journal of the Soil Mechanics and Foundations Division, ASCE, 92* (SM6), pp. 105-134.

Seed, H. B., and H. A. Sultan (July 1967). "Stability Analyses for a Sloping Core Embankment," *Journal of Soil Mechanics and Foundation Division, ASCE, 92* (SM4), pp. 69-84.

Sevaldson, R. A. (1956). "The Slide in Lodalen, Oct. 6, 1954,"

Geotechnique, 6, pp. 167-182.

Shields, D. H., and A. Z. Tolunay (December 1973). "Passive Pressure Coefficients By Method of Slices," *Journal of the Soil Mechanics and Foundations Division, ASCE, 99* (SM12), pp. 1043-1053.

Skempton, A. W. (1961). "Effective Stress in Soils, Concrete and Rocks," *Pore Pressure and Suction in Soils*, p. 4, Butterworths, London.

Smith, E. A. L. (August 1960). "Pile Driving by the Wave Equation," *Journal of the Soil Mechanics and Foundations Division, ASCE, 86* (SM4), pp. 35-61.

Sowers, G. B., and G. F. Sowers (1970). *Introductory Soil Mechanics and Foundations*, 3d Ed., Macmillan Publishing Company, New York, 556 pp.

Stern, O. (1924). "Sur Theorie der elektrolytischen Doppelschicht," *Zeitschrift fur Elektrochemie und Angewandte Physikalische Chemie*, Leipzig-Berlin, pp. 508-516.

Su, H. H., and R. H. Prysock (1972). "Settlement Analysis of Two Highway Embankments," *Proceedings, Specialty Conference, ASCE; Performance of Earth and Earth-Supported Structures* I, Part 1, pp. 465-488.

Taylor, D. W. (1948). *Fundamentals of Soil Mechanics*. John Wiley & Sons, New York, 700 pp.

Terzaghi, K. (1925). *Erdbaumechanik*, Vienna, F. Deuticke.

Terzaghi, K. (1936a). "The Shearing Resistance of Saturated Soils and the Angle Between the Planes of Shear," *Proceedings, 1st International Conference on Soil Mechanics, I*, Cambridge, Mass., pp. 54-56.

Terzaghi, K. (1936b). "Stability of Slopes of Natural Clay," *Proceedings, 1st International Conference on Soil Mechanics, I*, Cambridge, Mass., pp. 161-165.

Terzaghi, K. (1943). *Theoretical Soil Mechanics*, John Wiley & Sons, New York.

Terzaghi, K., and R. B. Peck (1948). *Soil Mechanics in Engineering Practice*, John Wiley & Sons, New York.

Terzaghi, K., and R. B. Peck (1967). *Soil Mechanics in Engineering Practice*, Sec. Ed., John Wiley & Sons, New York.

Turner, M. J., R. W. Clough, H. C. Martin, and L. J. Topp (September 1956). "Stiffness and Deflection Analysis of Complex Structures," *Journal of Aeronautical Sciences, 23* (9), pp. 805-823.

U.S. Department of the Interior, Bureau of Reclamation (1963). *Earth Manual*, Denver, Colorado.

Vesic, A. S. (1967). "Ultimate Loads and Settlements of Deep Foundations in Sand," *Proceedings of a Symposium: Bearing Capacity and Settlement of Foundations*, edited by A. S. Vesic, Duke University, Durham, North Carolina, pp. 53-68.

Vesic, A. S. (January 1973). "Analysis of Ultimate Loads of Shallow Foundations," *Journal of the Soil Mechanics and Foundations Division, ASCE, 99* (SM1), pp. 45-73.

Vidal, H. (1969). "The Principle of Reinforced Earth," Highway Research Record 282, Highway Research Board, Washington, D.C.

Westergaard, H. M. (1938). "A Problem of Elasticity Suggested by a Problem in Soil Mechanics: Soft Material Reinforced by Numerous Strong Horizontal Sheets," *Contributions to the Mechanics of Solids*, Stephen Timoshenko 60th Anniversary Volume, Macmillan Publishing Company, New York.

Winterkorn, Hans F., and H. Y. Fang (eds.) (1975). *Foundation Engineering Handbook*, Van Nostrand Reinhold Company, New York.

Worth, E. G. (1972). Personal correspondence with Dr. L. R. Anderson.

Wright, S. G., F. H. Kulhawy, and J. M. Duncan (October 1973). "Accuracy of Equilibrium Slope Stability Analyses," *Journal of Soil Mechanics and Foundation Division, ASCE, 99* (SM10), pp. 783-792.

Zienkiewicz, D. C. (1977). *The Finite Element Method*, McGraw-Hill Book Company, New York.

Author Index

Subject Index

compression curve, 124, 126, 131
consolidation, 121
double layer, 15, 32, 316
expansive, 9, 13, 318
fissured, 180, 236
footing settlement, 285
minerals, 11
normally consolidated, 9, 125, 126, 128, 169, 171, 173, 182, 189
overconsolidated, 9, 125, 129, 131, 175, 179, 185, 189
permeability, 52, 60, 132, 135, 318, 373
quick, 32
sensitive, 31
shear strength, 30, 162, 168, 175, 182, 378
stress-strain, 126, 131, 170, 172, 179, 180
Cobbles, 10
Coefficient, at rest pressure, 113, 195, 196, 203, 229
conductivity, 52
consolidation, 135, 140
curvature, 27, 37
lateral pressure, 299
permeability, 52, 53, 58, 60, 132, 135, 318, 373
uniformity, 26, 37
volume decrease, 134
Cohesion(c), 31, 182, 183, 185, 200, 202, 207, 226, 243, 247, 249, 253, 267, 293, 297, 325
Compaction, backfill, 230
expansive soil, 318
field equipment, 320
field-laboratory comparison, 323
granular soils, 22, 314, 320, 322
silts and clays, 30, 181, 315
test, 313
theory, 312
Compressibility, 119
compacted clay, 318
normally consolidated clay, 121, 123, 125, 126
overconsolidated clay, 125, 129
sand, 120
Compression curve, 124, 126, 131
Compression index, 128
Computer applications, 335, 336, 350, 363, 391
Computer programs
consolidation, 391, 394
seepage, 391, 392
Cone penetrometer, 159, 278, 376

Consistency, 27, 28, 30
Consolidated drained shear, 164, 169, 178, 181
Consolidated undrained shear, 164, 171, 178, 181
Consolidation, 119
coefficient, 135, 140
computer programs, 391, 394
construction period correction, 143
degree of, 137, 138
field curve, 126, 129, 131
finite difference solution, 347, 391, 394
finite element, 351
rate, 132, 136, 347, 350, 391
rheological model, 122
secondary, 139, 140
test, 123
theory, 121
time factor, 137, 139
Consolidometer, 123
Contact pressure, 91
Continuous footings, *see* Footings, strip
Coulomb
active pressure, 211
passive pressure, 221
Creep, 208, 230, 236
Critical void ratio, 166
Crystalline, 11
Culmann's graphical construction, 217, 221
Curvature, coefficient of, 27, 37
cv 135, 140

Darcy's law, 50
Darcy's velocity, 51
Degree of consolidation, 137, 138
Degree of saturation, 17, 19
Differential settlement, 264, 293
Dilation, 166, 169
Direct shear test, 159
Dispersed structure, 30, 316, 317, 318
Double layer, 15, 32, 316
Drainage, 73, 84, 230, 331
Drilling, 370
Dutch cone penetrometer, *see* Cone penetrometer

Earth dams, critical design states, 257
discharge face, 78
filter criteria, 84
flow nets, 73, 78
Earthquake, 168
Eccentric loading, 272
Effective particle size, 26

Subject Index

transported, 8
types, 8
use chart, 38
Soil forming process, 7
Soil mechanics, 1
Soil stabilization, admixtures, 324
blasting, 327, 333
compaction, 311
compaction pile, 328
drainage, 331
heavy tamping, 329
injection and grouting, 326
precompression, 329
reinforced earth, 331
summary, 332
terra-probe, 327
vibroflotation, 327
Soil structure, crystalline, 11
dispersed, 30, 316, 317, 318
flocculated, 30, 316, 317, 318
Specific gravity, 17, 19
Square footings, 103, 104, 105, 270, 271, 274, 276
Standard penetration test, 121, 159, 280, 291, 374
Stern double layer, 15
Strain, 125, 165, 169, 171, 178, 180, 202
Strength envelope, 170, 173, 174, 178, 181, 182, 186
Stress, effective, 89, 91
finite element, 350
history, 114, 187
Mohr's circle, 106
principal, 106
total, 91
Stresses due to surface loads, 93
approximate, 94
influence charts, 99, 102, 104
point load, 95, 97, 98
uniform load, 99
Stress paths, 114, 187, 274
Stress-strain diagrams, *see* Clay, Sand
Structure, 11, 30, 316
Subsurface engineering, 1
Subsurface exploration, 365, 367
exploration planning, 4, 366
field classification, 382
recording field data, 380
report preparation, 384
sample preparation, 384

site reconnaisance, 5, 366
soil sampling, 367
Superficial velocity, 51
Surface charge, 14
Swelling clay, 9, 13, 318
Systeme International d'Unites, *see* S.I. units

Terra-probe, 327
Thixotropy, 31
Time factor, 137, 139
Total stress, method of analysis, 173, 182, 185, 238, 293, 297
strength envelope, 173, 174, 182
Triaxial shear test, 159, 161, 163

Unconsolidated undrained shear strength, 163, 173, 182
Unconfined compressive strength, 30, 159, 162, 181
Unified classification system, 36
Uniformity coefficient, 26, 37
Unit weight, 19
buoyant, 17
dry, 17
moist, 17
saturated, 17
typical, 18
U.S. Department of Agriculture classification system, 33

Vane shear, 159, 163, 373
Vibrating roller, 322
Vibroflotation, 327
Virgin compression curve, 125, 126, 129, 131
Viscosity, 58, 168, 169
Viscous lag, 121
Void ratio, 16, 17
critical, 166
typical, 18

Wall friction angle, 211, 213, 221
Water, capillary, 47
pressure, 45, 122, 134, 205, 240, 253, 259
Water content, 17, 18, 19
Weathering, 8
Weighted creep ratio, 82
Weight-volume relationships, 16, 17
Wells, 58, 68, 373
Westergaard, area load, 99, 102, 103, 105
influence chart, 103
point load, 95, 98

414

Subject Index

> Everything will be included in the new concept of
> newsreels.
>
> . . .
>
> The "Film-Eye" is challenging the visual
> presentation of reality as seen by the human eye.
> The "Film-Eye" is proposing its own way of seeing.
> The *kinok*-editor is organizing a new perception of
> life's moments for the first time.[38]

The original version of this manifesto, published in *LEF* (no. 3, 1923),
includes an epigraph summarizing more emphatically Vertov's critical
view on Russian film: "I simply want to state that everything we have
produced in cinema until now has been one-hundred percent wrong,
that is to say, absolutely contrary to what we had and have to do."[39]
As one can see, together with Gan, Rodchenko, and Mayakovsky, Vertov was convinced that Soviet cinema needed to be revitalized with a
revolutionary vision, and that this should be accomplished by rejecting the theatrical conventions used to present reality on the screen.

Another important constructivist critic with an acute understanding
of both the theoretical and practical aspects of documentary cinema
was Viktor Pertsov, whose essays (published in *LEF*) established his
reputation as one of the most analytical Soviet film theorists of the
1920s. In his famous article " 'Play' and Demonstration" (1927), Pertsov compares Vertov's films to those of Esther Shub,* another important Soviet documentary filmmaker:

Vertov's film *One Sixth of the World* and Shub's *The Fall of the Romanov Dynasty* and *The Great Road* represent genuine agitational journalism expressed
in a cinematic language. The authentic non-aesthetic impact of these films de-
rives from the real facts which comprise the films' structures. If the cinematic
juxtaposition of facts stimulates emotions, it does not mean that these emotions in and of themselves transform facts in a non-authentic aesthetic structure. We know that in staged films the impact on the audience is provided by
various dramatic devices; however, in unstaged film journalism, the impact is
provided by the rules of rhetoric. It is not a coincidence therefore that Vertov
conceives his films on an oratorical principle. He is a true cinematic orator
who makes his point in the manner dictated by the intrinsic power of the se-

---

* Esther (Esfir) Shub began her film career as an editor (adaptor) of foreign films distributed in the Soviet Union. Her method of editing had a substantial influence on both
Vertov and Eisenstein. In turn, she claimed that although she had learned many things
from Eisenstein, she considered herself "Vertov's pupil, regardless of our disputes and
disagreements." Shub demonstrated great mastery in her three major compilation films,
The Fall of the Romanov Dynasty [*Padenie dinastii Romanovikh*, 1927], *The Great Road*
[*Velikii put'*, 1927], and *The Russia of Nicholai II and Lev Tolstoi* [*Rossia Nikolaia II i
Lev Tolstoi*, 1928]. Shub's memoirs and articles are collected in the book *My Life — Cinema* [*Zhizn' moia — kinematograf*] (Moscow: Iskusstvo, 1972). More information about
Shub and her work can be found in my article, "Esther Shub: Cinema is My Life," *Quart-
erly Review of Film Studies*, no. 4 (Fall 1978), pp. 429–56.

lected facts, thus building a structure which holds the attention of the audience.... To edit facts [montage] means to analyze and to synthesize, not to catalogue.[40]

Pertsov's explanation of Vertov's method demonstrates his understanding of the two crucial aspects of Vertov's work: first, the evident ontological authenticity of each separate shot ("the non-aesthetic impact of the shot"), and second, the montage organization of the footage ("building a structure"), by which the filmmaker reconstitutes the spatiotemporal aspect of reality and conveys his message ("makes his point") while "holding the attention of the audience." It is difficult to find a more precise and equally sensitive elaboration of Vertov's directorial style even in contemporary critical literature.

Vertov's films drew the attention of another constructivist critic affiliated with the *LEF* circle, Sergei Ermolinsky, who emphasized the different attitudes in early Soviet "cinematographic journalism" that stemmed from the methods of "montage analysis and synthesis" as practiced by Vertov and Shub. Ermolinsky noted that Shub assembled footage as she found it in film archives, whereas Vertov *de*constructed the footage in order to achieve a new meaning and to convey his viewpoint. Shub's editorial strategy, Ermolinsky concluded, can be described as a faithful preservation of original newsreels – hence the long and unbroken (continuous) sequential takes in her films accompanied by descriptive intertitles. According to Ermolinsky, "the documentary shot was for Shub the actual goal," while "for Vertov it was always a means."[41] A further distinction between the two filmmakers, Ermolinsky contended, was the difference in their attitudes toward the film image as recorded by the camera. Vertov "threw himself on the given material, cutting it into numerous pieces, thus subordinating it to his imagination, while Shub regarded each piece [shot] as to a self-sufficient, autonomous entity."[42] Ermolinsky's astute analysis substantiated his characterization of Shub and Vertov as the two most significant documentary filmmakers of the silent era. From today's perspective, it is clear that Shub was the initiator of the "compilation film" genre, whereas Vertov was the precursor of the modern cinéma vérité style in its most genuine form.

To encourage a greater critical appreciation of cinema, the journal *Novyi LEF* organized panels that addressed the specific role of film in society. The most exemplary of these discussions was published in the December 1927 issue of the journal, under the title "*LEF* and Film." The panel, which included Osip Brik, Sergei Tretyakov, Viktor Shklovsky, and Esther Shub, reached the consensus that the controversy between staged and unstaged film constituted "the basic issue of contemporary cinema."[43] They concluded that Vertov and Shub produced "true cinema of fact," as opposed to Eisenstein who moved toward "cinema of fiction." Tretyakov summarized the discussion, noting that ontological authenticity in cinema should be measured by the extent to which reality is presented and/or transformed on the screen:

Based on the degree of transformation, filmed material falls into three basic groups. First, candidly filmed material, i.e. facts caught unawares; second, arranged material, i.e. events found in reality and molded by the filmmaker to a certain extent; and third, staged material, i.e. dramatic and narrative situations that are pre-scripted and then entirely acted out before the camera. The candid shots are, without exception, caught "off guard." Such is Vertov's "Life-Caught-Unawares." One can find a minimum of deformation in it, which does not mean that candidly filmed material lacks its own gradations of transformation.[44]

This summary reveals the subjectivity with which even the most committed advocates of "factual art" viewed the cinematic "deformation" of life's raw material; yet, subjectivity was allowed so long as it helped enhance the film's ideological meaning. According to the *Novyi LEF* panelists, only Vertov and Shub presented reality on the screen candidly, whereas Ermler, Pudovkin, and Eisenstein distorted reality to the point of misrepresenting life, a practice that became, as Mayakovsky termed it, "disgusting" and "outrageous," referring to the quasi-documentary presentation of Lenin in Eisenstein's *October*.[45]

In another discussion published in *Novyi LEF* of April 1928, Vertov's film *The Eleventh Year* and Eisenstein's *October* were directly compared. The panel, which called itself "The *LEF* Ring" (implying an "arena" of ideological contest), composed of Brik, Pertsov, and Shklovsky, unanimously proclaimed Vertov's film "outstanding" for its straightforward rendering of "life facticity" and "historial accuracy." At the same time, *October* was characterized by Brik as "Eisenstein's hopeless effort to jump over his own head."[46] The remark was meant to sting, since Brik condescendingly referred to Eisenstein as the "young director," although he was thirty, and *October* was his third film. But neither was Vertov immune to the panel's harsh criticisms, especially in Shklovsky's remarks stating that Vertov's insertion of metaphorical intertitles in *The Eleventh Year* not only incurred redundancies ("doubling the data") but also obscured the film's ideological meaning ("blurring the message"). The panelists concluded their discussion by establishing guidelines for "cinema of fact" (the unstaged film) as opposed to "cinema of mesmerization" (the staged film).

Struggles with NEP

The *LEF* critics adamantly opposed entertainment cinema even after Lenin had introduced the NEP policy,* which encouraged commercial film production and promoted the import of foreign trivial melodra-

---

* NEP stands for "New Economic Policy" [*Novaia ekonomskaia politika*], defined during

mas. As the editor of *Novyi LEF,* Tretyakov stated that during the NEP period American entertainment movies had a "devastating effect" on Soviet film. While in the early twenties the great American classic filmmakers inspired Kuleshov, Pudovkin, and Eisenstein to develop new revolutionary montage concepts, near the end of the decade many Soviet directors began to imitate Hollywood's conventional production, thus "contributing to a Pickfordization and Fordization of the workers' way of life."[47] This metaphorical warning went unheeded, however, because Lenin, suddenly, had concluded that a "reasonable dose" of entertainment films could function as an effective panacea for the masses, adding variety to the film repertory while yielding greater profits. The formula for achieving an appropriate "balance" between entertainment films and newsreel propaganda became known as the "Leninist Proportion,"* which commercial distributors immediately took as an excuse for making concessions to popular taste. With its obvious eclecticism, the Leninist Proportion contained contradicting suggestions, one demanding that Soviet film production "should begin with newsreel," and the other stating that if "good newsreels and serious educational films exist, then it doesn't matter if some useless film of the more or less usual sort is shown to attract an audience."[48] Naturally, the producers and distributors grabbed the second suggestion while almost totally neglecting the first one. As a result, experimental filmmaking became extremely difficult, while the NEP audience [*nepmanovska auditoriia*] was coddled by cheap imported films, much to the chagrin of the *kinok*s and the *LEF*ists.[49]

Vertov's diaries reveal how he continued to urge his *kinok*s to make films with pertinent ideological substance in order to "give proletariats of all countries the opportunity to see, hear, and understand each other better."[50] Not surprisingly, the Leninist Proportion proved to be at odds with Vertov's ideals about revolutionary cinema, but his concept of documentary film was equally antagonistic to Stalin's promotion of blatant political propaganda. Consequently, Vertov put forth a new "proportion" in his 1939 article "In Defense of the Newsreel," emphasizing the necessity to establish legitimate "rights" for both staged and unstaged films, to support the production of films concerned with documentary presentation, and to proscribe "intermediating" [*promezhutechnoi*] and "typage" [*tipazh*] films.[51] Vertov intentionally used the

Lenin's famous speech at the Tenth Congress of the Communist Party (March 1921) when he declared: "We are in a condition of such poverty, ruin, and exhaustion of the productive powers of workers and peasants that everything must be set aside to increase production." Quoted in George Vernadsky, *A History of Russia* (New Haven: Yale University Press, 1954), p. 323.
* Anatoly Lunacharsky claimed that Lenin "emphasized the necessity of establishing a definite proportion between entertainment and educational movies" at the meeting they held in January 1922. *About Film* [*O kino*], (Moscow: Iskusstvo 1956), p. 4.

terms "intermediating cinema" and "typage" to focus his criticism of Eisenstein who had introduced them to film theory while applying basic principles of documentary cinema to his staged films. The *LEF* critics considered this method inappropriate for "cinema of fact," thus supporting Vertov's criticism of Eisenstein's films as "staged films in documentary trousers."[52] Mayakovsky attacked Eisenstein's concept of typage publicly, denouncing this "most shameful method" of depicting historical personalities on the screen.*

Vertov's contempt for the NEP policy is understandable: he saw it as detrimental to the development of Soviet documentary cinema for which he had fought all his life. Since the Leninist Proportion represented yet another impediment to the growth of the "cinema of fact," Vertov proposed another "balance" in which newsreels and documentary films should constitute forty-five percent of all films distributed, educational and scientific films should make up thirty percent, and entertainment films should account for no more than twenty-five percent.[53] Needless to say, this suggestion hardly received any attention; actually, it had little hope for realization because the NEP mentality, enforced by the state, prevailed in film companies. With unconcealed bitterness, Vertov wrote in his diary:

The official voice of our [state] cinematography, the *Proletarian Film,* issued the following order: either make a transition to the staged film, make your mothers and fathers cry and get rid of your documentary cinema, or we shall destroy you with administrative measures [*unichtozhim vas administrativnimi merami*].[54]

This entry, written one year after the completion of *Three Songs about Lenin* (1934), reflects the frustration Vertov felt during the production and distribution of the film, in spite of its topic and the workers' enthusiastic response immediately after the Moscow opening. In the same section of the diary, symbolically entitled "About My Illness" (1935), Vertov openly revealed the inconveniences endured in the course of making the Lenin film:

The actual shooting in Central Asia took place under the most abnormal circumstances. We worked in constant danger of typhus, with a lack of transportation, and with irregular wages. Often we did not eat for three days, and sometimes we had to repair clocks for the local villagers in order to earn a meal without bread. We had to function covered with flea powder from head to toe, or rubbed with a greasy, smelly liquid which irritated our skin but saved us from the flies. Yet, all the time, we preserved our patience, retaining a strong will, because we did not want to give up. We decided to endure and fight to the finish.[55]

* See also note 127.

The difficulties continued throughout the editing of the film, and although Vertov's diary does not provide details about it, there is no doubt that the film was severely censored, since all the archival footage with Trotsky, Zinoviev, Kamenev, and Radek were cut from the final print. Even three years after the film was completed, Stalin personally ordered that an additional 700 feet be inserted at the end, "showing how Stalin was continuing Lenin's work."[56]

All these difficulties, however, did not shake Vertov's confidence that *Three Songs about Lenin* would reestablish documentary film as the most suitable method for reconstructing the revolutionary past and for conveying "Revolutionary Truth." In reality, his expectations were met with a paradoxical resistance. In his article "The Last Experiment," Vertov expressed his fear that the Lenin film was the last experiment of the "Film-Eye" method, since the Soviet film repertory was dominated by "entertainment movies" [*uveselitel'nye kartiny*] and by "translated films" [*perevodnye fil'my*], that is, films that "imitate the languages of theater and literature." Consequently, all the creative attempts at making "original films" [*fil'my originali*] and "author's films" [*avtorskie fil'my*] were "neither financially supported nor artistically encouraged by the government."[57] Surprisingly, on January 11, 1935, one week after his article appeared in *Literaturnaia gazeta*, Vertov received the Order of the Red Star.* However, the fact remains that after *Three Songs about Lenin*, Vertov was not able to make a major film, although he continuously offered his scripts to the Ministry of Cinematography.

By the end of 1938, Vertov had resigned himself to the realization that cinematic experimentation was no longer possible in the Soviet Union and that his own work in the Soviet film industry was considered – by the officials – as "inappropriate." He wondered at one point whether "in this situation" he could reasonably justify "fighting for his personal principles unscrupulously" as most of his colleagues did in order to work in their profession:

Can one conform to whatever modes, the habitual abominable modes – shameful, humiliating, and disgraceful methods – which hypocrites and imposters use all the time? I obviously cannot. . . . As long as I search for truth by means of truth alone.[58]

Similar moral dilemmas haunted Vertov until the end of his life: his initial pursuit of "truth" as a conscious effort to reveal "Life-As-It-Is" had grown into an ethical issue concerning freedom of expression in a

* The Order of the Red Star is the third one in the hierarchy of Soviet state honors, the Order of Lenin being the highest of all. Vertov received the honor at the celebration taking place in the Bolshoi Theater, January 11, 1935. Eisenstein received an even less important honor (People's Artist). For more information about this event, see Leyda (*Kino*, p. 319), who participated in the ceremony.

society that was beginning to stifle all individual accomplishments that did not comply with political dictates. As with all true artists, Vertov could not sacrifice his artistic freedom for political pragmatism, and he found himself isolated from the film community.

Vertov and Mayakovsky

From his student days, Vertov was fascinated by Mayakovsky's poetry, his innovations in language and prosodic style — especially his experiments with auditory effects and the musical rhythm of poetic structure. While studying psychology at the Psycho-Neurological Institute in Petrograd, Vertov had expended great effort, examining in practice the effects of direct sound recordings of auditory signals. Much like Mayakovsky, Vertov explored the expressive possibilities of sound and its role in the creation of what he called "a new type of art — the art of life as it is — that could contribute to the unstaged documentary film and the newsreel."[59] Also at this time, Mayakovsky delivered public readings in various cities including Petrograd; hence, it is no surprise that Mayakovsky's revolutionary poetry became inspirational to Vertov's research on aural comprehension and visual perception.

A brief autobiographical sketch reveals how Vertov's experiments with sound led him to his interest in cinema, as he began to understand that the two media — radio and film — were closely related. Reflecting on those days of youthful experimentation, Vertov emphasized his fascination with the possibility of "capturing" the auditory and visual aspects of everyday life:

From my childhood, I was interested in various means for making documentary recordings of the exterior world of sound through montage, stenographic recordings, phonographic recordings, etc. In my "Laboratory of Hearing," I created documentary compositions as well as literary-musical montages of words. I was particularly interested in the possibilities of the motion picture camera, its capacity to register segments of life as a true chronicle and newsreel of vanishing and irrevocable events occurring in reality.[60]

The act of recording as well as rearranging the perceptual elements of the visual and auditory world was of paramount importance for Vertov. His early experiments, in which recorded sounds were cut up and restructured according to musical rhythm, bear a striking resemblance to the linguistic theories and experiments done by the Soviet formalists, especially their investigation of poetic structure, the musical function of words, and the subliminal impact of syntax. The formalists' discoveries and their subsequent theories concerning the auditory dynamics of prosody helped futurist poets to realize that the resonance of the arranged words could alter and expand the thematic meaning of a stanza so much that a poem could be fully grasped only when read aloud in an appropriate tone of voice and at a particular pace. Similarly, as the forthcoming analysis of the montage structure of *The Man with the Movie Camera* will prove, it is the tempo of visual changes occurring on the screen that makes this film such an exciting visual experience. Many of these optical beats are not directly perceptible because their impulses are below the threshold of consciousness, but even though the spectators react to them subliminally, these impulses contribute to the sequence's meaning.

The rhythmic organization of words to achieve a musical impact in poetry was of exceptional importance to Mayakovsky as well. In his essay "How to Make Verses" (1926), he emphasized the precedence not only of line length but also of the "transitional words" that connect one line with the next. Mayakovsky urged his fellow poets to take advantage of all the formal possibilities available to them, or, as he put it, to give "all the rights of citizenship to the new language, to the cry instead of the melody, to the beat of drums instead of a lullaby."[61] If a poem was intended to reflect the dynamism of the new technological age, then, Mayakovsky insisted, its style and, even more, its formal structure should be equally "energetic"; otherwise, the poem would merely echo the mawkish and oldfangled [*staromodnii*] conventions of a symbolist-romantic imagination, only to function on the thematic level. The opening stanzas of "Morning" (1912), an early poem in which Mayakovsky isolates various objects from the external world, clearly illustrates the poet's search for a rhythm that is expressive and musical at the same time. The structuring of the lines, some of which consist of only one or one-and-a-half words (!), is reminiscent of Vertov's use of a single frame as a shot or montage unit. In both cases the result is an intensified prosodic or cinematic rhythm.

The morose rain looked askance.	Ugriumyi dozhd' skosil glaza.
And beyond	A za
the well-defined	reshetkoi
grillwork	chotkoi
of the wires' iron through –	zhelezhnoi mysli-provodov
a featherbed.	perina.
And on	I na
it	nee
lightly rest	vstaiushchikh zvezd
the feet of awakening stars.	legko operlis' nogi.
But the per-	No gi-
dition of lanterns,	bel' fonarei
of tsars	tsarei
with the crown of burning gas	v korone gaza
for eyes	dlia glaza
made it painful to take	sdelala bol'noi

| the odorous bouquet of | vrazhduiuschii buket |
| boulevard prostitutes. | bulvarnykh prostitutok.[62] |

In his analysis of this poem, Edward Brown points out that Mayakovsky achieves his rhymes by matching the last two syllables in paired lines, thus building a unique metric beat.[63] Unlike traditional poets, Mayakovsky also creates rhymes by splitting words and placing each half in a different line; this draws the reader's attention to the vocalistic sounding of a word; it also energizes its meaning within the thematic context, which takes on a new, auditory significance. Rather than restrict himself to traditional poetic forms, Mayakovsky expanded the stylistic features of his poetry, much as Vertov was preoccupied with experimenting with image and sound to form his unique cinematic style. Instead of letting the narrative restrictions dictate the arrangement of shots and relationship between sequences in *The Man with the Movie Camera,* Vertov and Svilova based their editing decisions on graphic and visual features, juxtaposing shots in such a way that their compositions match graphically, and only when perceived as part of an overall cinematic structure do these shots reveal their full meaning within the thematic context. This principle of editing, developed from Vertov's "Theory of Intervals,"* draws a lot from the constructivist concept of film as a "building" comprised of many bits and pieces whose ultimate meaning depends on the interrelationship between various components. Applying this concept to silent cinema, Vertov achieved a high degree of cinematic abstraction through the "battle" of different visual structures and movements, producing a "kinetic impact," the basis of *kinesthesia,*† the most unique experience that cinema can provide. He later extended the idea of intervals to his sound films, especially *Enthusiasm,* in which one can detect a contrapuntal relationship not only between image and sound but also between various sounds juxtaposed to each other. Vertov's "Theory of Intervals," aimed at creating primarily visual impulses by "a movement between the pieces [shots] and frames,"[64] has many features in common with the "Rhythmicosyntactic Theory" as defined by Osip Brik in his famous essay "Contributions to the Study of Verse Language."[65] Both theories suggest that the rhythmic beat of a poem or a sequence

---

* Vertov formulated his "Theory of Intervals" as early as 1919, although it was published two years later as part of the "We" manifesto. Later, Vertov elaborated the theory in his article "From 'Film-Eye' to 'Radio-Eye' " (1929). The most important aspect of this theory is its emphasis on the perceptual conflict that occurs between two adjoining shots as the result of cutting "on movement," so that the sequence functions like a musical phrase, with its rhythmic ascent, peak, and decline.
† In his "We" manifesto, Vertov uses the term "kinetic resolution" [*kineticheskoe rezreshenie*], but he actually had in mind what is known today as the kinesthetic impact. For an explanation of this term, as well as the notion of "kinesthesia," see Slavko Vorkapich, "Film as a Visual Language and as a Form of Art," *Film Culture* (Fall 1965), pp. 1–46.

not only supports the thematic meaning but acts as an autonomous structure with its own impact and signification.

Among Vertov's papers in his Moscow archive, there is a "deconstructed" poem that testifies to Vertov's fascination with Mayakovsky's syllabic meter. Written in the mid 1920s, the poem employs a metric pattern typical of Mayakovsky's early poetry (especially "Morning"), with split words that enforce the musical structure of the rhymes. Titled "Mouths Are Gaping Through the Window," the poem is dedicated to the thousands of children who were dying of hunger in drought-stricken areas:

The field is	Pole
bare.	golo.
Bodies are everywhere	Tel
as after a snowstorm.	metelitsa.
The year	God
like a sacrificial lamb	tel'tsem
fell	v grob
into the grave	leg.
Hun-	Go-
-ger	-lod
long-	do-
-lasting.	-log.
Mountains of	Gory
grief.	goria.
While the city looks like	Gorod
a rainbow	radugoi.
I throw	Gorem
my burning pain	goriu
into the city's	gorodam
face.	v upor.[66]

The alliterative use of entire words ("*gorem*," "*goriu*," "*goro*dam") or only their separate syllables ("go-*lod*," "do-*log*"), the splitting of words ("hun-ger"; "long-lasting"), and their treatment as autonomous lexical units with pertinent signification, and especially the graphic distribution of verses on the page, are clearly influenced by Mayakovsky's metric patterns. The poem can also be related to the formalist principle by which the rhythmic flow of the lines should reveal per se the respective situation or the poet's emotional state, as each word or its severed part triggers an image linked to another image-word in the manner of montage. The aggressive tone of Vertov's poem is paralleled by the staccato pace of its deconstructed lines, a principle Vertov often uses in the climatic portions of his films, by "scattering" the segments of the same shot throughout the sequence.

Mayakovsky produced complex poetic images by breaking up common syntax and by forcing the reader to abandon the rational search for a sequential order and thematic progression in poetry. Similarly, Vertov relied on the intricate juxtaposition and inversion of filmed fragments (shots) with the intention of disrupting the film's linear development and thwarting the reader's narrative expectations. Yuri Lotman's analysis of Mayakovsky's technique emphasizes the innovative and rhetorical figures that make Mayakovsky a most individual poet, a statement one can apply to Vertov with respect to cinema.* Both artists used innovative "communicative structures," whether in words or in images, to transpose the "life-facts" into a new vision of external reality that corresponded to their subjective perception. Above all, they shared an uncompromising attitude toward artistic creation in general, struggling – to the end of their lives – for the legitimate right of the artist to experiment in all media, to search constantly for new expressive means, and to interpret reality according to a personal world view. Vertov saw this search for a novel poetic expression as an exciting confrontation of creative forces.

In his poem "Conversation with a Tax Collector about Poetry," Mayakovsky refers to his rhymes as "a barrel of dynamite" and calls the poetic line "a fuse that's lit" and when "the line smoulders, / the rhyme explodes – / and by a stanza / a city / is blown to bits."[67] In an almost identical fashion, Vertov's and Svilova's cutting produces optical "explosions" at the juncture of two shots. Their juxtaposition of different visual compositions and their insertions of unexpected light flashes (optical pulsations) correspond to Mayakovsky's arrangement of stanzas to shock the readers, to engage them in a novel and unconventional perception of the external world. Mayakovksy talks about the "extracted" and "distilled" tropes, just as Vertov talks about the "concentrated way of seeing and hearing." In his youthful poems, Vertov demonstrated great fascination with futurist rhetoric. In one of these humorous epigrams dedicated to cinema and in opposition to the exploitation of the medium by the commercial entrepreneurs, Vertov unabashedly imitated similar epigrams by Mayakovsky. Intended as a proclamation of the *kinoks*' view of contemporary cinema, this 1917 short poem also informs us about Vertov's early interest in the psychology of perception.

---

* Yuri Lotman is one of the leading contemporary Soviet semiologists involved in extensive research on the relationship between media and the message. His study of cinema, *Film Semiotics and Problems of Film Aesthetics* [*Semiotika kino i problemy kinoestetiki*] (Tatlin: Eesti raamat, 1973), has been translated into English by Mark Suino under the title *Semiotics of Cinema,* published in the series *Michigan Slavic Contributions,* no. 5 (Ann Arbor: University of Michigan, 1973).

Not "Pathé," nor "Gaumont,"	Ne "Pate," ne "Gomon,"
Not this, not about this.	Ne to, ne o tom
The apple should be seen as Newton saw it.	N'iutonom iabloko videt'.
Open eyes to the Universe.	Miru glaza.
So that the ordinary dog	Chtob obychnogo psa
By Pavlov's eye can be seen.	Pavlovskim okom videt'.
We go to the movies	Idem v kino
To blow up the movies,	Vzorvat' kino,
In order to see the movies.	Chtoby kino uvidet.[68]

Vertov's brother, Mikhail Kaufman, explained that this poem was written "in reaction to a popular ad for the 'Pathé Newsreel,' which claimed to know everything and to see everything. . . . Dziga made fun of this ad by indicating that the 'Pathé Newsreel' sees very little and thinks even less."[69] The last three lines of the poem, however, have a broader meaning: they reveal Vertov's revolutionary idea about the "cinema of fact" rising from the ashes of the old movie dramas. His mention of Pavlov, with whose work he was certainly familiar and whose classes he might have attended at the Psycho-Neurological Institute, indicate Vertov's interest in behaviorist concepts of art. One can argue that from the idea of "Pavlov's eye" emerged Vertov's "Film-Eye," with its power to unveil the external world not in a conventional manner (à la "Pathé" or "Gaumont"), but through an analytical penetration into the internal structure of visible reality where, he believed, the true meaning of things and events is concealed. As unsophisticated and simplistic as Vertov's ruptured lines may be, they can be compared to Mayakovsky's own "movie-poems" in which he mocked traditional art and the type of filmmaking the *kinoks* denounced. In the early 1920s, Mayakovsky wrote a short but emphatic poetic aphorism, "Film and Film," which glorified the revolutionary cinema and condemned bourgeois photoplays (film melodramas) by distinguishing between film as entertainment and as an engaged social force:

For you cinema is spectacle	Dlia vas kino – zrelische
For me – a view of the world	Dlia menia – pochti mirosozertsanie
Cinema – conductor of movement	Kino – provodnik dvizheniia
Cinema – innovator of literature	Kino – novator literatury
Cinema – destroyer of aesthetics	Kino – razrushitel' estetiki
Cinema – fearlessness	Kino – besstrashnost'
Cinema – sportsman	Kino – sportsmen
Cinema – distributor of ideas.	Kino – rasseivatel' idei.[70]

Mayakovsky wrote these lines in 1922, at the time when Vertov's *Film-Truth* series had already attained considerable popularity. At this juncture it should also be noted that Mayakovsky had shown quite a different attitude toward film *before* the revolution. In his first article on cinema, written in 1913, Mayakovsky questioned whether contem-

porary theater could survive competition from the cinema, and postulated that theater should learn from film how to rid itself of the artifi-
ciality and lifelessness of sets. Claiming that cinema can only
"imitatively register movements in real life," Mayakovsky believed that
film would "open the way to the theater of the future and to the actor,
unfettered by a dead backdrop of a painter's set."[71] Evidently, the
young Mayakovsky regarded cinema as a medium whose sole capacity
was to reproduce exterior events, as opposed to the theater which possessed the ability of interpreting reality with poetic license. Like many
other artists of the period, Mayakovsky failed to recognize the possibility of cinema as an autonomous medium, an art form with a unique
means of expression. Vsevolod Meyerhold adhered to a similar attitude
and, in his 1913 essay on film and theater, stated that cinema was a
"shining example of obsession with quasi-verisimilitude," and contended that film was of "undoubted importance to science, but when it
is put to the service of art, it senses its own inadequacy and labors in
vain to justify the label of art."* After the revolution, however, both
Meyerhold and Mayakovsky changed their attitude toward cinema and
involved themselves directly in film production. Their prerevolutionary
scorn for cinema agreed with the common belief that film was unworthy of the title "art"; it also attested to the atmosphere in which the
Soviet avant-garde filmmakers had to work and promote their unorthodox ideas and experimental works.

In his second article, "The Destruction of the Theater," published
the same year (early 1913), Mayakovsky continued to describe film as
"a cultural means for the emancipation of the theater."[72] Even in his
third essay on cinema (published in late 1913), Mayakovsky maintained a distrust of film as an art form:

Can cinema be an independent art form? Obviously no. . . . Only an artist can
extrapolate the images of art from real life while cinema can act merely as a
successful or unsuccessful multiplier of the artist's images. Cinema and art
are phenomena of a different order. . . . Art produces refined images while cinema, like a printing press, reproduces them and distributes them to the remo-

---

* Vsevolod Meyerhold, *Meyerhold on Theatre*, trans. and ed. by Edward Braun (New
York: Hill and Wang, 1969), pp. 34–5. In 1915, Meyerhold also claimed that "the film as
it exists today is entirely inadequate and my attitude toward it is negative" (Leyda, *Kino*,
p. 81). That same year, Meyerhold directed his first film, *The Picture of Dorian Gray*
[*Portret Doriana Greia*], based on Oscar Wilde's work, which is considered to be "the
most important Russian film made previous to the February Revolution" (Leyda, *Kino*, p.
82). Unfortunately, this and the other Meyerhold film, *The Strong Man* [*Sil'nyi chelovek*,
1916], are lost. In 1925, Meyerhold planned to direct a film based on John Reed's *Ten
Days That Shook the World* for Proletkino, but the project was never realized. In 1928,
he was assigned to direct the historical epic *Twenty-Six Commissars*, which was subsequently directed by Nikolai Shengelaia in 1932. As an actor, Meyerhold appeared in *The
Picture of Dorian Gray*, *The Strong Man*, and *The White Eagle* [*Belyi orel*, 1928], directed
by Yakov Protazanov.

test parts of the world.... Hence cinema cannot be an autonomous art form. Of course, to destroy it would be stupid; it would be like deciding to do away with the typewriter or telescope because they do not relate to theater, literature, or Futurism.[73]

In the same article, Mayakovsky addressed the question: "Can cinema provide an aesthetic experience?" This time his response was affirmative, yet without any theoretical substantiation for his position. After a brief discussion of what he considered essential to art, Mayakovsky reiterated his previous claim that "at its best, cinema can only have a scientific or merely descriptive function."[74]

Mayakovsky's failure to appreciate the new medium and to anticipate its creative potentials seems strange in retrospect, especially because his initial disdain for cinema was not much different from the acrid note Tsar Nicholas II wrote along the margin of a 1913 police report:

I consider cinema to be an empty, totally useless, even harmful form of entertainment. Only an abnormal person could place this farcical business on a par with art. It is complete rubbish, and no importance whatsoever should be attributed to such idiocy.[75]

However obsolete today, the above quotations illustrate to what extent cinema was considered inferior to the other arts in Russia, both by artists and laymen. Even more than in other European countries, film in Russia was reduced to the level of fairground entertainment, which discouraged the young artists from participating in its development and emancipation. Such was the climate in which Vertov arrived with his revolutionary ideas regarding cinema as "the art of inventing movements of things in space" and "the ordered fantasy of movement."[76] Initially contemptuous of cinema as an art form because of his practical and enthusiastic involvement in the theater – one of the most developed and highly esteemed art forms in tsarist Russia – Mayakovsky reconsidered his negative attitude toward film only after he became acquainted with the works of Soviet revolutionary filmmakers such as Shub, Vertov, and Eisenstein. He even embraced cinema as his own expressive means: in 1918, he wrote three scripts and played leading roles in the subsequent films, while his critical writing about cinema revealed an increasing awareness of the distinction between film as entertainment and film as an art form.*

* Mayakovsky wrote eleven scripts altogether, although only three were produced. These three are: *The Young Lady and the Hooligan* [*Barishnia i khuligan,* 1918], based on the novel *Coure* by Edmondo d'Amicis; *It Cannot Be Bought for Money* [*Ne dlia deneg rodivshisia,* 1918] based on Jack London's *Martin Eden* (with Mayakovsky in the title role); and *Shackled By Film* [*Zakovannaia filmoi,* 1918], with Mayakovsky again playing the lead. The first and most popular film was directed by Evgeni Slavinsky, the other two by Nikandr Turkin. All of Mayakovsky's scripts appear in *Complete Works,* XII, pp. 7–212, 481–7.

During Mayakovsky's editorship of *LEF,* numerous discussions were organized about the nature and function of documentary cinema. Mayakovsky supported Vertov and Shub in denouncing the tide of commercial production prevailing during the NEP era; the *LEF* group believed that the promotion of entertainment movies jeopardized the role of revolutionary artists in Soviet society and worsened the already low cultural level of the masses. Mayakovsky wholeheartedly joined Vertov in his intention to change the common audience's consciousness about cinema. His support was clearly demonstrated in his humorous movie-poem "Film and Film" placed right above a photograph of Vertov in *Kinofot,* the journal dedicated to revolutionary art and edited by the enfant terrible of constructivism, Aleksei Gan, who was equally outraged over the importation of trivial movies from Germany, France, and the United States during the NEP years.

The *kinok*s' most difficult problem was to find theaters for screening their films, since at that time distributors, misinterpreting Lenin's idea of "proportion," established a strictly economic policy for showing films, a policy favoring profit-making movies. Despite this obstacle, Vertov urged his *kinok*s not to be discouraged by the lack of understanding from politicians, producers, and film distributors, but instead to hold to their revolutionary ideas in fighting the prevailing bourgeois mentality in cinema:

Although our newsreels are boycotted by both the film distributors and the bourgeois or semi-bourgeois audiences, this fact should not force us to comply with the habitual taste of a philistine audience. To the contrary, it should prompt us to change the audience.[77]

Vertov seized every opportunity to denounce "the ruthless money-grabbing profiteers" of the Soviet import companies, just as Mayakovsky used the official film congresses (organized by *Narkompros*) to attack the "commercial mentality" of Soviet film producers and distributors. Under Mayakovsky's leadership, *LEF* became a stronghold of critical and often vitriolic reactions against NEP policy. In its April 1923 issue, *LEF* published Tretyakov's famous article "*LEF* and NEP," in which he pointed out that NEP existed for one of two reasons: either as a vehicle for "restoring the old bourgeois attitudes towards art," or (according to the official definition) as a "temporary step that will make true art possible in the future."[78] For Tretyakov, Mayakovsky, Shub, and Vertov, there was no doubt about which of the two evaluations of NEP's role in art was the more appropriate. With their revolutionary eagerness, they rejected any compromise that was against their artistic conviction and their love for "cinema of fact." Vertov was above all enraged at the economic contradictions brought about by the NEP policy, not only in the cinematic world but also in overall living conditions. This is evident from his unpublished poem, "Mouths Are Gaping Through the Window," whose first part has already been quoted. The closing stanzas even more vehemently cry out against social injustice by condemning the stores fully stocked in the cities while hungry children die in the country:

NEP! Screw thy cafés and catafalques,	NÈP! Tvoiu-kafe-katafalk,
Carriages full of candies and violets!	Konfekt i fialok fiakr!
Children's cries – be a knot in thy throat!	Detskogo krika – oskolok v kadyk!
Take the cakes from thy mouth!!	Torty-to-iz gorla vytashchi!!
Mouths are gaping through the window!	Rty u vitrin.[79]

The avant-garde's resistance to and criticism of the NEP was extremely risky, since it meant disagreement with the party line. But despite this danger, Vertov and Mayakovsky continued to be vocal partakers of the anti-NEP attitude, romantically believing that the artist's voice might change the situation. On October 15, 1927, Mayakovsky declared publicly that "Soviet cinema is utterly archaic and based on obsolete aesthetics which have nothing in common with contemporary Soviet life."[80] The following year he published the famous poetic epigram "Film and Wine" (in Russian the two words rhyme: *kino* and *vino*), poking fun at the state company Sovkino for producing films ("*kino*") like wine ("*vino*"), and declaring that the company's bosses possessed minds not unlike those of "wine traders."[81] Supporting the truly revolutionary filmmakers, Mayakovsky used both his prestige and his rhetoric to discredit the NEP officials who made decisions about film production. He urged "Communism . . . to free film from the hands of the speculators" and advised "the Futurists to purify the putrid water – the sluggishness and immorality in art," because "without this, we shall have their imported two-step from America or a perennial eyes-full-of-tears *à la* Mozhukin. The first is a bore. The second even more so."[82] In actuality, Mayakovsky and Vertov could do little to change the NEP practice of making profit on movies, even less to improve popular taste. Against their attitude stood not only the official policy but also the Soviet equivalent of the Hollywood "dream factory,"* which thrived under the NEP auspice and produced a new type of sociorealist melodrama.

The comparison between Vertov and Mayakovsky shows the extent to which these two artists were concerned with the formal aspect of the creative process and with the involvement of the masses in that

---

* "Dream factory" [*fabrika snov*] is the term Shklovsky used to deprecate Hollywood productions, contrasting it with the term "Film factory" [*kinofabrika*], which described the practice of Soviet revolutionary filmmakers.

process as well, be it literary or poetic production. Their preoccupation with the auditory-visual structure and specific expressive means was so profound that their political commitments, let alone their obligations to the party, became of secondary importance. For such an "individualistic" behavior they had to pay a price, which turned out to be more emotionally devastating than they could foresee. With all their ideological radicalism, neither Vertov nor Mayakovsky ever placed political dogma above their personal artistic visions and their humanistic attitude toward freedom of expression. And if there is a Soviet contemporary of Vertov to whom he should be compared, both ideologically and psychologically, it could only be Mayakovsky. To equate Vertov with Trotsky ("cinema's Trotsky"[83]) seems unjustifiable. Whereas Trotsky valued political doctrine above all else during *and* after the time he possessed political power, Vertov and Mayakovsky always held their artistic visions and art in general above politics, defending uncompromisingly freedom of expression. It is therefore more appropriate to call Vertov the "Mayakovsky of cinema."

Futurist and formalist expression

A mutual aversion toward the staged film (photoplay) led Vertov and Mayakovsky to arrive at identical definitions of "cinema of fact." The actual source of this concept was Mayakovsky's "factual poetry," which inspired Vertov to formulate his "Film-Eye" and "Radio-Eye" methods. His first article on Mayakovsky's poetry (written in early 1934) openly imitated the rhetorical style of futurist proclamations, emphasizing the infuence of Mayakovsky on the *kinok*s: "Mayakovsky is 'Film-Eye.' He sees what the human eye does not see. 'Film-Eye,' like Mayakovsky, fights against the clichés of the world's film production."[84] Vertov's next article (written near the end of 1934) reiterated his enthusiasm for Mayakovsky, especially the auditory impact of his poetic language. Pointing to Mayakovsky's prosody as the source of his cinematic vision, Vertov wrote: "The unity of form and content which dominates folk art is equally striking in Mayakovsky's poetry. Since I have been working in the field of poetic documentary cinema, both Mayakovsky's poetry and folk songs have had an enormous influence on me."[85] Vertov's emphasis on the unity of form and content explains his concern for rhythmic structure (dominant in folk songs) as well as poetic imagery (evident in Mayakovsky's verse), which are the most significant features of *The Man with the Movie Camera*, especially its symphonic montage structure.

After studying Mayakovsky's essays on poetic expression while trying to imitate his versification, Vertov strove for a "cinematic poetry" devoid of conventional narrative linearity and the theatrical presentation of reality. In his diary, Vertov refers to himself as a cinematic poet: "I am a writer of cinema. I am a film poet. But instead of writing on paper, I write on the film strip."[86] Reflecting back on his earlier work, he confirms that he "discovered the key to recording documentary sounds while analyzing the musical rhythms of Mayakovsky's poems." Eisenstein also recognized the Soviet avant-garde cinema's debt to Mayakovsky, particularly to his method of deconstructing rhymes in order to achieve stronger auditory and visual effects and new meanings. In his essay "Montage 1938," Eisenstein wrote that Mayakovsky "does not work in lines...he works in shots, verses... cutting his lines just as an experienced film editor would construct a typical film sequence."[87] Both Eisenstein's "Montage of Collisions" and Vertov's "Montage of Intervals," preoccupied with the shots' formal structure, parallel Mayakovsky's concern for the words' syntactical relation and their inflection. *The Man with the Movie Camera* went farther than any other silent film in linking shots with the intention to create a musical structure, to produce a poetic impact, and to provoke subliminal responses in the viewer.

Another device integral to Mayakovsky's poetic technique is the condensation of thematic elements in which one character represents an entire social class, or one fragment of reality evokes a set of associations in the reader's mind. In the "Friends" section of "About This," Mayakovsky depicts the decadent atmosphere of a reception given by the "new bourgeoisie" through a few select details: the raven-guests, trivial conversation, bubbling champagne, and the two-step (which in the Soviet Union at the time connoted Western licentiousness). In his *Strike* (1924), Eisenstein used symbolic details in depicting (unmasking) negative characters by superimposing the emblematic close-ups (the owl, the fox) over the faces of his *typages* (nonprofessional actors chosen according to their facial features). In contrast, Vertov did not adhere to such a direct – and quite histrionic – visual symbolism; instead, he always showed real people within their natural environment, engaged in their usual activities (workers in mines, bureaucrats in offices, peasants in markets or on farms, old people in churches, derelicts sleeping among garbage, ladies in beauty parlors, children in amusement parks, drunkards in pubs and streets, athletes on fields, and the omnipresent cameramen). What makes these people representative of their class, occupation, mentality, and psychological attitude is the fact that they are "caught" by the camera when their true nature exposes itself fully, even if they instantly realize they are being filmed. It is the juxtaposition between the people (mostly shown in close-up) and the action they are involved in or the environment in which they function that generates the authors' comment on the "characters" and their actions. Through montage, Vertov and Svilova succeed in communicating the complex meaning of "the boisterous ocean of life," into which the Cameraman "throws himself," capturing things "unawares" so that "life's chaos gradually becomes clear,"[88] and the essence of the visible world becomes apparent.

The "Film-Eye" method, whose basic goal was to make possible a "cinematic sensation of the world" [*kinooshchushchenie mira*] and whose basic technique was to "extract" (Vertov's term) the most revealing images from thousands of feet of film, is analogous to the process of "creative mining" (Mayakovsky's term) in poetry. Referring to this process in the "Conversation with a Tax Inspector about Poetry," Mayakovsky writes: "Poetry's / also radium extraction. / Grams of extraction / in years of labour. / For one single word, / I consume in action / thousands of tons / of verbal ore."[89] In fact, Vertov quotes these lines in his diary, stating that this was the style he wanted to follow in his films, and not "the path of the poets whose lines hit the viewer (!) as the arrow of a cupid-lyre chase."[90] This, "organic" repulsion of conventional, trivial, and entertainment art was both the source of Vertov's and Mayakovsky's creative energy and their critical acrimony.

Vertov's published – but unrealized – scripts (particularly those written between 1935 and 1940) most clearly reveal his concern for poetic impact and formal structure. Referring to them as "musical/poetic films without magic or talisman,"* Vertov expected his readers to "see" the described images, and even to "edit" them into an imaginary montage structure! The ruptured and seemingly isolated lines, the contrapuntal organization of words, the concatenation of terse sentences, the musical flow of phrases, and the specified photographic viewpoint (angle), all of these do indeed inspire the reader of Vertov's scripts to envision the described situations. The introductory paragraph of the script "A Girl Plays the Piano," with its unusual graphic arrangement of sentences, parallels Mayakovsky's technique of laying out lines (or shots) by underscoring graphically outstanding details. After a brief "overture" that explains the basic intention of this "scientific and fantastic cinematic poem," Vertov's script begins just like a poem:

---

* Vertov, "A Girl Plays the Piano" (1939) [*Devushka igraet na roiale*], *Articles*, p. 288. In several of his other scripts, Vertov shows a particular interest in various aspects of the life and the social position of women in the Soviet Union. Among these scripts are "She" [*Ona*, 1939], "The Girl Composer" [*Devushka-kompozitor*, 1936], "Song of a Girl" [*Pesnia o devushke*, 1936], and "The Letter from a Girl Tractor Driver" [*Pismo traktoristki*, 1940]. Also, his third sound film, *Lullaby* (1937), deals with women and is therefore subtitled "A Song to the Liberated Soviet Woman."

A girl is playing the piano	Devushka igraet na roiale
She is watched	Na nee smotrit
through the open windows of a	
terrace	skvoz' raskrytye okna terrasy
by a starry night.	zvezdanaia noch'.
The Moon illuminates her hands.	Luna osveshchaet ee ruki.
The Moon illuminates the keyboard.	Luna osveshchaet klavishi.
And to her it seems not sounds	I ei kazhetsia, chto ne zvuki,
but rays of distant, invisible worlds,	a luchi nevidannykh dalekikh mirov,
rays of glimmering stars	luchi mertsaiushchikh zvezd
that sing from under her fingers.	poiut iz-pod ee paltsev.[91]

Even without a profound analysis of this text and its structure, one can realize that Vertov's goal is to create a musical rhythm with words. Each sentence rhythmically correlates with the next, so much so that in order to be fully apprehended the phrases must be visualized. Prosodically, the structural arrangement of rhymes is reminiscent of Mayakovsky's technique, whereas the imaginistic presentation of nature and landscape may be related to other poets Vertov admired, Whitman among them.* The next passage from the same script requires the reader to participate even more imaginatively in "extracting" a series of images from a literary description:

> This could take place on a significantly lesser planet,
> let us say the Moon.
>   But the surrounding setting would not change.
>   This means only gravity would change.
>   Actually,
>   on the tennis court,
>   on the basketball court,
>   in the gymnastic compound,
>   on all the spaces which meet
>   her eyes,
>   strange things occur:
> all the players do not run,
>   but glide floating.

> Eto moglo by imet' mesto na znachitel'no men'shei planete,
> skazhem na Lune.
>   No obstanovka krugom ne izmenilas'.
>   Znachit, izmenilos' tol'ko tiagotenie.
>   Deistvitel'no,
>   na tennisnom korte,
>   na basketbol'noi ploshchadke,

* In his diary written during World War II, after describing the difficulties he faces in persuading the administration to approve his new project, Vertov suddenly begins to write in a highly emotional manner ("The poetry of science....The poetry of space.... The poetry of unheard numbers...."), and quotes several lines from Walt Whitman's *Song of Joys. Articles,* "Notebooks" (February 1, 1941), p. 234.

```
              v gimnasticheskom gorodke,
                   na prostranstve, kotoroe obkhvatyvaet
                        ee glaz,
                              proizkhodiat strannie veshchi:
vse igraiushchie ne begaiut.
      a plavno porkhaiut.⁹²
```

Because of their position within the line, the words themselves appear to "glide," thus inviting the reader to experience the line's "tonal glissando" as a stimulant in envisioning the events in "slow motion." The rhythm of the sentences is constructed in such a way as to enhance the mood of the event while at the same time suggesting a particular montage pace to be achieved in the process of editing. Indeed, it is a pity that Vertov was not allowed to realize this and other scripts he conceived in the later period of his career. The lexical use of the words and their syntactic order in these scripts indicate that he sought to expand his "Film-Eye" method as well as the "Film-Truth" principle by developing more symbolic montage "figures," and by allowing rather stylized mise-en-scène and shot compositions. One wonders how the last few "shots," as described in the script "A Girl Plays the Piano," would appear on the actual screen, if Vertov had been permitted to shoot them:

```
The universe                           Vselennaia
      looks                                  smotrit
           with the eyes of the stars             glazami zvezd
at the girl,

                                        na devushku,
           dreaming at the piano.            mechtaiushchuiu za roialem.⁹³
```

The futurists' and formalists' unprecedented concern for stylistic (formal) aspects and expressive modes of art stemmed from their belief that form and structure can produce their own meaning, which is as important as the narrative context. Vladimir Markov, who personally participated in the Russian futurist movement throughout the 1920s, explains that the main goal of the futurist poets like Mayakovsky was "to make the word the real protagonist of poetry, and, more importantly, to insist consciously and aggressively that poetry grows out of the word."⁹⁴ The cubo-futurists in particular were concerned with the value, function, and aesthetic impact of the "word as such," not only in poetry but also in "fine prose." In her analytical comparison of Vertov's work with Soviet futurist poetry, Anna Lawton points to A. Kruchenykh's 1913 futurist poem "Pomada" [*Pomade*/Lipstick], whose stanzas (e.g., "*dyr buv shchyl*") are composed of newly constructed words without any meaning, in the typical manner of the "Zaum language." According to Lawton, "the images [shots] in this poem are liberated from any kind of causal relationship [diegetic editing] and arranged in rhythmic segments" [sequences], thus "endowing the text [film] with a new and fresh meaning based on analogical relationships [associative editing] – a meaning which relies on the participation of the reader's [viewer's] intuition."[95] As one may see from the words suggested in the brackets, this statement can be applied almost verbatim to Vertov's film *The Man with the Movie Camera*, by replacing the terms describing literature-poetry with ones pertinent to cinema. Indeed, it is difficult to find a filmmaker who believed as strongly as Vertov in the importance of individual shots as the foundation of the film's ontological integrity. Like the futurist poets, Vertov knew that in order to fully exploit the potentials of the (cinematic) medium, it was necessary to prevent narrative and theatrical conventions from shackling the kinesthetic impact of the film. This does not mean that Vertov did not draw on other media whenever he felt they could contribute to the overall unity of a sequence; yet he did insist that these "borrowed" elements be fully transformed into new cinematic values that, when properly integrated, would become an organic part of the cinematic vision of reality.

With all their concern for the formal aspect of artistic creation, Mayakovsky and Vertov demonstrated an equal interest in political content whenever they found it necessary. They saw great possibility and appeal in the use of political cartooning as a means of visual communication. During the Russian Civil War, Mayakovsky personally designed hundreds of widely distributed political ROSTA posters for the Russian Telegraph Agency.[96] In many respects, the graphic style of these cartoons resembles animated segments that Vertov incorporated in his early *Film-Eye* series, especially in terms of the treatment of characters and the simplicity of their pictorial execution. The use of the instantly identifiable types (e.g., "the worker," "the bureaucrat," "the bourgeois imperialist") and characteristic settings (e.g., "the factory," "the office," "the mine") made Mayakovsky's cartoons accessible to largely illiterate audiences. Vertov's cartoon insets in the *Film-Eye* series were also intended to convey a political message to the masses in the most understandable manner, through simplified graphics in motion. Ironically, it is in these segments of his films (which are the least cinematic) that Vertov succeeded in being the most "accessible" to the masses, those same masses that – to his great disappointment – proved totally unresponsive to his idea of the "ultimate language of cinema" as demonstrated in *The Man with the Movie Camera*.

The futurists' admiration for the technological age is the major theme of Mayakovsky's "hymns" to factories and machines. He saw the steel constructions as a challenge to the pastoral scenes glorified by the traditional poets, especially the symbolists and ego-futurists* whom Mayakovsky ridiculed as being soaked in sentimental laments and sobs. Mayakovsky urged the revolutionary poets to write about common people, "those thousands of street folks," students, prostitutes, salesmen, and workers "who are creators within a burning hymn / the hymn of mills and laboratories."[97] His enthusiasm for the mechanical age erupted in its full fervor during his 1925 visit to the United States, where he wrote his ode to the Brooklyn Bridge which, together with his ode to the Eiffel Tower, best illustrates the futuristic preoccupation with technology:

What pride	Ia gord
I take	vot etoi
in that mile of steel,	stal'noiu milei,
and from it arose	zhiv'em v nei
my living vision	moi videniia v staliu-
the struggle	bor'ba
for construction	za konstruktsii
instead of style,	umesto stilei
the stern calculation	raschet survovoi
of steel	gaek
precision.	i stali.[98]

The relationship between the futurists and the formalists was on one level harmonious and on another antagonistic. They shared the identical view of art as a "device" artists must learn to use properly, that is, within the constraints of a given medium; they mutually disagreed with the proponents of socialist realism and especially with its insistence on politicizing art. Of course, futurists promulgated their own political views concerning artistic creation that considerably differed from the party's interpretation of Marx and dialectical materialism. In fact, there were many diverse ideological tendencies among futurists, all of them incompatible with the party's demand that its members follow the prescribed ideological trend regardless of personal opinion. In contrast, formalists were ideologically disengaged; they concentrated predominantly on the structural, linguistic, and formal aspects of artistic work. In this respect, they had much in common with constructivists who were equally preoccupied with form and structure in art. All three groups, however, were fascinated with the current technological revolution and modern industrial environment. Futurists wrote poems about factories and machines; constructivists painted abstract geometrical forms inspired by a mechanical world or

---

* Mayakovsky used to attack the ego-futurist poets publicly, especially their leader, Igor Severianin (Lotarev), who was prolific and extremely popular before the revolution. Severianin emigrated to Estonia in 1919, and even after the Soviets occupied that land, he continued to publish in the USSR.

built sculptures that were creative replicas of the machine; formalists helped poets discover the melody and rhythm of language, the dynamic structure of words and stanzas that can echo the mechanical beat of factories and machines. The widespread practice of photographic collage became even more dramatic when motion pictures were combined into sequences that generated optical rhythms and mechanical movement unattainable in still-photography.

From a structural position, *The Man with the Movie Camera* can also be seen as an ode to the industrial revolution reflected in each and every aspect of the daily lives citizens lead in a great city that is persistently watched and scrutinized by the "Film-Eye." Emphasizing the structural beauty of machines, extolling the workers' zeal, and empathizing with ordinary citizens performing their duties, this film is one of the truly artistic documents about a technological – as well as social – transformation. Yet, given the circumstances under which it was made, as a cinematic achievement this film is more a statement about the future than a record of actual socioeconomic conditions. What Vertov and Mayakovsky conceived and dreamed of as an ideal social order or as a novel art structure was largely suppressed by the major ideology in practice; restrained in their creative drives, they could only hope that future generations would recognize their works as genuine avant-garde achievements and revolutionary artistic visions.

The practice of producing a musical rhythm by selecting (or constructing) words according to their "sounding" was essential for the formalist poets (and critics) gathered around the OPOYAZ society.* Dedicated to the "scientific inquiry into literary technique," OPOYAZ introduced the concept of a transcendental (transrational) poetic language known as "Zaum,"[99] meaning "beyond the rational." Comprised of the combination of sounds that do not represent concrete words, but which are constructed on musical principles to evoke feelings and imply abstract ideas, Zaum poetry grew out of suprematist painting, particularly its concept of the emotional impact of colors. The basic premise of the Zaum movement was that all art should be free from its subservience to thematic meaning so that sounds, colors, movements, and nonrepresentational shapes sustain their autonomous associations and pure aesthetic function. In actual poetic achievements, the alternation of various auditory beats, according to a symphonic pat-

* OPOYAZ stands for "Society for the Study of Poetic Language" [*Obshchestvo izucheniia poeticheskogo jazyka*], founded in Petersburg in 1916. It included some of the most significant formalist poets and theorists, such as Yuri Tynianov, Osip Brik, Boris Eikhenbaum, Viktor Shklovsky, and Vladimir Zhirmunsky. The members of this society differed markedly in their literary preferences, although futurists considered its "Formalist method as the key element in the study of art." Victor Erlich, *Russian Formalism: History – Doctrine*, p. 47.

tern, approximates the structure of so-called concrete music and is associated with the sounds of working machines and mechanized factories. Examples of such a mechanistic rhythm, achieved through the repetitive use of words with one or two syllables, are poems by the greatest and most controversial of Zaumists, Velemir Khlebnikov, who inferred that a poem "is built of words as the constructive units of the edifice"; accordingly, in his dramatic poem "Snake Train," he made up words based on the Russian linguistic root *"um"* [mind], to evoke certain feelings and to achieve a musical beat:

Deum.	Deum
Boum.	Boum.
Koum.	Koum.
Soum.	Soum.
Poum.	Poum.
Glaum.	Glaum.
Noum.	Noum.
Nuum.	Nuum.
Vyum.	Vyum.
Bom! Bom! Bom!	Bom! Bom! Bom!
It's a loud toll in the bell of the mind.	Eto bol'shoi nabat v kolokol uma.
Divine sounds flying down from above	Bozhestvennye zvuki, sletaiushchiesia
At the summons of man.	sverkhy na prizyv cheloveka.*

One can find a direct relationship between certain nonrepresentational shots and optical pulsations in *The Man with the Movie Camera* and the "verbal flashes" or "syntactical drumming" achieved through the rhythmic beats of vowels and syllables in Zaum poetry. Explaining how he conceived his film *Three Songs about Lenin* "in the style of folk-song images," Vertov stated that he also created "verbal concatenations in the manner of Zaum,"[100] both as part of the spoken commentary and written intertitles. Vertov's practice of using the "flicker" (achieved by repeatedly inserting a single black, gray, or transparent frame between various shots) produces a unique perceptual sensation in the viewer. Given the irritating impact of the flickering effect employed even more emphatically in *The Man with the Movie Camera*, it is clear why this film, which challenges human perception and rejects a traditional approach to art, has been attacked by the orthodox critics as mere optical pyrotechnics. But the futurists immediately understood Vertov's experiment and wholeheartedly defended his film as an outstanding avant-garde achievement.

Aesthetically, Vertov's theory and practice in many respects comply

---

* Velemir Khlebnikov, *Snake Train,* trans. Gary Kern (Ann Arbor: Ardis, 1976), p. 77. Published in 1922 (written between 1920 and 1921), Khlebnikov's *Zangezi* was produced in the theater of the Museum of the Materialist Culture in Petrograd, 1923, under the direction of the constructivist painter and sculptor, Vladimir Tatlin.

with the futurist's, suprematist's, and formalist's belief that every artist should express his or her ideas within the specificity of the given medium and without making concessions to popular taste. Just as Mayakovsky dedicated his poetic talent to changing the traditional prosodic structure, Vertov decided to create an international visual language understood by all people regardless of national boundaries. In his article "The Forward Looking," Fevralsky discusses the conceptual relationship between Vertov and Mayakovsky, particularly the way they use "cinematic" devices to present the "characters" in their films/scripts:

Vertov's film-poems were conceived according to Mayakovsky's poetical tradition. The visual concept of *The Man with the Movie Camera* has many contiguous points with Maykovsky's script *How Do You Do?*, written in 1926. Mayakovsky's protagonist is "The Man with the Pencil," i.e. the poet, as Vertov's protagonist is "The Man with the Movie Camera." Like Vertov, Mayakovsky insists on using "specific cinematic devices" instead of replacing them by other means of expression. It is an attempt to apply documentary style to a narrative film script.[101]

Fevralsky also points to other thematic and formal similarities in the works of the two artists. Commenting on the treatment of Lenin in Vertov's film *Three Songs about Lenin* and Mayakovsky's poem "Vladimir Ilyich Lenin," Fevralsky emphasizes Mayakovsky's concern for the historical documentation of the presented events. This insistence on historical accuracy was in line with constructivist ideas about "art of fact," which required the artists not to falsify the actual matter, but to register it faithfully before putting it into a new context — requirements almost identical to the demands Vertov assigned to his *kinoks*.

Educating the masses

Accessibility to the masses was a problem Mayakovsky and Vertov tried to resolve in an uncompromising yet painful way. Believing that the masses could be gradually educated to understand and appreciate unconventional means of expression, both men eagerly experimented with a novel structural relationship between words and images. In their revolutionary romanticism, they tended to overlook the masses' inability to appreciate avant-garde art and experimentation. Even after they became aware of this fact, they did not yield to popular taste, but continued to demand that the masses change their attitude toward art! Mayakovsky expressed his criticism of popular taste openly — and often with aggressiveness — at many public meetings. Reacting rather disdainfully to the repeated accusation that his poetry was "difficult," he would instantly reply: "I agree that poems must be understood, but the reader must be understanding as well.... You must finally learn to appreciate complex poetry.... You cannot say, 'If I don't understand a poem, then it is the writer who is the fool.' "* Vertov, who claimed that he "has never considered Mayakovsky incomprehensible or unpopular,"[102] responded almost identically to the allegations that his films were "inaccessible" to the masses when he wrote: "If the NEP audience [*nepmanovskaia auditoria*] prefers dramas of kisses and crime [*potseluinye ili prestupnye dramy*], it does not necessarily mean that our work is inappropriate; it may well mean that the mass audience is still unfit and incapable of understanding it" [*ne goditsia publika*].[103]

Paraphrasing the Marxist view of religion in his "Simplest Slogans," Vertov stated that "film-drama is an opiate for the people,"† and insisted that "it is necessary to distinguish between being popular and pretending to be popular."[104] Given the political and cultural circumstances, Vertov and Mayakovsky proved to be excessively idealistic in their intention to educate the masses. Just as their vision of the ideal socialist state was utopian, their belief in the substantial change of the masses' artistic taste was impractical. However, true artists and humanists are expected to dream of far-reaching ideals; hence it seems scandalous to call such visions "idiotic and ugly,"[105] as Raymond Durgnat does in referring to Vertov's ideas of society.

The concern for the cultural sophistication and education of the masses was part of the constructivist's activism, which counted on the audience's participation in the creation of the artistic meaning rather than on the passive consumption of art. To encourage such a participation, the "artistic products" were intentionally made strenuous and self-referential, in keeping with the formalist device known as "making-it-difficult,"‡ thereby forcing the audience to search for a more

---

* Herbert Marshall, *Mayakovsky* (London: Dennis Dobson, 1965), p. 66. According to Marshall, only five days before committing suicide, Mayakovsky presented a lecture at the Plekhanov Institute of Economics in Moscow and, among others statements, responded to criticism of his poetry as being "difficult" in the following way: "In fifteen or twenty years, the cultural level of the workers will be raised so high that all my works will be understood.... I am amazed at the illiteracy of this audience. I never expected such a low cultural level from the students of such a high and respected educational institution." p. 71. Quoted from *Literary Chronicle of Mayakovsky,* ed. V. Katanyan (Moscow: The State Publishing House of Artistic Literature, 1956), p. 406.
† Vertov's aphorism appears as the third section (entitled "Simplest Slogans") of his 1926 statement "Provisional Instructions to 'Film-Eye' Groups" [*Vremennaia instruktsiia kruzhkam kinoglaza*], *Articles*, p. 96. The statement contains nine "instructions," and the first one reads: "Film-drama is an opiate for the people!" [*Kino-drama – opium dlia naroda!*].
‡ Among the three formalist concepts, the idea of making a work of art "difficult" to understand was most criticized by the proponents of socialist realism. For more information about "estrangement," "*otstranenie*," "*zatrudnenie*," and "*obnazhenie priema*," see Victor Erlich, *Russian Formalism: History – Doctrine*, pp. 145–63.

rarefied meaning in the given work, and to enjoy aspects of artistic creation beyond the narrative ones. The practice of making the structure of an artistic product "difficult" also meshed with the formalist principle of "baring-the-device" in order to reveal how reality is aesthetically transformed into art through the process of construction [*stoitel'nyi protsess*]. Vertov adhered to this principle throughout the production of *The Man with the Movie Camera,* hoping that it would "open the masses' eyes" toward film as an autonomous art. Unsurprisingly, his belief proved to be premature, as his most avant-garde films have had to wait more than half a century to be fully appreciated, and even then only by certain film scholars.

By adopting such an uncompromising attitude toward cinema, Vertov made himself vulnerable, especially to those critics who stood for the principles of socialist realism which catered to the popular taste in promoting political ideas. At the same time, there was a contradiction in Vertov's attitude: on the one hand, he believed that film should educate the masses; on the other, he constantly challenged popular taste. Almost every *Film-Truth* series featured a specific filmic device – from reverse to accelerated motion, from associative editing to jump cuts – often with the intention that the viewers use their full intellectual acuity if they wanted truly to understand the film. Thus, especially in his later work, Vertov alienated himself from the masses by being too demanding, an attitude he was aware of without feeling guilty or obliged to follow party orders:

One of the chief accusations leveled at our method is that we [the *kinok*s] are not understood by the masses. Well, even if you accept the fact that some of our works are difficult to understand, does it mean that we are not supposed to create any serious work? That we must abandon all exploration in cinema? If the masses need simple agitation and polemical pamphlets, does it mean that the masses do not need serious essays by Engels and Lenin? What if a Lenin of the Soviet Cinema appears among us and he is not permitted to work, because the products of his creation are new and inaccessible?[106]

This statement, reminiscent of Mayakovsky's response to the criticism concerning the "impenetrability" of his verses, parallels Shub's defense of the artist's right to produce "difficult" works. In her reply to Shklovsky's objection that Dovzhenko's film *Ivan* (1932) was not accessible to the masses, Shub wrote:

Not long ago Mayakovsky was also criticized for being incomprehensible to a large audience. Bearing this in mind, today's critics have no right to say: "This is an important work of art, but it is a failure as an achievement." This is unacceptable. The task of a true critic is to create the climate for such "difficult" works of art. Whenever an important work of art is created, the critic's duty is to help it with his pen so that a temporarily incomprehensible work soon becomes comprehensible to everybody.[107]

What a wise attitude toward avant-garde creation and its inaccessibility to the masses! Mayakovsky also used his authority to support the avant-garde artists whose works were criticized for not taking into account popular taste and current political needs. At a conference designated "Theater Policies of the Soviet Government," Mayakovsky publicly denounced the party's decision to ban Mikhail Bulgakov's play *The Days of the Turbins** as politically "inappropriate":

I consider the politics of suppression absolutely pernicious.... No, art must not be suppressed.... What shall we achieve by suppression? Only that such literature will be distributed around the corner and read with even greater satisfaction as I have read poems by Yessenin in manuscript form.[†]

However vocal, Mayakovsky's cry for creative freedom remained futile. A few days before his suicide,[‡] he sardonically admitted to the victory of bourgeois aesthetics in Soviet art. Exhausted and depressed, he saw numerous signs of the avant-garde's collapse in the government's repeated attempts to suppress unorthodox achievements and to promote consumer products for both political and commercial purposes. In what was his last public appearance, Mayakovsky attacked the bourgeois mentality in art, in a tone humorous and nostalgic at the same time:

All those Venuses of Milo with their lopped off arms, all that Greek classical beauty, can never satisfy the millions who are entering into the new life of our noisy cities, and who will soon be treading the path of revolution. Just now, however, our chairwoman offered me a sweet with the label Mosselprom on it. Above the label there is the same old Venus. So, what we've been fighting

---

* *The Days of the Turbins* [*Dni turbin*] is a stage adaptation of Bulgakov's famous novel *The White Guard* [*Belaia gvardiia*, 1925]. It was produced by the Moscow Art Theater in 1926 with great success. But after a short run, the performance was banned. Disappointed, Bulgakov personally appealed to Stalin, asking for permission to emigrate. Stalin told him (over the phone) that he could not leave the country, and promised to lift the ban on the play; this soon occurred, but only for a short period of time. A few years later, another great Soviet avant-garde writer, Yevgeny Zamyatin — the author of *We*, the fascinating and prophetic satire on a futuristic totalitarian society — was also forced to request permission to leave Russia, which Stalin granted him in 1931. Zamyatin emigrated to France, but lived in total seclusion and died in Paris six years later. His and Bulgakov's contemporaries, however — Valentin Katayev, Vsevolod Ivanov, and Veniyamin Kaverin — sacrificed their literary talents to become hacks, manufacturing whatever was required in the shape and style demanded by the party.
[†] Marshall, *Mayakovsky*, p. 45. Mayakovsky drew parallels between Bulgakov and Yessenin because Yessenin's poetry was also criticized as "defeatist" and was banned for a considerable period of time. Yessenin hanged himself on December 17, 1925, in a Leningrad hotel.
[‡] Mayakovsky shot himself on April 14, 1930, in Moscow.

against for twenty years apparently has become victorious. And the lopsided beauty is being circulated among the masses, appearing even on candy wrappers, poisoning our brain and destroying *our* idea of beauty all over again.[108]

Mayakovsky's bitterness reflects his inability to change the state of Soviet art and to improve the mental inertia of the masses for whose enlightenment both he and Vertov had devoted their talents. Disillusioned, they saw no future for their avant-garde experimentation, and therefore had no motivation for further contributing to a culture they found stale. The psychological burden of this realization took its toll in them: Mayakovsky soon ended his life, and Vertov became a recluse who gradually lapsed into a state of apathy.

The Vertov–Eisenstein controversy

It is clear today that, together with Eisenstein, Vertov's work as well as his ideas about film represent the most revolutionary contribution not only to Soviet but also to world cinema. Although their aesthetic and theoretical concepts differ in many respects, their writings are all relevant to the understanding of cinematic language. Vertov fought for authentic cinematic expression throughout his career, while Eisenstein — especially in the later part of his life — extended his theoretical research to film modes that embrace other arts, particularly literature and theater. In contrast, Vertov made no concession whatsoever to theatrical (staged) and fictional (literary) film, remaining both in theory and practice firmly committed to documentary, nonfictional cinema. Even in his unrealized scripts, that include prearranged mise-en-scène, Vertov made sure that every situation he envisioned would correspond as closely as possible to what he considered "a faithful copy" [*tochnaia kopiia*] of the given "life-facts."

That Eisenstein and Vertov adopted different approaches to film is hardly surprising: Eisenstein came to cinema from theater, whereas Vertov was involved in cinematic experimentation from the very beginning of his career. Their conflicting attitudes regarding the aesthetic value and social function of cinema reflect in many ways the conceptual and ideological discord characteristic of Soviet revolutionary art throughout the 1920s. A comparative examination of their theoretical disagreements provides further insight into the evolution of Soviet art in general. It reveals the extent to which the development of Soviet avant-garde film depended on experimentation in other media and was influenced by the theoretical and aesthetic concepts of the revolutionary critics.

Viktor Shklovsky, in his booklet *Their Genuineness* (1927), was the first to compare Vertov and Eisenstein. He praised Vertov's early *Film-Truth* series, but exhibited a clear preference for the shot composition and pictorial stylization of Eisenstein.[109] Reiterating his standpoint in

a more recent monograph on Eisenstein (published in 1973), Shklovsky supported Eisenstein's critique of Vertov's "Film-Truth" principle.
He agreed with Eisenstein's statement that Vertov used montage to
make "essentially static shots simulate movement,"[110] and although
admitting that Vertov's theories influenced Eisenstein's early films, he
maintained that Eisenstein "had gone much further" in his montage
experiments. Shklovsky's critique of Vertov is challenged by the modern Soviet film theorist Sergei Drobashenko, who infers that the two
filmmakers influenced each other equally, and that both materialized
their theoretical ideas in a unique way.[111] Another contemporary Soviet
theorist, Tamara Selezneva, draws a similar conclusion in her discussion of the problem of staged and unstaged cinema. In her book *Film
Thought of the 1920s* (1972), Selezneva states that Eisenstein and Vertov were equally important as creative figures within their respective
domains.[112]

Eisenstein expressed his initial disagreement with Vertov shortly
after he began to attend the *kinok*s' workshop.* Since he was preparing to redirect his artistic career from theater to film, Eisenstein tried
to learn about techniques of shooting and editing. In his memoirs,
Vertov recalls his encounters with Eisenstein before *Strike* (1924) was
made, at the time Eisenstein used to visit the *kinok*s' workshop:

At that time [1918 to 1924], Sergei Mikhailovich Eisenstein still respected the
*kinok*s; he used to attend every single screening at which we discussed the
Film-Truth series. But although we respected his mind and talent, we were not
grateful to Sergei Mikhailovich. We had constant fights with him, because we
felt that his concept of "intermediating cinema" [*promezhutochnoi*] (this term
is not mine, but Eisenstein's) impeded the advancement of documentary film.
In contrast, we considered that the application of the documentary method to
the staged film was unnatural [*protivoestestvennyi*].[113]

The term "intermediating cinema," as the core of the controversy, implies a mixture of "staged" and "unstaged" film, a hybrid scheme Vertov flatly rejected. Eisenstein, however, saw in the stylized mise-en-
scène and the expressive shot composition (especially the lighting) "a
conscious and active remaking [*perekraivanie*] of reality, not so much
reality in general, but every single event and each specific fact."[114] Vertov considered such a directorial method incompatible with the "recording of facts, classification of facts, dissemination of facts, and
agitation with facts."[115]

---

* Eisenstein's first film experience was associated with the *kinok*s' group. The cameraman Boris Frantsisson shot Eisenstein's movie *Glumov's Diary* [*Dnevnik Glumova*, 1922]
as part of the Proletkult Theater's production of Ostrovsky's *The Wise Man*, while Vertov
was assigned as the "artistic instructor" (although Eisenstein claims that Vertov soon
left the shoot). Conceived as a "parody on the idea of Pathé newsreels" (Eisenstein, *Selected Works* II, p. 454), this short film (about 160 feet) was later included in *The Spring
Film-Truth* [*Vesennaia kinopravda*], May 1923.

Eisenstein repeatedly stated that he was not influenced by the "Film-Eye" method, nor by any of Vertov's early newsreels. But his close friend Esther Shub recalls that Eisenstein admitted this influence, if not publicly, then at least under a particular circumstance:

Eisenstein, who otherwise disavowed the influence of Vertov on his film *Strike,* admitted with the frankness of a great artist – not only to me personally, but also to his students in VGIK – that one of the best sequences in *Battleship Potemkin,* the gathering of the people around Vakhulchik's dead body, was conceived under the direct inspiration of Vertov's *Leninist Film-Truth.*[116]

There is no doubt that the relationship between Vertov and Eisenstein was tense, despite various attempts to make it look incidental. Elizaveta Svilova, in her 1976 recollections about her husband, tried to illustrate the "friendly" relationship between the two men with an episode in which Eisenstein was depicted as "bursting with his impression" while "telling me what exceptional success *Symphony of the Don Basin* received in England"[117] (from which Eisenstein had just returned in early 1931). If this was true, then it seems strange that, having had such enthusiasm about the international success of Vertov's film, Eisenstein never mentioned it in his writings nor analyzed the film's unique audiovisual structure, which in many respects substantiated Eisenstein's own theory of sight and sound counterpoint. Svilova obviously wanted to tone down the existent controversy when she wrote:

In the later period, Vertov and Eisenstein argued a lot about creative problems, but they nevertheless preserved good personal relations. Their polemics, in fact, sharpened their theoretical views, and they actually enjoyed their discussions. After all, they rejoiced in each other's successes, and accepted them with pleasure, not with jealousy.[118]

This statement can be taken only as a circumstantial gesture, unsupported by what one finds in Eisenstein's writings about the *kinok*s, especially in his attitudes toward *The Man with the Movie Camera,* which to the end of his life he considered "mischievous."

The theoretical disagreement between Vertov and Eisenstein stems, by and large, from their divergent definitions of "ontological authenticity"* in cinema, that is, the extent to which the viewer accepts an event presented on the screen as actually taking place in the real world. In his book *The Phenomenon of Authenticity* (1972), Sergei Drobashenko finds Eisenstein's polemics with Vertov centered so much on this issue

---

* "Ontological authenticity" implies that the motion pictures projected on the screen have a unique associative power that evokes in the viewer's consciousness – by a code of perceptual recognition – a sensorial notion of reality as well as a strong feeling that the events, characters, and the environment exist as a real world, in spite of the fact that the cinematographic projection consists merely of light, shadows, and sound waves.

that "Eisenstein's entire theory and practice in the 1920s (and even much later) was subordinated to a single goal and to one overriding idea, which was to find the best expressive means and *stylized devices* capable of jarring the viewer's psychology"[119] [italics mine]. Pointing directly to the theoretical dispute between the two filmmakers, Drobashenko concludes:

Eisenstein accused Vertov's films of having insufficient purposeful intention in commenting upon the filmed event. . . . therefore, he considered Vertov's work as primitive impressionism and a dispassionate representation of reality.[120]

According to Drobashenko, this argument reveals two different if not irreconcilable concepts of "cinema of fact." Eisenstein did not believe that the camera – a mere instrument in the filmmaker's hands – was capable of penetrating reality or revealing the hidden meaning of everyday events, as Vertov, Shub, Gan, and other constructivists contended. Drobashenko further infers that Eisenstein rejected the idea that a film image per se could achieve any substantial impact on the viewer unless sufficiently stylized *before* and *after* the shooting. In contrast, Vertov had a complete trust in the camera's power to "unveil those aspects of the filmed event which otherwise cannot be perceived."[121] This penetrating and revelationary capacity of the camera was the basis of the "Film-Truth" principle, which Eisenstein considered an artistically useless, purely mechanical device.

Eisenstein defined the representational character of the film image as the "shot's tendency toward complete immutability rooted in its nature";[122] in order to counteract the shot's faithfulness to reality, Eisenstein opted for techniques – expressionist lighting, histrionic acting, exaggerated makeup, and symbolic mise-en-scène – which reduced the ontological authenticity of the shot. Vertov opposed all those "histrionic" interventions, believing that the true nature of cinema resides in the camera's capacity to uncover the hidden aspects of reality, which must be caught unawares before they are aesthetically reconstructed through montage. Whenever he abandoned this principle by rearranging the events in front of the camera (as in some segments of the *Film-Eye* series, *One Sixth of the World,* and *The Man with the Movie Camera*), Vertov never concealed the fact that the photographed people were aware of the presence of the camera. The candid recording of "life-facts" was Vertov's primary principle substantiated throughout *The Man with the Movie Camera,* which presents the actual making of the film as yet another "life-fact." Here, the Cameraman appears on the screen in a double "role": he is a worker (who makes a film), as well as a citizen participating in daily life, whether by shooting on location or by posing for another camera.

Rejecting the "Film-Truth" principle, Eisenstein criticized Vertov's shots as "static" and "lacking metaphorical implications," and expressed his intense dissatisfaction with the "Film-Eye" method as a "disguising" device:

By structuring images of authentic life (as the post-Impressionists use authentic colors), Vertov weaves a pointillist picture emphasizing the external dynamism of events in an attempt to disguise the virtually static presentation of events on the screen. . . . Thus by recording the outer dynamics of the facts, [Vertov] neglects the inner dynamism of the shot [*vnutrikadrovaia dinamichnost'*], which actually is nothing but the beautification [*grimirovanie*] of the stationary shots.[123]

Eisenstein made these remarks shortly after the completion of his *Strike* (1924), at the time he was experimenting with the ideological function of montage. Criticizing the "Film-Truth" principle, Eisenstein claimed that it did not have "enough ideological power" because Vertov "selected only those facts from the outside world which *impressed* the filmmaker, and not those with which he should plough up the viewer's psyche" [*perepakhivaiut ego psikhiku*].[124] This criticism was most painful for Vertov, whose ultimate goal was to change the viewers' conventional perception by "shaking" their minds, not merely by "impressing" them optically, but by challenging their customary way of seeing.

Eisenstein felt that only through the stylistic modifications of photographed objects and events could the viewers' attitudes be affected; therefore, from the very beginning, he found the "Film-Eye" method ideologically inadequate. He was joined by Viktor Shklovsky in devaluating Vertov's method, especially his way of presenting newsreel footage. In his 1926 article "Where is Dziga Vertov Marching?" Shklovsky tried to discredit the *kinoks*' rejection of actors and staged events:

"Film-Eye" rejects the actor, believing that by doing so it separates itself from art. However, the very act of selecting images is an artistic gesture and the result of conscious will. . . . In the *Film-Eye Series*, cinema is not achieving a new level of artistic expression, but merely narrowing the old one. In the films made by the *kinoks*, photographed events are impoverished because they lack an artistic bias in their relation to the objects.[125]

From today's perspective, it is difficult to understand Shklovsky's uncompromising rejection of Vertov's work, especially when one knows that the *kinoks* produced two outstanding newsreels, *Forward March, Soviet!* (1926) and *One Sixth of the World* (1926), which were far superior to the contemporary newsreel, and marked a turning point in the evolution of documentary cinema.

Politically, Vertov differed from Eisenstein in various subtle, yet important, respects. Both opposed the party's control of art, but when directly faced with official demands to alter the ideological content of their works, they reacted differently. Confronted with the choice between his own artistic integrity and subservience to the party, Eisenstein did agree to include an apotheotic (and Stalinistic) finale in *Ivan the Terrible, Part Two* (1946/58), while Vertov intrepidly refused to make any such concessions. When pressed by the party to reedit his 1937 film *Lullaby* (dealing with the emancipation of Soviet women), he firmly declined to do so, and subsequently a mutilated version of the film was released *without* Vertov's consent. The same occurred a few months later with the compilation film *Sergo Ordzhonikidze* (1937), whose completed version was "altered against [my] will," as Vertov stated in his "Diaries."* Furthermore, he demonstrated great courage to disapprove of the political appendix to *Three Songs about Lenin*, which was nevertheless attached to the film by the production company in order to present Stalin as equal, if not superior, to Lenin.

The use of the nonprofessional actor [*naturshchik*] was yet another theoretical point on which Vertov and Eisenstein disagreed. The concept of "typage,"[126] according to which the nonprofessional actors in a film should be chosen according to their physiognomy, their facial expressions, and physical stature, is an important aspect of Eisenstein's theory. In keeping with the strict distinction between documentary and fictional cinema, Vertov was against the inclusion of any *artificial* ingredients in documentary films and, conversely, any *quasi-authentic* elements in fictional films. He believed that the two film genres could develop fully only if each explores its own specific means of expression – a theoretical extension of his concept of autonomous cinema "without the aid of theater and literature." Consequently, he considered the use of nonprofessional actors in staged films as incompatible with the notion of "life-fact" because such a practice "impedes the evolution of the documentary film. . . . [Since] newsreels have their own path of development, [and] the staged films should not follow the same path, the inclusion of the newsreel technique into the organism of staged cinema is simply unnatural."†

In numerous articles and "collective discussions," the *LEF* journal fully supported Vertov's view, and when Eisenstein chose an ordinary worker to play the role of Lenin in *October*, all of the *LEF* critics de-

---

* In his diaries (at the beginning of 1945), Vertov states that *"Three Songs about Lenin"* was acknowledged one year after the film was completed. *Lullaby* was forcibly butchered during the editing and given very little coverage in the press. Undated, unpreserved, it was finally destroyed. *Sergo Ordzhonikidze* was ruined in the course of reediting and endless revisions. *Articles*, p. 263.
† Vertov, "In Defense of Newsreels," *Articles*, p. 153. In the same text, Vertov complains that at a 1939 film conference "all the speakers avoided the question about the newsreel film, [and] . . . talked only about 'artistic' [staged] cinema, [as if] we are not artists, not creative workers."

cried his decision. Osip Brik refuted Eisenstein for being "totally insensitive to the historical truth,"[127] and for using the "most shameful method" to betray "historical facts." Mayakovsky was equally scathing in his objection to Eisenstein's casting the worker Nikandrov for the role of Lenin, and – in contrast – praised Vertov's and Shub's methods as *the* appropriate way of presenting the October Revolution and its leaders on the screen. In his 1927 article "On Film," Mayakovsky attacked Eisenstein and his production company (Sovkino) for "dramatizing the Revolution" in a "disgusting manner," while pointing to the "workers' revolutionary newsreels" produced by the *kinok*s as the "correct approach to history." Referring to *October* as "the staging of Lenin by various Nikandrovs and the like," Mayakovsky wrote:

It is disgusting to see a person imitating poses and gestures resembling those of Lenin. Behind all these appearances one feels nothing but emptiness and a total lack of thought. A comrade was perfectly right in saying that Nikandrov may resemble Lenin's numerous statues, but not Lenin himself.[128]

In his 1928 public speech "Paths and Politics of Sovkino," Mayakovsky was even more sarcastic when he alluded to Eisenstein's film:

I promise that I will boo and throw eggs at this fake Lenin, even at the most solemn moment, whenever it may be. For this is outrageous! Under the name of Eisenstein, Sovkino is giving us a counterfeit of Lenin, i.e., a certain Nikanorov or Nikandrov – who knows his name! Sovkino must take blame for this, because it refuses – just as it did in the past – to understand the importance of newreels. No wonder we buy our own newsreel footage from America in dollars.*

Naturally, Mayakovsky came out in support of Vertov's use of archival footage as the proper alternative to Eisenstein's method; he felt Vertov's approach ideally suited the cinematic reconstruction of historical events. Objectively, Mayakovsky's criticism of the "outrageous" impersonation of Lenin's character prevented him from recognizing other unique features in *October,* especially its "intellectual" and "overtonal" montage, both of which have many touching points with Vertov's theories.†

Historically, the theoretical difference between Vertov and Eisenstein emerged from the time their first theoretical essays appeared in

---

* Mayakovsky, "Paths and Politics of Sovkino" (1927) [*Puti i politika Sovkino*], Complete Works, XII, p. 359. Referring to the newsreel footage that had to be bought "from America in dollars," Mayakovsky had in mind the fact that Shub, in order to complete her compilation films, had to purchase the original footage about prerevolutionary Russia from American distributors who initially bought it from various individuals during and immediately after the revolution.
† The forthcoming close analysis of the major sequences in *The Man with a Movie Camera* will demonstrate how, in their practical achievement, Vertov and Eisenstein arrived at similar montage resolutions.

the same June 1923 issue of *LEF*. Eisentein's article, "Montage of Attraction," dealt with his theatrical staging of Ostrovsky's *Enough Simplicity in Every Wise Man* (in the Proletkult Theater), and focused on the mise-en-scène as a theoretical issue.[129] Vertov's article, "*Kinoks*. Revolution," addressed only cinema by declaring "war" on staged films. Inspired by the constructivist concept of art, Vertov introduced the first sophisticated theory of montage[130] one full year before Eisenstein became involved in cinema. From that moment, the rivalry between the two men grew bitter; and although it was Eisenstein who resorted to political insinuations, Vertov was not above mudslinging either, even though his attacks were not as prompt or systematic.* Perhaps Eisenstein would not have been so critical of Vertov had Mayakovsky and the other *LEF*ists not been so steadily supportive of Vertov's cinematic concepts and achievements.

Eisenstein's most elaborate theoretical disagreement with Vertov appeared in his essay "On the Question of a Materialist Approach to Form" (1924), which begins as a formal analysis of *Strike* and develops into a dissection of Vertov's views of cinema. In the introduction, Eisenstein reluctantly admits the importance of the *Film-Truth* series in the development of Soviet cinema, while almost simultaneously stating that "the formal structure and method of realizing *Strike* are antithetical to the *Film-Eye* series."[131] The central theme of the essay can be read as an ideological denouncement of Vertov's failure both to explore the "act of seeing" and to avoid "impressionistic representation"[132] of reality. Concluding his essay, Eisenstein states ironically that cinema does "not need a 'Film-Eye,' but 'Film-Fists' [*kinokulaki*]!"[133] This was Eisenstein's direct political allusion in his criticism of the *kinok*s. Apparently, Vertov's article "The Factory of Facts," published two years later in *Pravda,* was a delayed response to Eisenstein's criticism. Accepting the play with words, Vertov launched neologistic counterterms – "fists of facts" [*kulaki faktov*], "lightning of facts" [*molnii faktov*], and "hurricanes of facts" [*uragani faktov*] – while urging the *kinok*s to treat "film witchcraft" [*kinokolodovstvo*] and "film mystification" [*kinomistifikatsiia*][134] as antithetical to cinema. True revolutionary cinema, Vertov stated in his article, is "neither FEKS, nor the 'Factory of Attraction' as conceived by Eisenstein, nor the factory of kisses and doves.... Simply, it is the Factory of Facts. Recorded facts. Sorting facts. Disseminating facts. Agitating with

---

* In two of his articles, Vertov directly claimed that his "Film-Eye" method and his "Film-Truth" principle influenced Eisenstein. In " 'Film-Eye' about *Strike*" ["*Kinoglaz*" o "*Stachke*"], printed in *Kino* (March 24, 1925), Vertov stated that *Strike* was the first attempt at applying the *kinok*s method to staged cinema. In "The Factory of Facts," printed in *Pravda* (July 24, 1926), he wrote that the *Film-Eye* series inspired Eisenstein to shoot the mass sequences in both *Strike* and *Battleship Potemkin.*

facts. Propagating with facts. Fists of facts."[135] Vertov also requested that staged films with actors and dramatized events "stay away" from "cinema of fact," while referring to Eisenstein's films *Strike* and *Battleship Potemkin* as "isolated examples" that "inappropriately" adopt the "Film-Eye" method to fictional cinema. Obviously, Vertov's intention was to identify Eisenstein with one of those orthodox film directors who exemplified what Vertov defined as "the factory of grimace, i.e., an association of all kinds of theatrical moviemakers, from Sabinsky to Eisenstein."* Undoubtedly, this was Vertov's most caustic remark.

The "Theory of Intervals," as a conceptual basis of the "Film-Eye" method, was the core of the Vertov-Eisenstein controversy. In "The Fourth Dimension in Cinema," Eisenstein attacked the "Theory of Intervals" as the failing point of *The Eleventh Year* (1928), a film "so complex in the way its shots are juxtaposed that one could establish the film's structural norm only with a 'ruler in hand,' that is, not by perception but only by mechanical [metric] measurement."[136] When *The Man with the Movie Camera* was released, Eisenstein used the same tactics to underrate Vertov's method by describing the accelerated shooting and slow motion used extensively in the film as "mere formalistic jackstraws and unmotivated camera mischief."[137] This can be seen as a latent political accusation in light of the party's current engagement in ridding art of "eccentricism" and "formalism." Whether Eisenstein's attacks were motivated by personal bias or whether they happened to be unconscious side effects of a professional debate carried out in a politically charged atmosphere is a matter for speculation. Nonetheless, the rumors about Vertov's stubbornness, craziness, and personal quirks – which were at odds with the party's idea of an obediant "socialist artist" – may have been supported by Eisenstein who, at that time, did not realize he would soon fall prey to a similar pressure from the condemnation of his "eccentric" directorial style, as well as his "personal" interpretation of the historical events and his "ideologically inappropriate" presentation of contemporary life circumstances.† It may be only a coincidence that Eisenstein's article "What

---

* Vertov, "The Factory of Facts" (1926), *Articles,* p. 88. Cheslav Sabinsky was a set designer at the Moscow Art Theater, who turned to film before the revolution and made a series of studio-made spectacles. He continued to direct photoplays, mostly taken from literary and dramatic texts such as *Mumu* (based on Turgenev's story) and *Katerina Izmailova* (based on Leskov's novel), and produced numerous films until he died in 1941.
† In his statement about the "mistakes" of *Bezhin Meadow* [*Bezhin lug*], 1937, Eisenstein explained: "Mistakes of generalization divorced from the reality of the particular occur just as glaringly in the methods of presenting the subject. . . . Philosophical errors lead to mistakes in method. Mistakes in method lead to objective political error and looseness. . . . A detailed scrutiny of all the consecutive scenes fully revealed to me my wrong approach to this subject. The criticism of my comrades helped me to see it." The entire statement is reprinted in *Sergei M. Eisenstein* by Marie Seton, pp. 372–7.

Lenin Gave to Me" (1932) appeared during the period of the ideological "purification" of Soviet art ordered by Stalin. Denouncing the *kinoks* as the "Talmudists of pure film form" and the "Talmudists of 'Film-Truth' documentary" [*talmudisti chistoi kinoformy i talmudisti 'kinopravdy' dokumentalizma*],[138] Eisenstein made an overt political allusion that, by its implication, could only aggravate the *kinoks'* standing because the term "Talmudist" belonged to the vocabulary used by Stalin and Zhdanov[139] in their attacks on the artists and their work considered ideologically incorrect and hence deserving of punishment.

But Eisenstein's attitude toward Vertov and his *kinoks* was inconsistent and unpredictable. In his essay "Pantagruel is Born" (1933), Eisenstein acknowledged the contribution Vertov's *Film-Eye* series had made to Soviet cinema by stating figuratively that "Soviet sound cinema was born not from the ear, but from the eye of the silent film (and perhaps even from the 'Film-Eye')."[140] Not long after this article, Eisenstein reversed his stance and condemned the "Film-Eye" method for its "formal excesses" [*formal'nye ekstsesy*], which he labeled a "peculiar type of film-schizophrenia" [*kino shizofreniia*].[141] Claiming that Vertov's cinematic style suffered from "a hypertrophy of montage which detracts from the representational aspect of the shot,"[142] Eisenstein totally dismissed Vertov's contributions to the theory of montage, and he often quoted *The Man with the Movie Camera* as an example of the "misemployment" of cinematic technique, a film unworthy of scholarly attention. Then, at the time *Three Songs about Lenin* became internationally acclaimed, Eisenstein suddenly showed considerable interest in Vertov's film, which Mayakovsky and the *LEF* critics used to counteract Eisenstein's method and practice of presenting historical figures on the screen. In his rather opportunistic article "The Most Important of the Arts" (1935), whose title was obviously "inspired" by Lenin's famous statement about cinema, Eisenstein wrote that "together with Kuleshov's experiments, Vertov's *Film-Truth* series and *Film-Eye* series offer profound views and insights into our contemporary society."[143] In the article "Twenty" (1940), Eisenstein retreated to his earlier critical position which objected to the newsreel footage of the *Film-Truth* series for "lacking a proper concept in its montage organization."[144] After World War II there was another vacillation in Eisenstein's attitude toward Vertov: the *Film-Eye* series, which he initially found "problematic," became "important" again! In the article "About Stereoscopic Film" (1948), he admitted for the first time that Vertov's newsreels successfully melded documentary authenticity with the maker's own view of the world: "What we see on the screen is the result of Vertov's personal view; hence his films represent not only objective reality, but also Vertov's own cinematic self-portrait."[145]

The theoretical differences between Vertov and Eisenstein are best exemplified in their attitudes toward the role of sound in cinema. From the beginning of his career, Vertov was fascinated with the possibilities of sound, particularly synchronous sound recording. His belief that the sound film camera is capable of unveiling the substance of reality is significant when one realizes that in the Soviet Union at that time the technical equipment for direct sound recording was extremely primitive.* Eisenstein, on the other hand, remained skeptical of sync-sound as a creative device for a long period of time. When Vertov was enthusiastically writing about the "auditory facts" out of which the "Radio-Eye" could compose a "cinematic symphony," Eisenstein was warning filmmakers about the "dangers" of sound cinema in his famous 1929 "Statement,"[146] which, in essence, was antagonistic to the inclusion of sound in cinema in general, and only on a secondary level discussed the concept of sight-and-sound counterpoint. In contrast, Vertov proved receptive to sound cinema long before it was technically feasible, and he produced the first Soviet full-length unstaged sound film, *Enthusiasm* (1930), with the intention of abolishing "once and for all the immobility of sound recording equipment...confined within the walls of the film studio."[147] At the same time, Eisenstein, lecturing at home and abroad, claimed that "the hundred percent talking film is nonsense."† *Enthusiasm* upheld Vertov's contention that the direct recording of sound on location had enormous expressive possibilities, especially in creating "the complex interaction of sound and image."[148] Most importantly, Vertov's experimentation with sync-sound took place *after* the appearance of Eisenstein's 1929 "Statement about the Future of the Sound Cinema." Released in the fall of 1930, long before Pudovkin's first sound film project, *Deserter* (1933), was completed, *Enthusiasm* has remained the most creative experiment in the early use of sync-sound.‡

---

* Vertov explained his concept of sound cinema in various articles. One of the earliest, " 'Film-Truth' and 'Radio-Truth' " (1925), urged the *kinok*s to "campaign with facts not only in terms of seeing but also in terms of hearing." In "Film has Begun to Shout" [*Kino zakrichalo*, 1928], he introduced the idea of applying the principle of "Film-Truth" to the recording of natural sound. In "Watching and Understanding Life" [*Bachti i chuti zhitiia*, 1929], Vertov anticipated the principle of sync-sound shooting at a time when it was considered impossible. In "The First Steps" [*Pervye shagi*, 1931], he acknowledged the principle of direct sound recording with portable sound equipment and expanded the idea of sight-and-sound counterpoint. At the Conference of Workers in Sound Film (held in the Summer of 1933), Vertov delivered a report urging the sound engineers (who resisted recording sound on location) to develop a portable sound recording system capable of "instantly capturing the sound in the street simultaneously with photographing the image of the event."
† In his lecture delivered at the Sorbonne (February 17, 1930), Eisenstein declared: "I believe that the hundred per cent talking film is nonsense, and I hope that everybody agrees with me." *Selected Works*, I, p. 553. The title of the lecture was "The Principles of the New Russian Film" [*Printsipy novogo ruskogo fil'ma*].
‡ Although both are based on the principle of asynchronism, *Deserter* and *Enthusiasm*

The fact that the release of the "Statement" preceded Vertov's preparatory work for *Enthusiasm* has led some critics to consider the film a response to Eisenstein's concept of sound cinema. Vertov, however, claimed that another statement, written also in 1929, motivated him to produce the first Soviet sound film, a project which he had looked forward to undertaking for quite some time.[149] He referred to an article written by the sound-film engineer Ippolit Sokolov, who insisted that sound should be "recorded entirely in the studio" to ensure the "aesthetic relationship between the sound and the image."[150] Discussing the auditory aspect of the sound track, Sokolov maintained that "auditory reality is essentially non-phonogenic."[151] On the basis of this contention, he claimed that it is "impossible to capture 'life unawares' with the microphone, because unorganized and random noises occurring in reality make a true cacophony of sound, literally a feline concert" [*bukval'no koshachii kontsert*].[152] Evidently, Solokov's conclusion was in direct opposition to Vertov's "Film-Truth" principle as well as his "Radio-Ear" method, both of which favored sound captured candidly as an auditory "life-fact."

Sokolov's statement "The Potential of Sound Cinema" in many ways echoes Eisenstein's rejection of sync-sound, warning the filmmakers that "sound recording on a naturalistic level, i.e., sound which faithfully corresponds to movement on the screen, will destroy the culture of montage."[153] This complied with most filmmakers and theorists of the silent era who intrinsically opposed sound, but at the same time it enraged those who tried to demonstrate that "sounds captured at random" *could* become expressive components of the cinematic structure, and even function as "music" of a different sort.* In his 1931 article "The First Steps," written immediately after the completion of *Enthusiasm*, Vertov explained:

The beginning of work on *Enthusiasm* was preceded by the "theory of the feline concert" [*teoriia koshacheo kontserta*], as defined by Ippolit Sokolov. It was also motivated by some foreign and domestic authorities who rejected the possibility of recording sound for the newsreel. Hence *Enthusiasm* resulted from the negation of this negation.[154]

Vertov's vanguard achievement in sound cinema was conditioned by a whole range of factors, both theoretical and personal: his longstanding experimentation with auditory perception as an inevitable part of the "montage way" of seeing *and* hearing was meant to func-

---

treat sound differently: while Pudovkin's sound track can stand apart from the picture as a musical structure of its own, Vertov's sound track is closely related to – and integrated with – the image, thus commenting on the shots emotionally and ideologically.
* See Lucy Fisher, "*Enthusiasm:* From Kino-Eye to Radio-Eye," *Film Quarterly*, no. 2 (Winter 1977), pp. 25–34. Fisher suggests some fifteen different uses of sound in Vertov's film.

tion as "a transition from silent to sound cinema in the domain of the newsreel and unstaged films."[155]

It is interesting to compare the ways in which the two most avant-garde Soviet films were received by the press and critics: *Enthusiasm* was attacked by Soviet critics for being "eccentric" and "anti-proletarian,"* but it received overwhelmingly favorable reviews in Europe and the United States; *The Man with the Movie Camera* was equally misinterpreted and rejected both in the Soviet Union and abroad. Eisenstein, who might have been expected to defend these films for their unorthodox structure on both visual and auditory levels, remained either antagonistic or reserved: he derided *The Man with the Movie Camera* as "formalistic" play, and failed to mention *Enthusiasm* in any of his numerous essays about the relationship between image and sound.

Vertov's difficult years

Excluding personal grudges that may have come into play, the Vertov–Eisenstein controversy testifies to the divergent interests and opposing theoretical views among the leading figures of the Soviet avant-garde movement in the 1920s. Official ideologists exploited these internal divisions to encroach on the freedom of artistic expression and to subordinate the artists' integrity to the pragmatic goals of the party. Those who opposed the dogma of socialist realism were pronounced "unsuitable" and denied full participation in the artistic community; those who adhered to the canonized doctrine were given special treatment. As a member of the former category, Vertov was made to feel abandoned and useless. He became aware of the support being accorded to other filmmakers, while he was consistently prevented from realizing his own projects. From 1935 to 1939 he submitted numerous manuscripts to the *Soiuzkino* film company, but received no favorable replies, a fact that left his mood cramped and forlorn:

It is impossible to assure my comrades of the necessity for me to work. . . . If they only would ask me to make films with the same persistence as they ask comrade Eisenstein, I would certainly be able to turn the world around. But nobody has asked me to do anything.[156]

* After criticizing *The Man with the Movie Camera* as "futile, self-infatuated trickery," Lebedev attacked *Enthusiasm* from the same orthodox position, in his article "For Proletarian Film Journalism" [*Za proleterskuiu kinopublitsistiku*], Literatura i iskusstvo, no. 1 (November 1931), pp. 15–16. Vertov responded to the Lebedev criticism in general in "The Complete Capitulation of Nikolai Lebedev" [*Polnaia kapitulatsiia Nikolaia Lebedeva*], Proleterskoe kino, no. 5 (May 1937), pp. 12–13.

The situation worsened in the ensuing years, and Vertov gradually withdrew into himself. "I do not isolate myself – I am isolated – nobody invites me anywhere any more," he complained in the early 1940s.[157] Later the same year, a lengthy documentary film to celebrate the twentieth anniversary of Soviet cinema (*Our Cinema*) was made without mentioning Vertov whatsoever!* The apparent futility of his attempts to make films finally exhausted him, and in 1941 he wrote: "I am sick and tired of everything, I do not understand anything that is going on. I have no energy left to cope with my enemies' intrigues, and I am afraid to look behind the screen."[158] By this time Vertov had reached the point of no return. He felt his "nervous system [had] been completely ruined by endless procedures of approval and disapproval, agreement and disagreement."[159]

Tragically for the history of cinema, he spent the remainder of his life supervising newsreel production at the Central Studio for Documentary Films,† amid working conditions that were antithetical to his youthful dreams about a "truly socialist" film production cooperative similar to his *kinoks'* workshop. To paraphrase Mayakovsky, everything Vertov and his *kinoks* had fought against and considered bourgeois suddenly appeared victorious, while their idea of unspoiled beauty and their vision of a revolutionary art had been destroyed all over again.‡ Vertov's and Mayakovsky's careers metaphorically reflect the destiny of the Soviet avant-garde: at the height of their creative energy they were scarred by disappointment, as the impossibility of putting into practice their revolutionary ideals became clear, and as their creative achievements were constantly rejected by the state as well as by the masses. Their initial unreserved faith in the new government "of" and "for" the working class was shattered by a subsequent realization that their achievements were judged only according to current political needs, and that their expectation of the masses' appreciation of the avant-garde art was naively idealistic. The precariousness of remaining faithful to his own beliefs and the impossibility of expressing

---

* *Our Cinema, Twentieth Anniversary of the Soviet Cinema* [*Nashe kino, dvatsatiletie sovetskogo kino*, 1940] was a full-length documentary made by Fedor Kiselev in honor of the twentieth anniversary of Soviet cinema. It was composed of sequences from "the most important Soviet films," but never mentioned Vertov's work. Vertov points to this injustice in his diary entry of February 12, 1940, *Articles*, p. 228.
† Vertov's last documentary film, *The Oath of Youth* [*Kliatva molodykh*, 1944], was completed just before the end of World War II. From that period until his death (1954), Vertov was entirely engaged in supervising the production of the official newsreel series *News of the Day* [*Novosti dnia*], altogether fifty-five issues, produced by the Central Studio for Documentary Films. Although the credits identify Vertov as "author-director," his actual role was rather editorial, since all the essential decisions had to be made by the official and political board of the production studio.
‡ See note 108.

freely his cinematic vision had a devastating effect on Vertov; he expressed this frustration in poetic terms through lines found among his notebooks and unfinished manuscripts:

Arrived as a youngster of medium height	Prishel iunoshei srednego rosta
To Malyi Gnezdnikovskii, no. 7	v. Malyi Gnezdnikovskii, 7.
Proposed the "Film-Truth,"	Predlozhil kinopravdu.
It seemed simple.	Kazalos' by, prosto.
No. Not at all.	Net. Ne sovsem.
Here begins a long story.	Zdes' nachinaetsia dlinnaia istoriia.
In this story History will be clarified.	V etoi istorii razberetsia Istoriia.[160]

Indeed, history did clarify Vertov's "story," albeit unfavorably for him, as well as for Mayakovsky and other Soviet avant-garde artists, many of whom were persecuted by Stalin for being too candid in voicing their personal views about life, society, and art.

Mayakovsky's life ended tragically: shortly after he publicly expressed his discontent with the NEP policy and its negative impact on popular culture, he committed suicide. But Mayakovsky was "dead" long before he destroyed himself physically on April 14, 1930. Ten years later, reflecting upon the relationship between the artist and society, Vertov cried out: "Is it possible to die not from physical but from creative hunger?" And immediately he gave the inevitable answer: "It is possible."[161] However, despite his irrevocable situation, Vertov never even contemplated the idea of leaving the Soviet Union. Like Mayakovsky, he felt that he must remain on his native soil – in life *or* death!

The same question haunted the entire Soviet avant-garde throughout the 1920s, as its creative potential – initially manifested without any limitation – became increasingly suppressed by political terror. Faced with hopeless prospects, most of the artists complied with the party's demands; others could not do so even when their very existence was threatened. Vertov belonged to those revolutionary artists who felt unjustly prevented from satisfying their "creative hunger," a hunger stronger than the survival instinct. Therefore, it is of no surprise that he considered himself "dead" before his actual passing.

The demise of the Soviet avant-garde

As the danger of engaging in any form of nonconformist artistic endeavor became more apparent, disillusionment crept into the Soviet art community. Anxiety reached a critical point in the late 1920s when the most prominent avant-garde artists were denouced by *Narkompros** as

---

* *Narkompros* stands for the People's Commissariat of Education [*Narodnyi komissariat prosveshcheniia*], headed by Anatoly Lunacharsky from 1917 to 1929.

"ideologically unsuitable." Consequently, traditional artists who had fortified within RAPP,* the stronghold of socialist realism, took absolute control of Soviet art, whereas those who disagreed with RAPP policy had little choice but to rescind their ideological stance or to lapse into silence. According to Vahan Barooshian, the demise of the Soviet avant-garde was accelerated by their attempt to "deemphasize the role of ideology in art[†] and their disagreement with the dogma of socialist realism. As the pressure to bend to the party line increased, schisms between the proponents of pluralism in artistic expression and the pragmatic followers of socialist realism became critical.

By the end of the 1920s, the internal ideological shakedown had taken its toll on the strength of the avant-garde movement, which gradually yielded to mounting political pressure. Even Mayakovsky shied away from his initial radicalism, and in 1929 he founded REF (Revolutionary Art Front), intending to merge *LEF* with RAPP and thus postpone the extinction of the avant-garde. Unlike his earlier statements, Mayakovsky's speech at the first REF meeting stressed the "primacy of [political] aims over content and form," emphasizing that "only those literary means which lead to Socialism are correct."[162] Although he did not specify what he meant by "correct" literary means, it is reasonable to assume that he referred to the directives stipulated by the party. However conciliatory, Mayakovsky's speech satisfied neither the RAPP ideologists nor his avant-garde friends who, like Brik, took his action as a betrayal of futurism.[163] Mayakovsky's failure to reconcile the two antagonistic ideologies exacerbated his disillusionment with an environment in which one group of artists was promoted at the expense of the other and evaluated exclusively on political grounds.

Vertov's diaries from that period reveal that he was increasingly confronted with obstacles that thwarted his creative energies. After returning from London, where *Enthusiasm* had received overwhelming praise (Charlie Chaplin described the film as "one of the most exhilarating symphonies"[‡]), Vertov encountered difficulty in acquiring distri-

* RAPP (founded in 1928) stands for the Russian Association of Proletarian Artists [*Russkaia assotsiatsiia proletarskikh pisatelei*]. With their traditional attitudes toward the arts, the members of this organization, by and large, believed in the central position of party ideology in literature. Many of RAPP's concepts were later developed into the dogma of socialist realism.
† Vahan Barooshian, "Russian Futurism in the Late 1920s: Literature of Fact," *Slavic and East European Journal,* no. 1 (Spring 1971), p. 43. Barooshian emphasizes that the most active contributors to *Novyi LEF* were both militant and exclusive in their demand that traditional art be abandoned altogether in the name of "art of fact."
‡ Vertov, "Notebooks" (November 17, 1931), *Articles,* p. 173. The original statement reads: "Never had I known that these mechanical noises could be arranged to sound so beautiful. I regard it [*Enthusiasm*] as one of the most exhilarating symphonies I have heard. Mr. Dziga Vertov is a musician. The professors should learn from him and not quarrel with him. Congratulations. Charles Chaplin." Initially, the statement was pub-

bution for his newly completed Lenin film. It may seem strange that a film that glorified the legendary communist leader was surreptitiously denied major distribution. The actual reason: Stalin was unsatisfied with the film because it ignored what he thought was his role in the October Revolution.* In his diary of May 17, 1934, Vertov describes the party's indifference regarding the promotion of his film, which had a demoralizing effect on him:

It is almost four months since *Three Songs about Lenin* was completed.... The agony of waiting. My entire being is tense – like a drawn bow. Anxiety – day and night.... They managed to exhaust me completely.[164]

Expressing his growing frustration at the government's reluctance to promote the distribution of his film, he wrote two days later: "I do not know whether I am a human being or a 'schema' invented by my critics."[165] On November 9, 1934, after the film had finally been released in a few neighborhood theaters in Moscow, Vertov learned about the film's overwhelming success in the United States. Yet this fact had little influence on the Soviet distributors, even after the official press conceded that *Three Songs about Lenin* was an international success. Consequently, Vertov's diary notes became increasingly grim and desperate, as he repeatedly accused the Soviet bureaucracy of purposely disregarding his new proposals. In a diary entry written at the end of 1939, Vertov quoted Saakov (a functionary at the Ministry of Cinematography) as warning him: "You must act as we tell you to do, or you will not be allowed to work in cinema anymore."[166] Unwilling to comply with orders, Vertov turned to script writing, vowing to "work with the pen if it is not possible for me to work with the camera."[167]

The Man with the Movie Camera was plagued with even greater problems. In spite of the acclaim it had received in the domestic and international press, the film was withheld from the broad audience Vertov had particularly wanted to reach. The paradoxical situation continued: following the premiere of the film in Moscow, *Pravda* wrote that "rarely was an audience so attentive as during the screening of this film,"[168] and recommended that *The Man with the Movie Camera* be shown widely among Soviet theaters. Again, one would expect that a recommendation coming from such an authoritative organ would automatically be followed and put into practice. However, although the positive review appeared in the party newspaper, the suggestion for the film's wide release went unheeded by the distributors. Their bureaucratic attitudes were fostered by the champions of socialist realism, among them Nikolai Lebedev who attacked Vertov's work as

lished in the German journal *Film Kurrier*. In addition to this acknowledgment of *Enthusiasm,* Chaplin included it in his list of the best films made in 1930 (responding to a questionnaire organized by *The London Times*).
* In 1938, the film was reedited to include Stalin's speech on Lenin and the October Revolution.

"confused, formalistic, aimless, and self-satisfied trickery,"* prompting the distributors to shelve the film. As a result, *The Man with the Movie Camera* was, on the one hand, praised by the official reviewers, and on the other, criticized by the party ideologists whose opinion was meant to be the "working policy" for the "money-hungry" distributors who readily withdrew *The Man with the Movie Camera* from the large theaters — a fact which distressed Vertov enormously, as his friend Sergei Ermolinsky recalls:

Most critics accused the film of being too formal and eccentric without ever realizing that its "eccentricity" was aimed at discovering a new and genuine cinematic expression. . . . I met Vertov after the first few screenings near Pushkin Square and found him depressed, which he did not attempt to conceal. . . . Generally, he was a reserved person who did not easily express his emotions, but this time he said to me: "It didn't work, it did not!" The situation was thus more difficult for him since the very concept of this film was so dear and important to him.[169]

Obviously, Vertov found himself in a most peculiar position since the leading newspapers and journals supported his film, emphasizing the originality and significance of his experiment. For example, *Vechernaia Moskva* (January 22, 1929) heralded Vertov as one of "the greatest innovators of Soviet and world cinema whose work is virtually unknown to Soviet audiences," and recommended that *"The Man with the Movie Camera* be circulated widely."[170] *Kino-Gazeta* (February 19, 1929) was critical of the fact that "the Moscow theaters have not as yet ordered a single copy of this film, although it was completed last November."[171] In the Ukraine, where the previews of *The Man with the Movie Camera* were particularly successful, the problems with the general distribution of the film were no different. *Vechernii Kiev* (March 9, 1929) making veiled references to party bureaucracy, asked the following embarrassing questions:

Who has the right to say, and who dares claim, that Vertov's film lacks social value, that it is incomprehensible, inaccessible, etc.? Whom did they ask? Did they conduct a poll? In what cities? With what kind of people? Clear the way for Dziga Vertov![172]

Kino i kul'tura (no. 4, March–April 1929) was most vehement in criticizing the narrow-minded bureaucracy, claiming that the "distribution companies continue a policy implemented by pre-Revolutionary petty

* Nikolai Lebedev's attack on Vertov and his *kinoks* appeared in his article "For A Proletarian Film Journalism" [*Za proleterskie kinopublitsistiki*], Literatura i iskusstvo, 1931, no. 9–10. Almost word for word, he reiterated the attack in his *Survey of the History of Silent Film in USSR: 1918–1934* [*Ocherk istorii kino SSSR: nemoe kino, 1918–1934*] (Moscow: Iskusstvo, 1965), claiming that "in *The Man with the Movie Camera*, it is virtually impossible to find any logic in the succession of the shots" (p. 206).

bosses," and that "to Vertov's disadvantage, they make profits for mediocre imitators, unaware that without Vertov, Soviet cinema would wither away."[173] Vertov personally took part in this struggle for the distribution of *The Man with the Movie Camera*. On March 12, 1929, he sent the following letter to *Kinogazeta:*

Instead of answering countless enquiries about the fate of *The Man with the Movie Camera,* let me inform the editorial staff, which promptly acclaimed the appearance of my film, as well as all of the friends who supported it on numerous occasions both in Moscow and the Ukraine, that *The Man with the Movie Camera* (released in November of last year) has not reached major theaters in Moscow or Leningrad so far (i.e., after five months) simply because it was shelved and boycotted by all means available to the film distributors, while you, comrades, at the same time were asking when the film would be shown. Evidently, this will cease to occur only when the opinion of the public and the press becomes more relevant than the opinion of those who run our distribution companies, whose taste is on the level of movies like *Six Girls Behind the Monastery Walls.* *

However, it soon became clear to Vertov that the large audience as well was incapable of grasping the complex structure of his film, and that the masses had tastes identical to those of the film distributors; hence, the labeling of *The Man with the Movie Camera* as "inaccessible" and "formalistic" was all the more painful for him. Yet he did not make self-debasing apologies for "political errors"[†] (as Eisenstein did in 1936 and 1937), even after the most virulent party spokesman, A. Fedorov-Davydov, discredited Vertov as an "idol" of foreign "decadent" critics:

It is not accidental that this film received the greatest recognition abroad by the aesthetes belonging to European avant-garde cinema. They welcome Vertov's complete retreat from a realistic reflection of reality to an empty and fruitless play with form underscored by the philosophy of rejection of an objective perception of the world.... However, [Vertov] does not understand the whole reactionary, anti-realistic essence of *The Man with the Movie Camera.*[174]

---

* Roshal', *Dziga Vertov,* pp. 200–4. *Six Girls Behind the Monastery Walls* or *Six Girls in Search of Night Shelter* [*Shest devushek kotorye za monasterskoi stenoi ili Shest devushek ishchut pristanishcha*] is a commercial German melodrama [*Sechs Mädchen suchen Nachtquartier,* 1927] directed by Hans Behrendt. It was shown in seven leading Moscow theaters during March and April of 1929, with extensive promotion. Vertov's letter, dated March 12, appeared in the March 19, 1929 issue of the journal *Kinogazeta.*
† Although Eisenstein had ideological problems with his earlier films, the real crisis occurred during the work on *Bezhin Meadow* [*Bezhin lug*], based on Turgenev's story. It was Boris Shumyatsky, a mediocre administrator of Soviet cinematography, who elaborated Stalin's criticism of Eisenstein's concept of peasant life. After several "corrections" of the script and inspections of the rushes, the entire project was canceled, and Eisenstein had to expose his "mistakes" publicly, See *Sergei M. Eisenstein* by Marie Seton (New York: A. A. Wyn, 1960), pp. 351–78. About Stalin's interference during the editing of *October,* see Herbert Marshall's *Masters of Soviet Cinema* (London: Routledge & Kegan Paul, 1983), pp. 191–200.

Following this type of defamation (in early 1936), Vertov's already precarious position in the Soviet cinema became even more aggravated. All his projects for new films and anticipated experiments were turned down by the officials: he was simply "written off" as a creative force in the Soviet cinema. Consequently, his diaries were filled with grim thoughts, and his articles lacked theoretical weight, dealing more and more with his growing depression:

Nobody requires anything of me. Therefore, it is difficult for me to talk. But, nevertheless, I will not give up. I will try once again to convince my comrades of the necessity for me to work.... They say that I do not know how to admit my mistakes. That is true. But I want to correct my mistakes by creating films which will be more accessible to the masses.[175]

Rather than a recantation of his works' ideological content, this desperate statement is a result of Vertov's frustration over the proletariat's inability to appreciate his avant-garde work. "There is nothing more terrible for an artist than to see his creative work misunderstood and destroyed,"[176] wrote Vertov in 1944, referring both to the masses and the officials who discarded his films as politically inept. Seen within the flood of other artists' ideological renunciations, Vertov's readiness "to correct [his] mistakes" seems pathetic. Unable to realize his numerous projects at the time of his creative peak, exhausted by humiliation and political pressure he could neither overcome nor surrender to, he was left at the mercy of the "obsequious" bureaucrats and political "mediators" [posredniki] who impeded his creativity, following the official directive to crush every filmmaker who did not meet current political dictates. The following note written during World War II at Alma-Ata in Kazakhstan (where the Soviet film industry had been relocated) summarizes Vertov's agony:

Most of my projects have not borne fruit, have not been finished. Nobody picked them up, hence they rotted. I have no powerful friends, nobody to go to for support.... Nobody wants to help me. Everybody is afraid of losing his head. But I am not afraid of losing mine.[177]

Unfortunately, individual courage was not a quality the political bureaucrats appreciated: they proclaimed him "stubborn" and treated him as an impediment to the "positive" evolution of Soviet cinema, while his work was disregarded by the film historians and theorists long before his death in 1954. Even when it became evident that Vertov's method exerted substantial influence on numerous documentary filmmakers such as Jean Rouch, Mario Ruspoli, Richard Leacock, Lionel Rogosin, and other representatives of the "cinéma-vérité," the "direct cinema," and the "candid-eye" concepts of shooting,[178] the Soviet film critics insisted that Vertov's place in world cinema was insignificant. Nikolai Abramov exemplifies this disdainful attitude toward Vertov, and *The Man with the Movie Camera* in particular, by stating: "This film, just like other pseudo-innovative films by foreign 'avant-garde' filmmakers, failed to exert the slightest influence on the development of the expressive means of cinema. Like all the films of this kind, it remained barren, with undeveloped characters that quickly pass over the screen, loaded with images of scandalous sensations, incapable of enriching, in any way, a living and growing art."[179] However, Jean Rouch, in his introduction to Georges Sadoul's 1971 monograph on Vertov, fully acknowledges Vertov's influence on the contemporary non-fictional cinema. Comparing Vertov with such great innovators in art as Marinetti and Apollinaire, Rouch metaphorically describes the avant-garde significance of Vertov's masterpiece:

The Man with the Movie Camera is an anxious provocation of the man who decided to make a film which we have not yet been able to understand completely and, for this very reason, which has continued to surprise the spectators over the past forty-two years.[180]

Faced with the political repression that expanded near the end of the 1920s, most of the Soviet avant-garde artists became cautious about expressing their nonconformist views. But for Vertov, to compromise his cinematic experimentation and free artistic expression was unthinkable, totally alien to his creative and ethical standards. This period of inactivity and frustration is reflected in his theoretical writings, which lost all the visionary style that characterized the *kinoks'* proclamations of the early 1920s. Stalin's crackdown on the avant-garde throughout the 1930s completely stifled creative exploration in all areas of Soviet art.* As Jean Laude explains, the party's initial ideological criticism of formalism turned into a general threat against the Soviet avant-garde:

In the USSR, the Zhdanov Report, the defeat of *LEF*, and the state takeover of ideology were to mark the moment from which the term formalism, which in the first place had been polemical, came to designate counter-revolutionary design.[181]

Following Zhdanov's notorious report at the Party Congress in 1936, "Zhdanovism" [*Zhdanovshchina*] took over – confirming that the dogma

---

* Ideological pressure in the Soviet Union had varying intensity and repercussions depending on the economic and political situation. The attack on constructivism and futurism was initially launched in 1922, when the Central Committee issued a letter entitled "About the Proletkult," which stated that "under the guise of 'proletarian culture' the artists grouped around *LEF* present to the people decadent bourgeois views of philosophy (Machism), while in the domain of art they offer poor taste and perverse ideas to the workers." *About the Soviet Party Press: Collection of Documents* [*O partiinoi i sovetskoi pechati: sbornik dokumentov*] (Moscow: Pravda, 1954), p. 221. Ernst Mach was an Austrian physicist and philosopher whose subjective idealism Lenin severely criticized in his work *Materialism and Empirio-Criticism* (1909).

of socialist realism was definitively accepted as the only "correct" way of dealing with art and culture. The party's stringent policy "to reduce all tendencies in art to a single one" engaged the entire bureaucratic apparatus "to institutionalize *its own* truth."[182] Such a dogmatic attitude was at odds with Vertov's life-long quest for truth, not only in art but in life as well.

Inevitably and ironically, one of the most innovative Soviet avant-garde artists was officially considered "counterrevolutionary" and "reactionary." Given these circumstances, a film like *The Man with the Movie Camera* would automatically become a target of the Zhdanovist and Stalinist critics who accused it of being a "formalist," "antirealistic," and "fruitless play" with images, thereby incompatible with what was considered to be "the straight road that leads toward Socialism."

The official criticism of his work and suppression of his numerous creative plans alienated Vertov from his colleagues, even his friends. He became totally withdrawn and depressed; his diaries and notebooks from that period are bleak, often bewildered. He refers to himself as if to another person, suspicious of everybody and everything: "What an injustice is concealed in the recent treatment of Vertov who is threatened with an ultimatum – either accept Sisyphean toil or face the accusation of being idle."[183]

Not surprisingly, he felt that "Sisyphean labor is torture – useless, aimless, painful, and grueling work,"[184] wholly adverse to his creative nature. He helplessly cried: "How can one explain intolerance of the talented and support of the mediocre?"[185] For Vertov, such a situation was a bitter personal blow; for the Soviet avant-garde in general, it was a clear sign of its eventual defeat.

Experiment in cinematic communication

From the beginning of his career, Vertov's goal was to break the literary/theatrical conventions of cinema in order to attain an autonomous film language based on specific visual devices unique to the nature of the medium. Considering the camera a technological vehicle capable of recording "life-unawares" and recognizing montage as the means by which one could create the genuine "film-thing," Vertov and his *kinoks* dreamt of a newsreel that would not merely present "Life-As-It-Is" but also stimulate viewers to participate in the "associative construction" of the projected images. The result of this attitude is *The Man with the Movie Camera,* a film so radically different from both the traditional fictional and documentary film genres that even those who sympathized with Vertov's experimentation could neither understand its extraordinary significance nor fully appreciate its unorthodox cinematic structure.

The film's introductory credits (see Appendix 1) describe *The Man with the Movie Camera* as a cinematic presentation of visual events "without the aid of intertitles, without a script, without sets and actors," an "experiment in cinematic communication" intended to develop "an ultimate language of cinema based on its total separation from the language of theater and literature." Vertov identifies himself as the "Author-Supervisor" of the experiment, and calls his film "an excerpt from the diary of a cameraman," referring to his brother, Mikhail Kaufman (the "protagonist" in the film). The third person cooperating on the production of the film is Elizaveta Svilova, Vertov's wife, shown at the editing table, classifying footage and putting shots together in montage units.

Functioning as the Council of Three* for the *kinoks,* Vertov, Kaufman, and Svilova could be considered the collective authors of *The*

---

* See Chapter I, note 1.

Man with the Movie Camera. But the credits clearly assign the role of "Chief Cinematographer" to Mikhail Kaufman, and the function of "Assistant Editor" to Elizaveta Svilova, implying that Vertov was in charge of the entire production and therefore should be regarded as the sole "author" of the film. Vertov emphasized that he prepared for this project "through all the previous experiments of the 'Film-Eye' method,"[1] and that his ultimate goal was to present "the facts on the screen by means of 100 percent cinematic language."[2] Hence, *The Man with the Movie Camera* should be studied as the expression of Vertov's essential theoretical views on cinema. For Kaufman, however, this film's complex structure was a failure and disappointment. In a 1979 interview, recalling how the Council of Three decided to make a film that would exemplify the *kinoks'* method, Kaufman only repeated his initial doubts:

We needed a film theory and film program expressed in a purely cinematic form. I suggested such an idea to Vertov, but it could not be realized at the time. . . . Well, we finally had to break off shooting, and Vertov started editing. I was very disappointed then. Instead of a film which had been thought out, what came out was actually only its first part. And it was terribly overloaded with events which were, from my point of view, very intrusive.*

Kaufman's disappointment proves that he failed to grasp the uniqueness of the film's "overloaded" structure even long after the project was completed. Furthermore, innovation and experimentation had never excited him, as one can see from his traditional documentary *In Spring*, produced immediately after the completion of *The Man with the Movie Camera*. Therefore, it seems unjustified to claim that "the structure of Mikhail Kaufman's 1926 documentary film *Moscow*... seems to have influenced both Walter Ruttmann's *Berlin: Symphony of a Great City* (1927) and Vertov's *The Man with the Movie Camera*," as Annette Michelson insists.[3] Even a superficial comparison of the two films reveals that the structure of Vertov's film is light-years ahead of the conventional manner in which Kaufman depicts a city.

To facilitate the close study of *The Man with the Movie Camera*, it is useful to break up the film's diegetic evolution into separate segments so that the narrative function of each thematic unit can be easily discerned. The film's overall montage structure, which Vertov called "the film's indissoluble whole," develops around a central theme – that of the Cameraman performing his daily routine in an urban environment – and is built by juxtaposing events occurring in di-

---

* In his interview published in *October,* no. 11 (Winter 1979), Mikhail Kaufman also implies that he "suggested" to Vertov in 1924 the idea of making a film that would express "a Kino-theory and a Kino-program in cinematic form" (p. 55). Vertov, however, never mentioned this in his writings.

verse locations and at different times, all unified by the ideological connotation of the photographed "life-facts."

The appearance of the Cameraman as the protagonist and the thematic grouping of the selected events lends to *The Man with the Movie Camera* what one might call a "surrogate narrative." However, although the viewer is temporarily prompted to follow certain diegetic threads, "there is little possibility of fantasizing about what happened before the film's beginning or after its end, only of thinking through the paradigm of its construction."[4] This observation implies that the thematic meaning of the sequences depends exclusively on the structural relationship of the connected shots and the filmic devices employed during shooting, editing, and laboratory processing. The film's overall meaning ought to be sought in the photographic execution and montage structure of its recorded and subsequently concatenated shots. Mikhail Kaufman states that he and Vertov "accumulated an enormous number of devices of all sorts which were supposed to be revealed in *The Man with the Movie Camera* ... as a means to an end,"[5] whereas Vertov describes *The Man with the Movie Camera* as a film that is not merely "a sum of recorded facts," but a "higher mathematics of facts,"[6] which permeated all stages of this production. Consequently, the ideological meaning of the film should be discerned from the relationship between the cinematic form and the thematic connotation of the film material (footage).

Segmentation of the film's structure

In his 1926 article "*Kinok*s and Montage,"[7] Vertov specifies different stages of montage (the organization of shots according to their basic function; an overall familiarization with the material in order to detect the correct connecting shots; the establishment of major themes and discovery of smaller motives; and finally, the reorganization of the entire material in the most appropriate cinematic manner), all of which can be discerned in the fifty-five segments of *The Man with the Movie Camera*.

Structurally, *The Man with the Movie Camera* consists of four parts: the Prologue (2 min. 45 sec.), Part One (33 min. 12 sec.), Part Two (27 min. 50 sec.), and the Epilogue (7 min. 5 sec.).

The Prologue introduces the film's major topic (the Cameraman with his camera) and the environment (the movie theater that reappears in the Epilogue). The Prologue concludes with the orchestra's performance, indicating that the screening of the film has started. Part One depicts an early morning urban setting; it follows the Cameraman's activity, introduces the Editor, and develops the themes of birth, marriage, and death. Part Two begins with various kinds of labor and culminates in sequences showing machines in operation, sports events, and a musical performance executed with spoons and bottles, all in the presence of the Cameraman as he witnesses "Life-As-It-Is." The Epilogue, a brief coda, brings the viewer back to the movie theater in order to recapitulate the diegetic as well as cinematic elements of the entire film recorded by the "Film-Truth" principle and restructured by the "Film-Eye" method.

The segmentation of a film in which narrative development is not conceived as the basis of its thematic structure can be nothing but arbitrary. The two existing (published) segmentations of *The Man with the Movie Camera* present filmic units according to different principles. Croft and Rose base their segmentation on a thematic/structural concept that divides the film text into seven sections:

"A Credo" (shots 1–4),

"Introduction: The Audience of the Film" (shots 5–67),

"Section One: Waking" (shots 68–207),

"Section Two: The Day and Work Begin" (shots 208–341),

"Section Three: The Day's Work" (shots 342–955),

"Section Four: Work Stops, Leisure Begins" (shots 956–1,399), and

"Coda: The Audience of the Film" (shots 1,400–1,716).[8]

Another segmentation of the film's thematic development, this one by Bertrand Sauzier, published in his essay "An Interpretation of *The Man with the Movie Camera,*"[9] contains ten segments noting 1,712 shots, as opposed to 1,716 shots specified in the Croft–Rose segmentation. This discrepancy in the number of shots indicates the difficulty of delineating cuts between shots often consisting of only a few – usually highly abstract – frames.*

The segmentation proposed here contains fifty-five units with 1,682 shots, excluding the "shots" that consist of transparent or black leader and blurred or decomposed images without any representational signification.

What follows is a specific thematic segmentation of the film (16mm print), indicating the number of shots and their duration within each sequence (based upon a projection speed of eighteen frames per second):

CREDITS: 1,765 fr., 1 min. 38 sec.

BLACK SCREEN: 16 fr. (1,766–1,781).

---

* There also exists an in-depth transcription of *The Man with the Movie Camera* in Seth Feldman's *Dziga Vertov: A Guide to References and Resources* (Boston: G. K. Hall, 1979), pp. 98–110; however, Feldman presents an extensive and detailed description of shots in the film, rather than an analytical segmentation into sections or sequences.

PROLOGUE: 5 sequences, 66 shots, 3,967 fr. (1,782–5,748), 3 min. 45 sec.

I.   *Introduction*: 4 shots, 502 fr. (1,782–2,283), 28 sec.
II.   *Movie Theater*: 17 shots, 1,280 fr. (2,284–3,563), 1 min. 11 sec.
III.   *Arrival of the Audience*: 10 shots, 879 fr. (3,564–4,442), 49 sec.
IV.   *Preparation*: 21 shots, 846 fr. (4,443–5,288), 47 sec.
V.   *Performance*: 14 shots, 460 fr. (5,289–5,748), 26 sec.

PART ONE: 19 sequences, 491 shots, 47,813 fr. (5,749–53,561), 44 min. 20 sec.

VI.   *Morning*: 62 shots, 5,555 fr. (5,749–11,303), 5 min. 15 sec.
VII.   *Awakening*: 37 shots, 3,277 fr. (11,304–14,580), 3 min. 1 sec.
VIII.   *Vagrants in the Street*: 14 shots, 1,248 fr. (14,581–15,828), 1 min. 10 sec.
IX.   *Washing and Blinking*: 24 shots, 1,798 fr. (15,829–17,626), 1 min. 40 sec.
X.   *Street Traffic*: 18 shots, 3,006 fr. (17,627–20,632), 2 min. 47 sec.
XI.   *Factory and Workers*: 36 shots, 3,276 fr. (20,633–23,908), 3 min. 2 sec.
XII.   *Travelers and Pedestrians:* 11 shots, 1,926 fr. (23,909–25,834), 1 min. 47 sec.
XIII.   *Opening Shutters and Stores*: 18 shots, 2,668 fr. (25,835–28,502), 2 min. 28 sec.
XIV.   *Vehicles and Pedestrians*: 14 shots, 1,874 fr. (28,503–30,376), 1 min. 44 sec.
XV.   *Ladies in a Carriage*: 22 shots, 2,726 fr. (30,377–33,102), 2 min. 31 sec.
XVI.   *Arrested Movement*: 6 shots, 936 fr. (33,103–34,038), 52 sec.
XVII.   *Editing Room*: 38 shots, 3,506 fr. (34,039–37,544), 3 min. 20 sec.
XVIII.   *Controlling Traffic*: 7 shots, 423 fr. (37,545–37,967), 24 sec.
XIX.   *Marriage and Divorce*: 35 shots, 7,726 fr. (37,968–45,693), 7 min. 9 sec.
XX.   *Death, Marriage, and Birth*: 16 shots, 1,890 fr. (45,694–47,583), 1 min. 45 sec.
XXI.   *Traffic, Elevators, and Cameraman*: 18 shots, 2,192 fr. (47,584–49,775), 2 min. 2 sec.
XXII.   *Street and Eye*: 77 shots, 500 fr. (49,776–50,275), 28 sec.
XXIII.   *Accident on the Street*: 15 shots, 1,418 fr. (50,276–51,693), 1 min. 19 sec.
XXIV.   *Fire Engine and Ambulance*: 23 shots, 1,868 fr. (51,694–53,561, 1 min. 44 sec.

PART TWO: 27 sequences, 855 shots, 40,021 fr. (53,562–93,582) 37 min. 10 sec.

XXV	*Various Kinds of Work*: 53 shots, 3,037 fr. (53,562–56,598), 2 min. 49 sec.
XXVI.	*Manufacturing Process*: 31 shots, 1,598 fr. (56,599–58,196), 1 min. 29 sec.
XXVII.	*Working Hands*: 73 shots, 1,876 fr. (58,197–60,072), 1 min. 44 sec.
XXVIII.	*Mine Workers*: 9 shots, 1,387 fr. (60,073–61,459), 1 min. 17 sec.
XXIX.	*Steel Workers*: 16 shots, 1,876 fr. (61,460–63,335), 1 min. 44 sec.
XXX.	*Power Plant and Machines*: 23 shots, 2,784 fr. (63,336–66,119), 2 min. 35 sec.
XXXI.	*Cameraman and Machines*: 152 shots, 875 fr. (66,120–66,994), 49 sec.
XXXII.	*Traffic Controller and Automobile Horn*: 28 shots, 989 fr. (66,995–67,983), 55 sec.
XXXIII.	*Stoppage of Machines*: 9 shots, 578 fr. (67,984–68,561), 32 sec.
XXXIV.	*Washing and Grooming*: 6 shots, 414 fr. (68,562–68,975), 23 sec.
XXXV.	*Recreation*: 24 shots, 2,667 fr. (68,976–71,642), 2 min. 28 sec.
XXXVI.	*Wall Newspaper*: 8 shots, 754 fr. (71,643–72,396), 42 sec.
XXXVII.	*Track and Field Events*: 37 shots, 3,236 fr. (72,397–75,632), 3 min.
XXXVIII.	*Swimming, Diving, and Gymnastics*: 14 shots, 1,279 fr. (75,633–76,911), 1 min. 12 sec.
XXXIX.	*Crowd on the Beach*: 24 shots, 2,323 fr. (76,912–79,234) 2 min. 9 sec.
XL.	*Magician*: 17 shots, 960 fr. (79,235–80,194), 53 sec.
XLI.	*Weight Reducing Exercises*: 18 shots, 1,488 fr. (80,195–81,682), 1 min. 23 sec.
XLII.	*Basketball*: 18 shots, 645 fr. (81,683–82,327), 36 sec.
XLIII.	*Soccer*: 30 shots, 1,402 fr. (82,328–83,729), 1 min. 18 sec.
XLIV.	*Motorcycle and Carousel*: 43 shots, 1,944 fr. (83,730–85,673), 1 min. 48 sec.
XLV.	*Green Manuela*: 3 shots, 648 fr. (85,674–86,321), 36 sec.
XLVI.	*Beerhall*: 16 shots, 1,391 fr. (86,322–87,712), 1 min. 15 sec.
XLVII.	*Lenin's Club – Playing Games*: 8 shots, 812 fr. (87,713–88,524), 45 sec.
XLVIII.	*Shooting Gallery*: 41 shots, 1,172 fr. (88,525–89,696), 1 min. 5 sec.

XLIX.   *Lenin's Club – Listening to the Radio*: 22 shots, 1,013 fr. (89,697–90,709), 56 sec.

L.   *Musical Performance with Spoons and Bottles*: 111 shots, 1,354 fr. (90,710–92,063) 1 min. 15 sec.

LI.   *Camera Moves on Tripod*: 21 shots, 1,519 fr. (92,064–93,582), 1 min. 24 sec.

EPILOGUE: 4 sequences, 271 shots, 8,722 fr. (93,583–102,304), 8 min. 5 sec.

LII.   *Spectators and the Screen*: 89 shots, 4,881 fr. (93,583–98,464), 4 min. 31 sec.

LIII.   *Moscow Bolshoi Theater*: 3 shots, 465 fr. (98,465–98,929), 26 sec.

LIV.   *Accelerated Motion*: 70 shots, 2,675 fr. (98,930–101,604), 2 min. 29 sec.

LV.   *Editor and the Film*: 108 shots, 699 fr. (101,605–102,304), 39 sec.

BLACK SCREEN WITH THE TITLE "END": 69 fr. (102,305–102,374)

TOTAL: 55 sequences, 1,682 shots, 100,523 frames, 93 min. 10 sec.

The Prologue begins symbolically with a shot of a tiny Cameraman climbing up a giant camera. This is followed by a series of shots of streets and a movie theater as the Cameraman and audience enter it. The projection booth and orchestra pit are prepared for the screening: medium shots and close-ups of the projectionist, the projector, and the reels precede the shots of the conductor, the musicians, their instruments, and the seated audience. From the outset, the notion of film-making, projection, and screening informs the audience that the "film-thing" they are going to see will involve both the medium of cinema as well as everyday life situations recorded by the camera.

Part One opens with a shot of the numeral "1" (numerals "2" and "3" do not appear – as one may expect – in the surviving prints*). The

---

* According to the 1935 shot list compiled by A. A. Fedorov and G. Averbakh (kept in the Dziga Vertov Archive, Moscow), in the original print of the film three additional numerals were included: the numeral "2" appeared after the "Arrested Movement" sequence (XVI), the numeral "3" after the "Fire Engine and Ambulance" sequence (XXIV), and the numeral "4" after the "Traffic Controller and Automobile Horn" sequence (XXXII). See Feldman, *Dziga Vertov*, pp. 98–110. However, the existing position of the numeral "1" does not stand at the end of the first reel, that is, after the "Washing and Blinking" sequence (IX), but after the "Performance" sequence (V). It is obviously placed there to introduce Part One, which clearly implies that Vertov used the numerals to divide different parts (segments) of his film on a structural/thematic level. Hence, my contention is that the original placement of the numerals appeared after the Prologue, Part One, Part Two, and before the Epilogue. Even the design of this numeral (which essentially differs from the conventional numbers used to mark the reels) indicates that its function was conceptual and not technical: it is simply composed as a vertical figure against a black background.

numeral "1" slowly rises toward the camera and is followed by several shots of empty streets at dawn as people awaken, wash, and prepare for work. Gradually, the traffic increases, and more people are seen walking the streets or riding in carriages, trains, and trolleys. The act of filmmaking is reemphasized by showing the Editor at work. At this point (in the sequences "Death, Marriage, and Birth" and "The Street and Eye") Vertov achieves cinematic abstraction by concentrating on formal compositional patterns prevailing in the shots and graphic designs of photographed objects and events. Part One concludes with a straightforward recording of an accident in the street: a man is hit by a car, an ambulance races across the screen in one direction and is repeatedly crosscut with a fire engine speeding away in another.

Part Two begins with several shots of the lens (in close-up) and the Cameraman (in medium shot) filming in the streets, power plants, factories, mines, workers' clubs, stores, amusement parks, beaches, sports stadiums, pubs, and theaters. After the day's work, workers prepare for their recreation; the central sequences show people playing games, listening to the radio in a workers' club, or attending a "concert" created by a man hitting ordinary objects with spoons. The Cameraman and the Editor are repeatedly shown involved in their everyday work on the film to be projected in the movie theater, and are both seen again in the Epilogue. At the end of Part Two, the camera assumes human characteristics: the tripod begins to move, and the audience, watching the film in the same movie theater introduced in the Prologue, laughs at it.

The Epilogue is comprised of shots of the audience in the movie theater intercut with shots of other people and objects moving in accelerated motion on the screen-within-the-screen. The closing sequence focuses on Svilova as she edits shots seen earlier in the film. The final shot – a close-up of the mirror image of the camera lens with a human eye superimposed on it – metaphorically concludes the film with the message that the "Film-Eye" method produces a "film-thing" in which reality is captured by the "Film-Truth" principle, which "observes and records life *as it is,* and only subsequently makes conclusions based on these observations."[10]

It is pertinent to note that *The Man with the Movie Camera* begins and ends with images related to the acts of filmmaking and film viewing: the camera and the Cameraman, the movie theater and the projector – all are associated with the spectators as they watch the screening accompanied by an orchestra. The central part of the film depicts various aspects of life in a Soviet city, intercut with images of the actual shooting, editing, and screening of the film. The closing segment summarizes the "film-thing" in a most dynamic visual manner: extremely brief shots — seen throughout the film — are repeated here, condensed into a symphonic crescendo cut off abruptly by the emblem of the "Film-Eye" method.

A "difficult" movie

Throughout the film, the process of filmmaking is presented as one of many working activities, with those involved in it shown as an integral part of a socialist environment. Following the constructivist tradition, Vertov acknowledges both the productive and creative aspects of cinema, believing that the masses will appreciate the *kinoks'* innovative approach to filmmaking. In his romantic revolutionary enthusiasm, however, Vertov overestimated his audience: not only common moviegoers but many of the avant-garde filmmakers, theorists, and critics were disappointed with the film. As already mentioned, Eisenstein accused Vertov of producing "trickery" and "unmotivated [formal] mischief,"[11] even though Vertov clearly stated that *The Man with the Movie Camera* was conceived and executed with one major intention: to rid the viewers of their conventional manner of watching movies. He maintained that the film's complex structure could be difficult in the same way "serious essays by Engels and Lenin are difficult."[12] Yet the critical attitude toward Vertov's work as being "difficult" and "inaccessible" has persisted over the last fifty years.

Only recently have some critics gone beyond regarding Vertov's film as a mere display of cinematic fireworks and instead have recognized its unique structural and cinematic importance.[13] Commenting on the structural difficulty of the film, Noel Burch writes:

This film is not made to be viewed only once. It is impossible for anyone to assimilate this work in a single viewing. Far more than any film by Eisenstein, *The Man with the Movie Camera* demands that the spectators take an active role as *decipherers* of its images. To refuse that role is to leave the theater or escape into revery.... One may safely say that there is not a single shot in this entire film whose place in the editing scheme is not overdetermined by a whole set of intertwined chains of signification, and that it is impossible to fully decipher the film's discourse until one has a completely topographical grasp of the film as a whole, in other words, after several viewings.[14]

If one wants to go beyond the "topographical grasp" of this film, it is necessary not only to see this film "several" times, but to examine each sequence and even individual shots on an analyst projector or editing table. The forthcoming analysis is based on such a close and rigorous "shot-by-shot" study of the film, including the segments that consist of only one or two frames and transparent or black leader.

This concentration on every aspect of the film's structure and its thematic function corresponds to the actual process by which *The Man with the Movie Camera* was made. Vertov and Svilova spent nights and days cutting out and adding in single frames of innumerable shots scattered throughout their workshop, arguing for hours whether a shot should be one frame longer or shorter. The analysis that follows parallels this working procedure, but *in reverse:* in order to determine the basic filmic units that generate thematic meaning, this analytical procedure deconstructs the film's montage framework, which consists of sequences, segments, shots, frames, and so-called subshots (montage units, which are not clearly demarcated).

The four parts of *The Man with the Movie Camera* (Prologue, Part One, Part Two, and Epilogue) build a circular structure that evokes the theme of Vertov's work – a Cameraman's daily activity in a Soviet city – reflecting the cyclic notion of cinematic creation. Continually reminding the viewers of the perceptual illusion of the projected image, Vertov demonstrates how "the film stock is transformed from the movie camera, through the laboratory and editing process, to the screen."[15] By disclosing the cinematic procedure, *The Man with the Movie Camera* confirms the educational function of the "Film-Eye" method; by focusing on ordinary people "caught unawares" in everyday situations, Vertov's film underlines the social aspect of the "Film-Truth" principle.

Because the film communicates through purely visual means, critics often highlight formal or technical aspects alone, instead of concentrating on the *role* technique plays in creating thematic meaning through intellectual associations. Once it becomes clear that this film has neither cogent narrative continuity nor developed dramatic characters, two choices remain: either reject Vertov's work as an "eccentric" display of optical tricks, or look for the meaning and interpretation in the structural relationship between form and content. This interaction can be fully understood only through a close scrutiny of the complex linkage between the representational aspect of each shot and the function of the cinematic devices employed in the process of shooting, editing, and laboratory development.

In film histories, Vertov's work is most often classified as a "city symphony" and is commonly compared to Walter Ruttmann's 1927 film, *Berlin: Symphony of a Great City.* This comparison, as Vertov himself notes, is "absurd," since Ruttmann's film, conceived within a traditional documentary framework, lacks genuine avant-garde characteristics. In his July 18, 1929, "Letter from Berlin," Vertov wrote: "Some Berlin critics – attempting to explain the cinematic features of the *Film-Eye* series – claim that my 'Film-Eye' method is merely a more 'fanatic' version of the concept Ruttmann used in *Berlin.* This half-statement, this half-truth is absurd."[16] Vertov's claim is even more pertinent to the innovative structure of *The Man with the Movie Camera.* The film does not merely depict a city but rather reflects life's dynamism – and its multifaceted contradictions – in Soviet society, developing a cinematic metaphor for human perception and the interdependence of life and technology. This metaphoric connotation is implicit in the film's diegetic structure, which follows the constructivist principle that the creative process is part of the work's meaning and aesthetic import. The constructivist preoccupation with the relationship between man and technology is also manifested in Vertov's title, *The Man with the Movie Camera,* as opposed to the title Keaton gave to his film with a similar thematic content, *The Cameraman* (1928).* Unlike Keaton's film, which implies a specific profession, Vertov's title suggests that his Cameraman is only an individual among numerous citizens-workers who participate in building a society using various tools – in this case, a camera. This idea permeates the entire movie by means of visual juxtapositions that create the analogy between the filmmaker and his camera, and the worker and his tools.

Thematically, one can discern five types of images in *The Man with the Movie Camera:* industrial construction, traffic, machinery, recreation, and citizen-workers' countenances. The construction shots are mostly stationary and of a broad scale, with either enhanced or curtailed perspective (depth of field). In the traffic shots, the camera, often with reduced cranking speed, catches glimpses of life. Machines are shown mostly in close-up, interrelated with workers' enthusiastic expressions and fervent movements. The images of citizens enjoying recreational sports and participating in various forms of entertainment are photographed in close-up or medium shot and are rhythmically juxtaposed to the shots depicting the sports competitions. The montage integration of these five types of images creates a "grand metaphor" about a society free of any capitalist exploitation of workers. At the same time, however, this cinematic trope discloses all the contradictions of an undeveloped and/or badly managed socialist state. The overall ideological message of the film is meant to work "at a distance," that is, it depends on the viewers' active mental participation during the screening, on their subsequent thoughts, repetitive viewing, and additional research. Isolated from this analytical framework, *The Man with the Movie Camera* is often dismissed as a "difficult" film

---

* Keaton's silent film *The Cameraman* (1928), directed by Edward Sedgwick, is considered a masterpiece; it includes several sequences that can be related to Vertov's film on a conceptual level (e.g., shooting a newsreel on the streets of New York, or accelerated motion to produce comic effects).

with an "obscure" meaning, "overloaded" with optical tricks. Only when seen within its structural context can this film reveal its full connotative potential and its ideological significance.

Reactions to the camera

The response of the people within the film to the act of shooting plays a substantial part in the film's self-referentiality. For the most part, the citizens, involved in their work, remain unaware of the camera and pay no attention to the shooting. However, there are several exceptions to this pattern, as in the shot of the poorly dressed woman who throws plaster on the wall and, unashamed of the dirty work she is engaged in, casts a flirting smile acknowledging the presence of the camera (Fig. 48). The majority of the workers in the factories do not pay attention to the Cameraman who perpetually runs around them, be it in the mine (Fig. 145), the factory (Fig. 147), or at the dam (Fig. 148). But when flames blast from furnaces (in front of which the Cameraman appears as a silhouette), a worker warns him to move away for his own protection (Fig. 146).

The female workers in the mill reveal rather indirectly their awareness – even pleasure – of being photographed: although they seem absorbed in their activity, their facial expressions and physical gestures acknowledge the presence of the camera. Similarly, the ladies on the beach subtly display their awareness of the camera as they coquettishly apply their lipstick (Fig. 162), cover themselves with mud (Fig. 161), or chat with one another (Fig. 163). The male workers seem to be more involved in their work, rarely paying any attention to the Cameraman, even when he shoots them from unusual positions (Figs. 56, 57). Kaufman and Svilova are equally immersed in their respective activities, with one possible exception: at the end of the first part of the film, Kaufman's assistant, while shooting from a moving fire engine, recognizes the presence of another camera (Fig. 120). Perhaps those most oblivious of the camera are the children (Fig. 165) watching the magician's tricks (Fig. 166), or both children and adults when involved in situations that capture their complete attention, as in riding a carousel (Fig. 174).

Those citizens who live in uncomfortable conditions exhibit a different attitude toward the camera's omnipresent eye: they consider it an intrusion upon their privacy and respond accordingly. For example, when the camera turns toward the young woman sleeping on the park bench and focuses on her legs, she indignantly jumps up and flees (Fig. 53). A milder reaction occurs in the shot of two ladies in a carriage: they amusingly respond to the Cameraman by mimicking the Cameraman's cranking of his camera, and then conceitedly gaze in the opposite direction (Fig. 69). Later in the same shot, a barefoot woman, absorbed by her duty of carrying the ladies' luggage, shows no awareness of the camera, while a boy on the sidewalk acknowledges the act of shooting with obvious amusement (Fig. 81). A young bum, ragged and lying on littered grounds, responds nonchalantly to the camera by making faces (Fig. 28). A woman, signing a divorce agreement in the Marriage Bureau, covers her face with a purse, as if to protect herself from being exposed to the camera, while her ex-husband laughs with overt contentment (Fig. 93). To emphasize the difference in the couple's reactions, the shot is cut at the point when the husband signs the paper with obvious satisfaction, as the wife dejectedly turns away and leaves. The intention is clear: the camera exposes "Life-As-It-Is," recognizing people's different reactions to being filmed. One might say that by so doing, Vertov, Kaufman, and Svilova violated one of the tenets of the "Film-Truth" principle, which states that the filmmaker must shoot people in such a way that his or her own work does not impede the work of others.[17] This may be only partially true because, although the camera "impedes" the communication between the bureau clerk and the couple, it functions as a "witness" to a "life-fact." By recording bureaucratic procedure, the camera captures the couple's individual responses toward the matter of divorce, and, in its own way, conveys the filmmaker's attitude with regard to the concept of marriage in a socialist country.

The citizens filmed on the street are engaged in their actions and respond to the camera only when it intrudes on their work or "threatens" them physically, as when the tracking camera makes its way through a crowd of startled pedestrians running on all sides (Fig. 110). Consequently, the audience is made "aware" of both the "life-fact" as it occurs in reality and of how such a "fact" is modified when the "subjects" become aware of the rolling camera. Nevertheless, the most common response to the camera is an unconcealed empathy toward the recording apparatus, a notion that contributes to the idea that filmmaking is like any other work, and that the Cameraman is in no way an exceptional citizen; rather, he is part of "a whirlpool of interactions, blows, embraces, games, accidents"[18] representing "Life-As-It-Is."

Self-referential associations

It is through associative editing and symbolic intercutting that Vertov conveys his commentary on the recorded "life-facts," especially when ordinary people are shown in conjunction with *quotidien* and yet connotative objects. There are numerous examples of this editing procedure. A shot of a young woman washing her face (Fig. 40) is followed by a shot of a sprinkler hosing down the street (Fig. 29); in the same vein, a shot of the young woman drying herself off with a towel (Fig. 40) is compared with a shot of another woman cleaning a window (Fig. 48). On a more self-referential level, the act of splicing film (Fig. 129) is associated with fingernail polishing (Fig. 128), stitching (Fig. 130), and sewing (Fig. 131). As a couple signs their marriage papers (Fig. 88), an insert of a manual traffic signal changing its position against the sky appears briefly (Fig. 33). When another couple is seen arguing over their divorce (Fig. 91), there suddenly appears on the screen a split image of two streets "falling" away from each other (Fig. 92). As the traffic fills the streets, a slapstick effect is created by accelerating the motion of the vehicles. After filming in a pub, the Cameraman is shown "drowning" in a magnified beer mug (Fig. 182), and the ensuing hand-held shots (of beer bottles carried on a tray) are unsteady (Fig. 183), as though photographed in an inebriated state. Similarly, a magician's performance is "outdone" by the camera's tricks, when frame-by-frame shooting produces the illusion that five sticks independently form a pyramid on the ground (Fig. 155). The magician himself "materializes" (through a dissolve) on the screen (Fig. 151), thereby implying that the camera is even more "magical" than the illusionist, or, on a more metaphorical level, that the Cameraman's work is not mechanical but creative.

At other points, the camera performs unexpected optical surprises by "catching" hurdling athletes in midair (Fig. 160) and freezing horses as they trot in a race or gallop through the street (Fig. 70). The prowess of a weight lifter is optically ridiculed: as he lifts weights, his legs and arms are severed and decomposed (Fig. 167). At the end of the film, the camera operates without the guidance of the Cameraman and attains human capabilities as it "walks" on its three "legs" and shoots independently, as indicated by the turning crank-handle (Fig. 197). This animated (through stop-trick) anthropomorphism (enhanced by the spectators' acceptance of the illusion) reinforces the already promulgated idea of the marriage of man and machine. The camera's "human behavior" is further emphasized when it is shown "standing" on its tripod as if on guard duty while the Cameraman, dressed in his bathing suit, takes a break and cools off in the shallow end of the beach (Fig. 164).

While the film's most humorous situations are created by unusual mise-en-scène, visual jokes are achieved through montage, by building an expectation in one shot and then disappointing the viewer with an improbable resolution of the situation in the following shot. For example, in the "Soccer" sequence (XLIII), shots of the soccer players are intercut with the shot of an athlete throwing a javelin (Fig. 170), followed by a slow motion shot of the goalkeeper catching a soccer ball coming from the same direction (Fig. 171). The most humorous effects, however, are achieved by the accelerated motion of the traffic and pedestrians (sometimes moving in reverse), which seems to ridicule the frantic tempo of urban life.

Within the context of the constructivist practice of "baring the device," the film repeatedly reveals the cinematic technique used in a given shot. In addition to enlightening the audience as to how the film was made, its self-referential aspects help the viewers to understand the process of human perception and its similarity to – and distinction from – the camera's "perception." Vertov's comments on various events are conveyed through particular cinematic devices: slow, accelerated, and reverse motion, superimposition, pixillation, overlapping, jump cuts, leitmotifs, and various other montage techniques. As will be demonstrated, these self-evident devices are always intrinsically related to the thematic aspect of the photographed object-event. Furthermore, in order to expand upon the notion of reflexivity, Vertov repeatedly points to the following: the motion picture as perceived by the viewer; the motion picture projected on the screen and simultaneously recorded by the camera (screen-within-the-screen); the "freeze-frame"; the film frames as part of the material (footage) handled by the Editor; the actual film moving through the editing table; and the film posters recorded by the movie camera and subsequently projected on the screen (like a slide). All these correspond to Vertov's directorial style, according to which the basic function of the "Film-Truth" principle is "not to rewrite life in terms of a literary script, but to observe and record 'Life-As-It-Is,' "[19] thus providing material (footage) for the construction of the film by the "Film-Eye" method.

Ideological implications

In *The Man with the Movie Camera,* parallel editing is intended to provoke the viewers to think about juxtaposed shots and to establish ideological connections between various events. For example, the shot of a nude woman on the beach covering her chest and neck with mud for salubrious purposes (Fig. 161) is followed by a shot of a woman (obviously belonging to the prerevolutionary bourgeoisie) in a fancy bathing suit and pearl necklace, applying heavy lipstick in an effort to make herself more attractive (Fig. 162). The similarity between the application of fingernail polish (Fig. 128), the act of splicing film (Fig. 129), and the act of stitching (Fig. 130) or sewing (Fig. 131) provides a basis for a metonymic comparison that works on both psychological and metaphorical levels: the parallelism between these procedures is essentially technical, but the implication subliminally alludes to the social differences between the activities of the bourgeoisie and those of the working class. Other shooting devices underscore the ideological distinction between the filmmaker's personal attitude toward the filmed events and things as they objectively exist in external reality. One of the most obvious examples is the split-screen view of the Bolshoi Theater: the classical columned façade of the old building suddenly breaks apart (Fig. 213), symbolizing Vertov's (and the constructivist's) break with bourgeois art and cultural tradition.

The interaction between montage pace and camera movement (panning, tracking, and tilting) reveals Vertov's attitude toward social changes occurring in postrevolutionary Russia. The "Lenin's Club — Playing Games" sequence (XLVII) consists of five shots (linked by invisible cuts) in which the camera glides in front of the religious store, tracks along a cobblestone street, tilts diagonally over the façades and church spires to refocus on the main entrance of one church where Lenin's portrait is exhibited above the door, then, with a swift pan, finally focuses on another church entrance bearing the inscription "Lenin's Club" (Fig. 184). These five shots are perceived as a continuous and fluid camera movement within a unified space that, through the relationship between the smooth cuts and the camera movement, provides a clear ideological message: on a cinematic level, the sensorial impact of the "swish" and "flick" pans "converts" the church environment into the workers' club, thus "sanctioning" the historical "life-fact" of the new social reality in which the religious stores and church buildings are turned into socially utilitarian environments. Due to the camera's fast panning of the sky and the buildings' façades, the transitions between the five shots are hardly perceptible; thematically, the ideological associations in this sequence are similar to the message conveyed by authentic footage showing people demolishing churches at the beginning of *Enthusiasm*.*

A close examination of this sequence's structure confirms that its ideological implication results from the integration of cinematic devices and the images' representational content. For example, the introductory shakey close-up of blurred beer bottles carried on a tray (by an invisible waiter) through a crowded beer hall (Fig. 183) in contrast to the closing shot of checkers and chess figures miraculously jumping into place on a board (Figs. 185, 186) humorously alludes to the roguish chaos of the beer hall as opposed to the orderliness of the workers' club. The two environments are connected by the camera's panning and tracking through space: camera movement selects "life-facts" from their respective milieux and prompts the viewer to syn-

---

* The introductory sequences in *Enthusiasm* include archival footage of the revolutionary masses knocking down the cross from the steeple and taking out icons from the church.

thesize ideologically the images of drunken people in a beer hall (Fig. 181), a religious store, and church buildings transformed into workers' clubs (Fig. 184) where the camera encounters only several workers relaxing (Fig. 187). These images are interrelated on a thematic level through crosscutting, thus creating a new cinematic whole with a message that metaphorically merges the two locations (beer hall–religious store) and two actions (demystification of churches–formation of the workers' clubs), suggesting that in the new society the pub and the church should be replaced by the cultural club, just as bottles have been replaced by the checker game, and church icons have been replaced by portraits of Marx and Lenin. At the same time, the sequence's ideological message reflects a paradoxical situation with ironic overtones: although the pub is filled with many customers, inside the workers' club there are only a few people, Vertov among them (Fig. 188). Obviously, the masses do not readily change their cultural and spiritual habits despite the revolution. Following the "Film-Truth" principle, Vertov refrains from misrepresenting that reality, let alone falsely embellishing it to satisfy propaganda. Instead, he acknowledges "life-facts" as they appear in front of the camera at a given moment, and only in the process of editing adds to them his personal comments.

Some of Vertov's allusions require that the viewers be acquainted with the specific contemporary circumstances, as in the shots, for example, depicting three film posters that, within the montage context, signify more than just a technical reference to the current movie repertory. One of the posters advertises a popular foreign film by depicting a young woman standing next to a man who holds his finger to his lips, as if warning her to keep silent (Fig. 26). This poster is shown in different scales, four times throughout the first part of the film, compelling the viewer to wonder why so much attention is focused on it. Was Vertov so fascinated by this poster's graphic execution that he wanted his audience to see it in various contexts? Or was the advertised film chosen for its particular content in order to discredit commercially oriented contemporary film production/distribution in the Soviet Union?

The poster is first seen (sequence VI) inserted between two close-ups of a sleeping woman, thereby automatically suggesting "keep quiet – the woman is sleeping" (Fig. 25). On a strictly graphic level, its design can be related to postrevolutionary posters of the period that warned the Soviet citizens to remain vigilant and closelipped in fighting the "enemies" of socialism. It is not until "Street Traffic" (sequence X) that a shot appears in which the entire poster is revealed, advertising the popular German film *The Awakening of a Woman** (or

---

* *The Awakening of a Woman* or *The Awakened Sex* [*Probuzhdenie zhenshchini ili Pro-*

The Awakened Sex), launched at that time as "an artistic drama of passionate love" (Fig. 26). Familiar with this type of film, contemporary viewers were expected to make their own judgment of the sleeping young woman (obviously of bourgeois origin) and the bourgeois lady appearing in the German movie known for its trivial treatment of sex. The postponed revelation of the film's title on the poster, in conjunction with the sleeping young woman, suggests a humorous double entendre, especially for those spectators who are familiar with the German movie and know what kind of awakening is implied. In the second instance (sequence VII, 1), the close-up of the movie poster follows the shot in which Kaufman, toting his camera, leaves an apartment building (or the *kinoks'* workshop), gets into a car, and hurries off to work (Fig. 34). Again the difference between romanticized staged films (photo plays) and documentary cinema (newsreels) is brought to the fore: instead of going to a studio – where movies like *The Awakening of a Woman* are made – the Cameraman veers into everyday life, just as Vertov urged his *kinoks* to do.[20] In the third instance (sequence X, 8), there is a long shot of the same poster exhibited at the entrance of a movie theater, clearly a direct allusion to the current entertainment repertory promulgated by the New Economic Policy and the "Leninist Proportion." The viewers are enlightened as to the poster's actual scale because they see it in relation to the passing Cameraman (from right to left) who, carrying the camera on his shoulder, now looks relatively small (Fig. 55). The spatial relationship between the enormous advertisement and the tiny Cameraman running to his daily assignment suggests that the latter is engaged in shooting "Life-As-It-Is," as opposed to the dramatized situations exploited by entertainment movies. The graphic display implicit in the decorative style of the huge poster highlights the phenomenological disparity between the two worlds – one fabricated by a "dream factory," the other reflecting the real world from which the Cameraman selects significant "life-facts." The fictional cinema epitomized by the huge film poster and the newsreel represented by the tiny Cameraman are shown as opposite facets of the same environment in which *The Man with the Movie Camera* was made and shown. As Kaufman bustles in front of the fanciful poster without paying any attention to it, one realizes that the antithetical worlds of the *kinoks* and commercial cinema coexist indifferently. Yet, there is a practical interdependence between business and perception, as in the ensuing shot which focuses on an inscription, decoratively exhibited in the window of an optician's shop as the blinds are rolled up and reveal the words "GLASSES – PINCE NEZ" (Fig. 62).

buzhdennyi pol] are the Russian titles for the German film, *Das Erwachsen des Weibes* (1927), directed by Fred Sauer.

The final appearance of the poster (sequence XIII, 17) is a summation of all the contradictions. The poster, reflected in the revolving door of a movie theater, looks even more paradigmatic of the world it represents. As the door begins to rotate, the reflection of the poster slowly exits to the left of the frame and is gradually replaced by that of a distant city moving from right to left. This optical mutation, the transformation of one image (the fictionalized world of a photoplay) into the other ("Life-As-It-Is"), suggests a triumph of Vertov's "Film-Truth" principle over entertainment cinema. The optical blending of different images reflected in the window hypostatizes the self-referential aspect of the film while confirming the illusory nature of cinematic projection and the ambiguity of human perception. The shot is composed of four visual phases in which one composition replaces the other: (1) the movie poster, (2) the skyline of a distant city, (3) the Cameraman cranking his camera, and (4) the door frame obliterating the reflection. The fascinating metaphorical peak of the shot is reached when the mirror image of the Cameraman cranking his camera (in front of the rotating glass) gradually replaces the view of the city until the Cameraman himself is obliterated by the wooden frame of the door (Fig. 63). This shot, by contrasting the reflection of a fictional movie poster with that of the outside world in which the Cameraman captures "Life-As-It-Is," metaphorically confirms the "life-fact" that the screening of an entertainment film supercedes the screening of a documentary film and newsreel. Contemporary viewers could easily make such a conceptual connection because the film genre advertised on the poster represented ninety-five percent of the Soviet film repertory,[*] and it exemplified the distributor's commercial credo: "The more conventional – the better" [*chem shablonnee chem luchshe*].[21] Indeed, this was the attitude against which the *kinok*s had to struggle in order to establish a new revolutionary cinema, and to abolish the production of bourgeois photoplays, which, according to Mayakovsky, "have nothing in common with contemporary Soviet life."[†]

Another movie poster appears in the second half of the film, and depicts the citizens' recreation. The film's title, *Green Manuela* or *Where They Trade in Bodies and Souls*,[‡] refers to another German "adventure spectacle" from the early twenties. The first shot of the "Green Manuela" sequence (XLV) is a long shot of several trees in a park. The following shot (a long take) begins with a pan of the sky and then moves to the façade of a building on which the film poster is

---

[*] See Chapter I, note 49.
[†] See Chapter I, notes 80–2.
[‡] *The Green Manuela or Where They Trade with Bodies and Souls* [*Zelenaia Manuela ili Gde torguiut telom i dushom*] are the Russian titles for the German film *Die Grüne Manuela* (1923), directed by E. A. Dupont.

exhibited; the pan continues diagonally over the sky and across several trees to the roof of a movie theater where a large horizontal panel, marked "Proletarian Movie Theater," is displayed; finally, the pan scans past several telephone poles and rests upon a cluster of trees (Fig. 179). The third shot is, actually, comprised (along the horizontal axis) of two images in which the Cameraman, kneeling beside his tripod, reloads his camera on the roof of a tall building towering over the city (Fig. 180). The subsequent shot, a close-up of the sign "Beerhall/ Bierhalle" (Fig. 154), introduces the "Beerhall" sequence (XLVI). The message cannot be more clear: thanks to the NEP policy, trivial melodramas have invaded the "proletarian" movie theater. By preserving the spatial connection between the two phenomena − in a concrete environment and at a specific historical moment − the camera movement *(trajectoir)* makes an ideological gesture-statement that cannot be conveyed by a cut. Resolute diagonal upward panning functions as a kinesthetic metaphor for the *kinoks*' cry against bourgeois photoplays, the main target of Vertov's struggle against traditional staged cinema.

Finally, there is an insertion of a third movie poster, this time advertising a Soviet contemporary film, *The Sold Appetite,** directed by Nikolai Okhlopkov in 1928, and based on a political pamphlet by Paul Lafargue. Promoted at the time as "a social satire on millionaires," this film was advertised by a poster depicting a group of elegant ladies and gentlemen set against the background of a modern city (Fig. 52). A medium shot of the poster is inserted within a segment depicting early traffic on Moscow streets (Fig. 54), and is followed by the shot of the young woman sleeping on a park bench (Fig. 53). This juxtaposition ideologically brings together the "life-facts" of (1) everyday events captured "unawares" (traffic, pedestrians), (2) historical documents (film posters), (3) social conditions (citizens sleeping on the streets), and (4) the technological device (the camera's lens). The phenomenological interaction of these elements is intended to confront

---

* *The Sold Appetite* [*Prodannyi appetit*], produced by VUFKU (Odessa) in 1928 and directed by Nikolai Okhlopkov, was based on a satirical pamphlet, *Un apetit vendy,* by Paul Lafargue. Some sequences of this dramatic film contain original archival footage of Lenin's funeral (January 24, 1924), intercut within the main story of the film. The action is situated in a "capitalist country." A multimillionaire suffers from a mysterious disease: his stomach cannot digest food. Doctors recommend that he replace his stomach and try to find someone who is willing to sell his own. A poor bus driver, pressed by his girlfriend, agrees to sell his stomach to the millionaire. The operation is performed, and the re-stomached millionaire begins to consume enormous quantities of food; in turn, however, the bus driver becomes weak, falls into depression, and attempts to commit suicide. His girlfriend tries to revive his interest in life, and slowly he recovers. He continues to drive the bus, but, after several attacks of depression, he drives into a wall on the street and kills himself; at the same time, the millionaire dies of obesity in his fancy home.

the viewers with the paradoxes of Soviet reality, just as, on the other hand, the necessity of cooperative relations between various industries is alluded to in the sequence "Power Plant and Machines" (XXX) through parallel montage: shots of factories in full production are juxtaposed to shots of the Cameraman shooting a dam (Fig. 148) that provides the power to drive the machines (Plate 1).

Through the careful selection and interaction of "life-facts" as found in everyday life, Vertov and Svilova use the "Kuleshov effect"* in a highly sophisticated manner to achieve more complex associations and ideological implications that, in most sequences of *The Man with the Movie Camera,* emerge not only from the Kuleshovian juxtaposition of different images but also from the ideological connotation intrinsic to the execution of the shot. For example, the famous split-screen shot of the Bolshoi Theater (Fig. 213) is associative in and of itself, that is, by its representation of a prerevolutionary architectural monument being "slashed into halves," and equally so by its position among other shots. The direct ideological implication emerges from the shot's composition and its optical distortion. Because the shot of the Bolshoi Theater already carries a particular cultural and ideological implication, when it is intercut between two close-ups of a big pendulum, its ideological connotation becomes multifaceted: it links the past with the present. In the first close-up, the pendulum swings at normal speed, as if to denote the regular passage of time (Fig. 212). After the split-screen shot of the theater, the same pendulum (by now a death knell for the Bolshoi and all traditional art) begins to swing faster, metaphorically signifying the dynamism of the new society in which classical art (symbolized by the decorative Bolshoi architecture) appears anachronistic and doomed to be replaced (even abolished) by new forms of art and communication. (A similar idea is conveyed in Vertov's *Enthusiasm* through the multiple exposure of the crucifix and church cupolas splitting apart in an optical distortion.) Finally, after the splitting of the Bolshoi Theater is completed, all shots of traffic and pedestrians are shown in accelerated motion, foreshadowing the end of the film and its dynamic montage sequence in which Svilova's eyes (Fig. 227) and repeated brief shots of the swinging pendulum (Fig. 231) are interspersed among numerous close-ups of the film spectators (Fig. 198) sitting in the movie theater (Fig. 204), apparently the same one seen at the film's beginning (Fig. 4). Thus, the notion of *temporal progression,* symbolizing the dynamism of the revolutionary era (as opposed to the stagnancy of bourgeois art), culminates in the

---

* The "Kuleshov effect" is based on the montage principle according to which the ultimate meaning of a shot is qualitatively affected by the preceding or following shot. See Lev Kuleshov, *Kuleshov on Film,* p. 200.

film's final montage crescendo where representational shots merge into a semiabstract kinesthetic and symphonic vision characteristic of pure cinema.

Negation of narrative

On the diegetic level, *The Man with the Movie Camera* defies the narrative as a means of drawing the viewer's attention to its meaning. Yet certain events are presented in a sequential order which fosters the expectation of linear development. However, each time such a narrative core becomes apparent, it is immediately thwarted. The initial shots of the Cameraman carrying his camera and entering the movie theater suggest that we are about to follow the film's "protagonist" throughout different stages of his daily work. After the shot of the camera (Fig. 1), the top of a tall building (Fig. 2), and a streetlight pole (Fig. 3), there appears the empty movie theater (Fig. 4), with its light fixtures (Fig. 5) and a cordon at its entrance (Fig. 6). The Cameraman enters the theater (Fig. 7); then the projector is seen (Fig. 8) and is prepared for the screening by the projectionist (Figs. 9–11). As the audience fills the auditorium (Figs. 12–14), the conductor and his orchestra get ready to accompany the film (Figs. 15–22). With this, the "mininarrative" is completed, and any further appearance of the Cameraman is related to situations that take place elsewhere. Actually, he does reappear in the movie theater at the end of the film, this time not among the spectators, but rather as the object of their perception – on the screen-within-the-screen (Figs. 217, 226).

The most evident example of the negated narrative core occurs in the "Awakening" sequence (VII) in which a young woman, upon awakening, washes and dresses herself as she prepares to leave her apartment. This diegetic embryo could be developed into a full narrative story, but the moment it reaches a thematic statement (in which the young woman's blinking is compared to the opening and closing of shutters), the "character" portrayed by the young woman no longer appears. However, other components of this sequence (e.g., the blurred shots of the speeding train, which illustrate the young woman's sleeping consciousness) reappear as associative inserts at the end of the film, contributing to the montage crescendo of the finale.

The first shot of the young woman shows her asleep in bed (Fig. 25); this and the following close-up are intentionally composed in a fashion that suppresses the "reading" of its representational content: the young woman's head is cut off by the upper line of the frame, and only after repeated intercutting of the same images does the viewer detect a female hand, chin, and neck under the sheets. The setting in which she sleeps is then compared – through parallel editing – to the dirty streets in which derelicts, a handicapped man, and a young woman sleep (Figs. 28, 53), alluding to the state's inability to provide shelter for all its citizens. The sleeping young woman is reintroduced after a shot in which the Cameraman leaves a building (possibly the *kinoks'* quarters) carrying a camera, and getting into a car (Fig. 34). On his way he crosses the railway tracks with his convertible (Fig. 43) and shoots the on-coming train from below by placing the camera alongside the tracks (Fig. 37). Completing his assignment, the Cameraman leaves the railroad (Fig. 42), and the sequence comes to a close with a long shot of the convertible speeding along a dusty road, telephone poles occupying the left side of the frame (Fig. 44). By means of crosscutting between the two places (exterior/railroad and interior/bedroom), the sleeping young woman is shown again, this time with her right hand resting on her head (Plate 5). Immediately thereafter, a series of brief shots of the young woman's head moving from left to right is juxtaposed to a set of brief shots of a speeding train (Fig. 38). The semiabstract composition of these shots, intensified by accelerated and reverse motion, parallels the woman's transitional phase of sleep prior to awakening. This oneiric mood is further enhanced by the flickering effect: as the young woman wipes her face with a washcloth and begins to blink (Fig. 49), an exchange between brief tracking shots of the railway tracks (Fig. 41) and black frames creates an optical pulsation akin to one's visual perception for the first few moments while emerging from sleep.

The "story" of the young woman continues to develop on a diegetic level: after she gets up from her bed, she moves around the room and prepares for work (Figs. 39, 40). Suddenly, there appears a close-up of the Cameraman's hand replacing the regular lens of his camera with a telescopic one (Fig. 45). Then, the camera swiftly turns toward the right side of the frame, and the subsequent shot reveals a derelict lying on a pile of garbage. The medium shot of the derelict is intercut by two close-ups of the camera being cranked, the second of which is seen upside down (Fig. 134), as the young man reacts to the act of filming by making faces, arranging his cap, and scratching his bare chest (Fig. 28). This uninhibited exchange between the derelict and the Cameraman informs the viewer about the candidness of the "Film-Truth" principle: the camera is neither avoiding nor covering up the unpleasant "life-facts," and the derelict — freed from moral or political restrictions — is not intimidated by the camera's presence. The derelict shots are then related to the shots of a female worker sweeping the streetcar tracks (Fig. 48) and another (older) bum (with an amputated leg) lying on a street bench (Fig. 28). While perceiving all these situations, the spectators are made aware of the Cameraman's being an eyewitness to these "life-facts," as he is reflected in the lens (Fig. 46).

The shots of the young woman blinking (Fig. 49), edited to produce the flickering effect, are compared to the shots of opening and closing venetian blinds (Fig. 50), while the shot of the contracting and expanding photographic iris (Fig. 47) is compared to the shot in which a blurred image of flowers gradually comes into focus. Both of these cinematic effects function self-referentially as they draw the viewer's attention to the nature of the film's projection. At the same time, the crosscutting between the contracting iris and blooming flowers points to the technical aspect of the cinematographic apparatus as well as the nature of human perception. Once again, the *kinoks'* idea of the merger between human being and machine is brought to the fore: the eye "armed" with the camera lens yields an "extracting" way of seeing, by which one can unearth the essence of the photographed objects.

The "Ladies in a Carriage" sequence (XV) is another example of a diegetic core that transforms into a cinematic metaphor devoid of linear development. First, by crosscutting a close-up of locomotive wheels and a medium tracking shot of a horse-drawn carriage, two means of transportation are related. At the outset of the sequence, the Cameraman is seen standing behind his camera mounted upon its tripod in a moving convertible (Fig. 65). But the Cameraman remains inactive as several horse-drawn carriages pass before him. Then a shot of three well-dressed ladies sitting in a stationary carriage (Fig. 66) is followed by a tracking shot of the Cameraman's convertible moving through traffic; this time, two ladies with an open parasol ride in front of him. In the ensuing shot, photographed from behind the convertible, the Cameraman stands poised but does not yet use his apparatus. Toward the end of this shot, the Cameraman takes his right hand off the camera in order to adjust his hat. At this point, a low-angle close-up of stationary locomotive wheels (Fig. 67) is inserted. The next shot is the second part of the previous shot in which the Cameraman and his assistant prepare to shoot. In another insert of the locomotive, the wheels finally begin to turn and, as if "triggered" by this movement, in the following shot, the Cameraman begins shooting: positioned behind the tripod, he cranks the camera and directs the lens at the ladies in the carriage riding parallel to his convertible (Fig. 65). The third insert of locomotive wheels shows the train picking up speed as white steam fills the right corner of the frame. The succeeding shot is a rear view of the Cameraman in the convertible, zealously cranking the camera as he films the ladies in the moving carriage (Fig. 66). A fourth insert of the locomotive – an extreme low-angle shot of railway tracks over which a dark train rushes from foreground to background as pedestrians move in accelerated motion on the right side of the moving train – concludes the sequence (Fig. 68). Because of the Cameraman's determined pursuit of the ladies, the viewer expects further development of their "story." However, after the ladies reach the apartment house where a barefoot maid waits to carry their luggage into the building, they no longer reappear.

On a thematic level, the intercutting of the locomotive wheels reiterates the theme of traffic and general transportation shown throughout the film. On a denotative level, the modern means of transportation, the locomotive – a symbol of the technological age – is contrasted to the outmoded form of the carriage which, with its (bourgeois) passengers, belongs to the prerevolutionary era. The juxtaposition of these two sets of images is repeated several times, as the initial stirring of the locomotive is compared with the starting of the camera, both products of modern technology. By associating the camera with a locomotive, Vertov creates another cinematic metaphor for his "Film-Eye" method influenced by the constructivist vision of the world and the futurist admiration for the technological age, suggesting that the machine should dominate all areas of human activity – transportation, industry, as well as cinema.*

In other instances, Vertov and Svilova use specific cinematic devices to emphasize the idea of technological revolution as an essential feature of the contemporary world. In the "Accident on the Street" sequence (XXIII), the image of a telephone (Fig. 114) becomes a decisive communication channel between the site of the accident (Fig. 118) and the approaching ambulance (Fig. 115). After a montage sequence (achieved by intercutting the eyeball and street traffic), the viewer's perception is "shaken" by an unexpected optical jolt: a close-up of the ambulance rushing toward and over the camera (Fig. 116) is inserted between two medium shots of the operator answering an emergency call (Fig. 113); this optical shock sensorially foreshadows the actual accident, which is not shown until later. The message (anticipating the accident) is conveyed in a purely visual manner: the order of the shots and the camera position create the illusion that the ambulance is running over the operator – and the spectator – who responds to the emergency call. This perceptual "warning" is followed by the shot of the Cameraman jumping into his convertible, his camera ready and mounted on its tripod (Fig. 117). The automobile exits frame left, while the operator's hand is shown hanging up the receiver (Fig. 114), confirming the emergency call and dispatching help. The apparent "story" related to the accident is underscored sensorially by the Cameraman's subjective point of view: a fast tracking shot (taken from the moving vehicle as it penetrates the crowd) forces several surprised and fright-

---

* About the futurists' adoration of machines and the technological age, as well as the constructivists' preoccupation with metallic structures, see Chapter I, "Futurist and formalist expression."

ened pedestrians to jump aside (Fig. 110). This is followed by a symbolic oblique shot of the traffic with pedestrians seen blurred in the foreground (Fig. 111). The sequence ends with multiple crosscuts between the ambulance going in one direction and the fire engine speeding off in another (Fig. 119).

In response to the camera's physical movement (particularly its aggressiveness toward the pedestrians), the ensuing shot reveals a wounded young man lying on the pavement, his bleeding head wrapped in bandages (Fig. 118). This is an example of Vertov's metonymic "visual" syntax: the tracking camera is identified with the vehicle that "runs over" the pedestrians (Fig. 110) *before* the actual accident is shown. The two additional shots, an oblique view of the street "tilting" the moving cars (Fig. 111) and a shot of a trolley running over the camera (Fig. 112), enhance the meaning and sensorial impact (the "danger") of the previous aggressive tracking shot. Together these three shots complement the function of the "Street and Eye" sequence (XXII) which ends elliptically, by visually conceptualizing the tragic consequences of the accident (Fig. 118). The tension is accordingly built by means of a genuine cinematic device (montage) and resolved by a straightforward depiction of the accident, which emotionally attracts the viewers' attention to the injured man. (A shot-by-shot breakdown of this sequence is included in Chapter III.)

Disruptive–associative montage

The metaphorical implications of *The Man with the Movie Camera* result from the apposition of often unrelated and contradictory themes. The emergence of new imagery at the end of a sequence invites the viewer to establish an ideational connection between two topics, one fully developed, another just emerging. Once introduced, the new theme takes over in the forthcoming sequence, only to be replaced by another subject as the sequence approaches its end. This editing procedure is based on "disruptive–associative" montage that develops through several phases: a sequence establishes its initial topic and develops its full potential through an appropriate editing pace until a seemingly incongruous shot (announcing a new topic) is intercut, foreshadowing another theme that, although disconcerting at first glance, serves as a dialectical commentary on the previously recorded event. The metaphorical linkage between the two disparate topics occurs through an associative process that takes place in the viewer's mind. Through such dialectical intercutting, the initially presented topic acquires an additional meaning that complicates the already achieved thematic integrity of the sequence. But this apparent complication is only momentary: the instant the inserted "disruptive" shot is perceived, it begins to function retroactively, providing more information about the surrounding shots than about itself. The "disruptive" shot(s) are repeated (often through different pictorial composition) until their content begins to dominate the screen, developing a new meaning that will then suffuse the ensuing sequence.

Disruptive–associative montage can be seen as a cinematic parallel to the "dialectical unity of opposites," which Friedrich Engels defines in the following way:

The two poles of an antithesis, like positive and negative, are just as inseparable from each other as they are opposed.... Despite all their opposition, they mutually penetrate each other.... Dialectics grasp things and their images, ideas, essentially in their interconnection, in their sequence, their movement, their birth and death.[22]

One of Vertov's most dialectical implementations of montage is based on Engels's belief of criticizing and challenging all things, including Marx's own teaching. Vertov and Svilova used this principle to confront the overt meaning of a sequence by inserting "subversive" shots that function as thematic antitheses which prompt a dialectical grasp of the sequence's message. Eisenstein also considered the dialectical thesis–antithesis conflict as part of his theory of intellectual montage. Drawing from Hegelian dialectics, Eisenstein developed his concept of dialectical montage to "resolve the conflict-juxtaposition of the psychological and intellectual overtones"[23] through "conflicting light vibrations" that affect the viewer on a physiological level and, at the same time, "reveal the dialectical process of the film's passage through the projector."[24] According to this definition, Eisenstein's dialectical conflict is more comparable to Vertov's "Theory of Intervals" (the dialectical juxtaposition of the various movements within the connected shots) than to his disruptive–associative montage (the dialectical function of a disruptive shot on a thematic level).

Unlike Eisenstein's concept of dialectical conflict that produces emotional/intellectual overtones due to the optical pulsation of images projected on the screen, Vertov's cinematic dialectics result from unexpected insertions of shots whose thematic connotation contradicts the already established meaning of the entire sequence. Vertov's disruptive-associative montage is best exemplified in the "Editing Room" sequence (XVII), the "Controlling Traffic" sequence (XVIII), the "Marriage and Divorce" sequence (XIX), and the "Death, Marriage, and Birth" sequence (XX). Altogether, these four sequences are made up of ninety-six shots, with a total duration of approximately four minutes. The shot scales vary as indicated in the shot-by-shot breakdown (below), which includes the following specifications: high angle (HA), low angle (LA), extreme close-up (ECU), close-up (CU), medium close-up (MCU), medium shot (MS), medium long shot (MLS), long shot (LS), and extreme long shot (ELS). At the end of each shot description, it is often noted that "the first frame is black," pointing to an optical "pulsation" within the sequence (discussed in Chapter III). The length of the shots, marked by the number of frames, indicates the montage pace of the sequence.

XVII *Editing Room:* 38 shots, 3,506 fr. (34,039–37,544), 3 min. 20 sec.

1. MS. Two shelves of film rolls, one above the other; their tails extend upward against a lit background so that the frames can be easily identified (128 fr.).

2. HAMS. Seven parallel shelves with rolls of labeled film; three signs in Cyrillic (*zavod* – plant; *mashiny* – machines; *bazar* – market) indicate that the rows are arranged thematically (71 fr.).

3. CU. Empty circular film plate rests on editing table (51 fr.).

4. ECU. A filmstrip with two frames of a peasant woman, head wrapped in white scarf; frames are positioned horizontally with image on its side (3 fr.).

5. CU (same as VII, 6). The film plate rotates with filmstrip winding around central spool rotating counterclockwise (77 fr.). Note: first frame black.

6. HAMS. Svilova, in profile, working at her editing table, examining a piece of filmstrip and watching it against the light table while cranking handle with her right hand (46 fr.).

7. HACU. A dark piece of filmstrip moving rapidly and diagonally from upper right to lower left against lit screen; the movement is reversed, and a bright piece of film strip (with image of an infant) moves from lower left to upper right; a silhouetted hand holding a pair of scissors enters from the right and cuts the strip where the light and dark segments join; hand withdraws frame right, leaving two pieces of film strip on the lit screen (143 fr.).

8. HAMS (same as XVII, 9). Svilova takes a roll of filmstrip from a plate on her right and labels it (63 fr.).

9. ECU (same as XVII, 7). The frame of the peasant woman in a white scarf is now seen right side up (74 fr.).

10. LAMS. Svilova in right foreground working at editing table; she puts a roll of film on the top shelf in front of her, below a hanging lamp, and then looks at the exposed frames lit from behind; the rest of the frame is in darkness (94 fr.).

11. CU (same as XVII, 7 and 12). The peasant woman in white scarf, smiling and moving her head to the right (94 fr.).

12. HAMS (same as XVII, 11). Svilova examines the filmstrip, looking at it against the lit screen (37 fr.).

13. ECU. A filmstrip with bands of perforation on both sides; one and a half frames of the face of a laughing boy with skull cap facing left and girl behind him (102 fr.).

14. HACU. The boy laughing, seen on the filmstrip in previous shot (102 fr.).

15. MS (same as XVII, 15). Svilova examines film, leaning over the lit screen (39 fr.).

16. ECU. Faces of two children, girl in center and boy at right of frame, with bands of transparent perforation on both sides (105 fr.).

17. CU. Faces of children from the filmstrip in previous shot, now seen directly on the screen (77 fr.).

18. HACU (same as XVI, 6; freeze-frame). An old peasant woman (50 fr.).

19. HALS (same as XVI, 5; freeze-frame). Street and square filled with people (31 fr.). Note: the first frame is black.

20. CU (same as XVII, 20). Two children's faces (29 fr.).

21. HALS (same as XVII, 22). Street filled with moving pedestrians (137 fr.).

22. HAMCU (same as XVII, 18). An old peasant woman arguing and moving forward and backward diagonally from lower left to upper right (106 fr.).

23. MS (same as XVII, 18). Svilova examining filmstrip on the light table (33 fr.). Note: the first frame is black.

24. CU (similar to XVII, 7). A strip of film moving diagonally from lower left to upper right across the light table; strip moves until it reveals a new shot with darker images (52 fr.).

25. HAMS (same as XVI, 2; freeze-frame; motion). Two ladies sharing parasol and man in straw hat; the shot is frozen, but after twenty-eight frames the freeze-frame turns into a motion picture, with pedestrians moving in the background (127 fr.). Note: the second part (i.e., motion picture) of the shot is composed slightly differently from the freeze-frame; it is a jump cut created by the exclusion of a few frames from the continuity of the action from freeze-frame to motion. (243 fr.). Note: the first frame is black.

26. CU (same as XVII, 2). Face of a middle-aged peasant woman laughing (53 fr.).

27. CU (similar to XVI, 1). The head of the dappled horse (seen earlier) in profile facing left as it pulls vehicle; houses, pedestrians, and trees in background (77 fr.).

28. MS (similar to XVI, 4). Two ladies in black hats in profile facing left in carriage as it moves from right to left (158 fr.). Note: the first frame is black.

29. CU. A filmstrip moving vertically across the light table from bottom to top, with hand silhouetted against light table (51 fr.).

30. HAMS. The back of the carriage driver in left foreground, as the lady in white dress pays him; the woman in the black hat and the maid in the background walk forward, while the boy in the left middle ground stares at the camera (95 fr.). Note: the first frame is black.

31. CU (same as XVII, 8). A small roll of film spinning on spindle in the center of the film plate (53 fr.). Note: the first frame is black.

32. HAMS (same as XVII, 30). Barefoot maid comes between the two women and accepts a big trunk from the carriage driver; the carriage occupies the left horizontal foreground; in middle ground the boy who was staring at the camera exits frame right; the maid puts the trunk on her left shoulder

and walks into the background, behind one of the women, while the other takes her coat from the carriage and walks toward the background (243 fr.).

33. HAMS (lateral tracking). With camera over his left shoulder the Cameraman walks toward the left, in shadows of buildings and trees; at intervals, vertical tree trunks appear in the foreground, creating the illusion of moving from left to right (68 fr.). Note: the first frame is black.

34. HAMS. The carriage with dappled horse in the left half of the frame; the carriage driver unloads suitcases and hands them to the two ladies and a child, one lady already carrying a small suitcase; they exit frame right (154 fr.).

35. LAMCU. Rotating upper edges of the panels of a revolving door moving clockwise, projecting shadows on the ceiling (80 fr.).

36. HALS. Street square and park; traffic moves from left to right in the middle ground, pedestrians mill in the background (159 fr.). Note: the first frame is black.

37. MS. The revolving door as silhouetted people pass through it and exit right foreground (151 fr.).

38. HALS. An intersection, traffic moving perpendicularly and passing each other, from upper right to lower left and middle left to lower right corner of the frame (162 fr.).

XVIII *Controlling Traffic.* 7 shots, 423 fr. (37,545–37,967), 24 sec.

1. HAMS. A woman sits at desk and picks up a telephone receiver from frame right (55 fr.).

2. MS. Lower segment of a traffic signal, with a car passing in background; two hands and a torso of a uniformed traffic policeman shown as he touches the control's handle (53 fr.).

3. MS (same as XVIII, 1). The woman on the phone takes notes (51 fr.).

4. LAMS (tilted). The traffic policeman, seen from chest up, blows his whistle; he then puts his left hand on the control lever and looks away (58 fr.).

5. HALS (same as XVII, 39). Square and traffic (99 fr.). Note: the first frame is black.

6. LACU (tilted). The traffic signal turns, silhouetted, against the gray sky, changing its diagonal position from lower left–upper right to upper left–lower right corners of the frame (52 fr.).

7. HALS. The camera with telescopic lens in right foreground tilts downward toward street and receding tall buildings and traffic (105 fr.).

XIX *Marriage and Divorce.* 35 shots. 7,726 fr. (37,968–45,693), 7 min. 9 sec.

1. MS. A man in dark suit at left and a woman in white blouse at right leaning over the counter in the middle ground; a clerk sits at left reading from a book; the man and woman are listening to him (156 fr.). Note: the first frame is black.

2. ECU. Marriage application, with typed and written words positioned diagonally (parts of words cut off); text is in Ukrainian and in Cyrillic: "Record of Marriage. Place – Veresna. Year – 1928" (80 fr.).

3. MS (same as XIX, 1). The couple converses with the clerk sitting in the foreground (87 fr.). Note: the first frame is black.

4. HALS (same as XVIII, 7). The camera slowly pans to right, revealing more of the street and traffic in the background (82 fr.).

5. MS (same as XIX, 3). The clerk in the foreground stands up and hands the couple the marriage certificate; the woman signs it (140 fr.).

6. HALS (same as XIX, 4). The camera continues panning to right (51 fr.).

7. MS (same as XIX, 4). The man signs the marriage certificate (65 fr.).

8. LACU (same as XVIII, 6; tilted). The traffic signal in its changed position, as the top of a trolley passes diagonally from lower right to upper left (42 fr.).

9. MS (same as XIX, 7). The man smiles and folds the certificate (95 fr.).

10. LACU (same as XIX, 8). The traffic signal returns to its original position (124 fr.). Note: the first frame is black.

11. MS (same as XIX, 9). The man and woman leave the room, while the clerk sits down and continues to work at his desk (63 fr.).

12. HACS (same as XIX, 6). The camera, facing toward the background, suddenly turns 180 degrees; the street and traffic are seen in background (58 fr.).

13. MS (closer view of XIX, 11). A couple, woman on left and man on right, with the clerk's head in the foreground; both look grimly at the clerk, who is looking down and apparently writing (74 fr.).

14. ECU. A divorce certificate with printed and written information in Ukrainian (78 fr.).

15. LAMCU. The woman in pearls and modern dress (seen in the shot before last) with right hand on her cheek looking down with a worried expression on her face (67 fr.).

16. LAMS (same as XIX, 13). The couple (woman on left, man on right) listen to the clerk (47 fr.).

17. LAMCU (same as XIX, 15). The woman moves her head to left (60 fr.). Note: the first frame is black.

18. MS (same as XIX, 16). The couple listens to the clerk reading (64 fr.).

19. LAMCU (same as XIX, 17). The woman with right hand on her cheek looks down (32 fr.). Note: the first frame is black.

20. LAMCU. The man talks while gesticulating with left hand in the foreground (28 fr.).

21. LAMCU (same as XIX, 19). The woman looks toward the man and speaks (44 fr.).

22. LAMCU (same as XIX, 20). The man speaks, gesticulating with left hand in the foreground (31 fr.).

23. LAMCU (same as XIX, 21). The woman with right hand on her cheek lifts her head (18 fr.).

24. LAMCU (same as XIX, 22). The man looks down, his hand in front of his mouth (20 fr.).

25. LAMCU (same as XIX, 21). The woman begins to speak; top of the clerk's head appears in left foreground (106 fr.).

26. LS (split screen; tracking). Two tilted views of street with traffic as the camera moves perpendicularly forward; produces feeling that the screen is falling apart with traffic going in opposite directions in either half (159 fr.).

27. MS. The woman, in profile, facing right, signs a document, leans forward and finally moves out of lower right frame to sign the paper (31 fr.).

28. LAMCU (same as XIX, 24). Man looks down while speaking (66 fr.). Note: the first frame is black.

29. MS (similar to XIX, 27). Woman receives a paper handed to her by the clerk below the frame; she reads it while the hands of her husband enter frame right (115 fr.).

30. HALS (similar to XXI, 5). Two trolleys passing each other in opposite directions (91 fr.).

31. MS (similar to XIX, 18). The man on left and the woman on right hiding her face with her left hand, while the clerk sitting in left foreground writes and talks; as the woman turns her head away from the camera, the man looks at the camera and smiles (82 fr.).

32. HAMS. In the center of the screen, a weeping woman wearing a dark scarf on her head, leans over a fence. In the background there is a tombstone revealing the family name "Brakhman" (103 fr.).

33. MS (same as XIX, 31). There is a man on the right and a woman in a white scarf with a purse over her face to hide herself while signing a document handed to her by the clerk. The man smiles and looks on as she signs (169 fr.).

34. HAMS. An old woman in white scarf weeping over a grave (131 fr.).

35. MS. The woman finishes signing certificate and leaves while the man signs it; a hand enters from frame right with a piece of paper and then retracts frame right (67 fr.). Note: the first frame is black.

XX *Death, Marriage, and Birth.* 16 shots, 1,890 fr. (45,694–47,583), 1 min. 45 sec.

1. HAMS (parallel tracking). A dead man in an open casket. In the foreground, strewn flowers are seen moving from left to right. People in the background move in the same direction (131 fr.).

2. MS (panning). A bride in white, carrying flowers in her left arm, is helped out of a carriage. A man and woman follow her (57 fr.).

3. HAMCU. The face of a woman screaming in pain; a hand enters frame right and arranges white towel on the woman's forehead. (200 fr.).

4. HAMLS (tracking). A limousine carrying a casket as people move toward the screen; the windshield of the car is seen in the foreground, the casket in the middle ground and the crowd in the background (152 fr.). Note: the first frame is black.

5. MS. The bride and groom get into a horse-drawn carriage; she is carrying

flowers and an icon; in the background peasants stand and watch; the driver sits in front (96 fr.). Note: the first frame is black.

6. HAMS (similar to XX, 3). The woman, lying in a bed, screaming in pain (94 fr.).

7. MS–LS. The bride and groom at left, sitting in the carriage; he carries an icon, she carries an icon and flowers; as the moving carriage exits frame right, there is a long shot of a road with peasants in the background (96 fr.).

8. HAMCU (same as XX, 3). A woman screaming because of intense pain (214 fr.).

9. LS. Two rows of cars on a street lined with dark trees silhouetted against clear sky. The vehicles move toward the foreground with the trees receding into the distance, and the sky in the middle (251 fr.).

10. MS. The woman's legs spread with open vagina from which a nurse receives newborn infant; she unwinds the umbilical cord from around the infant's neck (59 fr.).

11. LAMLS (split screen; tilted; superimposition). Two different buildings recede diagonally from lower right to upper left, forming a triangular strip with sky above; the two shots are joined along a diagonal line; in lower left corner the superimposed right hand of the Cameraman moves the camera toward upper right, then he looks through the lens and begins to crank (122 fr.).

12. HAMS. A nurse's hands are shown washing the infant in a tub. (119 fr.).

13. LAMLS (same as XX, 11). The Cameraman continues to crank his camera (36 fr.).

14. HAMS. Having given birth, the woman, covered with a white blanket, lies still in bed; the nurse enters from right with the infant wrapped in a labeled blanket; the mother reaches for the infant (105 fr.).

15. LAMLS (same as XX, 13). The Cameraman continues to crank his camera (54 fr.).

16. HAMCU. The mother, with the aid of the nurse, holds her infant in bed; she kisses it and laughs (104 fr.).

A close examination of these four sequences confirms that Vertov and Svilova used disruptive–associative montage to create metaphors about life, society, class distinction, and human behavior. In the "Editing Room" sequence (XVII), the idea of mechanical motion and the idea of a frantic urban tempo are expressed in a triple range of shots: a revolving glass door through which people continuously rush (Fig. 83), the ladies in a carriage arriving at their apartment (Fig. 81), a crowded square with traffic moving in various directions and pedestrians milling about (Fig. 84). The "Controlling Traffic" sequence (XVIII) begins with the operator picking up the receiver (Fig. 85) and continues with a shot of the policeman manipulating a mechanical traffic signal (Fig. 86). The juxtaposition of a telephone and a traffic signal emphasizes the common theme of communication (one aural, another visual). The symbolic link between street traffic and the city life's panopticum is visually enhanced by a split image in which two views of street traffic, each leaning toward the opposite side, are merged so that it looks as if the cars are falling apart (Fig. 92). Although organically belonging to the previous sequence, this shot reappears in the middle of the "Marriage and Divorce" sequence (XIX), functioning on a thematic level as a disruptive—associative image. Consequently, this shot can be read in two ways: as a reminder to the viewer that a different world exists outside the Marriage Bureau, and as a comment on the actual situation (divorce) shown in the previous shots. The most symbolic shot (no. 6) of the sequence is a low-angle close-up of the silhouetted traffic signal against the gray sky (Fig. 33). Graphically, this shot is abstracted to the extent that if it were to be shown isolated and severed from its diegetic context, identifying it as a traffic signal would be difficult if not impossible. As a metaphor of communication — both mechanical and human — the traffic signal is inserted (intercut) according to the disruptive—associative principle, and it assumes various connotations depending upon its position within the sequence's montage structure.

Another constructivist technique known as "baring the device" (revealing the respective creative process) is used to acquaint the viewers with the immediate creators of the film — the Cameraman and the Editor. The Cameraman is first seen (in the "Editing Room" sequence) walking along the street with the tripod over his shoulder (Fig. 82). Then, at the end of the "Controlling Traffic" sequence and in the middle of the "Marriage and Divorce" sequence, the camera alone is shown with its telescopic lens (Fig. 87), while in the following "Death, Marriage, and Birth" sequence, the Cameraman is superimposed over the compounded images of the buildings he photographs (Fig. 105). Consequently, the self-referential nature of the film manifests itself in different contexts: (1) the Cameraman performing his daily routine; (2) the omnipotent "eye" of the shooting apparatus ready to capture "life-unawares"; and (3) the magical capacity of cinema to capture "invisible" aspects of reality that are inaccessible to the naked eye.

The "Editing Room" sequence introduces several new topics: the montage process as part of filmmaking, the perceptual distinction between motion pictures and the freeze-frame, and urban activity as part of social stratification represented by the contrast between the peasant women and the bourgeois ladies. Near the end of the sequence, two additional inserts of the revolving door (Fig. 83) are associated with traffic and city commotion, creating a metaphor for the repetitiveness and limitation of life's circles. Similarly, in the "Controlling Traffic" sequence, the events on the street (traffic, pedestrians) are directly related to mechanical aspects of communication, as exemplified by the linkage of the operator answering the telephone (Fig. 85) and the policeman handling traffic (Fig. 86). In this context, the operator functions as a liaison between the accident on the street and the ambulance, while the sequence ends with a self-referential shot of the camera, armed with a telescopic lens ready to catch "life-unawares" from a distance (Fig. 87). Though invisible, the Cameraman's presence is evident as the camera swings around to capture something on the other side of the shooting axis that has obviously drawn his interest – the Marriage Bureau.

Once the physical relationship between the camera and the environment is established, the "Marriage and Divorce" sequence begins with a seemingly narrative event. A young couple stands in front of a desk while being interviewed by a Marriage Bureau clerk (Fig. 88). The next shot is a close-up of a white piece of paper bearing the inscription "Marriage Certificate" (Fig. 89), followed by the same medium shot of the couple conversing with the clerk (Fig. 88). Then comes the shot already seen at the end of the previous "Controlling Traffic" sequence (XVIII), that of the camera panning to the right and overlooking the street from a high angle (Fig. 87). This shot is inserted within the shots of the couple signing the marriage certificate (Fig. 88), thus preparing the viewer for the symbolic shot of the traffic signal (Fig. 33) already seen in close-up in the previous sequence (XVIII, 6). In this context, the traffic signal connotes the marriage approval. The bridegroom smiles and folds the certificate as, in the ensuing shot, the traffic signal returns to the "go" position, humorously implying that the road to life is open for the new couple! The comic overtone of the situation has already been anticipated by the shot of the camera equipped with a telescopic lens that abruptly turns to the street as if to catch up with the outside world (Fig. 87): as soon as the couple leaves the bureau, the camera hastily turns forward (as if "facing" the audience), and the ensuing shot reveals the second couple standing in the same spot while listening to the clerk as he reads a divorce certificate (Fig. 90). The next ten shots of the couple arguing are edited on the shot–countershot principle (Fig. 91), culminating with a notable metaphorical image (split-screen) composed of two tracking shots of the street, each half tilted to the opposite side (Fig. 92). The alternate movement of the traffic adds to the graphic symbolism of this shot by creating the impression that the screen is being rent. The visual metaphor (implying a broken marriage) is intensified on the level of mise-en-scène in the following shot depicting the depressed wife signing the paper, the husband arguing with her, and the woman receiving the divorce certificate (Fig. 91). Finally, the communicative aspect of traffic is reintroduced to create a pun on the speed and facility with which a divorce can be obtained in the Soviet Union: the marriage's instant termination is indicated by a series of shots depicting the third couple (Fig. 93) and an insert of two trolleys passing each other in opposite directions (Fig. 94). But the camera does not merely focus on the bureaucratic procedure; it is particularly engaged in recording the human condition, as in the shots of the spouse who hides from the camera while the husband appears pleased with the outcome of the case (Fig. 93). Near the end of the sequence, an unexpected disruptive–associative shot is inserted: an old woman stands weeping over a tombstone (Fig. 95). The final shot returns to the divorce procedure: as the wife signs the certificate, she covers her face with her purse (Fig. 93). Then an additional disruptive–associative insert of another old woman crying over a grave (Fig. 96) foreshadows the next sequence, which begins with a public funeral (Fig. 97).

The "Death, Marriage, and Birth" sequence (XX) opens with a shot of a public open-casket funeral (Fig. 97), followed by a shot of a bride in white (Fig. 98), a shot of a woman in labor (Fig. 99), another shot (a frontal view) of a limousine bearing the casket (Fig. 100), a shot of the bride and groom climbing into a horse-drawn carriage (Fig. 101), another shot of the woman in labor (Fig. 99), a shot of the bride and groom holding religious icons (Fig. 102), and the closing long shot of the cars in the funeral procession (Fig. 103). The sequence culminates with a shot of the woman giving birth to a child (Fig. 104), generally considered the most provocative image in the entire film. In the next compound shot, the figure of the Cameraman cranking his camera is superimposed over two tall buildings (their positions askew), presumably the hospital (Fig. 105). This shot is repetitively inserted among the shots of the woman giving birth (Fig. 104), of two hands washing the infant in a tub (Fig. 106), of the mother lying in bed and reaching for the baby (Fig. 107), and of her kissing her child (Fig. 108). The "Film-Eye" is alert at all times: while the midwife assists the woman in her delivery, the Cameraman is involved in recording "life-facts," without closing his eyes even in front of the most "shocking" occurrences. But the "Film-Eye" uses this event as an opportunity not to develop a plot, but rather to document "Life-As-It-Is."

On the structural level, these four sequences support the ideological implications initiated throughout Part One: the interrelationship between various aspects of "Life-As-It-Is" and the unique power of technology to expand human experience and perception. The "Editing Room" sequence (XVII) develops the theme of people (mostly in close-ups) and traffic (mostly in long shots) as the two most dynamic aspects of urban life, a milieu to which the Cameraman, Editor, and Filmmaker belong. In the "Controlling Traffic" sequence (XVIII), the traffic signal becomes a symbol of a more personal human relationship. Inserted frequently in the "Marriage and Divorce" sequence (XIX), the signal parallels the second couple's conflict. As the sequence reaches its thematic peak, the characteristic gestures and expressions of the divorcing couple are alternated in ten shots, while the ironic comment on the divorce procedure is rendered by the insertion of a compound shot of two streetcars passing in opposite directions. Then a third couple appears in front of the Marriage Bureau clerk, and it is at this point that a disruptive–associative shot of an old woman crying over a gravestone is inserted. To emphasize this parallelism, after the couple signs the divorce papers, another shot of the old woman crying in the graveyard is inserted to complete the "Marriage and Divorce" sequence. Although the ensuing sequence, "Death, Marriage, and Birth" (XX), begins with the shot of a corpse in an open coffin carried in a limousine, the marriage shots dominate by a ratio of seven to three; and even though the death theme prevails emotionally and visually in the tracking shots of the funeral, the sequence ends with a celebration of life through a series of shots in which the happy mother cuddles and kisses her newborn baby.

Other examples of disruptive–associative montage in *The Man with the Movie Camera* (based on a similar principle of intercutting shots seen earlier in the film) prod the viewer to search for the images' metaphoric meaning within the context of the respective sequence. For example, the shots of machines and gears are inserted in sequences completely unrelated to industrial production; in this associative context, they are meant to symbolize the movement and progress of a new society. Similarly, the shots of a train, first inserted at the beginning of the "Awakening" sequence (VII), reappear in "Travelers and Pedestrians" (XII), "Ladies in a Carriage" (XV), and "Spectators and the Screen" (LII). Distinct types of labor, shown in the "Various Kinds of Work" sequence (XXV), are reinserted in "Working Hands" (XXVII), "Traffic Controller and Automobile Horn" (XXXII), and in "Washing and Grooming" (XXXIV). "Disruptive" inserts appear among the sports events to disrupt the montage flow of the games; for example, the shot of a somersaulting athlete (Fig. 172) is inserted at the beginning of the "Motorcycle and Carousel" sequence (XLIV), between a shot of a ball in flight (Fig. 169) and the shot of cyclists on the motorcycle track (Fig. 173). The same insert appears among other shots of the racing motorcyclists; the third time, however, the somersaulting athlete is shown jumping backwards, associatively alluding to the danger of motor racing. Similarly, an insert of a female athlete doing the high jump (Fig. 159) appears among the shots of the rotating carousel. In the same vein, several blurred shots of the speeding train (Fig. 38) introduced in the "Awakening" sequence (VII), reflect the sleeping young woman's consciousness and appear within the "Vehicles and Pedestrian" sequence (XIV) to emphasize the hectic tempo of city life. Finally, in the "Spectators and the Screen" sequence (LII), the same blurred shots of the passing train (photographed from an extreme low angle) are inserted among images projected on the screen-within-the-screen, just as the shot of the young sleeping woman is inserted among the shots of the "Morning" sequence (VI) to illustrate this sequence's main theme (the dawn) and to anticipate the next sequence's theme (awakening). In general, the "disruptive" shots − whose associative meanings depend upon the diegetic context of each separate sequence − contribute to the film's rhythmic pulsation, stimulating the viewer to think about montage as *the* determinator of thematic meaning.

Vertov and Svilova build a dialectical unity of thematic opposites on the structural basis of disruptive−associative montage, as the selected "life-facts" are related to each other not through narrative continuity but through an ideological juxtaposition of presented events. Because the connection between the inserted shots and the main thematic blocks depends upon the viewer's capacity and willingness to search for and establish the ideological implication of the montage concatenation, the sequences based on disruptive−associative montage pose a substantial demand on the viewer both during and after screening *The Man with the Movie Camera*.

Social and political commentary

Social criticism in *The Man with the Movie Camera* can easily be overlooked or misread. Due to the metaphorical function of disruptive−associative montage, the full ideological meaning of the intercut images emerges only retroactively. Note, for example, the shot of a giant bottle dominating the park landscape (Fig. 27) preceding the shot of a derelict asleep on a park bench surrounded by rubbish (Fig. 28). By virtue of instant association, the shot of the giant bottle (shaped like a common wine container) implies that it contains alcohol. At the same time, the composition of the shot with the decorative bottle in the center of a park connotes abundance, leisure, or pleasure, and as such, it is in ideological conflict with the derelict's living conditions (dirty streets). Obviously, Vertov expected his audience to ask questions pertinent to the social circumstances that surround and identify these two environments.

More relevant social criticism is generated by relating shots of sweating workers carrying ore in primitive carts (Fig. 56) and shots of activity in a dirty factory yard (Fig. 57) to those of well-dressed women strolling down the street and riding the trolley (Fig. 209), alongside poorly made horse carts in the middle of a public square (Fig. 64). Different social strata − the working class and the bourgeoisie, the poor and the well-to-do − are juxtaposed, alluding to the contradictions in their coexistence within the new socialist regime. A similar ideological implication is conveyed by juxtaposing images of a rich display of various (some of them quite luxurious) consumer products in city store windows (Fig. 31) with images of poorly dressed female workers engaged in heavy physical labor (Fig. 48), or juxtaposing well-to-do ladies being made up in the beauty salon (Fig. 121) with the haggard faces of peasant women at the market and on the street (Figs. 59, 71). A less ideological, more thematic association is achieved by crosscutting the shot of women doing laundry (Fig. 122) with the shots of women having their hair shampooed and dried (Fig. 121).

At some key junctures, entire sequences are juxtaposed for social purposes. The sequence in which a group of citizens is shown drinking beer in a *crowded* pub (Fig. 181) precedes the sequence of a workers' club dedicated to "V. I. Ulyanov – Lenin" (Fig. 195), where only a *few* individuals-workers are shown reading papers, playing checkers and chess (Fig. 187). In other sequences, Vertov's social criticism is more satirical, assuming the form of a political joke. The shaving of a man's face (Fig. 123) is humorously associated with the sharpening of a razor (Fig. 124), which in turn is juxtaposed to the sharpening of an ax on a large grindstone (Fig. 125). This triple comparison, in addition to its thematic meaning, implies the shortage of high-quality razors and the general lack of consumer goods in the postrevolutionary period.

Montage juxtaposition sometimes lends itself to a direct political comment which stems from Vertov's statement that "either one serves the members of the new bourgeoisie, or one works to eliminate them."[25] This belief shines through poignantly in another juxtaposition of the bourgeois ladies enjoying the attention lavished upon them in a beauty parlor (Fig. 121) with the close-up of the ax being sharpened (Fig. 125). Moreover, in the latter shot, the worker's hand is missing the middle and index fingers, as opposed to the beautifully manicured hands of the bourgeois ladies (Fig. 128). Other shots make their sociopolitical statements through a more direct contrast, as, for example, when a traditional peasant wedding, with a horse and buggy carrying the formally dressed bride and the groom holding religious icons (Fig. 102) is juxtaposed with the bureaucratic and businesslike nature of a civil wedding hastily performed and, with even greater expedience, later invalidated in the Marriage–Divorce Bureau, a grim place devoid of any ritualistic or ceremonial atmosphere (Fig. 88). Vertov compares these two "life-facts" without explicitly conveying his personal opinion, allowing the viewers to make their own judgments according to their socioreligious inclinations.

Perhaps the most blatant social criticism is made by the shot depicting elegant bourgeois ladies arriving at their apartment building in a carriage, while a barefoot and shabbily dressed maid obediently waits out on the street to carry their heavy suitcases on her shoulders (Fig. 81). The characteristic reactions of the individuals in this shot reveal their different socioeconomic backgrounds and attitudes. The maid, earnestly involved in her work (carrying the burden), does not pay any attention to the camera, while the well-dressed ladies either flirt with the camera or disdainfully pretend to be uninterested. To emphasize the spatial aspect of the situation, Kaufman found the optimal position for his camera: he photographed the ladies sitting comfortably in the carriage from a high angle so that they appear in the foreground of the shot, while beneath them the barefoot maid moves deferentially in the background. The inquisitive boy (standing on the side) is positioned in the middle ground facing the camera, and the coachman (who places the suitcases on the maid's shoulders) fills the upper left foreground. As the ladies and the maid withdraw into the background, the driver pulls out of the left foreground, and the boy exits to the right, leaving the shot composition "open," in contrast to its being "closed" at the beginning. The integration of the camera angle, mise-en-scène, depth of field, and the spontaneous reaction of the people points to class distinction in the Soviet Union, a contrived irony revealed by the "Film-Truth" principle. Although it is not explicitly stated whether the ladies belong to the old bourgeoisie or represent the "new class" emerging from the growing socialist bureaucracy, it is obvious that Vertov disapproves of one group of people exploiting another. This is corroborated in a humorous way by the next (and last) shot of the sequence, in which the Cameraman is shown walking along the street with the tripod and camera over his shoulder (Fig. 82), unequivocally allying himself with the barefoot maid carrying the suitcase.

In the early 1920s, most of the truly revolutionary Soviet artists were profoundly disappointed with the consumer mentality of the NEP era that perpetuated a bourgeois worldview. Vertov, Kaufman, and Svilova responded to such a mentality by recording "life-facts" from this period while identifying themselves with ordinary citizens and workers who suffered because of the economic situation. Vertov compares the Cameraman cranking his camera (Fig. 134) to the shoeshiner brushing shoes (Fig. 127), and shows Kaufman reflected in the mirror on which a shoeshine sign is painted (Fig. 126), visually elaborating his claim that the *kinok*s should be proud of equating themselves with the "shoemakers of cinema," rather than the fictional film directors, whom Vertov labeled as "film-mensheviks," "directors of enchantment," "priests-directors," and "grandiose directors."* Reacting against all these, Vertov and his *kinok*s sided with ordinary working people as opposed to the old bourgeoisie and the "new class."

* These terms can be found throughout Vertov's *Articles,* as, for example, "shoemaker of cinema" [*sapozhnik kinematografii*], p. 71; "film-mensheviks" [*kinomen' sheviki*], p. 79; "director-enchanter" [*volshebnik-rezhiser*], p. 92; "director-priest" [*zhrets-rezhiser*], p. 96; "grandiose directors" [*grandioznye kinorezhisery*], p. 99.

The nature of the film image

Vertov's ideological allusions are further enhanced by cinematic techniques that call into question the ontological authenticity of the motion picture image. Particularly, the appearance of freeze-frames within the motion pictures emphasizes the perceptual distinction between still and motion photography. The perceptual "clash" created by juxtaposing freeze-frames and motion pictures projected consecutively on the screen points to the fact that the cinematic vision can be expanded by modern technology in order to provide a deeper insight into the external world. On the other hand, by pinpointing the perceptual distinctions among still photography, freeze-frames, and motion pictures, Vertov acknowledges his commitment to "cinema of fact," inviting his audience to experience "Life-As-It-Is" from the standpoint of the "Film-Truth" principle and "Film-Eye" method, which can achieve, as no other medium, "condensed time" (accelerated motion), a "negative of time" (reverse motion), "paralyzed time" (freeze-frame), "fractured time" (stop-trick), and a "close-up of time" (slow motion).*

Trick processing is also used in the film's introductory shot of the tiny Cameraman who, carrying his camera on a tripod, appears on the top of the giant camera – a metaphor which, at the very outset, generates a surreal image, even though composed of authentic "life-facts." The same effect of compounding two images is accomplished in the shot of the Cameraman looming over the city and loading his camera on a building's roof (Fig. 180). Some of the most pictorially attractive compound shots, dissolves, and multiple exposures of the city traffic and factory scenes are achieved by complex trick effects (Plates 1–4).

Stop-trick (frame-by-frame shooting) is extensively used in the "Shooting Gallery" sequence (XLVIII) to promulgate another political statement: the young woman (Fig. 189) repeatedly shoots (aims the gun at) the swastika, the symbol of nazism (Fig. 190). The political innuendo becomes clear at the moment the bullet hits the swastika on the figure's hat: the target flips over, and on its rear side there is the inscription "Death to Fascism!" (Fig. 191). Later in the same sequence, Vertov uses this technique to draw an ideological association by cross-cutting between another young woman shooting (Fig. 192) and a set of bottles arranged in a wooden box (Fig. 193). As the young woman fires, the bottles disappear one after another, creating the illusion that the bullets are hitting their "target" (Fig. 193); it becomes obvious that the young woman is "shooting" alcoholism.

Vertov does not only distinguish between the image and the written word but also delineates specific techniques of projecting images. The

---

* "Close-up of time" is, actually, the term introduced by Vsevolod Pudovkin. See also notes 27 and 28.

phenomenological difference between the (static) freeze-frame and (dynamic) motion pictures is fully demonstrated near the end of the film in several shots that combine – by means of matched frames – the motion picture screen-within-the-screen and the frozen and moving images of the theater. In the first shot, both segments move, as the audience watches the abstract rotating spool of wire on the screen-within-the-screen (Fig. 201). In the next four shots, the audience is seen "frozen," while the images on the screen-within-the-screen (planes, the Cameraman shooting from the motorcycle, a train) are moving (Figs. 216, 224). In the last two shots of the auditorium, again both segments of the image move (Fig. 226). The third combination, however – the "live" audience and the "frozen" screen-within-the-screen – never takes place, which may symbolically imply that, for the *kinoks*, the *cinematic illusion* cannot be destroyed. The "Film-Eye" never sleeps, the "Film-Eye" is always active, shaking the audience to life! This message is reiterated in the shot where a locomotive runs directly over the "petrified" audience in the movie theater (Fig. 208). To transmit the feeling of being awakened from mental and perceptive inactivity, Vertov uses the same shot of the rushing locomotive, this time over the real screen (Figs. 207, 223), perceived by the contemporary viewers (Fig. 198). Symbolically, the ensuing shots of the milling pedestrians and the traffic on the street are shown in extremely accelerated motion, as if in reaction to the lifelessness of the audience sitting in front of the screen-within-the-screen. Then the Cameraman, zealously cranking his camera in the speeding convertible, appears on the screen-within-the-screen. After several additional shots of the street crowd in accelerated motion, the shot of Svilova editing the footage marks the beginning of the montage crescendo that will end the film.

Reduced to a small rectangle within the real screen, the moving image on the screen-within-the-screen looks like the theater's "window" facing reality; although it is not always possible to identify what is projected on the screen-within-the-screen (e.g., the ambiguous shot associated with the rotating spool of wire, whirling strips of light, or spinning cones on wheels), the images on the screen-within-the-screen clearly represent the external world. In the "Spectators and the Screen" sequence (LII), two "realities" interact, producing an exciting cinematic experience through the amalgamation of images projected on the real screen with those appearing on the screen-within-the-screen. In this context, the three shots of the rotating spool of wire are extremely significant. First, we see an abstract close-up of a rotating structure (a spool of wire) on the real screen (Fig. 200). The shot is positioned between the "Camera Moves on Tripod" sequence (LI) and the "Spectators and the Screen" sequence (LII), thus marking the "border" between Part Two and the Epilogue. Then there appears the same rotating structure projected on the screen-within-the-screen in front of the seated audience (Fig. 201). Finally, the previous abstract close-up of the rotating spool reappears on the real screen and is followed by a high-angle shot of the sparse audience conversing and ignoring the screen (Fig. 202). In contrast, after the next two shots (composed like a typical constructivist collage) which depict dancers performing to piano accompaniment (Fig. 203), the audience's interest is recaptured, and its attention is redirected toward the screen (Fig. 204). Comparing these two sets of shots — one abstract and the other representational — Vertov reveals the variety of responses elicited by the cinematic image, demonstrating at the same time the distinction between (1) the film image as part of the diegetic world and (2) the film image as a graphic structure with purely abstract/kinesthetic significance/impact.

In accordance with his decision to make a film "without the aid of intertitles," Vertov uses verbal inscriptions only when they function as organic parts of the image, as, for example, in the shots of the "Beer-hall/Bierhalle" (Fig. 154); the Cameraman's reflection in the mirror with the sign "Shoemaker" (Fig. 126); the inscription on the optician's store, "Pince-nez" (Fig. 62); the "Proletarian Movie Theater" (Fig. 179); the target in the shooting gallery, "Death to Fascism" (Fig. 191); the workers' club, "V. I. Ulyanov" (Fig. 195); the mailbox (Plate 2); the marriage and divorce certificates (Fig. 91); the ambulance (Fig. 115); the film posters (Figs. 26, 52, 179); the street banner announcing the anniversary edition of Maxim Gorky's collected works (Fig. 32); and the religious store (Fig. 184). In the "Wall Newspaper" sequence (XXXVI), the written text is used as a thematic "bridge" between two different topics within the segment which shows a worker arranging hand-printed articles, drawings, and photographs on a board known as "wall newspapers" (Fig. 156), widely used in the Soviet Union at the time of the paper shortage and reduction of the official newspaper's circulation. The last shot of this segment is a close-up of an article's hand-written title that reads "About Sports," followed by the introductory shot of the sequence depicting various sports activities (Figs. 157–9).

The variety of perceptual experiences in *The Man with the Movie Camera* is summarized in the finale, which points to the various natures of the projected images, to the distinct methods of subverting the illusion of reality on the screen, to the technical aspect of cinematic creation *before*, *during*, and *after* shooting, and to the multiple levels of visual interpretation inherent in the cinematic process. Through the interaction of these four aspects of the film medium, Vertov invites his viewers to reconsider their perception of both cinema and reality or, as he put it, to "clarify their vision" in order to see the world from a fresh perspective that will then allow "the proletariat of the entire world to

determıne their position in relation to the circumstances which surround them."[26]

Cinematic illusion

Perhaps the most important feature of Vertov's method is that it scru-
pulously examines the ambiguity between the objective and subjective
aspects of human perception. This is most evident in the shot of the
Cameraman standing on the ground floor while being photographed by
another camera from inside an ascending/descending elevator. As the
elevator rises, the Cameraman slowly disappears beneath the bottom
edge of the frame (Fig. 109). This deceptive "motion" produces perceptual ambiguity due to the illusion that the Cameraman − who in reality
remains stationary − has been moving! More precisely, as a result of
the eye's natural tendency to accept every movement on the screen as a
faithful record of external reality, the viewers are under the impression
that the Cameraman and camera are "descending." Due to this optical
illusion, the shot calls the viewer's attention to the unseen cameraman
(Vertov?) located in the moving elevator. This deceptive effect unset-
tles the viewer's common perception of movement: as the abrupt plum-
meting creates a kinesthetic shock, the elevator shaft momentarily
plunges the entire screen into darkness, suspending every illusion and
revealing only the black screen (as shown on the final frame enlarge-
ment in Fig. 109).

Other shooting strategies in *The Man with the Movie Camera* subvert the conventional perception of space and time, and challenge the
locus of movement as it is experienced in external reality. The viewers
often cannot be sure of what is actually moving on the screen, and
their capacity to discern where the "real" motion occurs and at what
point their eyes are deceiving them is thwarted. Part One ends with
several lengthy shots filmed by the omniscient camera tracking along
a moving fire engine on which the Cameraman and his assistant operate their apparatus (Fig. 120). The two parallel movements create the
impression that the Cameraman and his assistant, carried by the fire
engine, appear stationary in contrast to the rapid movement of pedestrians, traffic, and the apparent movement of building façades "rushing" in opposite directions behind them. Again the attention of the
viewer is shifted to the invisible camera which the assistant acknowledges; consequently, the lengthy tracking shot is cut off by the insertion of a static close-up of the camera lens, this time without the
superimposition of a human eye or the reflection of the Cameraman on
it (Fig. 47). Such subversion of the spatiotemporal continuity and expansion of the diegetic space established by the wide tracking shot in-

tercut with the stationary close-up (the lens) has a structural function: it punctuates the screen both symbolically and graphically at the border between the two parts of the film, one dealing mostly with working activities, the other with recreation.

In the second portion of the "Motorcycle and Carousel" sequence (XLIV), another enigmatic shot (no. 42) disrupts both the action and the illusion of reality on the screen. After the shot of the Cameraman riding a motorcycle with his camera mounted on the handlebars (Fig. 175), and after the shot of the rotating carousel (Fig. 176), there appears a shot of dark and white waves photographed from a moving boat, with a metal railing occupying the lower left corner of the frame (Fig. 177). Unexpectedly the shot "freezes," as if the projector has suddenly stopped! A close examination of the frames reveals that they contain identical halves of two "frozen" frames combined into one horizontally split image. Both segments of the frame represent a more detailed view of the previous shot – the water – seen only in the lower right segment of the frame, while the rest of both segments remains transparent (Fig. 178). This freeze-frame is mechanically reproduced fifty-four times (lasting about three seconds when projected at eighteen frames per second) before the tracking shot of the water continues to move, as if the projector or the camera – after malfunctioning for a while and tearing the film stock – continues to run again. It seems legitimate to ask whether this "torn" piece of film is a mistake created during shooting or processing. The print that is preserved in the Moscow State Film Archive contains the identical shot, as do those distributed in other countries. But since the original negative of *The Man with the Movie Camera* no longer exists, there is no evidence that this camera "mischief" is deliberate or that it occurred after the final printing of the positive.* Intentional or unintentional, the impact of this insert is shocking: it "lays bare" the technological aspects of filmmaking (shooting, projection, film stock, movie screen) while subverting the illusion of reality by abruptly replacing the image of moving water with a piece of fractured film. In addition, it makes the viewer aware of the

---

* In the course of my research on *The Man with the Movie Camera* at the Museum of Modern Art (New York), I suggested that the Film Study Center compare the print kept in its collection with the original print held in the Moscow State Film Archive. As a result of this enquiry, the following letter arrived (dated March 26, 1976), addressed to Eileen Bowser (then Associate Curator) from the director of Goskfilmfond, with the following information: "The print of Dziga Vertov's *The Man with the Movie Camera* which we hold in Gosfilmfond fully corresponds to the copy you have in your archive. The number 'one' appears on the screen after the introduction just as it does in yours, i.e., without, however, any other number appearing before the following sequences. The repetition of the frozen shot (of the torn film) also takes place in our copy. Unfortunately it would be impossible now to establish whether these two details existed in the original negative, since unluckily the original negative has not been preserved."

physical existence of the projection beam thrown on the bare screen, and points to the distinction between projected slides and motion pictures.

The "overlapping effect" is used sparingly in *The Man with the Movie Camera*. It appears for the first time in the sequence showing a young woman in a white dress as she opens wooden blinds seen from outside of the building. The shots of the shutters being opened are repeated five times, each time photographed from different angles, and connected by four slow dissolves that further underscore the temporal duration of the action (Fig. 61). This extended duration is appropriate to the theme of the forthcoming sequences which depict the solitude of city streets at dawn (Fig. 32) and the awakening of the young woman (Figs. 39, 40).

Overlapping is used again in the "Basketball" (XLII) and "Soccer" (XLIII) sequences, where fragments of the players' moving bodies – especially their arms – are repeated to create kinesthetic tension. In this case, the overlapping works against the basic tempo already established by editing because it slows down the actual movement of the soccer players and creates the illusion that they are dancing (Fig. 168) or that the ball is "flying" in the air (Fig. 169). As a result, the movements of the players become "balletic" and are intended to glorify – in a kinesthetic way – the physical nature of sports. In other instances, overlapping is used within otherwise "logically" edited shots, as, for example, in the sequence where the Cameraman walks toward the workers' club ("V. I. Ulyanov"). In the first shot, the Cameraman is seen moving toward the club's entrance; the shot is cut at the point where he reaches the bottom of the building's stairs; the next shot repeats the second portion of the Cameraman's walk, but on a smaller scale (Fig. 194). One can speculate as to the meaning of this spatiotemporal inconsistency, together with its facetious tone, which could be read as the evolutionary "distance" between the overcrowded pub seen earlier and the solitary workers' club. Yet it seems its primary function is to enhance the mise-en-scène visually through the repetition of portions of the same movement (action).

There are several explanations for Vertov's tempered use of the overlapping effect. He probably considered overlapping too contrived a rhetorical device that did not suit the nature of "life caught unawares," particularly at the time when Eisenstein used it extensively in his silent films, repeating many times over the closing segments of the same action to create a metaphor for extended temporal duration of a dramatic event (e.g., the endless opening of the draw bridge in *October*, ordered by the tsarist officer to prevent the demonstrators from entering the city). Or perhaps it was Svilova who did not appreciate this type of editing, which is dependent upon the available footage, especially if combined with a dissolve (achieved, at that time, by rewinding the filmstrip in the camera).

The dissolve is most successfully employed in so-called optical metamorphoses where the subjects of the camera are transformed. For example, the magician "materializes" in front of a cluster of trees before he himself makes things disappear (Fig. 151); a group of exercising women suddenly appears on an empty concrete platform (Fig. 150); swimmers "emerge" on the water's surface (Fig. 152); the covered carousel is magically unveiled so that the wooden horses are seen on their platform (Fig. 153); and the Russian title *"Pivnaia"* turns into the German *"Bierhalle"* (Fig. 154). In another instance, the stop-trick is used to provide the illusion of physical transformation, as, for example, with the disappearance, one by one, of beer bottles (Fig. 193) shot at in a target range by a young woman (Fig. 192). The same technique achieves the effect of the camera "walking" on its tripod (Fig. 197), and the sticks moving and forming themselves into a pyramid (Fig. 155).

The illusion of slow or accelerated motion (produced by cranking the camera faster or slower than sixteen or eighteen frames per second) in *The Man with the Movie Camera* is mostly associated with sporting events or traffic movement, in order to express the elegance of the athletes' motion or to intensify the hectic tempo of urban life. This technique is essential to Vertov's "Film-Eye" method, as it makes the viewer aware of the fact that the human eye, as opposed to the mechanical "eye," cannot halt ongoing motion in reality. The same holds for freeze-frames: the moment a horse, an athlete, a soccer player, or a human face appears frozen on the screen, action is isolated from the temporal flow of its natural movement, which shifts the cinematic illusion into the realm of still photography. These stationary images (equivalent to slides) are graphically relegated to the foreground (thus giving the impression of pictorial flatness), and contrast with the moving images that surround them. At the same time, freeze-frames permit a close view of the presented events or objects, revealing those visual features that are otherwise imperceptible when shown in tiny fractions of time. By their immobility, freeze-frames add to the self-referential aspect of the film by reminding the viewer of the illusionistic nature of motion pictures and their technological context. In contrast to the freeze-frame, slow motion functions as a "close-up of time," in Pudovkin's words,[27] or, as Vertov described it in his article "The Birth of the 'Film-Eye,' " slow motion is the "microscope and telescope of time, the negative of time, [thus providing] an opportunity to see things without limits and without spatial distance."[28]

Accelerated motion is used most often in the traffic sequences, particularly near the finale, which reinforces the montage pace of the symphonic coda of the film. In the last two sequences, "Accelerated Motion" (LIV) and "Editor and the Film" (LV), Vertov and Svilova insert numerous accelerated shots seen earlier in the film so that the viewers can immediately identify them and be fully immersed in the kinesthetic experience afforded by the optical beat of a montage cadence. The ontological authenticity of the accelerated final shots is considerably undermined through montage as their spatiotemporal condensation approaches cinematic abstraction. Consequently, in the closing portion of the film, the accelerated shots attain a level of "estrangement" [*ostranenie*] by "making it difficult" [*zatrudnenie*]* to discern what is actually moving on the screen. However, it is perfectly clear that the movement reflects "Life-As-It-Is" transformed into a "filmthing" by means of the "Film-Eye" method – the cinematic rebuilding of everyday reality.

Reverse motion is employed several times in *The Man with the Movie Camera*. At the beginning, through cinematic technique, a surreal mood is produced by the shots of pigeons flying backwards onto a roof (Fig. 36). Then, even a more surreal situation is achieved when checkers and chess figures arrange themselves on two playing boards (Figs. 185, 186); and, although barely visible during screening, the reverse motion is utilized in the two shots of the runner who is seen, though only for an instant, somersaulting both forward and backwards (Fig. 172). In this last case, the technique acts as a perceptual accent within the montage juxtaposition of various track and field events ending with a speeding motorcycle (Fig. 173) and a rotating carousel (Fig. 174). Lastly, reverse motion is employed in the finale of the film, which shows people walking backwards on the street (LV, 49; LV, 57). Combined with accelerated motion, reverse motion adds to the frenetic dynamism of city life. In particular, by deviating from the phenomenological norm of movement in reality, reverse motion exemplifies Vertov's belief that the "cinematic way of seeing" penetrates beneath the actual appearance of external reality, and conveys the filmmaker's personal attitude toward "life-facts."

At this point it would seem appropriate to discuss Vertov's and Svilova's use of jump-cuts, but since *The Man with the Movie Camera* is structurally based on the subversion of narrative linearity, jump-cuts appear almost at every instant of its diegetic development, contributing to the overall dialectical deconstruction of the film's spatiotemporal unity.

The Man with the Movie Camera also subverts the continuity of space and time by the use of pixillation, the blurring of images,

---

* See footnote on p. 11.

and the stroboscopic flickering effect. Of all these devices, the flickering effect is most apparent, since it is achieved by an abrupt exchange of bright and dark impulses on the screen (produced by intercutting black frames within a shot). For example, in the "Awakening" sequence (VII), the shot of the railway tracks (with the camera running over them) is interspersed with dark glimmerings (at the rate of two to four frames each time) that disrupt the natural flow of perception while assuming a metaphor for cinematographic projection and its intrinsically stroboscopic nature (Fig. 41). Likewise, in the "Accelerated Motion" sequence (LIV), the shot of two women and a man in a white suit being driven in a carriage (Fig. 214) is cut off by twenty black segments (each segment two or three frames long) as the actual movement of the carriage is gradually accelerated, thereby intensifying the deconstruction of the action's diegetic spatiotemporal continuity. Naturally, the viewers are inclined toward the representational continuity of the photographed event (the undisturbed motion of the carriage on the street), but as the recurrent flickering effect attains a disturbing level of intensification, the exchange of light and darkness on the screen assumes a hypnotic function that parallels the whirling of city traffic – the representational content of the sequence.

Finally, there are hundreds of dark frames, appearing mostly at the beginning of shots, produced by a delayed opening of either the shutter or the iris when the cranking begins. The "optical interference" of these frames is visually exciting as it intensifies the stroboscopic pulsation of the projected image. The function of the dark frames becomes obvious when an almost equal number of semidark frames at the very end of the shots draw attention to themselves and disrupt the illusion of spatiotemporal continuity on the screen. Less frequently, the last frame of the shot is lightened to the point of transparency, thus further contributing to the overall optical pulsation of the film. From a perceptual standpoint, the entire film can be seen as a challenge to the illusion of real time and space, both of which are taken for granted in conventional films. Vertov "warns" the viewers to question ceaselessly their perception of reality in general and as it is experienced in the movie theater. In this respect, *The Man with the Movie Camera* is, indeed, a dialectical negation of its own "being" since it deconstructs the very base of its existence – the "Film-Truth" principle. Consequently, the perception of reality is questionable per se: it consists not only of what one is accustomed to seeing but also of subliminal data that modify our cognitive grasp of the visible world and intensify our emotional response. Vertov demonstrates that these invisible components of perception exist in the materialistic sense (as one-frame shots) and that they considerably affect our understanding of reality in an unconscious way, creating a new vision of and attitude toward "Life-As-It-Is," not "Life-As-It-Appears."

Because the ontological authenticity of *The Man with the Movie Camera* is preserved in individual shots, even the most deconstructed sequences are not divorced from reality. The viewers are constantly reminded of the actual existence of events within an environmental context, even though the nonlinear progression of these events forces the spectators to establish for themselves a new spatiotemporal relationship among the objects and characters. On a psychological level, this self-referential aspect of the film generates a feeling of "estrangement" (defamiliarization) by showing everyday objects and events from unusual perspectives, thereby making the viewers more consciously aware of them. This is in harmony with the futurist attitude that encourages the audience to assume a critical position toward the art work: the disclosure of various techniques of filmmaking expands the viewers' awareness of the creative process, which is as important as the product itself. From such a duality of the presented images, there emerges a dialectical collision between the illusion of reality depicted on the screen and the "laying bare [of] the device" [*obnazhenie priema*]* that makes the projected illusion possible. This perceptual collision, initiated at the film's beginning and reiterated throughout, culminates in the finale as the montage pace becomes extremely aggressive. Kinesthetically compressed, the finale turns into a metaphor of the "Film-Eye" method: Svilova's eyes (Fig. 227) and the projection beam (Fig. 228) fuse with numerous shots retrieved from earlier parts of the film, as the film closes with the emblem of the "Film-Eye" method, exemplified by the image of the camera and the Cameraman's eye superimposed over the lens. The final image – just like the opening one – relates to the cinematic process: as the iris slowly closes, the superimposed eye continues to stare at the viewer (Fig. 233), perceptually ready to "arm" itself with the camera lens and capture "life-unawares."

Point of view

In a narrative film, the viewer is usually informed about the source of the camera's point of view, as each shot represents either a third-person's (the storyteller's) standpoint or a character's (actor's) view. In a traditional documentary film, point of view almost always resides in the third person, be it the filmmaker or the cameraman. *The Man with the Movie Camera* incorporates various perspectives, as it basically shifts between the third-person (author's) point of view and the subjective (Cameraman's) point of view.

* See footnote on p. 45.

The first five shots of the film clearly imply that matching points of view, typical of traditional cinema, will not be observed. The point of view of the opening shot, that is, the tiny Cameraman carrying the tripod and camera over his shoulder as he climbs a giant camera (Fig. 1), provides the audience with the instant expectation that the next shot will represent either the giant camera's point of view or that of the Cameraman's own tiny camera. Instead, what follows is a shot photographed from an extreme low-angle showing the top floors and the decorative façade of a tall building with the gray sky above (Fig. 2). The third-person point of view in the third shot is identical to the first one: the Cameraman toting the tiny camera over his shoulder descends from the giant camera that continues to "gaze" at the spectator (Fig. 1). The fourth shot is a low-angle view of a street light with clouds swiftly drifting behind it (Fig. 3). These four shots forewarn the viewer that there will be no diegetic continuity between angles of observation, and they anticipate the unorthodox cinematic structure of the entire film. Once it becomes obvious that the point of view cannot be justified by a conventionally valid logic, the connotative link between shots assumes a metaphorical signification, as the various points of view demonstrate the filmmaker's personal attitude toward reality and comment upon the nonnarrative relationships between the photographed objects.

From time to time, the isolated narrative cores in *The Man with the Movie Camera* provide the basis for a diegetic interpretation of the point of view in the sequences they occupy. However, each interpretation is unique to its corresponding sequence and cannot be applied to the following one, even if it deals with the same event or involves characters and locations seen in previous sequences. The relationship between the Cameraman's position and the photographed object is the sole basis on which the viewer establishes his comprehension of the (cinematic) world shown on the screen. With this in mind, one can distinguish six points of view, on the basis of (1) objective, narrative continuity; (2) the "protagonist's" vision of reality; (3) the position of the cinematic apparatus; (4) the relationship of the *montageur* with the film footage; (5) the interaction between people and objects or events; and (6) images that can be attributed to neither representational denotation nor diegetic connotation:

1. The third-person (author's) point of view
2. The Cameraman's point of view
3. The diegetic camera's point of view
4. The Editor's point of view
5. The character's point of view
6. The ambiguous point of view

The film begins with the third-person (author's) point of view of the Cameraman as he shoots with his camera positioned on a giant camera, and ends with an amalgamation of the Cameraman's and camera's points of view as conceived by the "Film-Eye" method – the human eye superimposed over the camera's lens. But it is the third-person point of view that dominates throughout the Prologue and 120 shots of Part One.

In Part One, two objective shots of the Cameraman – first, a full shot of him as he kneels on the railway track and positions himself to shoot the approaching train by placing the camera between the two tracks (Fig. 37); then, a close-up of his face in profile as the train speeds by – precede the Cameraman's direct point of view of the side of the blurred train. One would expect that, rather than this objective close-up, the next shot would be a subjective view of the oncoming train. Any such narrative development, however, is subverted by this and the succeeding objective close-up of the Cameraman seen again in profile, this time facing to the right as the train rushes behind him. The subjective shot of the train running over the camera (Fig. 68) does not appear until right before the second shot of the sleeping young woman moving restlessly in her bed (Fig. 25). A variation of this subjective shot reappears in the final sequence, "Editor and the Film" (LV), both on the real screen and the screen-within-the screen (Figs. 223, 224). On a psychological level, the abrupt intercutting of these objective shots (the Cameraman as he sets up his shot, as the train passes behind him, as he gets his foot caught in the track) with subjective ones (the blurred train seen both from the side at an oblique angle and directly oncoming) relate not only to the state of mind of the Cameraman – his fear of the approaching train – but also to the hypnagogic state of mind of the sleeping young woman introduced at the beginning of the "Awakening" sequence (VII), and shown throughout (see graphic display of shots in "Oneiric Impact of Intervals," in Chapter III).

The first time the Cameraman's and the camera's points of view are shown together is in sequence VIII ("Vagrants in the Street"), in which the Cameraman's hand and the camera, with its telescopic lens, rotate ninety degrees and face forward (Fig. 45). The ensuing high-angle medium shot shows vagrants reacting (Fig. 28) to the camera seen in the previous shot. Throughout this sequence, the Cameraman himself is not exposed, although the shot–countershot of the vagrant and the cranking camera is repeated several times, as the camera's point of view is enhanced by the extreme close-up of the lens in which the camera and the Cameraman's hand are reflected (Fig. 46). Symbolically representing the camera's point of view, this shot is highly self-referential: it informs the viewer about the act of shooting, and simultaneously reveals both the technological means of cinema and the object that has been "captured" on the camera lens.

In the "Washing and Blinking" sequence (IX), a character's (subjective) point of view is introduced more emphatically than in any other shot in the film. After the shots of the young woman awakening and getting out of bed, putting on her stockings, dressing, washing, drying herself off with a towel (Figs. 39, 40), and blinking (Fig. 49), a series of shots are inserted of venetian blinds opening and closing (Fig. 50), followed by a close-up of the camera lens contracting its iris (Fig. 47). Here two points of view are interrelated: the one directly associated with the young woman, the other metaphorically linked with the camera. Then a close-up of the young woman (as she begins to blink) is repeatedly intercut among a series of medium shots of the blinds alternately opening and closing. This intercutting compares the mechanism of the human eye with that of the camera lens. The parallelism is underscored by an inserted compound shot in which the cranking camera and the lens itself are reflected in a close-up of another lens with a superimposed close-up of a human eye (Fig. 46), serving as the emblem of the "Film-Eye" method. This compound shot is preceded by a shot of a rather well-dressed woman lying on a bench in the middle of a city square (Fig. 53) and is followed by a close-up of the same woman's legs. Another compound shot of the "Film-Eye" emblem reappears after this close-up, whimsically implying the camera's point of view and humorously alluding to the Cameraman's interest in women. Similarly, in the "Awakening" sequence (VII), when the young woman gets up from her bed, the Cameraman conspicuously focuses on her legs also. On an ideological level, the juxtaposition of the shots of one woman sleeping on a comfortable bed and another on a street bench confirms Vertov's critical attitude toward the inequality in living conditions. In reaction to the camera's intrusion of the citizen's private life, the young woman lying on the park bench – antagonized by the act of shooting – makes a discontented gesture and leaves (Fig. 53). On a more metaphorical level, the Cameraman hastily acknowledges his having overstepped the bounds of "life caught unawares," as the ensuing shots revert to the depiction of street traffic.

In the "Factory and Workers" sequence (XI), the Cameraman's point of view is indicated by a very low-angle view of the yard as the workers pull their carts over the (invisible) camera (Fig. 56). This point of view is subsequently corroborated by the succeeding high-angle (third-person) point of view that discloses Kaufman cranking his camera on the ground as a worker pulls his cart over the Cameraman (Fig. 57). Of course, the camera's point of view would be recognized even without the subsequent shot that uncovers the actual position of the camera; but, by underscoring the high angle, Vertov and Svilova make a clear distinction between the two points of view (the third person and the Cameraman's). In another case, the Cameraman's point of view is implied not by an exaggerated angle but by the camera's unsteady movements through space. For example, a street shot is executed with the hand-held camera that follows two peasant women walking along the street (Fig. 59). The vibration of the image instantly suggests that the camera is carried by the Cameraman as he walks, a notion already descriptively conveyed by the previous shot of the Cameraman holding his camera as he passes over an elaborate steel bridge (Fig. 58), and especially by the shot in which the Cameraman, with the camera over his shoulder, makes his way through the crowd (Fig. 60). Consequently, all the hand-held tracking shots are associated with the Cameraman's point of view, while also demonstrating his physical exertion and his active participation in everyday life.

The most equivocal sequence concerning point of view is the one in which ladies are driven in a moving carriage while the Cameraman photographs them from a speeding convertible (Fig. 65). In one of the shots, the lady in the striped blouse looks directly into the lens and moves her hand in a circle to imitate the cranking of the camera (Fig. 69). This self-referential gesture provokes the viewer's awareness of the cinematic process, as the ensuing shots turn into freeze-frames of horses, ladies in the carriage, traffic, the crowd on the street, and children (Fig. 71), drawing a phenomenological distinction between motion pictures and still photography. Freeze-frames, by their nature, can be considered neither the Cameraman's nor the motion picture camera's point of view. It is also obvious that these are not stationary stills photographed by the motion picture camera, but rather produced in the lab. As such, they inform the viewer about yet another phase of the cinematic process – that of laboratory printing. Because of these features, freeze-frames do not belong to any of the basic points of view; they stand instead outside of the film's diegetic structure, thereby implying an ambiguous point of view.

The self-referential notion is further enhanced by the inclusion of perforations on both sides of the filmstrip handled by Svilova in her editing room. The image on the filmstrip represents a little girl attentively looking ahead (Fig. 72); this shot is succeeded by a shot of numerous reels of film arranged on shelves (Fig. 73). Another set of shots reiterates the ambiguous point of view, some showing classified film reels (Fig. 74), others depicting an empty plate on Svilova's editing table (Fig. 75). Partially third-person and partially self-referential, these shots prepare the spectator for a new point of view, that of Svilova (the Editor) working at the editing table (Fig. 80). Since she looks at the filmstrips from above, all the shots of the editing table match the Editor's (high-angle) point of view: the film plate (Fig. 75), filmstrips (Fig. 76), scissors (Fig. 77), the lighting board (Fig. 78), and Svilova's hand as she writes numbers on index cards (Fig. 79). The

Editor's point of view is conceptually reinforced near the end of the film by a close-up of Svilova's eyes intercut among numerous brief and accelerated shots of traffic, pedestrians, pendulums, and the screening in the movie theater, metaphorically reflecting her creative vision.

In the puzzling elevator shot, the movement of the elevator confuses the viewer as to whether or not the Cameraman (standing behind the camera in the hall) is actually moving (Fig. 109). The point of view in this shot is equally perplexing: first it appears to be a shot taken by a stationary camera, but in fact it is the *omniscient* camera that is moving, not the object being shot (the Cameraman standing behind his tripod and camera). This perceptually deceptive point of view (trompe l'oeil) is displaced by the next shot in which the camera tracks along the street as pedestrians get out of its (or the vehicle's) way (Fig. 110). There is no doubt that in this shot the camera is physically moving and that the people are reacting to it. In contrast, the illusory movement of the ascending/descending Cameraman/elevator creates an impression similar to the experience of a traveler watching a moving train from his or her stationary position within another train.

In the "Street and Eye" sequence (XXII), a stationary close-up of a blinking eye (Plate 6) crosscut with the blurred swish-pan of a street photographed from an oblique angle (Fig. 111) generates a "psychological" point of view associated with the "frustrated" perception of pedestrians caught within hectic traffic. The dialectical clash between the contrasting graphic forms (circular/diagonal/horizontal/vertical), for the most part appearing superimposed one over the other, is reinforced by the vibrating camera movement (often blurring the image) and a frenetic montage pace (approaching the threshold of imperceptibility). The tumultuous tempo of this sequence contributes to the disturbing impact of the "life-fact" presented in the next sequence, in which the actual "victim" of the frantic traffic is shown lying injured on the street (Fig. 118).

In Part Two, the emphasis on the Camerman's or character's point of view is not as strong as it is in Part One. Instead, the physical dynamics of the photographed events (such as recreation and various sports) determine the montage pace, while the use of different photographic devices, including extremely oblique angles, supplies the sequence with optical expressivity rather than signifying a particular point of view. Only after more than 400 shots can one discern a character's point of view in the "Track and Field Events" sequence (XXXVII). Numerous shots of sports fans (Fig. 158) are matched by point-of-view shots related to the spectators' field of vision – the sports events themselves (Fig. 159). Similarly, in the "Magician" sequence (XL), close-ups of the children's astonished faces (Fig. 165) are matched by the shot of the magician performing his tricks (Fig. 166). This sequence begins with a close-up of a boy looking down to the right, followed by a high-angle medium close-up of the magician, the latter shot clearly implying the boy's subjective point of view. In this segment, point-of-view interaction creates a geographical ambiguity: by establishing a *spatial* relationship between shots obviously photographed in *different* locations,* Vertov and Svilova create the impression that people are looking at one another (according to the "Kuleshov effect"). In the "Basketball" (XLII) and "Soccer" (XLIII) sequences, the Cameraman's point of view is reinforced by the panning and tracking after the "flying" soccer ball photographed by a hand-held camera, which constantly changes its position, angle, and speed of movement through space (Fig. 169). Here the camera's point of view and motion are identified with that of the "dancing" soccer players (Fig. 168); the hand-held camera movement contributes to the viewer's motor-sensory experience of the event.

Both objective and subjective shots in the "Motorcycle and Carousel" sequence (XLIV) are sensorially exciting, as the viewer experiences the parallel circular motions (the rotating of the carousel and the movement of the motorcycles as they arc around a track) from a variety of angles. Shots of children and adults enjoying their ride on wooden horses (Fig. 174) represent the third-person point of view, as do the shots of the Cameraman shooting while on the carousel (Fig. 176) or while riding a motorcycle himself, his camera mounted on the handlebars (Fig. 175). The blurred images of spectators around the carousel (Fig. 174) obviously represent the subjective points of view of the Cameraman and the other riders. In addition, a male athlete turning a somersault (shown previously and here both forward and in reverse) and a female doing the high jump are intercut within a series of objective shots of the racing motorcycles and rotating carousel as they turn in opposite directions, apparently comparing the speed and movements of the carousel and motorcycles to physical human feats. The sequence ends with a low-angle view of the Cameraman as he rides his motorcycle and hits the objective camera. As if in response to this aggressive action, a brief semiabstract shot of water is quickly succeeded by a shot of torn film (Fig. 178), indicating in an unusual manner that an "accident" has taken place, both on the race track and in the projection booth – a self-referential shot that can be read as the ambiguous point of view and/or the point of view of the film's spectators.

In the "Shooting Gallery" sequence (XLVIII), the young women's points of view (Figs. 189, 192) correspond to the close-ups of the targets they aim at (Figs. 190, 193). In the "Musical Performance with Spoons and Bottles" sequence (L), the characters' points of view are underscored by montage, as close-ups of people looking at and listening to the musician as he performs are alternately exchanged (Plate 9).

---

* A close examination of the shots reveal that their backgrounds do not match.

The fact that the characters' points of view are inconsistent (since the listeners look in various directions) contributes to the emotional impact of this sequence, which achieves an abstract musical rhythm. The character's point of view prevails again toward the end of the film, in the "Spectators and the Screen" sequence (LII), which shows spectators attending the screening in the movie theater. The notion of film spectatorship is already touched upon in the previous sequence, the "Camera Moves on Tripod" (LI), which makes it clear that the people laughing at the animated tripod and camera (Fig. 197) are, in fact, the same spectators attending the screening of *The Man with the Movie Camera* in the movie theater (Figs. 198, 199), apparently the same one shown in the introductory sequence of the film (Figs. 13, 14).

There are ninety shots of the "Spectators and the Screen" sequence (LII) interspersed with arrhythmic insertions of the spectators (in close-up) engaged in watching the film projected on the screen-within-the-screen. At the beginning of the sequence, the movie audience watches (on the screen-within-the-screen) two ladies and the man in a white suit in a moving carriage (Fig. 215). Then the carriage is shown on the real screen (Fig. 214) in a shot that represents the third-person point of view, reiterated by the next shot of the Cameraman on a motorcycle, appearing on the miniature screen-within-the-screen (Fig. 217), and followed by a shot of him as he shoots from the convertible now seen on the real screen (Fig. 225). At this point, the existence of a diegetic point of view (related to the real screen) and a self-referential one (related to the screen-within-the-screen) is evident. After a series of crosscuts between these two shots, the shot of the two women and the man in a white suit reappears on the real screen: this can be read both as the Cameraman's point of view and the movie audience's point of view. The actual spatial relationship as well as the geographic connection between the Cameraman and the ladies in the carriage is specified by the earlier sequence (XV) in which several shots of the Cameraman (driven in a convertible in the foreground as he shoots the ladies driven in a carriage in the background) realistically depict the situation as it occurs in external reality (Fig. 65). All these conflicting points of view inform the viewer about the two levels of perception in cinema: one associated with the film's various themes, the other related to the filmmaking process. V. V. Ivanov refers to this collision of different points of view in Vertov's film as the "reverse connection" between the movie audience and the real audience, both engaged in watching images recording "life caught unawares," later to be restructured by means of montage into a "film-thing" from which selected shots are subsequently extrapolated and shown on the screen-within-the-screen. Ivanov particularly emphasizes the relationship between the shots that are part of the film's diegetic structure and these same shots that appear on the screen-within-the-screen – a "reverse connection" that contributes to the "the perceptual tension between the cinematic illusion and the audience's critical judgement."*

The Editor's point of view within the montage structure saturates the film's last two sequences (LIV and LV). After twenty-four shots of the Cameraman and street traffic in full commotion, a close-up of Svilova at work appears on the screen; the Editor's point of view prevails because of the repeated close-up of her eyes (Fig. 227) as she looks down at the editing table. During the central portions of the final sequence (shots 50–126), Svilova's close-up (composed of only two or three frames) recurs among equally brief shots of pedestrians and traffic, underscoring the notion that all these shots *conceptually* imply her editorial vision. As the frames whisk in front of her eyes, the phi-effect (illusion of double exposure of Svilova's eyes and various views of traffic) takes over, reflecting the Editor's ideational point of view, as well as the filmmaker's creative fantasy at work and the suprematist concept of film as a complex structure.

The Epilogue reintroduces the "Film-Eye" emblem – a fusion between the human and mechanical eye (Fig. 233) – signifying Vertov's method and exemplifying the *kinoks*' relationship to the camera: "I am eye.... I am a mechanical eye.... I, machine, show you the world."[29] The film's end recapitulates the interaction between machine, traffic/communication, and the cinematographic apparatus with the following shots: a traffic signal (Fig. 229), an automobile mounted on the railway tracks and running over the camera (Fig. 230), swinging pendulums (Fig. 231), a streetcar running over the camera (Fig. 232), ten black frames, and, finally, the superimposition of the eyeball and the gradually closing iris over the lens (Fig. 233). The eyeball remains visible throughout the iris's contraction, which lasts for eighty-six frames. The film ends as it begins – with the camera – but the final point of view is attributed to the human eye, which continues to stare at the viewer, even after the iris completes its contraction.

Taken in its entirety, *The Man with the Movie Camera* contains nearly an equal number of shots implying the Cameraman's and the Editor's points of view. Added together, these approximate the number of shots representing the third-person point of view, which provides the viewer with a balanced proportion of opposing perspectives, simultaneously confirming and challenging the objectivity of human perception. The lesser number of shots relates to the character's (subjective)

* V. V. Ivanov, "The Categories and Functions of Film Language" [*Funktsii i kategorii yazyka kino*], *Trudy po znakovym sistemam*, no. 7 (1975), pp. 170–92 (unpublished translation by Roberta Reeder). Analyzing the relationship between the two different points of view at the end of Vertov's film (i.e., real audience versus the audience seen on the screen), Ivanov writes: "Vertov uses extensively what could be termed a 'reverse connection' with the auditorium. In several sequences, the auditorium appears together with the screen-within-the-screen on which there is the same image seen earlier in the film, hence indicating a 'reverse connection' " (p. 3).

point of view, as in the shot where we see the object at which a young woman aims her rifle – the fascist symbol (connoting a political message); the shots following the young woman's blinking eyes (illustrating human perception); the shots following the close-ups of sports spectators (revealing the source of their excitement); and the shots appearing on the real screen directly after the close-up shots of the audience looking at the screen-within-the-screen (establishing the viewers' points of view in the auditorium).

Point of view is an essential structural component in *The Man with the Movie Camera:* it acknowledges the extent to which Vertov's film departs from traditional filmmaking as well as from conventional documentary film. The interaction of the three self-referential points of view (the camera's, Cameraman's, and Editor's) with those of the third person and of various "characters" generates a unique distancing effect sustained throughout the film. At the same time, this causes a shift of the viewer's perception from objective to subjective, with recurrent emphasis on the cinematographic creative process. Underscoring the perceptual distinction between reality as it appears in the exterior world and as it is presented on the screen, *The Man with the Movie Camera* proposes a unification of the human eye with the "Machine-Eye," in order to create a more substantial, more dynamic, and more revealing vision of reality. As Vertov astutely claimed: "If an artificial apple and a real apple are photographed in such a way that one cannot distinguish between them, this demonstrates not the skill of shooting but rather the lack of it."[30]

Glorifying technology's capacity to aid the naked eye, examining the nature of human perception, presenting reality with a critical attitude, *The Man with the Movie Camera* does not limit itself to one particular point of view. Rather, it affords the viewer with the privilege of having looked at life from all possible perspectives through dynamic cinematic means – a vision that is the ultimate goal of the "Film-Eye" method: not only to view life differently but above all to provide a more profound vision of reality than conventional observation can allow.

A mathematics of facts

The cinematic complexity and unorthodox structure of *The Man with the Movie Camera* prevented most critics of the period (including some of the avant-garde filmmakers) from recognizing the significance of the formal execution in the film's aesthetic impact. While most Soviet critics attacked Vertov's film for being "formalistic" and "inaccessible" to the masses, American critics by and large found it confusing and superficial, as exemplified by Raymond Ganly's 1929 review, which stated that *The Man with the Movie Camera* was "a hodge-podge . . . just a titleless newsreel embellished with trick photography."* Even a sophisticated critic such as Dwight Macdonald ridiculed "Vertov's extremist school," which "believes that it is possible to make a movie out of any shots that happen to be lying around the laboratory."† This attitude still prevails among traditional American film critics and historians, including Arthur Knight who, demonstrating his inability to perceive the cinematic features and aesthetic values of Vertov's work, continues to claim that Vertov's film is "increasingly mechanical and trick-filled."[1]

A close analysis of the film's montage structure reveals that *The Man with the Movie Camera* is far from being a "trick-filled" "hodge-podge," and that the film's unique value lies in the integration of its filmic devices (particularly its shot composition and montage) with all other elements

---

* Raymond Ganly, "Man with a Camera," *The Motion Picture News* (October 26, 1929), p. 2. The film was premiered in New York, September 16, 1929, and in San Francisco at the beginning of October, i.e., only five months after the Moscow opening (April 9, 1929).
† Dwight Macdonald, "Eisenstein and Pudovkin in the Twenties," *The Miscellany,* no. 6, March 1931, p. 24. This is the second part of the original article, which is – symptomatically – omitted from the book *Macdonald on Movies* (New Jersey: Prentice Hall, 1969). At the Film Conference held at the City University of New York (July 1975), Macdonald reconfirmed his misunderstanding of Vertov's film: at the panel discussion on Vertov, he stated that *The Man with the Movie Camera* "is but a small film about a photographer who runs around a city" (the quotation is taken from my personal notes kept during the conference).

to form a self-contained cinematic whole. Vertov emphasized that his film is not a mere demonstration of cinematographic technique or the craft of editing, but an attempt to create a "film-thing" that will convey a particular meaning and generate enough kinesthetic energy to affect the viewer's consciousness. He disregarded not only those reviewers who attacked the film as a "formalist attraction," but also those critics who discussed the film's montage structure from a journalistic perspective calling it a "mesmerizing visual concert." Vertov proposed instead that *The Man with the Movie Camera* be analyzed in concrete cinematic terms without the mystification characteristic of pretentious abstract theorizing. He insisted that his film be seen as neither a merely "fascinating collection" of "life-facts," nor just a "symphonic interaction" of shots, but as an integration of both:

The film is the sum of events recorded on the film stock, not merely a summation of facts. It is a higher mathematics of facts. Visual documents are combined with the intention to preserve the unity of conceptually linked pieces which concur with the concatenation of images and coincide with visual linkages, dependent not on intertitles but, ultimately, on the overall synthesis of these linkages in order to create an indissoluble organic whole.[2]

Earlier in this essay Vertov refers to his film as "a theoretical declaration on the screen," and emphasizes his concern with the interaction between the formal aspect of montage ("visual linkages") and the thematic meaning of juxtaposed shots ("conceptually linked pieces") in order to reach an ultimate cinematic integration ("overall synthesis of shot juxtapositions"). This implies that in order to grasp the full and appropriate meaning of *The Man with the Movie Camera*, it is necessary to undertake an in-depth analysis of the film's formal structure and its relationship to the content of each shot.

Adhering to the constructivist principle that a work of art is like a "building" unified by the cinematic integration of its numerous components, Vertov regarded every aspect of his film as inherently significant because each part would (1) affect the viewer by its own force and (2) acquire its proper meaning through its interaction with other elements and their combined or total relation to the photographed event. Thus, the shot's formal (pictorial) outlook becomes the "aesthetic fact" contributing to the diegetic meaning of the film's "indissoluble organic whole," its overall montage structure, which integrates its thematic/ideological meaning and its formal/cinematic execution.

Shot composition and visual design

Since shot composition is the basic perceptual unit of cinematic structure, we shall begin our analysis by examining the pictorial execution of the shots in *The Man with the Movie Camera*. There are three major characteristics of Vertov's and Kaufman's pictorial treatment of individual shots: (1) the black and white contrast of the image, (2) the graphic pattern dominating the frame, and (3) the abstract or semi-abstract outlook of the shot.

The black and white contrast is underscored mostly in wide-angle shots depicting factory machines, moving vehicles, traffic, mine shafts, bridges, and other metal constructions. This is most visible in the shots of the steel foundry, where the Cameraman and his camera appear silhouetted against the glowing light of the furnace or in those shots that depict bridges and trains photographed against the sky (Plate 1). Sometimes the distribution of black and white within the frame is carefully balanced, but for the most part it changes throughout the duration of the shot. Usually, early on in the shot, the composition depicts representationally various objects and events with a depth of field created by natural lighting; then, gradually, a sharp black and white contrast overcomes the screen, reducing the perspective and making the shot look "flattened." This is achieved either by changing the position of the camera while it continues to shoot against the light coming from behind the object (hence producing high contrast) or by narrowing the aperture so that less light is permitted to expose the film. Without depth of field, the photographed objects assume a symbolic meaning; deprived of detailed representational features, the shots subvert their own diegetic world. This shooting procedure is exemplified in the shot of the Cameraman climbing up a metal construction and looking much like an acrobat or a tightrope walker with the camera as his "balancing pole" (Plate 3). Similar visual symbolism is achieved with other silhouetted objects, such as the traffic signal photographed in various positions against a gray sky (Fig. 33) or the lower body of the train's wagon rushing over the camera and filling the foreground as it seems to "seal" the Cameraman back into the underground hole from which he is shooting (Fig. 68). The tilted angle of these shots (indicating the Cameraman's point of view) simultaneously enhances the thematic meaning (traffic's dynamism) and functions self-referentially. Numerous other shots taken from unusual and exaggerated angles are also associated with the Cameraman's activity or with the camera and its lens. These are mostly high-contrast shots (Plate 3), which invite comparison with Rodchenko's photographs; because of their pronounced graphic symbolism, these shots were often reproduced on the posters, programs, and leaflets promoting the film.*

The pictorial composition of split-screen images, double exposures, dissolves, and compound shots is particularly significant in that it illustrates Vertov's decision to use tricks not merely as attractive optical effects or picturesque images but as cinematic devices capable of con-

---

* See also p. 11.

veying messages through the very technique by which they are produced. The manner in which the Bolshoi Theater "falls apart" on the screen reveals the idea of "opening a new road" – the narrow black pathway that leads toward deep perspective – beneath the two parts of the severed building (Fig. 213). Compared to the shot of the same building that appears at the beginning of the film (Fig. 30), with its solid imperial façade, its classic columns, and decorative pediment photographed from a frontal low angle, this split-screen image of the Bolshoi Theater reinforces the ideological meaning implied by the special effect. The famous double exposure of the spinning wheel and the female worker's face is visually and metaphorically expressive because of the graphic balance of the head's position within the spinning wheel, both in its relation to the circular (or, to be more precise, oval) shape of the wheel and to its rotating movement (in opposition to the movement of the head), as white threads spread like sun rays from the center of the woman's forehead, connoting workers' enthusiasm (Fig. 206).

Dissolves, in general, are executed by the juxtaposition of similar graphic forms. The best example of this is the photographic presentation of Vertov's "Film-Eye" method: the interrelationship of two circular forms, the eye and lens (Figs. 46, 233). In contrast to this, the idea of "Radio-Ear" is conveyed through the juxtaposition of the circular form of a loudspeaker and the various forms of the ear, mouth, accordion, and piano keyboard (Fig. 196).

Perhaps the most fascinating compound shots are those of traffic, which combine two or more images of vehicles moving in various directions (Plate 2). This is most evident in the shots of streetcars as they move perpendicularly in different spatial zones, so that the combination of the vehicles' movements and their positions within the frame produces a surreal vision. Vertov and Kaufman executed these shots with a remarkable sense for merging the medium's tendency toward the ontological authenticity of the photographed object and its unique capacity to create fantastic situations by optical means. The composition of the compound shots usually begins as a faithful presentation of a street scene; then – although hardly perceptible – the positions of vehicles change, and one suddenly realizes the improbability of the newly developed situation. To achieve such deceptive and visually complex imagery, Vertov and Kaufman made sure that the "seams" between the compounded and/or superimposed shots would remain invisible: only after meticulous examination of these shots can one discover that the various vehicles are both compounded *and* superimposed. Even without such detailed scrutiny, the common viewer subliminally experiences the optical transformation of the shot – from a naturalistic presentation of "Life-As-It-Is" to a graphic abstraction — which immediately calls attention to cinematic technique.

The majority of compound shots depicting street crowds forces the viewer's perception to fluctuate. For a moment, the viewer has the impression of watching a large group of people teeming around a spacious square; then, a streetcar appears (from the left or right side of the frame) and penetrates the crowd, cutting it horizontally into two distinct segments. At this point, the viewer realizes the shot's collage structure — a self-referential act in itself — and begins to *examine* the shot as a pictorial composition rather than simply perceive it as a visual recording of a "life-fact." At other times, a shot's compounded composition is indicated by purely photographic means: as the area between two assembled images gradually turns darker or transparent, the border connecting the two shots gradually reveals itself on the screen. For example, the compound shot of the working typists is initially perceived as a long shot of numerous typists seated in several rows of desks (Fig. 221); as the shot continues, the flatness of the image becomes apparent, clearly indicating that the shot combines several images of typists lined up horizontally (this composition is irresistibly reminiscent of modern photographic collages).

One of the many shots depicting switchboard operators particularly manifests Vertov's and Kaufman's photographic technique. Upon seeing this shot for the first time, it looks as if the two groups of operators are sitting at parallel desks; but the moment one of the operators extends her hand and "plugs" a line "into" the neck of a fellow worker sitting in front of her, the representational illusion instantly vanishes (Fig. 222). A similar effect is repeated in numerous traffic shots (almost always taken from a high angle) where people and vehicles move in various directions. The viewer is often uncertain whether a street is actually oblique/curved and traffic chaotic, or, instead, photographically distorted by the compounding of two angles (views) of the same street and/or several images in which trolleys and cars run against and over each other (Plate 2). It is these semiabstract shots particularly that confuse and disturb common audiences as well as conventional critics who still refer to the film as "trick-filled" and "visually excessive." To serious theorists, however, these same shots confirm the sophistication of their creators, inspired by constructivist photographs that represent life situations in a slightly surreal manner, yet never completely cut from their organic ties with reality.

The plates "Factories and Machines" (1), "Traffic and Means of Communication" (2), "Graphic Shot Composition" (3), and "Visual Abstraction" (4) illustrate Vertov's and Kaufman's cinematic treatment of their subjects. Many of the visually exciting shots depicting the industrial environment are backlit, so that only silhouetted workers can be seen, thus unifying the human figures with the surrounding steel constructions. Shots of traffic are often either compounded or taken from high and/or oblique angles, which reveal the photographers' ideological and emotional attitudes toward their subjects. Industrial equipment and consumer goods are usually shot in close-up, with emphasis on their (horizontal, vertical, circular, square, rectangular, and triangular) shapes. In general, visually abstracted shots are executed in two ways: through the distortion of the representational depiction of the object by the proximity of the camera to the photographed surface, and through the meaningful graphic interaction of superimposed images.

The film's overall visual design is generated by three basic graphic patterns that dominate the entire film: the circle and vertical and horizontal lines. The circular form is most notable because of its natural association with the human eye and the camera lens, whereas the vertical and horizontal forms are generally related to industrial constructions, traffic, and communication. All three graphic forms are present in the very first shot of the film: the circular lens occupies the center of the frame, the horizontal position of the large camera (box) dominates the frame, and the vertical figure of the Cameraman carrying his tripod (which changes its position and becomes diagonal) appears on top of the large camera (Fig. 1). The viewer's attention is promptly drawn to the circular form of the lens by the mirror image of the camera manufacturer's logo "Le Parvo" ("Le Parvo") reflected in the lens' rim, forcing the viewer to read the words counterclockwise. The same circular form is reiterated in the closing shot of the film, showing a round camera lens with a human eye superimposed over it (Fig. 233). Between these two paradigmatic shot compositions, all other shapes and designs develop and change throughout the film according to their *visual* association with the respective themes and ideas prevailing in a given sequence.

After the first shot, the circle reappears in different contexts and with specific graphic variations. Initially associated with the camera's lens (shown in close-up), it is subsequently related to the actual screening of the film. In the projection booth, the emphasis is on close views of the circular film cores, cans, and projector reels (Figs. 8–11), whereas in the movie theater, the circular form is discerned in the various orchestra instruments (e.g., French horn) that accompany the screening (Fig. 17). As the film progresses, the circular form is underscored in traffic (wheels, headlights), in typewriter keys, in factories (gears, metal hoops, spindles, valve controls), in recreation (the wheels of bicycles, carriages, cars, a carousel), in sports (balls, ball and chain, motorcycles), in specific detailed areas of the locomotive, in the radio loudspeaker, in the pendulum, in the magician's hoops, in various targets at the shooting gallery, in the cranking of the camera, in the stacked reels of unedited film, and in the rotating plates on Svilova's editing table (Figs. 75, 80). It is important to note that the emphasis on the circular form is achieved not only by the careful selection of objects but also by photographing them in close-up (thus extracting the circular shape) and by contrasting them with different designs.

The vertical pattern (including squares and upright rectangles) appears in most of the shots depicting architectural constructions, buildings, scaffoldings, windows, machines, posters, mailboxes, checker boards, newspapers, street carts, and cigarette packets, as well as in those shots depicting the auditorium screen, the cable car moving across the dam, and the camera body. Verticality is most notable in the smokestacks, telephone poles, columns, bottles, pedestrians, and street lights. The horizontal pattern is associated with stretches of landscapes and sky (in the long shots), bridges, railroad tracks, trolleys and trains moving horizontally, as well as the filmstrip laid out across the editing table (Fig. 78). These three graphic forms often merge at climactic points into a pictorial composition of multiple crossing patterns without any single form being dominant. This is most evident in the superimposed shots of the sequence "Musical Performance with Spoons and Bottles" (L), which merges different forms, such as hands, keyboards, spoons, dancers, plates, and feet (Plate 9), or in the shots that reinforce one particular pattern by presenting variations of the same form, as in the superimposition of the woman's round face positioned in the center of threads moving around a spinning wheel (Fig. 206). In contrast, the "Street and Eye" sequence (XXII) merges two different forms through the "phi-effect," which juxtaposes a close-up of an eye (circle) with the sharp horizontal, diagonal, and vertical movements of the camera and street traffic (Plate 3). The "phi-effect" is also used in the "Cameraman and Machines" sequence (XXXI), where the juxtaposition of the image of the Cameraman (who appears basically as a vertical shape) with the circular designs of machines and wheels creates the illusion of superimposition, although none of the shots contains two images overlaying each other (Plate 8).

The semiabstract design of shots in *The Man with the Movie Camera* commonly serves as the transition from one shot composition to another. At these junctions, the fluctuating graphic and pictorial transformations in a particular shot turn each and every aspect of the "mise-en-scène" (the arrangement of elements in front of the camera) into the "mise-en-shot" (interaction of the camera movement with the mise-en-scène), eliminating any trace of the stationary composition characteristic of a still photograph. Periodically, and then only briefly, these transitional and transformational graphic patterns become highly kinesthetic moments that, while affecting the viewer on a subliminal level, can be examined only with the aid of an analyst projector. Such is the nature of the famous "enigmatic shot," composed of rotating horizontal and curved lines that form zigzag patterns projected on the screen-within-the-screen (Fig. 201); it is impossible to figure out the representational signification of this shot even after repeated screenings. Only when examined frame by frame does it become clear that the photographed object is a rotating spool of wire (Figs. 200, 201). Vertov obviously does not want the viewers to be able to identify immediately the object on the screen; rather, he wants to challenge the viewers to "decipher" the representational denotation of the shot positioned at the "border" between the two sections of the film (the end of Part Two and the beginning of the Epilogue). Similarly, the shot composed of a myriad undulating lines (or rays) crisscrossing the screen (Fig. 205) looks like a semiabstract luminogravure with shining threads diagonally intertwining all over the screen. And only by examining the shot frame by frame does one realize that it is composed of multiple superimpositions of glowing silken strands on a spinning loom that form a starlike composition.

Graphic abstraction in *The Man with the Movie Camera* occurs most often when objects or persons are positioned very close to the lens so that their representational appearance is either blurred or lacks sufficiently recognizable features for the entire object to be identified. For example, in the "Recreation" sequence (XXXV), a gray surface fills the entire frame and forms a triangular shadow in the upper right side of the screen with a slightly tilted line toward the lower part of the screen. The shot is instantly perceived as an "abstract tableau" reminiscent of a Malevich painting, until it dissolves into a gray concrete platform with bathers exercising in front of an instructor positioned in the lower part of the frame (Fig. 150). In the same sequence, the shot of the swimming pool opens as an abstract undulating composition before the swimmers gradually appear (through a dissolve) on the water's surface and swim in a geometrically arranged pattern (Fig. 152). In other shots, the representational composition appears first, and then gradually evolves into an abstract design, as in the "Shooting Gallery" sequence (XLVIII), in which a carton of beer bottles is slowly emptied (through stop-trick) so that the white box with square compartments (photographed against a black background) becomes a semiabstract composition reminiscent of constructivist paintings often reduced to simple geometric lines which interact with the rectangular shape of the frame (Fig. 193).

This manner of pictorial abstraction should be credited to Kaufman's meticulous use of the camera and its various – especially wide

– lenses, as in the "Swimming, Diving, and Gymnastics" sequence (XXXVIII). For example, one of the shots opens with two blurred vertical forms separated by a gray strip appearing at the right side of the screen, while the rest of the frame is black; as the form moves away from the camera, it becomes clear that it is a tripod carried by the Cameraman (Fig. 164). In other instances, soft focus blurs the physical appearance of the object on the screen, as in the "Spectators and the Screen" sequence (LII), in which the side of passing trolleys is constantly blurred and decomposed so that the white lines on them appear as parallel horizontal stripes vibrating on a black background; as the trolleys withdraw from the foreground, the semiabstract composition of the shot becomes a representational image clearly depicting the passing vehicles (Fig. 210). However, most often the representationally identifiable parts of objects are intentionally excluded from the frame, as in the "Crowd on the Beach" sequence (XXXIX). The shot, composed of a white surface with dark spots on the lower left and right sides of the screen, is equally reminiscent of a Malevich abstract painting, until hands enter the frame and smear mud on what becomes recognizable as a woman's back (Fig. 161). Finally, the deconstruction of the object's representational appearance in *The Man with the Movie Camera* is achieved by the rapid operation of machines, as in the "Cameraman and Machines" sequence (XXXI), where the shots of a rotating flywheel, photographed in extreme close-up, reach the point of abstraction and produce an intense interplay between diagonal and star-shaped graphic patterns (Plate 3).

The variety of graphic patterns, visual designs, and, especially, their kinesthetic interaction ought to be credited not only to Kaufman, who photographed most of the film's footage, but also to Svilova, who edited the shots in such a way that their formal features are instantly perceptible either by the contrast or similarity between the juxtaposed images. But it was Vertov who conceived the overall formal structure of the film, with the ultimate goal of demonstrating the medium's unique visual capacity. The coexistence of different visual designs is conceptually foreshadowed in the first shot of the film (the tiny Cameraman standing on the giant camera), and is restated in the final shot of the film (a close-up of the camera's lens). Of course, many of the shots' formal elements affect the viewer on a subliminal level; these abstract shot compositions constitute an "absolutely visual" and "universally accessible" means of communication, to use Vertov's terminology, and convey messages in purely cinematic terms. As explained in Chapter I, this concept is homologous to the structure of futurist poetry. Emphasizing the shots' visual design and creating kinesthetic links between them on the principle of "intervals," Vertov "destroys both the conventional semantics of the shots (by means of unusual

frame compositions and camera angles) and the conventional syntag-
matic relationship that would advance a narrative (by means of a
striking use of montage)."[3] By considering the shot as a "montage
phrase," Vertov creates a "film poem" in which "shots-phrases rhyme
with one another,[4] attesting to the influence of Mayakovsky on Vertov's
film method.

Following both the futurist's and constructivist's concepts of "art of
fact," Vertov urged his *kinok*s to "observe people and objects in motion" and to achieve the "condensation and deconstruction of time"
through a "concentrated way of seeing."[5] In many of his theoretical essays written during the making of *The Man with the Movie Camera,*
Vertov defines his cinematic tactics as being primarily concerned with
the correlation of planes, shot scales, foreshortenings (by choice of angles and lenses), movements within the frame, light (reflections,
shades), and speed (both of the events photographed and of the camera). His preoccupation with the formal aspects of filmmaking is evident in his manifestos, which describe cinema as "an art of organizing
the various movements of objects in space,"[6] and which hail "dynamic
geometry, the race of points, lines, and volumes."[7] Perhaps the most
important claim in Vertov's declarations is his contention that "the
material and elements of the art of cinematic movement are the *intervals* (transitions from one movement to the next), and by no means the
movements themselves," that is, the optical clash occurring at the
juncture where one movement "touches" another, and thus "pushes the
movement toward a kinetic resolution."[8] Obviously, Vertov had in
mind here the *kinesthetic* impact of a film, although he used the term
"kinetic" (more appropriate for mobile sculptures). To demonstrate his
concept of kinesthetic orchestration (which he also called "organized
fantasy of movement" and "geometric extraction of movement"[9]), Vertov decided to produce a film whose structure would generate a thun-
dering impact on the viewer. The result of this experimentation with a
"dynamic geometry" of recorded and consequently rearranged "life-
facts" is *The Man with the Movie Camera.*

Vertov's "Theory of Intervals" is fundamental to the understanding
of the formal aspect of his film. Theoretically, it can be related to Eisenstein's "Montage of Conflicts," which juxtaposes the shots according
to the direction of movement within the frame, its scale, volume,
depth, graphic design, and its contrast of dark and light.[10] In his essays, Eisenstein attributes his use of the same term, "interval," to music's influence on his theory, but fails to mention that it was Vertov who
first applied it to cinema. Eisenstein's contention was that "the quantity of intervals determines the pressure of the tension,"[11] especially
the perceptual stimulation of the viewers. Depending on the sequence's thematic meaning, Eisenstein used "intervals" to *intensify*

montage structure, whereas Vertov considered them the "elements of the art of movement" that contribute to the "poetic impact of the sequence," provided that the intervals are conceived as a "geometrical extraction of movement through the exchange of images."[12] Such an attitude is purely constructivist in its recognition of the physiological impact of the cinematic image, as will be shown in the analysis of those sequences whose montage structure is based on a "geometric" interaction of shots. Vertov's constructivist attitude is reflected in his claim that "the film's goal was not to *conceal* cinematic rhetoric but to *acquaint* the viewers with the grammar of cinema's expressive means."[13] It is of no surprise then that such an attitude was proclaimed "formalistic," and that the majority of viewers proved insensitive to the latent signification of the film's formal structure.

Phi-effect and kinesthesia

Vertov's interest in the theory of perception is evident in his use of the phi-effect* to achieve a highly kinesthetic impact on the viewer. Due to the fact that our eyes retain a perceived image on the retina for one one-hundredth of a second *after* an actual perception is completed (i.e., persistence of vision), the viewer experiences an illusion of double exposure (through a coalescence of two or more images on the same screen). As a result, the impression of a previous image merges with the perception of the succeeding image, and if the sequence contains a series of different images (shots) following one another in rapid succession, the film viewer will experience the stroboscopic illusion of these images being superimposed. Yet it is essential to note that this optical deception differs substantially from the mechanical superimposition of two shots (i.e., the *simultaneous* appearance of the two images on the same screen). While the mechanical superimposition – whether produced by the camera or through a laboratory process – affects the viewer's perception on a purely optical level, the phi-effect generates a stroboscopic pulsation that has a hypnotic impact on the viewer. Instead of perceiving an actual combination of two (or more) objects projected on the screen, the viewer sensorially experiences "intervals," which pulsate between the exchanging shots. In the former case, the impact is *soothing* and optically *impressive,* while in the latter it is *irritating* and optically *aggressive.* One can understand

---

* The phi-effect is determined by the stroboscopic nature of the cinematic projection. If two objects or graphic forms are projected alternately on the same screen, the viewer will have the illusion that one form is transformed into another, and/or that they exist simultaneously (which is a different effect from superimposition). Correspondingly, if the image of the same object is projected alternately on different areas of the same screen, the viewer will have the illusion that the object is jumping back and forth.

why Vertov opted for the latter: it conformed to his concept of revolutionary cinema as a way of "reshaping" the viewers' perception and forcing them to participate in exploring the external world through the "penetration" of its internal structure.

The following diagrams indicate the collision of intervals in the "Street and Eye" sequence (XXII), which generates the phi-effect in order to enhance the viewer's sensorial experience of the traffic accident. The diagrams include the scale, duration, content (objects, events), and the direction of movement in each shot. Assisted by frame enlargements (Plate 6), the reader can figure out how, through a "battle of movements," kinesthetic energy is built up in this sequence, and to what extent the impact of intervals is "geometrically" increased by gradually reducing the shots' duration. The graphic scheme of the basic movements in the alternating shots explains the "orchestration of the movements" in this sequence:

1. CU: 5 fr.:   EYE:

2. ML: 25 fr.:   STREET:

3. CU: 6 fr.:   EYE:

4. ML: 22 fr.:   STREET:

5. CU: 6 fr.:   EYE:

6. ML: 17 fr.:   STREET:

7. CU: 5 fr.:   EYE:

8. ML: 19 fr.:   STREET:

9. CU: 4 fr.:   EYE:

10. ML: 23 fr.:   STREET:

11. CU: 4 fr.:   EYE:

12. ML: 27 fr.:   STREET:

13. CU: 6 fr.:   EYE:

14. ML: 19 fr.:   STREET:

15. CU: 5 fr.:   EYE:

16. ML: 19 fr.:   STREET:

17. CU: 7 fr.:   EYE:

18. ML: 17 fr.:   STREET:

19. CU: 4 fr.:   EYE:

20. ML: 21 fr.:   STREET:

21. CU: 5 fr.:   EYE:

22. ML: 21 fr.:   STREET:

23. CU: 6 fr.:   EYE:

24. ML: 11 fr.:   STREET:

25. CU: 5 fr.:   EYE:

26. ML: 11 fr.:   STREET:

27. CU: 6 fr.:   EYE:

28. ML: 16 fr.:   STREET:

29. CU: 6 fr.:   EYE:

30. ML: 8 fr.:   STREET:

31. CU: 6 fr.:   EYE:

32. ML: 7 fr.:   STREET:

33. CU: 5 fr.:   EYE:

34. ML: 8 fr.:   STREET:

35. CU: 5 fr.:   EYE:

36. ML: 7 fr.:   STREET:

37. CU:	4 fr.:	EYE:	
38. ML:	6 fr.:	STREET:	
39. CU:	2 fr.:	EYE:	
40. ML:	6 fr.:	STREET:	
41. CU:	3 fr.:	EYE:	
42. ML:	5 fr.:	STREET:	
43. CU:	3 fr.:	EYE:	
44. ML:	4 fr.:	STREET:	
45. CU:	2 fr.:	EYE:	
46. ML:	4 fr.:	STREET:	
47. CU:	2 fr.:	EYE:	
48. ML:	3 fr.:	STREET:	
49. CU:	2 fr.:	EYE:	
50. ML:	3 fr.:	STREET:	
51. CU:	2 fr.:	EYE:	
52. ML:	3 fr.:	STREET:	
53. CU:	2 fr.:	EYE:	
54. ML:	3 fr.:	STREET:	
55. CU:	2 fr.:	EYE:	
56. ML:	3 fr.:	STREET:	
57. CU:	2 fr.:	EYE:	
58. ML:	3 fr.:	STREET:	

59. CU:	1 fr.:	EYE:
60. ML:	2 fr.:	STREET:
61. CU:	1 fr.:	EYE:
62. ML:	2 fr.:	STREET:
63. CU:	1 fr.:	EYE:
64. ML:	2 fr.:	STREET:
65. CU:	1 fr.:	EYE:
66. ML:	2 fr.:	STREET:
67. CU:	1 fr.:	EYE:
68. ML:	2 fr.:	STREET:
69. CU:	1 fr.:	EYE:
70. ML:	2 fr.:	STREET:
71. CU:	1 fr.:	EYE:
72. ML:	2 fr.:	STREET:
73. CU:	1 fr.:	EYE:
74. ML:	2 fr.:	STREET:
75. CU:	1 fr.:	EYE:
76. ML:	2 fr.:	STREET:
77. CU:	1 fr.:	EYE:

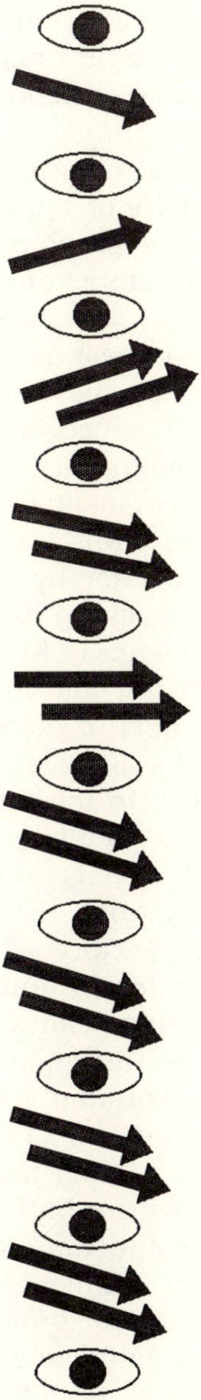

Characteristically, the first shot of the ensuing "Accident on the Street" sequence (XXIII) is entirely representational: it shows an operator talking on the phone, apparently responding to an emergency call (Fig. 113). As such, this shot has a transitional function, providing the necessary explanation for the unusual montage tension built up by the previous sequence. The kinesthetic energy released in the "Street and

Eye" sequence represents a cinematic equivalent of the subsequent street accident whose tragic aftermath is shown *consecutively* on the screen in straight newsreel manner (the arrival of an ambulance and treatment of the victim).

The phi-effect is used here to "shock" the viewers perceptually and to warn them *sensorially* of the approaching danger. In addition, this sensorial stimulation is prefigured by the three closing shots of the previous "Traffic, Elevators, and Cameraman" sequence (XXI). The first of these shots is a long, shaky traveling take in which the camera penetrates through the street as pedestrians scatter in all directions to avoid being run over by the vehicle on which the camera is placed. The spatial dynamism, sensorial impact, and the fluctuating pictorial composition of this lengthy tracking shot is reinforced by the aggressive perpendicular movement of the camera through the street (Fig. 110). At the same time, this shot's impact differs from the notion of "kinesthesia"* created later by the "Street and Eye" sequence (XXII) because the camera's tracking *through space,* in the "Traffic, Elevators, and Cameraman" sequence (XXI), preserves the spatial unity of the environment, thus affecting the viewer's sensory-motor centers. On the other hand, the phi-effect of the street and eye oscillation generates a stroboscopic pulsation (flashes) that excites the viewer's retina on a purely perceptual level. In the second shot, the street is photographed from an extremely tilted angle so that the trolleys and buses appear to move in opposite (diagonal) directions across the screen, while the blurred pedestrians obstruct the foreground (Fig. 111). The third and final shot of this introductory segment is an extremely low-angle (stationary) view of a trolley running over the camera and, presumably, the Cameraman shooting from a manhole (Fig. 112). This shot opens with the sky occupying the center of the frame and trees on either side receding into the background (twenty-five frames long); a trolley appears from the bottom center of the frame and moves aggressively forward (completely obstructing the view) until it disappears, revealing again the sky and trees (for only five frames). In addition to its sensorial impact, this shot functions as an overture to the "Traffic, Elevators, and Cameraman" sequence (XXI). In contrast, the introductory shot of the "Street and Eye" sequence (XXII) is perceptually unaggressive: it depicts an eyeball looking downward, and it appears on the screen for only an instant (five frames). Thus, at the very transition between the circular composition of the shot with the eyeball (occupying the greater portion of the frame) and the shot of the oncoming trolley, which then darkens the entire frame, an interval is created that pro-

---

* See footnote on p. 27.

duces a perceptual stimulus and acts as an overture to the succeeding "Accident on the Street" sequence (XXIII).

The "Accident on the Street" sequence, with its apparent narrative implication, has a thematic function: after a cinematic "announcement" of a disaster (in the "Street and Eye" sequence), an authentic "life-fact" is shown directly – undisturbed by montage. The thematic link between the two sequences is reiterated by the shot of the operator as she receives an emergency call that, naturally, causes the viewer to question what has happened at the end of the "Street and Eye" sequence. The answer is gradually revealed. First, the ambulance and the Cameraman's convertible, photographed from above and obviously heading toward the site of the accident, are shown speeding through crowded streets (Fig. 117). The expectation is cinematically intensified and only subsequently presented in narrative terms: the viewers are initially stimulated on a sensory-motor level and – once charged with the kinesthetic tension – they perceive the event as it occurs in external reality. The viewers' direct experience of the "life-fact" is preceded by a sensorial frustration created by the "Film-Eye" method. All key montage sequences in *The Man with the Movie Camera* can be read as cinematic "transpositions" of the previous or succeeding segments, which depict events in a direct documentary manner. They comment both ideologically and emotionally on "life caught unawares," by employing particular shooting techniques that "decode" the visible world and "extract" a deeper meaning from it.

The "Street and Eye" sequence (XXII) consists of "intervals" that collide through different shot compositions and durations (e.g., the eyeball moving in various directions and the camera panning and drifting over the street). The motion of the eyeball occurs from right to left (or vice versa) and up and down (or vice versa), whereas the camera movement mostly slants sideways and jumps diagonally. The chaotic confrontation of divergent movements is intensified by the variable and ever-increasing editing pace, so that near the end of the sequence the kinesthetic tension becomes perceptually aggressive, as the length of each shot is consecutively shortened until the image of the street (with its horizontal and diagonal lines) is seen "superimposed" over the eye, cutting across its circular shape. Probably more than in any other case, the formal structure of this sequence is the actual source of its conceptual meaning which transcends a mere illustrative depiction of the hectic traffic. At its perceptual and kinesthetic culmination, the sequence reaches a point of cinematic abstraction composed of the two basic graphic patterns: a circle and a line, the former representing the act of seeing, the latter the act of physical movement (traffic).

At first, the image of the eyeball is kept on the screen long enough (between five and six frames) to be perceived as a real human eye engaged in watching something (the street traffic). Gradually, the shots of the eye become shorter, until near the end of the sequence when the shot duration does not exceed more than one or two frames, thereby achieving the phi-effect. Although the eye is wide open throughout the sequence, in the final close-up the eye is almost totally closed, so that the eyeball, clearly visible in all preceding shots, can no longer be seen. Metonymically and symbolically, the sequence ends with a "closed eye," implying "nonseeing," a rejection of the foreshadowed disaster (accident), as the complete impression of the "superimposition" of the two forms takes place in the final forty shots (out of a total of seventy-seven shots). This kinesthetic buildup is unexpectedly halted by a stationary shot of a nurse answering the telephone; consequently, the following shots depict an ambulance rushing through the streets to rescue a young man hit by a car, the Cameraman (with his camera) participating also in that "rescue chase."

On an interpretive level, the development of the "Street and Eye" sequence (with several surrounding shots) can be described in the following way:

1. Urban streets filled with traffic.
2. Hectic and confused vehicular movement.
3. The camera penetrates the street.
4. A trolley runs over the camera (and spectators).
5. A human eye gazes restlessly (at street activity).
6. Traffic becomes frantic and accelerated.
7. Various movements (camera's, traffic's and eye's) collide.
8. Traffic (horizontal lines) "severs" the human eye.
9. A nurse answers the emergency call.
10. An ambulance races to the scene of the accident.
11. The Cameraman follows the ambulance.
12. The injured young man lies unconscious on the pavement.

These twelve events can be seen as part of a latent "story" expressed through visual symbolism and brought to a climax by means of a specific cinematic device (the phi-effect). Vertov's idea of developing "an ultimate international language" is thus conveyed via pure cinematic means generating an impact that no other medium can achieve.

There can be other ways of interpreting the formal structure of the "Street and Eye" sequence, but the attempt to associate the shot of the "severed" eye with the "slicing" of the eye in Buñuel's and Dali's *Un Chien Andalou* (1928) seems farfetched. The purpose of Buñuel's razor *physically* slashing the eyeball is to produce an "eccentric shock" (in a common dadaist/surrealist manner), while it may also be graphically associated with the preceding shot of a thin cloud passing across a full moon in a dark sky. Contrary to this, the viewer of Vertov's film gets the impression that the street traffic *deconstructs* the eyeball, which has an ideological implication both within the context of the ensuing sequence (the actual accident on the street) and in relation to other sequences which depict hectic traffic. It is important, therefore, to note that the "severance" of the eye in Vertov's film is *built into* the sequence's kinesthetic montage structure, and not presented merely as a flash of only one representational shot (as in Buñuel's film). On a structural level, this sequence yields multifaceted meaning: not only does it foreshadow the actual accident but its climactic portion also contributes to the sensorial experience of an urban environment with all its inherent contradictions. But above all, this segment is "a complex experiment" that sharply contraposes "Life-As-It-Is" seen "from the point of view of the human eye armed with the camera's eye" ("Film-Eye") and "Life-As-It-Is" seen "from the point of view of the unaided human eye."[14]

Vertov obviously desired maximum control over the viewer's conscious and unconscious perception of the projected images. In order to achieve this goal, he chose two basic graphic patterns – the circle and the line – associated with two objects (the eye and the street). It is probable that Vertov was acquainted with the phenomenon of the fovea, "a sphere smaller than a pinhead near the center of the eye's retina which appears to be the major source of consciously induced visual information."[15] The fovea permits only a small portion of visual data to be filtered into awareness, whereas the greatest amount of the picture's detail – especially areas on the periphery of vision – is routed to the unconscious. Most of the meaning generated by the brain, however, is based on *both* conscious and unconscious data, although modern psychology claims that "greater significance and meaning are derived, apparently, from the unconscious, i.e., from the enormous quantity of subliminal information."[16] This is especially pertinent for artistic perception. The manner in which Vertov edits this sequence indicates that he was particularly interested in the subliminal impact of shots whose duration approaches the awareness threshold.

Vertov's "Theory of Intervals" is based on the subliminal effect of the instantaneous conflict between two opposing movements. In the "Street and Eye" sequence, the same concept is applied to the conflict between different graphic shot compositions and light distributions. The brief close-ups of the eye show the eyeball's extremely subtle motions (micromovements) in a series. These conflicts cannot be perceived directly because they are registered on a subliminal level and, as such, stimulate sensory-motor activity in the viewer. The variety of movements and their repetitive patterns generate a kinesthetic power that has a great psychological impact with emotional overtones. In his work entitled *Subliminal Perception,* N. F. Dixon acknowledges that "there are enough scientific studies which have successfully employed psychological concomitance of emotion as indicators of subliminal effect."[17] Careful analysis of the key sequences in *The Man with the Movie Camera* supports the claim that Vertov was aware of the possibility of producing such an effect through subliminal montage: with it, he could arouse fear in the audience *before* the street accident actually occurs in the film or compel the viewers to empathize with the Cameraman when he "floats" in the "Cameraman and Machines" sequence (analyzed later in this chapter).

Intervals of movement

Although Vertov saw a great cinematic potential in the merger of "the eye plus the camera lens," he allowed the eye to be "severed" by the camera's tracking through space. As previously elaborated, the illusion of that severance is created by a "battle" of intervals from which the kinesthetic power of the "Film-Eye" method stems. In other instances, the interaction of intervals functions on either an ideological level, as in the "Working Hands" sequence (XXVII), or a poetical one, as in the "Cameraman and Machines" sequence (XXXI).

Both sequences are photographed with an emphasis on motion within the frame and are edited with maximum attention to the linkage of opposing movements (intervals). Yet, at the same time, the representational aspect of the photographed subject is preserved, while the intervals accumulate great kinesthetic energy that has a metaphorical implication: the "Working Hands" sequence symbolizes human work in general, and the "Cameraman and Machines" sequence presents the filmmaker as an integral part of the workers' society. In spite of the optical subversion of reality (through rapid editing, lens distortion, accelerated or decelerated motion, reverse screening, overlapping, split-screen, multiple exposure, camera movement, flickering, pixillation, and other filmic devices), both sequences retain a phenomenological semblance of photographed objects and events; their cinematic abstraction, however, begins at the point where the intervals approach the awareness threshold, creating an oneiric vision that makes an impossible situation probable.

The "Working Hands" sequence (XXVII) is positioned in the film's structural center, between two other sequences that depict various working activities: it is preceded by "Manufacturing Process" (XXVI) and succeeded by "Mine Workers" (XXVIII). The former is introduced by the Editor, shown arranging shots on the editing table; also in this sequence, other working activities are added to illustrate different aspects of human labor, represented by a steel worker, engineer, miner, shoeshiner, hairdresser, magician, entertainer, dancer, musician, driver, trolley conductor, waiter, athlete, traffic controller, barber, operator, and film editor. Thus the "Manufacturing Process" sequence (XXVI) functions as a thematic introduction to the "grand metaphor" built up cinematically in the "Working Hands" sequence (XXVII).

The "Working Hands" sequence (XXVII) is composed of seventy-three shots that accomplish maximum optical condensation during twenty-three shots (shots 47–69), focusing on hands engaged in various jobs. The sequence begins with a series of rhythmically arranged shots (mostly two or three frames long) depicting a woman's hands packing cigarettes in a factory (Fig. 132). These twenty-eight brief shots are followed by a stationary close-up of a typist's hands flying across the keyboard (Fig. 133) and a close-up of the Cameraman's hand cranking the camera (Fig. 134). Once the gestural similarity among these three activities is established, the next segment introduces the following shots depicting various manual activities: applying mascara (Fig. 135), calculating on an abacus (Fig. 136), putting on lipstick (Fig. 137), and operating a cash register (Fig. 138). Unexpectedly, there appears a close-up of a pistol being loaded (Fig. 139), followed by a close-up of a hand plugging in an electric cord (Fig. 140), a close-up of a hand pressing the buzzer next to a wall telephone (Fig. 141), a close-up of an operator's hands hanging up the receiver (Fig. 142), the operators' hands at work on a switchboard (Fig. 143), and the playing of the piano (Fig. 144). All these brief shots continue the sequence's thematic preparation for its climactic segment composed of twenty-six shots (47–73), which are juxtaposed on the principle of intervals, as described in the following shot-by-shot breakdown:

1. HACU. Hand enters from right and picks up the receiver from an ornate telephone (16 fr.).
2. CU. Cameraman's left hand cranks the camera clockwise (9 fr.).
3. HACU. Pair of hands playing piano on a diagonally positioned keyboard (13 fr.).
4. HACU. Pair of hands in lower frame typing on a Cyrillic typewriter in upper frame (9 fr.).
5. HAM (same as XXVII, 49). Hands play on piano keyboard (13 fr.).
6. HACU (same as XXVII, 50). Hands continue to type (10 fr.).
7. HAM (same as XXVII, 51). Hands continue to play on keyboard (11 fr.).
8. HACU. Razor being sharpened on leather strap positioned diagonally lower right to upper left (9 fr.).
9. CU. Gloved hands reach up vertically in the center of the frame for a cord and pull it down against a dark background (10 fr.).
10. HACU. Ax being sharpened, held by two workers' hands, index finger missing on right hand; the ax is sharpened horizontally on a grinding wheel (9 fr.).
11. MS. Lateral view of a group of silhouetted workers standing face right and pulling rods to the left; right side of frame in darkness (8 fr.).

12. LAMCU. Lateral view of woman's right arm operating the wheel on the right side of a sewing machine; the arm moves diagonally from upper right to lower left against a dark background (10 fr.).

13. CU (same as XXVII, 55). Gloved hand continues to pull cord down vertically against a dark background (7 fr.).

14. CU (same as XXVII, 35). Two female hands packing cigarettes seen horizontally from middle left to middle right against a white background (9 fr.).

15. MS (similar to XXVII, 57). Silhouetted workers push rods to the right against a dark background (10 fr.).

16. LAMCU (same as XXVII, 58). Woman's right arm seen diagonally against a dark background and turning the wheel of a sewing machine (6 fr.).

17. MS (similar to XXVII, 61). Workers push rods horizontally to the right against a dark background (11 fr.).

18. LAMCU (same as XXVII, 62). Woman's right arm moves diagonally, first to upper right, then to lower left against a dark background (6 fr.).

19. CU. Two blurred hands moving a flat chisel back and forth against a white background (11 fr.).

20. LAMCU (same as XXVII, 64). Woman's right arm moves diagonally from upper right to lower left against a dark background (8 fr.).

21. CU (same as XXVII, 65). Two blurred hands move chisel horizontally back and forth against a white background (11 fr.).

22. CU. Machine bar moves downward in center foreground; hand-held oil can stands behind it against a dark background (11 fr.).

23. CU (same as XXVII, 67). Two blurred hands move the chisel back and forth against a white background (10 fr.).

24. CU (similar to XXVII, 68). A closer view of metal bar moving up and down on left edge of the frame, with a worker's torso in the middle ground, and the right half of the frame dark (42 fr.).

25. HAMS (same as XVII, 2). Seven parallel shelves with rolls of labeled film; three signs (plant, machines, market) in Cyrillic indicate the rows are classified thematically; Svilova enters frame right and puts labeled reels on the upper shelves (85 fr.).

26. CU. Svilova's right hand places a reel of film on an upper shelf in the center of the frame; then her hand exits lower left; her hand reappears from below, placing another labeled reel on the left side of the shelf (49 fr.).

27. HAMS (same as XXVII, 71). After placing reels on upper shelves, Svilova exits to the right, leaving the horizontal shelves with classified and labeled reels of film (31 fr.).

Reduced to their basic graphic schemes and subordinated to the dominant montage beat, the first twenty-two shots are compressed through the reduction of their temporal duration, thus creating a symphony of optical movements (Plate 7). This kinesthetic orchestration illustrates what Vertov terms "a superior way of organizing documentary material, a method by which shots organically interact, enriching each other and unifying their forces to form a collective body." Vertov particularly emphasized that this method "is by no means formalistic but something completely different, an inevitable development that must not be avoided in the process of montage experimentation."[18]

The actual temporal duration (i.e., metric length) of these twenty-three shots is meant not only to produce the phi-effect but also to fuse optically different images of working hands by interrelating their gestures. The first four shots and the last four shots are of longer duration and thus function as frameworks for the symphonic interaction of intervals. In this interaction, two visual components – the pictorial composition of the shot and the hands' movements – collide with each other, thereby producing a sensorial as well as emotional impact that heightens the poetic meaning of the entire sequence.

Although there is great thematic variety in the chosen professional activities – ranging from heavy physical work in the steel foundry, through skilled labor (such as sewing, carpentry, and packing), to the kind of work involved in typing and playing the piano – it is the pictorial composition, graphic form, and direction of movement within each shot that make the sequence cinematically exciting. The shapes of the photographed objects are continuously underscored by their movement (e.g., the diagonal motion of a razor being sharpened or the horizontal glide of wood planing), often contrasted by an opposite movement (e.g., the circular gesture of the hand in contrast to the horizontal position of the sewing machine or the perpendicular position of the pianist's hands against the keyboard displayed diagonally). By building the sequence on the intervals of mechanical and gestural movements that complement or contradict each other, Vertov and Svilova achieve a fascinating choreography of physical activities, juxtaposing various forms and shapes characteristic of the nature of each respective activity. As a result, the kinesthetic orchestration produces a perceptual impact simultaneously disturbing and pleasing to the viewer, while never obscuring the thematic meaning of the sequence – the human hand at work.

The structural framework of the sequence enhances its ideological implication: the first shot is a close-up of the switchboard operator's hand hanging up the receiver (Fig. 142); the last two shots are a close-up of Svilova as she places a reel of film on a shelf (Fig. 73), followed by another medium shot of her arranging reels on the shelves in preparation for final editing (Fig. 74). These two functions of the hand (the operator's and the Editor's) are associated with two aspects of communication: one conveys messages verbally (by answering the telephone), another visually (by relating shots as image-signs). Activities not related directly to the hand are chosen rather arbitrarily, yet always with a distinct emphasis on the ritualistic nature of the work, as well as the movements characteristic of an industrial environment. Vertov's principle of choosing and juxtaposing "life-facts" is radically different from the one Walter Ruttmann demonstrated in his film *Melody of the World* (1929), in which he provides a topical survey of similar working activities occurring all over the world and at the same time of day, linking the shots either through thematic associations or through contrast, but with only marginal – if any at all – emphasis on the formal interaction between the movements (intervals) and graphic similarities/distinctions within the connected shots.*

In the climactic segment of the "Working Hands" sequence, the prevailing shot movement varies from horizontal (typing, sharpening an ax, planing a board with a chisel) to vertical (pulling a wire, lifting steel) to diagonal (playing a piano, sharpening a razor) to circular (sewing at a machine, cranking a camera). All these movements are interrelated on the graphic level as well, so that their juxtaposition (as demonstrated in Plate 7) produces a montage dynamism governed not only by the intervals' beats but also by the optical transformation of moving shapes. Pictorially, each shot incorporates both a graphic design intrinsic to the nature of the work activity shown and elements of a graphic design prefiguring the shape of the ensuing shot. For example, the cranking of the camera is basically circular except for the element of vertical design established by the square shape of the camera itself. And the operator's hand reaching for the telephone's base begins as a diagonal movement which terminates vertically, but also contains elements of horizontal design due to the position and shape of the receiver.

The beginning of this sequence is conceived in a semiabstract manner, as the viewer is aware only of the movement per se, with very little if any revelation of its actual purpose. As the sequence evolves, the precise function of the movements is emphasized rhythmically by the intervals' beats. Yet in spite of this substantiation of movements, the kinesthetic abstraction of the sequence is not weakened; rather, it becomes even more powerful as the graphic variations of different movements are persistently tied to the notion of the specific work the hand performs. Consequently, the sequence assumes a symphonic structure, developing into a "cinematic ode" to the *hand that produces*, as illustrated by the following diagrams:

---

* Although Ruttmann produced his two major films, *Berlin, Symphony of a Great City* [*Berlin, die Symphonie einer Grossstadt,* 1927] and *Melody of the World* [*Die Melodie der Welt,* 1929], before *The Man with the Movie Camera,* Vertov had not seen these two films before he completed his. Actually, Ruttmann was influenced by Vertov's early work, particularly the *Film-Eye* and *Film-Truth* series. In his *Weekend* [*Wochende,* 1929], a "sound film without images," Ruttmann follows directly Vertov's concept of "Radio-Ear" and his principle of radiophonic montage.

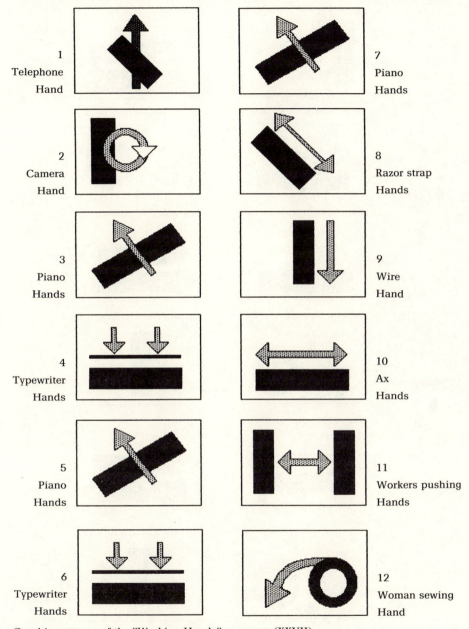

1
Telephone
Hand

2
Camera
Hand

3
Piano
Hands

4
Typewriter
Hands

5
Piano
Hands

6
Typewriter
Hands

7
Piano
Hands

8
Razor strap
Hands

9
Wire
Hand

10
Ax
Hands

11
Workers pushing
Hands

12
Woman sewing
Hand

Graphic patterns of the "Working Hands" sequence (XXVII)

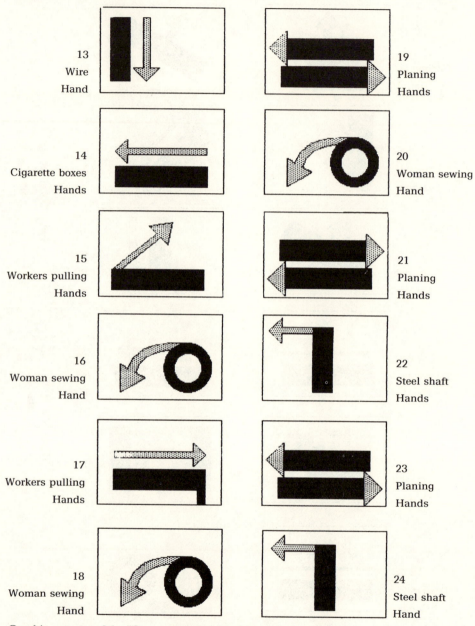

13
Wire
Hand

19
Planing
Hands

14
Cigarette boxes
Hands

20
Woman sewing
Hand

15
Workers pulling
Hands

21
Planing
Hands

16
Woman sewing
Hand

22
Steel shaft
Hands

17
Workers pulling
Hands

23
Planing
Hands

18
Woman sewing
Hand

24
Steel shaft
Hand

Graphic patterns of the "Cameraman and Machines" sequence (XXXI)

The position of this sequence within the film's structure justifies its semiabstract outlook: it is the second sequence in Part Two, following the "Manufacturing Process" sequence (XXVI) and preceding the "Mine Workers" sequence (XXVIII), two relatively descriptive segments of the film. Symbolically, the latter sequence concludes with Svilova as a worker-editor, with the focus on her right hand placing a small reel on a shelf (Fig. 73). Within such a structural framework, the "Working Hands" sequence functions as one of the major poetic, emotional, and ideological points of the entire film. In the broader context of international avant-garde experimentation, it is necessary to note a surprising correspondence between Vertov and the American filmmaker Maya Deren. In her theoretical pronouncements about cinema, Deren compared the function of "poetic illuminations" in film to that of dramatic monologue:

Every once in a while, the filmmaker arrives at a point of action where he wants to illuminate the meaning of *this* moment of drama, and at that moment he builds a pyramid or investigates it "vertically," if you will, so that you have a "horizontal" development with periodic "vertical" investigations, which are the poems, which are monologues.[19]

Deren's concept of "vertical" investigation/penetration seems relevant to the intervals of movement in Vertov's film. Like Maya Deren, Vertov and Svilova "vertically" investigate the phenomenological substance of a specific event and "poetically" intensify the ideological implications of its kinesthetic dynamism. The "Working Hands" sequence is one such "vertical" investigation, based on Vertov's "Theory of Intervals" and fully imbued with the ideological meanings that the film's middle section conveys, with neither explanatory intertitles nor explicit political allusion.

Subliminal propulsion

The "Working Hands" sequence (XXVII) is succeeded by three relatively brief sequences (composed of 9, 16, and 23 shots) that depict miners (XXVIII), steel workers (XXIX), and a power plant (XXX). In each of them, the Cameraman shoots as he runs tirelessly through mine shafts and factories. At the power plant he films the dam from a cable car as he hangs over the sloping water (Fig. 148). Then, an unexpected and "disruptive" split-screen image of traffic and pedestrians (Plate 2) introduces the "Cameraman and Machines" sequence (XXXI). The "Cameraman and Machines" sequence contains 152 shots and lasts less than fifty seconds, achieving another kinesthetic condensation of repeated visual data that turns into a cinematic metaphor about the Cameraman and the productive labor process of which he is a part. Numerous shots of machines are compiled from various factories (just as the traffic shots are photographed in different cities within the Soviet Union), so that their fusion (on the screen) assumes a broader environmental connotation related to the industrial environment essential for the building of a socialist society.

The interaction among shots in the "Cameraman and Machines" sequence is based on movement that occurs at two levels: the representational (the Cameraman's body moving in a semicircle) and the more abstract (visually distorted wheels and gears rotating in various directions). The juxtaposed shots include the three basic directions of motion – circular, vertical, and, most frequently, horizontal – supported by the complementary movement (of the tripod) progressing along the diagonal of the frame. Some shots of the machines are composed with the clear intention to underscore their linear design (bars and shafts), others their circular construction (wheels and gears). In the climactic portion, eighty-nine shots with linear composition are progressively "bombarded" by sixty-three medium shots of the moving Cameraman and the tripod placed diagonally over his shoulder. Here Vertov and Svilova create what may be termed "subliminal propulsion" of movements and shapes by intercutting discordant mechanical locomotion with the Cameraman's smooth and ethereal movement. The machines and wheels turn from left to right (clockwise), whereas the Cameraman and his tripod revolve incrementally in a diagonal from the upper right toward the upper left corner of the frame, forming a counterclockwise arc and creating a resplendent counterpoint of spiral movements that suffuse the entire frame with a kinesthetic syncopation.

The sequence begins with ten close-ups of rotating wheels, gears, and the cranking camera, establishing a pattern of repeated circular motion. Then a large factory chute (with black smoke billowing downward) suddenly breaks the circular pattern, followed by another vertical composition of two tall smokestacks. These inserts reinforce the vertical pattern that is subsequently superseded by the diagonal pattern of the shots with the Cameraman: seventy-six one-, two-, or three-frame shots of smokestacks, gears, wheels, and valves intercut with sixty-one shots (of the same duration) of the Cameraman. The differing pictorial composition of these two sets of shots is manifested not only through their graphic design but even more so by means of lighting, as can be seen in the frame enlargements (Plate 8). The shots are predominantly dark with bright contrasting accents created by the Cameraman's hand and tripod or bright reflections produced by the metallic surface of pistons and gears.

The entire sequence consists of 152 shots (lasting only 49 seconds), with the climactic segment containing only 55 shots that create an intense kinesthetic pulsation due to the frequency of cuts. The first 10 shots are relatively lengthy (ranging from 15 or 17 frames, to 52 and up to 108 frames), representing various parts of machines, with an emphasis on the triangular rods moving up and down (nos. 4, 6, 8, and 10). Inserted among these images are three shots of a lateral view of the camera, with the Cameraman's hand cranking it. The next three shots are: (no. 11) the vertical composition of the factory chutes (15 fr.); (no. 12) the triangular composition of the rods (2 fr.); and (no. 13) the lateral view of the Cameraman with tripod and camera over his shoulder (1 fr.). Then a strongly vertical composition reintroduces the dominant graphic pattern against which the diagonal composition of the Cameraman is juxtaposed. The climactic section of this sequence can be divided into three parts, based on the physical position of the Cameraman and the graphic design of machines within the frame. The first part contains twenty shots (14–33), as can be seen in the following shot-by-shot breakdown:

1. LS. Two tall, vertical smokestacks silhouetted against white smoke and gray clouds; other silhouetted factory constructions appear in the lower foreground; the upper and right side of the frame is dark (1 fr.).
2. LAMS. Diagonal (upper right to lower left) view of the Cameraman, against a dark background, with tripod and camera over his left shoulder (1 fr.).
3. LS. Vertical smokestacks (2 fr.).
4. LAMS. Diagonal position of Cameraman (1 fr.).
5. LS. Vertical smokestacks (2 fr.).
6. LAMS. Diagonal position of Cameraman (1 fr.).
7. LS. Vertical smokestacks (2 fr.).
8. LAMS. Diagonal position of Cameraman (1 fr.).
9. LS. Vertical smokestacks (2 fr.).
10. LAMS. Diagonal position of Cameraman (1 fr.).
11. LS. Vertical smokestacks (2 fr.).
12. LAMS. Diagonal position of Cameraman (1 fr.).
13. LS. Vertical smokestacks (2 fr.).
14. LAMS. Diagonal position of Cameraman (1 fr.).
15. LS. Vertical smokestacks (2 fr.).
16. LAMS. Diagonal position of Cameraman (1 fr.).
17. LS. Vertical smokestacks (2 fr.).
18. LAMS. Diagonal position of Cameraman (1 fr.).
19. LS. Vertical smokestacks (2 fr.).
20. LAMS. Diagonal position of Cameraman (1 fr.).

It is important to note that Vertov and Svilova chose two compositionally distinct shots, one with a stable vertical structure (smokestacks) and another slowly moving, diagonal composition (the Cameraman) to mark the first climax of the kinesthetic progression. The second part of the sequence's climax contains twenty-two shots. It begins with a medium shot of five steam valves moving against a black background; this shot is then juxtaposed to the shot of the Cameraman (shots 34–55), as specified in the following breakdown:

1. MS. Five vertical steam valves on black background, two in left foreground, three in right middle ground, moving rapidly against the dark background (2 fr.).
2. LAMS. Diagonal position of Cameraman (1 fr.).
3. MS. Vertical steam valves (2 fr.).
4. LAMS. Diagonal position of Cameraman (1 fr.).
5. MS. Vertical steam valves (2 fr.).
6. LAMS. Diagonal position of Cameraman (1 fr.).
7. MS. Vertical steam valves (2 fr.).
8. LAMS. Diagonal position of Cameraman (1 fr.).
9. MS. Vertical steam valves (2 fr.).
10. LAMS. Diagonal position of Cameraman (1 fr.).
11. MS. Vertical steam valves (2 fr.).
12. LAMS. Diagonal position of Cameraman (1 fr.).
13. MS. Vertical steam valves (2 fr.).
14. LAMS. Diagonal position of Cameraman (1 fr.).
15. MS. Vertical steam valves (2 fr.).
16. LAMS. Diagonal position of Cameraman (1 fr.).
17. MS. Vertical steam valves (2 fr.).
18. LAMS. Diagonal position of Cameraman (1 fr.).
19. MS. Vertical steam valves (2 fr.).
20. LAMS. Diagonal position of Cameraman (1 fr.).
21. MS. Vertical steam valves (2 fr.).
22. LAMS. Diagonal position of Cameraman (2 fr.).

In the next part, the image of the Cameraman is overpowered quantitatively by the image of technology – namely, steam valves – in the ratio of 2:1 (the image of technology occupies twice as many frames as the image of the Cameraman). A certain compositional balance is achieved on a purely photographic level: the shots of valves are photographed through diffusion, whereas the shots of the two prominently positioned vertical smokestacks (seen in the first part of the segment) are composed sharply against a dark background with pulsating bright spots.

The third part of the segment first alternates a three-frame shot of machine rods and gears moving up and down with a two-frame shot of the upright Cameraman. The world of machines is represented through various details: after ten close-ups of rods and a gear, other elements of machinery gradually take over. Two of them are particularly notable: horizontally rotating spindles and a blurred spinning wheel. They are intercut with shots of the Cameraman at a rate of three frames (of spindles and wheel) to two frames (of the Cameraman). After sixty-five shots, the Cameraman (who maintains a more or less stable diagonal position in the shot) begins to move slowly – at first counterclockwise and then in various directions, sometimes jumping from one position to another. The second section of this third part contains twenty-four shots (from 129 to 152) and brings the intervals' beats to the height of optical propulsion by maximally reducing the shot duration (one frame equals one shot).

As in the "Street and Eye" sequence, this segment of the film demonstrates Vertov's awareness of the fovea phenomenon.* The function of one-frame shots is to filter minimal visual data to the viewer's consciousness, simultaneously emphasizing the specific pictorial design of each shot by focusing on details that subliminally stimulate the viewers in a particular way. For example, the curved, spiral, and diagonal motions of the Cameraman's body are intended to incite the viewer's emotional identification with the "floating" Cameraman, as one can see from Plate 8 and the following shot-by-shot breakdown:

1. LAMS (double exposure). Cameraman with goggles (1 fr.).
2. HAECU. Rods and wheels spinning (1 fr.).
3. LAMS. Cameraman with goggles facing left (1 fr.).
4. CU. Spindles on a rotating wheel (1 fr.).
5. LAMS. Cameraman with goggles (1 fr.).
6. CU. Machine positioned diagonally (from lower right to upper left) with wheel turning at right and valves rotating in upper background (1 fr.).
7. LAMS (double exposure). Cameraman with goggles (1 fr.).
8. CU. Blurred wheel spinning (1 fr.).
9. LAMS (double exposure). Cameraman with goggles (2 fr.).
10. HAECU. Rods and wheels spinning (1 fr.).
11. CU. Spindles on rotating wheel (1 fr.).
12. HAECU. Rods and wheels spinning (1 fr.).
13. CU. Blurred wheel spinning (1 fr.).
14. CU. Spindles on rotating wheel (1 fr.).
15. CU. Blurred wheel spinning (1 fr.).
16. CU. Spindles on rotating wheel (1 fr.).
17. CU. Blurred wheel spinning (1 fr.).
18. CU. Spindles on rotating wheel (1 fr.).
19. CU. Blurred wheel spinning (1 fr.).
20. CU. Spindles on rotating wheel (1 fr.).
21. CU. Blurred wheel spinning (1 fr.).
22. CU. Spindles on rotating wheel (1 fr.).

* See also p. 147.

23. CU. Blurred wheel spinning (1 fr.).

24. CU. Spindles on rotating wheel (1 fr.).

XXXII. *Traffic Controller and Automobile Horn.* 28 shots, 989 fr., 55 sec.

1. ECU. The mirror image of the camera and the Cameraman reflected in the lens (1 fr.). Note: this shot is the same as shot 43 in the "Various Kinds of Work" sequence (XXV).

The montage crescendo of the final part of the "Cameraman and Machines" (XXXI) sequence is composed of twenty-four one-frame alternating shots of the Cameraman and rotating spindles. The two-frame duration of the shot of the Cameraman with goggles (no. 9) is arbitrary; it cannot be explained by any structural principle. There is no substantial reason for this unexpected disruption of an otherwise consistent pattern of shot duration, unless the single two-frame shot is seen as a reinforcement of the fovea phenomenon with regard to the Cameraman's vertical position versus the amorphous background formed by wheels and gears.

The emphasis on the graphic pattern of the movement within the shots creates a kinesthetic abstraction subliminally perceived by the viewer as a progressive spiral rotation. Thematically, this upward movement is associated with the representational aspect of the shots (the Cameraman, wheels, cylinders, spools, dynamos). Subliminally, the denotative meaning of the photographed subjects is linked with the kinesthetic experience of the physical transformations undergone by the Cameraman and tripod on the screen. Thus, when the Cameraman appears to float over the machines, the viewer's sensorial experience reaches a peak. As a result of this thematic/formal interaction, the vision of an elevated human body unified with the working machines can be interpreted as Vertov's celebration of the technological power controlled by the workers.

The collision of two directions of movement in this sequence provides a unique kinesthetic experience stimulated by the intervals as constructed by Vertov and Svilova. The rotating wheel introduces circular and elliptical patterns to the vertical and diagonal formations, as illustrated in the diagrams on p. 161.

The juxtaposition of circular–elliptical and vertical–diagonal patterns creates – through their continuous rotation – a surreal ambiance in which the Cameraman appears to be flying among and over machines. Gradually, the Cameraman's "flight" transforms his body's vertical position into a rotating circular structure similar to the structure of rotating machines. The abrupt "stop" to this kinesthetic choreography is produced by a one-frame close-up of the camera with the Cameraman reflected in its lens, an image that marks the beginning of the next sequence, "Traffic Controller and Automobile Horn" (XXXII).

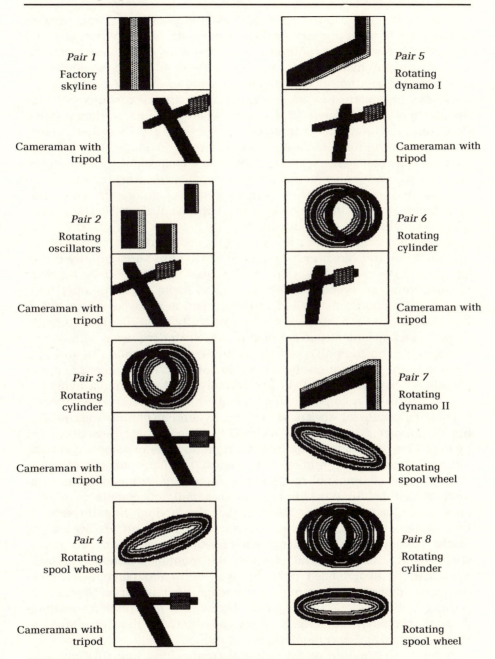

Pair 1
Factory
skyline

Cameraman with
tripod

Pair 5
Rotating
dynamo I

Cameraman with
tripod

Pair 2
Rotating
oscillators

Cameraman with
tripod

Pair 6
Rotating
cylinder

Cameraman with
tripod

Pair 3
Rotating
cylinder

Cameraman with
tripod

Pair 7
Rotating
dynamo II

Rotating
spool wheel

Pair 4
Rotating
spool wheel

Cameraman with
tripod

Pair 8
Rotating
cylinder

Rotating
spool wheel

Graphic patterns of the "Cameraman and Machines" sequence (XXXI)

Vertov's juxtaposition of man and machine can be compared with Eisenstein's "Cream and Separator" sequence in his 1929 film *Old and New,* which depicts peasants at a demonstration of the cream separator recently purchased by a cooperative farm. In order to emphasize the peasants' distrust of technology (in spite of Marpha Lapkina's excitement), Eisenstein's montage preserves the spatial distance between the machines and farmers in the barn, whereas Vertov's shots exclude the factory setting in order to achieve a spatial abstration that perceptually fuses the Cameraman and the machines. Ideologically, in Eisenstein's sequence, the two worlds – human and technological – contradict each other, whereas in Vertov's sequence, the two worlds are organically unified, complying with Vertov's vision of a socialist enterprise that depends upon the absolute unity between the workers and their technological means of production.

The variety of shapes in the shots that constitute the climax of the "Cameraman and Machines" sequence (XXXI) are substantially different from those in the "Street and Eye" sequence (XXII), where a steady circular form (an eyeball) is juxtaposed with the predominantly horizontal and diagonal movement of the camera as it sweeps over buildings and streets. This sequence is perceptually more disturbing due to its phi-effect and its structural position (after the sequence with people drinking beer in the pub), as well as its diegetic function (as an introduction to the accident on the street). In contrast, the "Cameraman and Machines" sequence (XXXI) is perceptually soothing, almost hypnotic; it is meant to be a constructivist statement about the expressivity and beauty of mechanical work as well as the dynamism of machines, reflecting the *kinoks*' admiration of technology. The graphic design created by the forms and movements in individual shots is extremely compact, although it consists of numerous elements. The light patterns continue to be vertical or diagonal and sharply angular, with unsteady and undistinguishable circular–elliptical shapes and movements that overtake the sequence's finale. The fact that the "flying" human being is identifiable as *the Cameraman* – a worker armed with his own set of "tools" – supplies this sequence with poetic reverberations. As Eisenstein stated,[20] the sensorial effect of the montage's interacting overtones can be experienced only through the actual perception of the projected images. Integrating all basic graphic patterns (horizontal, vertical, and circular), the sequence becomes a metaphor for communication, industry, and creativity, its constructivist elements functioning on both formal and thematic levels.

A close examination of the pictorial composition and rhythmic progression of the shots in the "Cameraman and Machines" sequence reveals the particular attention that Vertov and Svilova paid to the formal aspect of the shots. One should keep in mind that the technology of editing at the time was undeveloped and even primitive, particularly in the Soviet Union where a limited number of archaic editing tables existed. However, Svilova managed to splice numerous one- or two-frame shots in order to achieve a subliminal propulsion on the screen, producing the illusion that the Cameraman is superimposed over the rotating machines. The photographic execution of these shots is equally sophisticated: the vertical position of Kaufman's body within the frame consistently matches the diagonal position of his camera, so that the tripod appears as an extension of the Cameraman's body – a mechanical tool inseparable from the worker who uses it.

The first shot of the Cameraman in the sequence is a lateral view of his left profile with the tripod extended behind his back in the center of the frame; then, he appears on the right with a bright triangle of smoke from burning steel rising in the background. Wearing goggles like the other factory workers, the Cameraman moves further to the left (almost exiting the frame), as a sudden burst of bright smoke appears in the upper right corner over the camera. These eighteen one-frame shots form a flickering effect, introducing a bright accent to the predominantly dark image. Unlike the shots of machines, wheels, and gears photographed in a more abstract fashion, the shots of the Cameraman never lose their representational configuration. The accelerated movement of gears and wheels produces yet another blurring effect that enhances the graphic pattern of white lines within circular and diagonal movements. Integrated with the balletic motion of the Cameraman's body and his tripod, through a rhythmic alternation of images, these shots create the illusion of the Cameraman "floating" through and among the machines. An integral part of the technological environment, he is identified with the workers engaged in building the new society.

The last shot of the "Cameraman and Machines" sequence (XXXI) is a close-up of a circular spindle, followed by the close-up of the camera lens reflecting the camera. The ensuing "Traffic Controller and Automobile Horn" sequence (XXXII) begins with a tracking shot of the street and traffic signal. In the third shot of this sequence, the Cameraman is seen shooting as he sits on the cobblestone street amid milling pedestrians and cars (Fig. 147). With their relatively long duration (128 and 147 frames), these two shots structurally "break" the accelerated montage tempo (reached at the end of the "Cameraman and Machine" sequence), which is now replaced by the camera's own physical exploration of space. Consequently, the viewer's attention shifts from the stroboscopic "pulsation" of the subliminal montage to a sensorial experience of the film image as a reflection of the external world.

The kinesthetic choreography of the "Cameraman and Machines" sequence builds a metaphor of the Cameraman as a worker and an indispensable part of the industrial world. Overtaken by the mighty whirl of machines, the Cameraman appears free from the pressure of gravity as he floats in the factory milieu, dancing and hovering with his camera as a balancing pole. Gradually losing his earthbound identity, he becomes a "superman," a truly omnipotent "cinematic magician." Due to the hypnotic effect of the stroboscopic pulsation, the Cameraman's figure appears flat, therefore contributing to the oneiric effect of the entire sequence. The transition to the following sequence is abrupt, as if one had suddenly awakened from a vivid dream. As reality intercedes, the Cameraman is again seen performing the routine duties of his occupation, busily running along the streets as he "veers into the tempestuous ocean of life" from which he extracts details to build a new, more significant vision of the external world.

The dreamlike setting within which Mikhail Kaufman is presented in the "Cameraman and Machines" sequence (XXXI) can be interpreted as a futuristic poetic vision of the ultimate unification of the workers and their productive means. The fact that the Cameraman (wearing the proper protective gear and the worker's cap) is identified with the factory worker extends the cinematic metaphor further, in that it complies with Vertov's alliance of the *kinok*s with workers and engineers trained to use modern technology (the camera) in order to contribute to the building of a socialist country. The kinesthetic impact of the concentrated intervals is employed here to substantiate – in a poetic way – the filmmaker's ideological position in relation to technology, whereas, in other instances, the intervals' impact has a more direct oneiric implication (as in the "Awakening" sequence).

Oneiric impact of intervals

In addition to the kinesthetic and hypnotic power of the intervals (as demonstrated in the "Street and Eye" and the "Cameraman and Machines" sequences), Vertov and Svilova also employ the principle of cutting on intervals to parallel an anxious consciousness (the Cameraman's) and a sleeping consciousness (the young woman's).

The "Awakening" sequence (VII) is intended to induce a hypnopompic sensation* in the viewer by the inclusion of representationally am-

---

* There are two transitional states of consciousness: one preceding and the other succeeding the four stages of dreaming. Charles Tart describes them in the following way: "When we lie down to sleep at night, there is a period of time in which it would be difficult to say with any certainty whether we are awake or asleep. This borderline pe-

biguous shots that become clarified only within the context of their surrounding shots. The sequence begins with two such shots of the sleeping young woman reduced to semiabstract imagery that is difficult to identify because the shots include only the woman's neck and a fragment of the bed in which she sleeps (Fig. 25). The third shot is a long shot of a sunny terrace lined with trees and tables against an undefined horizon in the background (Fig. 27); the fourth shot is a long shot of the Cameraman placing his camera between the tracks as a train emerges from deep perspective (Fig. 37). After this introduction, the conflict of intervals begins. By intercutting shots of the sleeping woman among shots of a speeding train, a preawakening tension is generated, which at the same time matches the Cameraman's nervous tension. The following schematic presentation of the sequence illustrates the principle of editing on intervals. It also reveals Vertov's concern for diverse directions of movement within the same shot. (Compare the schematic drawings of these shots with the actual frame enlargements in Plate 5.)

1. HACU. A woman's chin and neck under sheets. A stationary shot forming a semiabstract composition in which the white area prevails (40 fr.). Note: The difficulty of recognizing what the photographed object represents intensifies the viewer's desire to find out what is excluded from the frame.

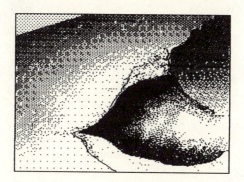

2. MLS. Part of a terrace with empty tables and two trees in the foreground, silhouetted against empty sky; branches of trees move slightly in the wind

riod has been termed the 'hypnagogic' period. The transitional state that occurs when we awake from sleep is called the 'hypnopompic' period." See Charles Tart, "Between Waking and Sleeping: The Hypnagogic State," *Altered States of Consciousness* (New York: Anchor Books, 1972), p. 75. In film theory, however, the term "hypnagogic" is used as a general term to indicate both borderline periods. Also, some contemporary film theorists use this term to describe the oneiric aspect of nonnarrative films, or to single out a specific movement in avant-garde cinema, such as "hypnagogic films" made by certain new American filmmakers.

(79 fr.). Note: The brightness of the composition and tranquility of the environment create a dreamy mood.

3. LS–MS. Railroad tracks with a train approaching from deep background and steadily moving toward the foreground; the Cameraman, kneeling on the tracks in the right foreground, prepares his camera (200 fr.). Note: The position of the camera in relation to the approaching train enhances the threatening situation.

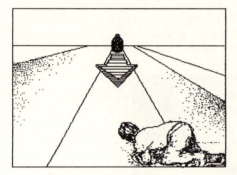 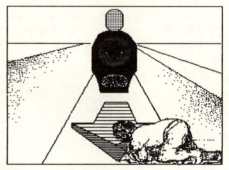

4. CU. The Cameraman's head in lower right portion of the frame, with tracks in the background; the turning wheels are blurred (7 fr.). Note: The optical shock produced by a jump cut (the position of the Cameraman's head does not match his position in the previous shot) marks the beginning of the oneiric part of the segment.

5. LACU. One side of the speeding train, as the windows move from upper right to lower left (22 fr.). Note: The camera's tilting and panning contribute to the semiabstract appearance of the speeding train.

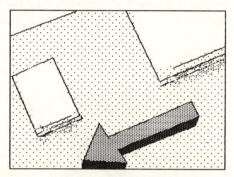

6. CU. The Cameraman's head in lower left portion of the frame, with tracks in the background and the wheels in blurred motion (5 fr.). Note: Another optical shock is produced by the reverse position of the Cameraman's head – identical but opposite to shot 4 (the flipped shot).

7. MCU. Train cars with windows move diagonally across the frame from the upper right to lower left corner; image is blurred due to the speed of the movement (22 fr.). Note: Simultaneous movement of the camera (panning and tilting) and the speeding train generate a kinesthetic impact on the viewer.

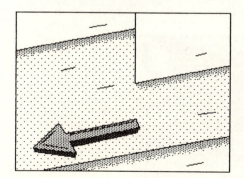 

8. CU. The Cameraman's foot caught on a railroad track (13 fr.). Note: As in shot one, the unusual composition of this shot enhances the viewer's desire to see the excluded part of the photographed object (the Cameraman's body). The tension is considerably heightened by the viewer's awareness of the approaching train and the position of the railway tracks within the frame, emphasizing the Cameraman's perilous position.

9. MS. Low angle (from beneath the train); the locomotive rapidly approaches the camera (and the spectators) and runs over it (40 fr.). Note: This is the most aggressive optical shock, intensified by the exclusion of frames (pixillation) which mechanically increases the train's advance.

 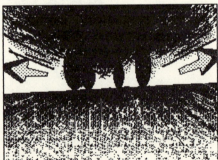

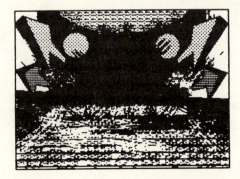

10. HACU. The woman's hair and forehead moving slowly from left to right, revealing the entire arm of the sleeping person (24 fr.). Note: This shot provides necessary information for a contextual denotation of shot 1. The head's movement anticipates a semiabstract appearance of the train in the following shot.

 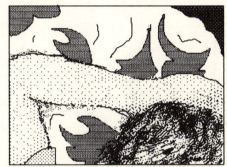

11. MS–LACU. The train at full speed moving diagonally from center background toward left of the frame, as the camera pans to the left, focusing on the windows, which first move diagonally, then horizontally, and again diagonally from upper right to lower left (26 fr.). Note: As in shot 5, the camera's tilting and panning intersperse with the movement of the speeding train and produce an even more abstract pattern on the screen.

 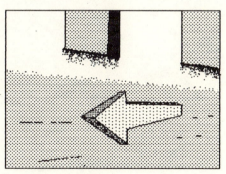

12. LAMS–CU. Blurred windows moving in different directions, constantly changing the composition and graphic patterns of the shot (68 fr.). Note: This is a central shot that achieves the greatest oneiric impact through its ten conflicting semiabstract and abstract movements/shapes, so that both the representational and spatial aspects of the photographed object are subverted and deconstructed.

 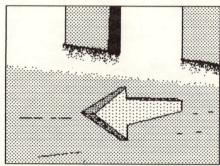

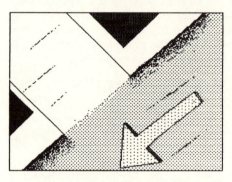 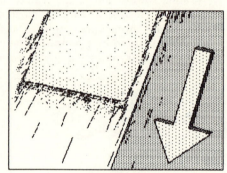

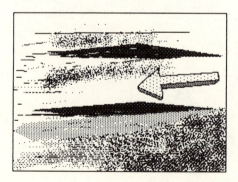 

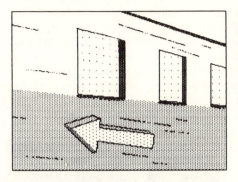
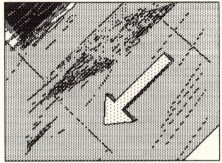

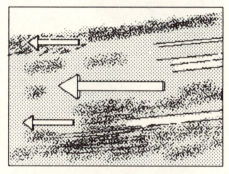

13. CU. The woman's head with her hair visible in the lower right corner; slowly, her arm extends horizontally to the left as her head makes a horizontal movement from right to left; in the upper part of the frame, decorative patterns of the bed sheet are prominent (18 fr.). Note: Perceptually, this shot "halts" the movement dominating previous shots and redirects the viewer's attention to the thematic aspect of the sequence (sleeping).

14. LACU–MS. Blurred windows of the train moving from upper right to lower left; the caboose becomes clear as it moves from a tilted to a horizontal position and then tilts again as the now silhouetted train disappears into the lower left corner; in the frame a clear sky and a telephone pole occupy lower right center (28 fr.). Note: With its changing graphic pattern, this shot visually unifies the abstract and the representational aspects of the sequence. It clearly implies the end of an optical dynamism, a sense of disappearance and departure.

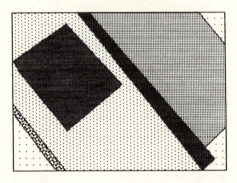
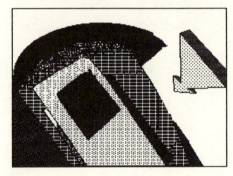

15. HAMCU. The arm and torso of the woman as she sits up in the bed; gradually her lower body and legs are seen as she gets out of the bed (57 fr.). Note: The clear representational outlook of this shot brings the thematic aspect of the sequence into full focus (wakening).

16. MCU. High angle of railroad tracks beneath the camera, which moves over them (168 fr.). Note: Although objectively the camera moves forward, it appears as if the tracks move in the opposite direction. A flickering effect is created by alternating black frames with the images of the tracks, according to the following scheme:

1. 28 fr. of track	9. 14 fr. of track	17. 5 fr. of track
2. 2 fr. of black	10. 2 fr. of black	18. 2 fr. of black
3. 25 fr. of track	11. 9 fr. of track	19. 5 fr. of track
4. 4 fr. of black	12. 2 fr. of black	20. 2 fr. of black
5. 19 fr. of track	13. 7 fr. of track	21. 4 fr. of track
6. 2 fr. of black	14. 2 fr. of black	22. 2 fr. of black
7. 17 fr. of track	15. 5 fr. of track	23. 4 fr. of track
8. 2 fr. of black	16. 2 fr. of black	24. 2 fr. of black

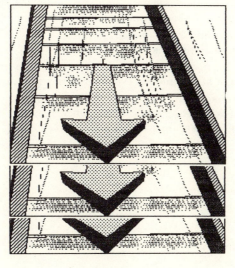

STRUCTURAL RECAPITULATION OF THE ONEIRIC SIGNIFICANCE OF THE "AWAKENING SEQUENCE"

1. *Semiabstract image: sleeping young woman*
   Function: introduction of oneiric state
   (ambiguous shot composition)

2. *Representational image: terrace with trees*
   Function: juxtaposition of exterior
   (reference to reality)

3. *Representational image: Cameraman and approaching train*
   Function: introduction to simultaneous event

4. *Representational image: Cameraman's head and turning wheels*
   Function: further amplification of dramatic tension

5. *Semiabstract image: speeding train*
   Function: extension of expectation – danger

6. *Representational image: Cameraman's head and turning wheels*
   Function: further amplification of dramatic tension – danger

7. *Semiabstract image: train moving diagonally*
   Function: further extension of expectation – danger

8. *Representational image: Cameraman's foot caught on railroad track*
   Function: peak of dramatic tension

9. *Representational image: approaching train*
   Function: visual aggression – strong motor-sensory impact

10. *Representational image: sleeping young woman*
    Function: linkage to oneiric state

11. *Semiabstract image: train moving diagonally*
    Function: intensification of oneiric effect

12. *Abstract image: train moving in various directions*
    Function: peak of oneiric effect – strong kinesthetic impact

13. *Representational image: sleeping young woman*
    Function: thematic linkage to oneiric state

14. *Representational image: disappearing train*
    Function: conclusion of oneiric state

15. *Representational image: young woman getting out of bed*
    Function: end of simultaneous event – return to reality

16. *Semiabstract image: railroad tracks*
    Function: cinematic abstraction of sequence's oneiric signification
    (tracking shot – flickering effect)

As one can see, at the beginning of this segment, Vertov relates the sleeping young woman's hypnopompic state of mind to the instinctive reaction of the Cameraman, who avoids the train by abruptly jerking his head. In fact, it is the Cameraman's response that triggers a series of semiabstract shots of the speeding train, which are then associated with the sleeping young woman. In turn, the slow-moving gestures of the young woman's head are contrasted to the abrupt jerking of the Cameraman's head. At the same time, her slow gestures are juxtaposed with the reckless movements of the train whose representational appearance is reduced to a few horizontal, diagonal, and square patterns. In shots 12 and 14, the pictorial compositions change rapidly, due to the camera's tilting and panning, or because of the variation in angle and scale of the shot. Although these two shots are relatively brief (sixty-seven and twenty-eight frames) and are without clear representational meaning, they consist of numerous visual elements, each introducing a new direction of movement and different composition. For example, in shot 12 one can discern ten different "subshots," each consisting of a particular graphic design. It is, of course, questionable whether these conflicting patterns and movements are produced by cuts (in the editing room) or during shooting. What is important here is the fact that on the screen they appear as separate pictorial units (subshots). It is through the interaction of these ten conflicting patterns and corresponding movements that optical tension ultimately affects the viewer on a subliminal level, thus intensifying the oneiric impact of the entire sequence. After shot 14 (which contains six subshots), the young woman gets out of bed as if "awakened" by this kinesthetic shock! Again, the duration of the patterns and movements in these subshots is irregular (the length of the juxtaposed pieces is not arithmetically determined), paralleling the uncomfortable feeling one customarily experiences upon awakening.

In a visual analogy to the sleeper's gradual emergence from the dream state, the young woman is not shown in full (through a long shot) as she gets up from bed, but rather in fragments. First, we see her legs slowly moving out of the bed and touching the floor (Fig. 39); then the diegetic progression of the action taking place in the bedroom is broken by an exterior shot that can be related to the earlier shot of the Cameraman shooting the train (Fig. 37). This disruptive shot is, actually, the concatenation (through montage) of twenty-four pieces (subshots) of railroad tracks (Fig. 41) crosscut with frames of black leader. The swift and optically irritating exchange of bright and dark surfaces flashing on the screen is spontaneously linked with the somnolent state of the young woman's consciousness as she awakes. With its stroboscopic pulsation, this shot (perhaps it is more legitimate to call it an optical effect) represents another example of a cinematic "illumination" of the human psyche. A more obvious connection between this nondiegetic shot (especially its perceptual impact) and the narrative core of the sequence is established later, when the young woman is shown in close-up, blinking her eyes as she dries her face (Fig. 49). This action is subsequently compared to the opening and closing of venetian blinds (Fig. 50), and also is related to the contraction of the camera's iris (Fig. 47). Once again, the eye's function is equated with the function of the cinematographic apparatus (the Camera-Eye).

Optical music

Many avant-garde filmmakers of the silent era were preoccupied with the idea of creating musical sensations by purely visual means. The French *cinéastes* of the twenties extensively experimented with cinematic images as components of musical structure. In an attempt to relate cinema to music, they introduced various metaphorical terms, such as *"la musique de la lumière"* (Ricciotto Canudo), *"l'inscription de rhythme"* (Leon Moussinac), or *"symphonie optique et visuelle"* (Germaine Dulac). Inspired by this concept, René Clair and Fernand Leger created two "musical" silent films, *Entr'acte* (Clair, 1924) and *Le Ballet mécanique* (Leger, 1924), in which shots interact – through editing – like notes in a musical composition.* Upon seeing another of Clair's experimental films, *Paris qui dort* (1923), Vertov explained that he had always wanted to experiment along these lines, particularly with accelerated and decelerated motion, but had been unable to pursue his vision.[†]

As a student, Vertov's interest in auditory perception and its mechanical recording stemmed from his experiments in the "Laboratory of Hearing" where he explored the "possibilities for making documentary recordings of the world of sounds through montage."[21] While

---

* In spite of many similarities in the use of specific cinematic devices to create visual abstraction on the screen, there is no evidence that Vertov actually saw *Entr'acte* or *Le Ballet mécanique* (both 1924) during his visit to Paris.
[†] Vertov, "Notebooks" (April 12, 1926), *Articles,* p. 161. The entry reads: "I saw the film *Paris qui dort* in the Art Theater. It upset me. Two years ago I had designed an identical project in which I wanted to use the same technical devices that have been employed in this film. All the time I have been searching for a chance to realize it. But that opportunity has never been given to me. And now, they have made it abroad. Film-Eye missed another position for attack. It is such a long road from the idea, conception, plan to its final realization. Unless we are given the opportunity to realize our innovative projects in the course of their actual execution, there will always be a danger that we shall never materialize our inventions." See also Annette Michelson, "Dr. Craze and Mr. Clair," *October,* no. 11, Winter 1979, pp. 31–53. Michelson infers that "Clair's ironic gloss upon the dynamics of alienation finds its extension in Vertov's systematic use of reverse motion ('negative of time') in the visual trope of hysteron proteron [which] thus becomes the pivotal element in the elaboration of his [Vertov's] Marxist propadeutic" (p. 52), which is a less apprehensible and more convoluted explanation about the ideological significance of Vertov's cinematic execution than he himself could ever have intended.

working on his early *Film-Truth* and *Film-Eye* series, Vertov often drew the viewers' attention to the sensation of sound by including shots that triggered aural associations. In *The Man with the Movie Camera,* the most evident example of this editing principle can be found in the "Traffic Controller and Automobile Horn" sequence (XXXII), in which a close-up of a horn (Fig. 149) is inserted twelve times among long shots of traffic and medium shots of the Cameraman shooting traffic on the street (Fig. 147). The idea of sound is alluded to earlier in the film (at the end of the Prologue) through the depiction of the conductor and the orchestra as they accompany the film screening (Figs. 15–22). The rhythmic exchange between various instruments and the conductor's gestures creates the illusion of a musical orchestration. A more kinesthetic representation of sound appears in the "Awakening" sequence (VII), through the intercutting of numerous blurred shots (of the speeding train) with close-ups of the sleeping young woman. The optical pulsation of these blurred shots functions as an equivalent to sound waves and, as if in response to them, the sleeping woman wakes up (as analyzed previously in this chapter).

But the most "auditory" sequence is "Musical Performance with Spoons and Bottles" (L), near the end of Part Two, with its 111 shots that build to a montage climax paralleling the structure of a musical score and its orchestration (Plate 9). At the end of the previous sequence, sound is denoted rather illustratively, that is, by a close-up of a radio with a large circular loudspeaker that fills the entire frame. The notion of sound is enhanced through multiple double exposures of details (implying sound) within the loudspeaker's circle: (1) a human ear, (2) two hands playing on a keyboard, and (3) a singing mouth (Fig. 196). These symbolic images replace one another by superimposition, slowly moving within the stationary circular loudspeaker attached to the radio, next to the drawing of a tsarist officer on horseback (in the left portion of the frame). Then follows the first shot of the musician's hands playing with spoons on a table, on top of which a cup, a plate, a tray, an ashtray, a washboard, and seven bottles are displayed. As soon as the musician begins to hit the "instruments" with two spoons, a "higher mathematics of facts" takes over, interrelating 111 shots of the "musical performance" and its listeners. Their order and duration, as shown in the following shot-by-shot breakdown, provide an idea of Vertov's and Svilova's strategy in building a montage glissando to stimulate auditory sensation through visual associations.

L. *Musical Performance with Spoons and Bottles.* 111 shots, 1,354 fr., 1 min. 15 sec.

1. HAMCU. Row of seven dark bottles with white labels arranged diago-

nally from lower left foreground toward upper right background, ending with a jar, and wood and metal scrubboard on dark table; behind is a row consisting of a light pot lid, six spoons, an overturned metal tray; from upper left corner two white hands with a spoon in either hand, blurred, move around, hitting various objects, lifting and banging pot lid on its surface (97 fr.).

2. LACU. Woman on left with white scarf looking down and laughing; then she turns to right and begins to move out of frame against wooden door. She is identified as Woman 1 (54 fr.).

3. HACU (similar to L, 1). Musician's two hands hitting objects with two spoons (57 fr.).

4. LACU. Profile of woman with long dark hair, her face in shadow, facing left, laughing; then she turns forward and laughs, turns back to left, becoming silhouetted against white sky. She is identified as Woman 2 (57 fr.).

5. HACU (same as L, 3). Hands hit objects with spoons (41 fr.).

6. CU. Woman with short hair on left, face brightly lit, facing right; then she turns left, moves back, then looks forward and laughs; she is against dark background. She is identified as Woman 3 (50 fr.).

7. HACU. Hands with spoons, tilted to the right against a dark background, tap on washboard (43 fr.).

8. CU. Two women, one facing the camera, the other with head turned in the opposite direction. Their foreheads meet (81 fr.).

9. HACU (same as L, 5). Hands gather up spoons and throw them on the table; hands then hit the tray with a spoon (43 fr.).

10. CU. Profile of unshaven man smoking, facing left; right side of his face highlighted, the rest in shadow against dark background (27 fr.).

11. HACU (same as L, 9). Hands throw the spoons on surface, then pick them up (15 fr.).

12. MCU. Several workers with caps; laughing worker in center and part of face of another worker on left also laughing and looking down toward right (15 fr.).

13. HACU (same as L, 11). Hands hit metal tray with spoons (14 fr.).

14. LACU. Three-quarter view of Woman 3 looking left; her face in shadow against light sky in background (15 fr.).

15. HACU (same as L, 13). Hands continue to hit metal tray with spoons (15 fr.).

16. LACU (same as L, 4). Woman 2, laughing in shadow, looks left against light background (10 fr.).

17. HACU (same as L, 15). Hands hitting objects with spoons (15 fr.).

18. CU (same as L, 14). Against dark background, Woman 3 with brightly lit smile; she is at frame left and looks down (10 fr.).

19. HACU (sames as L, 17). Hands hit objects with spoons (15 fr.).

20. LACU. Face of Woman 1 on left in white scarf, laughing and looking to lower right (9 fr.).

21. HACU (same as L, 19). Hands hit objects with spoons (11 fr.).

22. CU (same as L, 10). Profile of man facing and looking down left; he laughs and smiles; right part of his face highlighted, rest of his face in shadow, against dark background (8 fr.).

23. HACU (same as L, 21). Hands hit objects with spoons (12 fr.).

24. MCU (same as L, 12). Several workers laughing and looking down to right (8 fr.).

25. HACU (same as L, 23) Hands hit objects with spoons (10 fr.).

26. LACU (same as L, 18). Woman 3 faces left (7 fr.).

27. HACU (same as L, 25). Hands hit objects with spoons (9 fr.).

28. LACU (same as L, 16). Woman 2 with face in shadow, laughing, looking left, against light background (7 fr.).

29. HACU (same as L, 27). Hands hit objects with spoons (8 fr.).

30. CU (same as L, 26). Woman 3 on left, smiling and looking down (6 fr.).

31. HACU (same as L, 29). Hands hit objects with spoons (8 fr.).

32. LACU (same as L, 20). Face of Woman 1 on left in white scarf looking to right and sticking out tongue (6 fr.).

33. HACU (same as L, 31). Hands hit objects with spoons (8 fr.).

34. CU (same as L, 22). Profile of man facing left, right part of his face highlighted, rest of his face in shadow against dark background (66 fr.).

35. HACU (same as L, 33). Hands hit objects with spoons (8 fr.).

36. MCU (same as L, 24). Workers laughing and looking down to right (5 fr.).

37. HACU (same as L, 35). Hands hit objects with spoons (6 fr.).

38. LACU (same as L, 26). Woman 3 in shadow looking left against light background (4 fr.).

39. HACU (same as L, 37). Hands hit objects with spoons (6 fr.).

40. LACU (same as L, 28). Woman 2, laughing in shadow, looking left against light background (4 fr.).

41. HACU (same as L, 39). Hands hit washboard with spoons (6 fr.).

42. CU (same as L, 30). Woman 3 on left looking down, then raising head and turning right against dark background (4 fr.).

43. HACU (same as L, 41). Hands hit washboard with spoons (6 fr.).

44. LACU (same as L, 32). Woman 1 on left in white scarf laughing and looking right against dark background (3 fr.).

45. HACU (same as L, 43). Hands hit washboard with spoons (4 fr.).

46. CU (same as L, 34). Profile of man smiling and looking left, right part of his face highlighted, rest of his face in shadow against dark background (3 fr.).

47. HACU (same as L, 45). Hands hit washboard with spoons (4 fr.).

48. MCU (same as L, 36). Worker in center, his right profile in shadow, leans back and raises right hand (3 fr.).

49. HACU (same as L, 47). Hands hit washboard with spoons (4 fr.).

50. LACU (same as L, 38). Woman 3 in shadow facing left against light background (2 fr.).

51. HACU (same as L, 49). Hands hit washboard with spoons (3 fr.).

52. LACU (same as L, 40). Woman 2, face in shadow, laughing and looking left against light background, then turns her head slightly down (2 fr.).

53. HACU (same as L, 51). Hands hit washboard with spoons (3 fr.).

54. CU (same as L, 42). Woman 3 on left looking to right with open mouth against dark background (2 fr.).

55. HACU (same as L, 53). Hands hit washboard with spoons (3 fr.).

56. LACU (same as L, 44). Face of Woman 1 on left in white scarf laughing and looking right against dark background (2 fr.).

57. HACU (same as L, 55). Hands hit washboard with spoons (3 fr.).

58. CU (same as L, 46). Left profile of smiling man, right part of his face highlighted, rest of his face in shadow against dark background (2 fr.).

59. HACU (same as L, 57). Hands hit washboard with spoons (3 fr.).

60. LACU (same as L, 50). Woman 3 in shadow facing left against a light background (2 fr.).

61. HACU (same as L, 59). Hands hit washboard with spoons (3 fr.).

62. LACU (same as L, 52). Woman 2 with face in shadow against light background looks left (2 fr.).

63. HACU (same as L, 61). Hands hit washboard with spoons (3 fr.).

64. LACU (same as L, 62). Woman 2 with face in shadow against light background looks left, face turned more toward camera (1 fr.).

65. HACU (same as L, 63). Hands hit washboard with spoons (3 fr.).

66. LACU (same as L, 64). Woman 2 with face in shadow against light background looks left; face turned to right (1 fr.).

67. HACU (same as L, 66). Hands hit bottles with spoons (3 fr.).

68. LACU (same as L, 66). Woman 2 with face in shadow against light background looks left (1 fr.).

69. HACU (same as L, 67). Hands hit bottles with spoon (3 fr.).

70. LACU (same as L, 68). Woman 2 with face in shadow against light background looks left (1 fr.).

71. HACU (same as L, 69). Hands hit objects with spoon (3 fr.).

72. LACU (same as L, 70). Woman 2 with face in shadow against light background looks left (1 fr.).

73. HACU (same as L, 71). Hands hit objects with spoon (3 fr.).

74. LACU (same as L, 72). Woman 2 with face in shadow against light background looks left (1 fr.).

75. HACU (same as L, 73). Hands hit objects with spoon (2 fr.).

76. LACU (same as L, 74). Woman 2 with face in shadow against light background looks left (1 fr.).

77. HACU (same as L, 75). Hands hit objects with spoon (2 fr.).

78. LACU (same as L, 76). Woman 2 with face in shadow against light background looks left (1 fr.).
79. HACU (same as L, 77). Hands hit objects with spoon (2 fr.).
80. LACU (same as L, 78). Hands hit objects with spoon (1 fr.).
81. HACU (same as L, 79). Hands hit objects with spoon (2 fr.).
82. LACU (same as L, 80). Woman 2 with face in shadow against light background looks left (1 fr.).
83. HACU (same as L, 81). Hands hit spoons with spoons (2 fr.).
84. LACU (same as L, 82). Woman 2 with face in shadow against light background looks left (1 fr.).
85. HACU (same as L, 83). Hands hit spoons with spoons (2 fr.).
86. LACU (same as L, 84). Woman 2 with face in shadow looks left against light background (1 fr.).
87. HACU (same as L, 85). Hands hit objects with spoons (2 fr.).
88. LACU (same as L, 86). Woman 2 with face in shadow against light background looks left (1 fr.).
89. HACU (same as L, 87). Hands hit objects with spoons (2 fr.).
90. LACU (same as L, 88). Woman 2 with face in shadow against light background looks left (1 fr.).
91. HACU (same as L, 89). Hands hit objects with spoons (2 fr.).
92. LACU (same as L, 90). Woman 2 with face in shadow against light background looks left (1 fr.).
93. HACU (same as L, 91). Hands hit objects with spoons (2 fr.).
94. LACU (same as L, 92). Woman 2 with face in shadow against light background looks left (1 fr.).
95. HACU (same as L, 93). Hands hit objects with spoons (2 fr.).
96. LACU (same as L, 94). Woman 2 with face in shadow against light background looks left; she moves her head back and laughs (4 fr.).
97. HACU (multiple exposure, superimposition). With two spoons, the two blurred hands hit six spoons placed on table; after three frames, shoes appear tapping on floor with bottom of striped pants visible in the center, while the hands move back and forth in upper right; on lower edge of screen is lateral view of arm and snapping fingers (74 fr.).
98. LACU (same as L, 96). Woman 2, with face in shadow against a light background, laughs and looks left (4 fr.).
99. HACU (similar to L, 97; multiple exposure, superimposition). With two spoons, the two blurred hands hit six spoons on table; three piano pedals appear, then right foot steps on right pedal; keyboard appears in upper center extending toward upper diagonal right; left foot steps on left pedal (35 fr.).
100. LACU (same as L, 98). Woman 2 with face in shadow against light background looks and turns left (3 fr.).
101. HACU (same as L, 99; multiple exposure, superimposition). Keyboard ex-

tends diagonally from lower left to upper right across entire frame, while two hands with spoons move in various directions (30 fr.).

102. LACU (same as L, 100). Woman 2 turns more to left (3 fr.).

103. HACU (same as L, 101; multiple exposure, superimposition). Another keyboard superimposed horizontally, cutting the frame into two parts, as hands move in all directions; feet and pedals gradually disappear (68 fr.).

104. LACU (same as L, 102). Woman 2 turns left (3 fr.).

105. HACU (same as L, 103; multiple exposure, superimposition). Continuation of L, 103 (13 fr.).

106. LACU (same as L, 104). Woman 2's left profile; she turns to right (2 fr.).

107. HACU (same as L, 105; multiple exposure, superimposition). Diagonal keyboard repeatedly alternated between being lit and being in shadow; snapping fingers appear in center (92 fr.).

108. LACU (Same as L, 106). Woman 2 turns to right (2 fr.).

109. HACU (same as L, 107; multiple exposure, superimposition). Six spoons and a white lid are exposed against a dark background and positioned in the lower part of the screen (10 fr.).

110. LACU (same as L, 108). Woman 2 looking at camera (2 fr.).

111. HACU (same as L, 109; multiple exposure, superimposition). After twenty frames, two hands grab the spoons, lift them up and throw them down; two other hands on large keyboard are more visible (59 fr.).

This sequence of 111 shots (all close-ups) lasts just a little over a minute and is one of the most dynamic and visually expressive sequences in the film. It perfectly demonstrates Vertov's concept of organizing film-pieces into a film-thing, where the duration of the shot progressively decreases in established increments. While the decimal alternation of shots and the repetition of specific scales or lengths contribute to auditory sensation, the divergent, often totally opposed, movements within individual shots and their literal superimposition through multiple exposure produce an optical glissando associated with the musical performance. Here, Vertov created a *cinematic* variation of music, even though he believed that shots – however abstract – cannot replace notes, "just as literary descriptions of Scriabin compositions cannot express the true sensation of his music."[22]

The eight introductory shots of this sequence are relatively lengthy and stationary, with little perceptible movement of the women's heads. The cuts between close-ups of the listeners' heads and the medium shots of the musician's hands occur when each listener's movement becomes perceptible, and often just before the head leaves the frame (by "cutting on movement"). As the shots' lengths decrease, the physical movement of the head is considerably reduced, so much so that the viewer perceives the close-ups of the three young women – each with different facial features – as dominant circular shapes juxtaposed against the rectangular shape of the table with diagonally arranged bottles and the musician's hands playing the spoons. The medium shots of spoons and bottles are in fact segments of a continuous take broken up into separate shots and intercut among close-ups of the womens' heads; it is because of their strong representational outlook that these segments are perceived as a continuous action, in spite of the fact that the original take is broken up by the intercutting.

The musical overtones of this sequence are generated by the arrhythmic – or rather syncopated – reduction of the length of the shots that depict listening to the performance. If one traces the number of frames in shots 50–95, it becomes clear that Vertov and Svilova followed a specific arithmetic pattern: the women's heads in close-up are held for one to three frames, while the spoons and bottles are held for two or three frames. The overall montage pace of this segment is determined by: (1) duration of shots, (2) their scale, (3) their pictorial composition, (4) the conflict of divergent movements between the shots (intervals), and (5) the tension between the continuity of the physical action (musical performance) shown on the screen and its "metric" breakdown into montage units (individual shots). The connotation of musical sound arises from medium shots of spoons and bottles as they are intercut with close-ups of women "listening to the music," while the rapid alternation of the two sets of shot compositions (circular versus horizontal; vertical versus diagonal) reflects, in visual terms, the structure of a musical score.

Vertov's application of "metric editing" to intervals in order to create the notion of music is unique in the theory of silent cinema. He uses the same device in *The Eleventh Year* (1928), "linking montage with sound" by intercutting details of machines, hammers, axes, and saws on a "metric principle," thus achieving a musical "mechanical heart beat."[23] At the time when Eisenstein, Pudovkin, and Alexandrov launched their "Statement"[24] calling for the contrapuntal use of sound, Vertov was experimenting with his own concept of "visually conveyed sound," which, according to him, was an expansion of his "Film-Eye" method into the "Radio-Eye" method. Had Vertov realized his intention to "construct" an accompanying musical score for *The Man with the Movie Camera,* this sequence would probably be based on sight and sound counterpoint.* Soon after the completion of the film, Vertov wrote the essay "From 'Film-Eye' to 'Radio Eye,'" explaining that his film was conceived as a transition from "Film-Eye" to "Film-Radio-Eye."[25] In this respect, *The Man with the Movie Camera* can be seen as

---

* See Chapter I, note 149. Vertov's sound score for *The Man with the Movie Camera* would probably consist of ordinary sounds (recorded in reality) and musical inserts from various compositions and songs (mixage).

Vertov's conceptual exploration of the relationship between montage and sound, an idea fully realized in his next work, *Enthusiasm* (1930), the first avant-garde Soviet sound film.

To demonstrate the difference between the illusion of double exposure (created by the phi-effect) and the real superimposition (produced by the camera or in the laboratory) of two or more images that appear simultaneously on the screen, Vertov includes two sets of mechanical superimpositions in the "Musical Performance with Spoons and Bottles" sequence (Plate 9). One is composed of the two hands playing with spoons and snapping fingers, the other of a Jew's harp and dancing legs seen "through" keyboards. Placed next to each other, these superimpositions point to the difference between the phi-effect and mechanical double exposure, the latter lacking genuine kinesthetic energy, however visually attractive. This difference is even more perceptible in the multiple exposure of these and other objects, which appear in the sequence eight times (shots 97, 99, 101, 103, 105, 107, 109, 111). Shots 103–11 incorporate the additional image of two piano keyboards diagonally crossing each other and piano pedals with feet on them. Functioning as a "visual dominant" within the montage structure, the repetition of the multiple exposure (lasting ten to ninety-two frames) is disrupted by seven close-ups of Woman 2 (only two to four frames long) perceived as optical flashes that appear at various intervals. The smooth flow of these multiple exposures is rhythmically "bombarded" by close-ups of the woman's head as she listens to the performance, thus creating an arrhythmic beat of alternating compositions and durations: one contains numerous visual motifs and various designs perceived for a relatively long period of time (multishaped superimpositions); the other contains only one visual motif, the woman's head (the circle) pulsating on the screen. With its pronounced "musical" beat achieved through "metric editing," the "Musical Performance with Spoons and Bottles" sequence (L) foreshadows the symphonic finale of *The Man with the Movie Camera,* culminating with the "Editor and the Film" sequence (LV).

Movement versus stasis

In the "Musical Performance with Spoons and Bottles" sequence (L), kinesthetic tension is built up by juxtaposing stationary close-ups of the listeners with dynamically composed and superimposed shots of a "musical performance." In the "Spectators and the Screen" sequence (LII), even greater kinesthetic tension is created by juxtaposing similar close-ups of the movie audience, photographed by a stationary camera, with medium and long shots of various moving events photographed by a tracking camera.

In this sequence, the spectators in close-up watch the trolleys, cars, and pedestrians moving before their eyes on the screen-within-the-screen. A similar kinesthetic effect is achieved in the "Basketball" (XLII), "Track and Field Events" (XXXVII), and "Soccer" (XLIII) sequences whenever athletes' movements are "halted" by stationary close-ups of spectators or fixed objects. In all these sequences, the interaction between movement and stasis is repeatedly related to the photographed object's particular direction of movement. Movement versus stasis in the basketball shots is predominantly vertical (due to the players' jumping). In the soccer shots, it is mostly horizontal; in the track and field shots, the direction of the movement varies depending upon the sport. In the street shots, horizontal and curved movements interfuse, creating a perceptual obstruction that subverts the representational aspect of the juxtaposed images (vehicles and pedestrians).

The most kinesthetic interaction between movement and stasis is produced in the "Spectators and the Screen" sequence (LII) through the collision between the motionless audience and the moving vehicles captured by the camera as it pans from left to right. The camera then moves further to the right, revealing a deep perspective of the street (Fig. 210). In turn, the photographed objects are first perceived as representative images, then as semiabstract (blurred) silhouetted close-ups, which move further to the right as the camera follows them (at comparable speed and without adjusted focal lengths), so that near the end of the shot the back of one moving vehicle is clearly seen in long shot at the right middle ground, with traffic and buildings on the left side. These shots last only between forty and sixty frames (two to four seconds) and are repeated ten times in the same segment, continuously interrupted by close-ups or medium shots (twelve in all, each ten to fifteen frames long) of pedestrians' heads, which pop into the shot to occupy the entire frame. The result of this clash between the camera motion (panning) and the stationary shots (close-ups) produces a unique perceptual conflict that affects the viewers sensorially. Although this principle is used throughout the film, it is most apparent in the sequences that show spatial interaction between people and objects. It can also be detected – although to a lesser degree – in all the phi-effects, as, for example, in the "clash" between tracking shots of objects or buildings and relatively stationary close-ups of the eyeball in the "Street and Eye" sequence (XXII).

Static versus dynamic tension suffuses the entire structure of *The Man with the Movie Camera*. In a broader sense, the film begins with stationary images (the camera does not move throughout the entire Prologue). The first tracking shot, in which the camera moves perpendicularly toward a closed window, does not appear until the beginning of Part One. The numerical relationship between stationary and moving shots varies, depending upon the theme of the sequence. For example, all sequences of factories and machines consist of stationary shots juxtaposed to each other graphically, whereas in the sequences dealing with traffic and sports events, the majority of shots are photographed with the camera gliding through space or circulating around people and objects. Symbolically, the film begins with *stasis* and ends with intense *kinesis,* corresponding to the cycle of a working day, which starts slowly after wakening and then gradually increases its tempo through various urban activities that reach an explosive finale. As one might expect from Vertov, his film does not close with relaxation and composure, but rather with an ever-increasing montage crescendo that ends not with a "whimper," but with a "bang"!

Movement versus stasis is particularly emphasized after the "Musical Performance with Spoons and Bottles" sequence (L), which is built on the conflict between stationary close-ups of listeners and dynamic shots incorporating various objects. In the "Camera Moves on Tripod" sequence (LI), the first apparent juxtaposition of movement versus stasis is achieved through the apposition of static close-ups of faces – later recognized as the audience – and medium shots of the animated tripod and camera (Fig. 197). Once it becomes clear that the faces watching the camera "walking" on its three "legs" are the seated viewers in the movie theater, the movement versus stasis conflict is enhanced by freeze-frames of the auditorium (photographed from a high angle), incorporating the screen-within-the-screen on which the abstract image of the rotating spool of wire is projected (Fig. 201). This conflict is continued by alternating medium shots of the viewers (Fig. 204) with composite images of ballet dancers and double exposures that merge various entertainers and a piano keyboard (Fig. 203). Then, after a shot of the spectators seen from above (Fig. 199), there are shots of the Cameraman shooting planes (Fig. 219) with a telescopic lens (Fig. 218), followed by several composite shots of traffic (Fig. 220) and a compound shot of numerous diligent typists (Fig. 221); this segment ends with the abstract shot of the glittering spinning loom (Fig. 205) and the shot designed in a markedly constructivist manner – the superimposition of a rotating wheel carrying spools of thread and a close-up of a smiling female worker (Fig. 206).

The climax of the motion versus stasis conflict begins, in the "Spectators and the Screen" sequence (LII), with the juxtaposition of freeze-frames of the auditorium and a tracking shot of the Cameraman as he shoots from a motorcycle speeding toward the audience. This shot alternately appears on both the screen-within-the-screen (Fig. 217) and the real screen (Fig. 216). Then, a speeding locomotive (shot with the camera attached to its side) appears on the screen-within-the-screen (Fig. 208), followed by a composite shot of the Cameraman superimposed above the milling crowd (Fig. 211). The climax is further intensified by yet a greater perceptual separation of moving and stationary shots. This distinction is achieved by an arithmetic exchange of stationary close-ups of the film spectators with moving objects photographed by a panning camera, while the editing pattern is established by a progressive reduction in the number of alternating groups of shots, in the following manner:

Four close-ups of the spectators versus a carriage moving in front of the camera

Seven close-ups of the spectators versus a trolley moving in front of the camera

Four close-ups of the spectators versus a woman getting out of a trolley and moving in front of the camera

Three close-ups of the spectators versus a bicyclist passing in front of the camera

One close-up of a spectator versus two female pedestrians walking in front of the camera

Two close-ups of the spectators versus one pedestrian (the lady with a white hat) moving in front of the camera

Two close-ups of the spectators versus a car speeding in front of the camera

One close-up of a spectator versus a motorcycle driven in front of the camera

One close-up of a spectator versus a cabriolet speeding in front of the camera

One close-up of a spectator versus the front upper part of a trolley moving in front of the camera

One close-up of a spectator versus rear upper part of a trolley moving in front of the camera

One close-up of a spectator versus a carriage with women and children moving in front of the camera

The pattern is broken by the inclusion of a single medium shot of pedestrians (two ladies) walking in front of the camera from left to right, followed by one close-up of a spectator juxtaposed to a shot in which a speeding train moves laterally from right to left. This right–left movement of the train is then repeated four times on the screen-within-the-screen. The disruption of the montage pattern is therefore heightened by a reversed direction of movement: while all the previous movements occur from left to right (forming a curve that approaches the camera and obscures the lens' view as it pans in the direction of movement), the train moves from right to left as it crosses the screen-within-the-screen, first horizontally, then diagonally. The auditorium is reintroduced, this time with the spectators fidgeting in their seats, while the motion picture images projected on the screen-within-the-screen show the Cameraman shooting from a moving convertible (Fig. 226), which later appears on the real screen (Fig. 225); a carriage with two ladies and a man in a white suit, on both screens (Figs. 214, 215); and numerous shots of the traffic. A long shot of the Cameraman superimposed over a milling crowd (Fig. 211), a split image of the Bolshoi Theater (Fig. 213), and a close-up of a swinging pendulum (Fig. 212) introduce the "Editor and the Film" sequence (LV).

Structurally, the film builds step by step, from an apparently linear presentation of events (preparation for a film show, with the audience filling the auditorium and an orchestra beginning to play for the screening) toward a montage condensation of innumerable visual components, creating a kinesthetic abstraction of "Life-As-It-Is" that is more powerful – and more meaningful – than the representational content of each individual shot. The opening of the film is explanatory, as it provides the audience with basic information about cinema as a means of communication, while the end reaffirms cinema as a means of artistic expression, a unique medium capable of constructing its own time–space integrity.

The Man with the Movie Camera ends majestically, like the final chord of a great symphony. Its abrupt ending indicates that the true concern of its creators lay not so much in the mechanical recording of reality as with its cinematic transposition. Obviously, they chose to leave the viewer not with a thematic – nor even an ideological – statement about the concrete environment (a Soviet city) but with an exciting emotional experience that acknowledges cinema as a unique means of *aesthetic* construction. The "Film-Truth" principle of observing and recording the external world is thus integrated with the "Film-Eye" method used by the *kinoks*-engineers to deconstruct the recorded images and reveal the substantial truth about everyday reality, about "Life-As-It-Is," about those who produce the "film-thing," and about those who perceive it in the movie theater – both the real one and the one within the film.

Structural recapitulation

The "Camera Moves on Tripod" sequence (LI) serves as a preparation for the Epilogue, in which the audience attends a film screening. The first shots of this sequence are close-ups of men and women responding to the camera as it "walks" on its "legs"; hence the impression that they are amused by an "animated fact" incompatible with reality – an illusion further enhanced by the placement of the "walking" camera in an exterior. Suddenly, among the close-ups of the same people, there are inserted long shots of the audience sitting in front of the screen-within-the-screen on which appear various moving images of real events. These shots, marking the transition to the "Spectators and the Screen" sequence (LII), employ the composite image technique whenever both the audience and the screen-within-the-screen are seen simultaneously. The images on the screen-within-the-screen, in these composite shots, are always in motion, while the audience in front of it is sometimes "alive" and sometimes "frozen," thus further obscuring the border between reality and cinematic illusion.

The "Editor and the Film" sequence (LV) concludes the film by reemphasizing – both thematically and cinematically – the prominent role of montage in filmmaking. This sequence contains 108 shots in accelerated forward and reverse motion. The first shot of a dark film-strip moving diagonally over the brightly lit lighting board (Fig. 78) is followed by a close-up of Svilova as she looks downward (Fig. 80). These two shots are intercut four times before a close-up of Svilova's eyes appears on the screen, marking the beginning of the climactic part of the sequence (Fig. 227). Throughout the segment, the close-up of Svilova's eyes is repeated thirty-eight times with a progressive reduction of frames. Toward the end, brief shots of her eyes are contrasted with longer shots depicting traffic on the street; sporadically inserted, they produce another subliminal effect associated with the light pulsation of the film projection. The initial duration of the eye shots is between two and three frames (as, e.g., in nos. 20–45); however, the ensuing close-up of a moving trolley (no. 46) is twelve frames long, and the medium shot of the crowded theater (no. 47) is only eight frames long. The climax of the segment is composed of fifty-five shots, all of which belong to one of two groups: shots of two or three frames (close-ups of Svilova's eye) and shots of six or seven frames (traffic, pedestrians, the projector beam in the auditorium, a traffic signal, and a pendulum). Of all these flash shots, the projector beam (Fig. 228), the traffic signal (Fig. 229), and the pendulums – both the large one, which makes minute movements (Fig. 212), and the smaller one, which sweeps rapidly from side to side (Fig. 231) – carry the greatest weight in the montage rhythm. After the nine-frame shot of a speeding train photographed from below (no. 54), the duration of the shots alternates at a rate of three frames (Svilova's eyes) to six frames (trolley, traffic, auditorium). This geometric reduction of the shot length creates a visual abstraction that is abruptly terminated by the penultimate twenty-frame shot of the trolley running over the camera (Fig. 232) and a seventy-four-frame closing shot of the camera lens with the human eye superimposed over it (no. 108), functioning as a symbolic conclusion to the entire film (Fig. 233).

The montage structure of the "Editor and the Film" sequence (LV) is based on the juxtaposition of two different graphic forms: close-ups of Svilova's eyes (perceived as circles) intercut with close-ups of a pendulum (perceived as an unstable circular design) variously scaled shots of trolleys, the projector beam and the auditorium, a traffic signal, and a busy intersection – all of which fill the frame with vertical or diagonal patterns (shots 55–91). In this introductory section, the close-ups of the traffic signal (silhouetted against the gray sky) and the medium long shots of the projector beam (horizontally piercing the dark auditorium) contrast sharply with the circular forms of Svilova's eyes. In the closing section, Svilova's eyes are excluded, and the film ends with eighteen flash shots of the traffic signal, pendulum, auditorium, city market, railway cars, railroad tracks, trolleys, and the camera lens (shots 92–108).

What follows is a shot-by-shot breakdown of the last sequence of the film, its final 108 shots (out of a total of 1,682 shots in the film). From the specification of each shot's duration (given in frames) and the cranking speed (acceleration/deceleration), one can sense the fervent pace that distinguishes the film's finale:

LV. *Editor and the Film:* 108 shots, 699 fr. 39 sec.

1. HACU (similar to XXV, 50; accelerated motion). Dark filmstrip on light table moving from lower left to upper right (13 fr.).

2. LACU (similar to XVI, 11; accelerated). Right profile of Svilova looking down, face lit from below against dark background (15 fr.).

3. HACU (same as LV, 1; accelerated). Filmstrip continues to move diagonally from lower left to upper right (10 fr.).

4. LACU (same as LV, 2; accelerated). Svilova looks down (10 fr.).

5. HACU (same as LV, 3; accelerated). Filmstrip moves diagonally (5 fr.).

6. LACU (same as LV, 4; accelerated). Svilova looks down (4 fr.).

7. LAMLS–MS (similar to LIV, 4; tilted frame, accelerated). Dark strip of railway track extends from middle left to lower right with telephone poles silhouetted against gray sky and receding into distance; dark locomotive appears passing to right through foreground so that only lower part of coaches are seen; train moves quickly over the camera positioned below the track (28 fr.).

8. HAMLS (accelerated). Crowd milling about a market and filling the screen; in the lower right corner of the frame are brightly lit rectangular booth tops (55 fr.).

9. LAMLS (same as LIV, 64; accelerated). Train moves from background to foreground with landscape of a hill on either side. There is a shack on the left and a few trees silhouetted against the sky. As the train moves forward it fills the screen with its dark shape (10 fr.).

10. ECU (accelerated). Svilova's eyes and nose lit from below; the eye on the left is cut in half by the frame line; the eye on the right looks to the left, down center, then left, then center again (9 fr.).

11. LAMS (similar to LV, 7; accelerated). Dark train coaches pass over the camera positioned below the tracks; they are moving toward the audience; hill, trees, and sky are seen in upper right (9 fr.).

12. ECU (same as LV, 10; accelerated). Svilova's eyes move down center, then left, and back to center (6 fr.).

13. LAMS (same as LV, 11; accelerated). Dark train coaches move forward over the camera (6 fr.).

14. ECU (same as LV, 12; accelerated). Svilova's eyes move down center, left, then close and open again (5 fr.).

15. LAMS (same as LV, 11; accelerated). Dark train coaches move over the camera (9 fr.).

16. ECU (same as LV, 14; accelerated). Svilova's eyes move left, then center and down (4 fr.).

17. LAMS (same as LV, 15; accelerated). Dark train coaches move forward over camera (7 fr.).

18. ECU (same as LV, 16; accelerated). Svilova's eyes move down center, then left (3 fr.).

19. LAMS (same as LV, 17; accelerated). Dark train coaches move forward over the camera and pass it, then exit, revealing track and landscape with a hill on either side and a few trees silhouetted against sky (8 fr.).

20. HACU (same as LV, 1; accelerated). Dark filmstrip on the light table moving from lower left to upper right (5 fr.).

21. ECU (same as LV, 18; accelerated). Svilova's eyes look down center (2 fr.). Note: light from below is stronger than in previous shot.

22. HACU (same as LV, 20; accelerated). Filmstrip continues to move diagonally (4 fr.).

23. ECU (same as LV, 21; accelerated). Svilova's eyes look down center, then right (3 fr.).

24. HACU (same as LV, 22; accelerated). Filmstrip continues to move diagonally (5 fr.).

25. ECU (same as LV, 23; accelerated). Svilova's eyes look right, then move left (3 fr.).

26. LACU (similar to XVIII, 10; accelerated). Traffic signal, silhouetted against gray sky, rotates from right to left, changing diagonal position from lower left–upper right to upper left–lower right (65 fr.).

27. MLS. Trolley stands in center of screen; "No. 25–Saratovo-Leningrad-skaya" is written on top of it, with blurred pedestrians crossing from left to right in foreground; part of trolley on right seen moving toward background with buildings on left (7 fr.).

28. ECU (same as LV, 25; accelerated). Svilova's eyes look down, in center (2 fr.).

29. MLS (accelerated). Pedestrians passing in front of moving trolley in right foreground (7 fr.).

30. ECU (same as LV, 28; accelerated). Svilova's eyes look down, then move left (2 fr.).

31. MLS (accelerated). Pedestrians pass in front of moving trolley in right foreground (6 fr.).

32. ECU (same as LV, 30; accelerated). Svilova's eyes move from right to left (2 fr.).

33. MLS (same as LV, 31; accelerated). A few pedestrians obstruct background as they pass from left to right, extreme foreground; trolley moves forward (6 fr.).

34. ECU (same as LV, 32; accelerated). Svilova's eyes look right (2 fr.).

35. MLS (same as LV, 33; accelerated). One person, passing from left to right extreme foreground, obstructs background; trolley moves forward (5 fr.).

36. ECU (same as LV, 34; accelerated). Svilova's eyes look left (2 fr.).

37. MLS (same as LV, 35; accelerated). Person obstructs background while passing from right to left, extreme foreground; trolley moves forward (6 fr.).

38. ECU (same as LV, 36; accelerated). Svilova's eyes look right (2 fr.).

39. MLS (same as LV, 37; accelerated). Person obstructs background while passing from right to left, extreme foreground; trolley moves forward (6 fr.).

40. ECU (same as LV, 38; accelerated). Svilova's eyes look left (2 fr.).

41. MLS (same as LV, 39; accelerated). Person obstructs background while passing from right to left, extreme foreground; trolley moves forward (6 fr.).

42. ECU (same as LV, 40). Svilova's eyes look down right (2 fr.).

43. ECU–MS (accelerated). Dark blurred side of trolley in extreme foreground moves right and reveals tall buildings in left background; pedestrians and another trolley (No. 25) in center middle ground moving forward (6 fr.).

44. MLS (accelerated). Projector beam coming from small square aperture in upper left corner and extending diagonally toward center right against completely dark background; a thick black strip breaks the beam near the aperture, and another thin white vertical strip breaks it in the center; the intensity of the beam is diffused in the right; it dims for a moment, then reappears (6 fr.).

45. MLS (accelerated). Trolley in center moves forward as blurred pedestrians pass to right in foreground (2 fr.).

46. ECU–MLS (accelerated). Side of trolley in extreme foreground moves left, revealing trolley in center, part of trolley on right, and buildings and sky in distant background (12 fr.).

47. HAMLS (accelerated). Dark theater auditorium with audience facing left; pulsating projector beam positioned from upper center background toward upper left foreground (8 fr.).

48. ECU–MLS (same as LV, 43; accelerated). Dark blurred side of trolley in extreme foreground moves right, revealing tall buildings in left background and another trolley (No. 18) in center middle ground; pedestrians milling about (6 fr.).

49. ECU (same as LV, 42; accelerated). Svilova's eyes look down, then move left (2 fr.).

50. MLS (same as LV, 27; accelerated). A few pedestrians pass to right foreground in front of standing trolley (5 fr.).

51. ECU (same as LV, 49; accelerated). Svilova's eyes look left, then move right (3 fr.).

52. HALS (same as LII, 89; accelerated). Street filled with crowd moving in various directions and occupying lower part of the screen; dark landscape and gray sky above in background; on right, shadowed row of telephone poles receding from foreground to center background; dark trolley moving from left foreground to center background (6 fr.).

53. ECU (same as LV, 51; accelerated). Svilova's eyes look down, then move left (3 fr.).

54. LAMCU (similar to LII, 87; accelerated). Bottom of train car moving diagonally from middle right to lower left with blurred vertical white stripes flashing in the foreground (9 fr.).

55. ECU (same as LV, 53; accelerated). Svilova's eyes look down, then move right (3 fr.).

56. HALS (same as LV, 52; accelerated). Trolley continues to move toward center background in the midst of a crowd (6 fr.).

57. ECU (same as LV, 55; accelerated). Svilova's eyes look down (3 fr.).

58. HAMLS (same as LVI, 46; accelerated). Dark theater auditorium with audience facing left; pulsating projection beam positioned from upper center foreground to upper left foreground (6 fr.).

59. ECU (same as LV, 57; accelerated). Svilova's eyes look down (3 fr.).

60. HALS (same as LV, 56; accelerated). Trolley moves further into background (6 fr.).

61. ECU (same as LV, 59; accelerated). Svilova's eyes look right and open wide (3 fr.).

62. ECU (same as LV, 2; accelerated). Round brightly lit pendulum swings back and forth against dark background (6 fr.).

63. ECU (same as LV, 61; accelerated). Svilova's eyes look down, then she tilts face to left (3 fr.).

64. HALS (same as LV, 60; accelerated). One trolley disappears while another trolley enters blurred, left foreground (6 fr.).

65. ECU (same as LV, 63; accelerated). Svilova's eyes look down center (2 fr.).

66. LACU (same as LV, 26; accelerated). Traffic signal, diagonally positioned (from upper left to lower right) and silhouetted against gray sky rotates to right and comes into a vertical position, with black wire on left (5 fr.).

67. ECU (same as LV, 65; accelerated). Svilova's eyes look left, then move down (2 fr.).

68. HALS (same as LV, 64; accelerated). Trolley in left foreground continues to move right, obstructing the background; through the windows of the trolley one sees pedestrians on the other side of the street (6 fr.).

69. ECU (same as LV, 67; accelerated). Svilova's eyes look down center (2 fr.).

70. LAMCU (same as LV, 54; accelerated; panning, tilted frame). Blurred side of train moves upper right to lower left; as shot moves left, more windows and a strip of sky are revealed (6 fr.).

71. ECU (same as LV, 69; accelerated). Svilova's eyes look left (2 fr.).

72. HALS (same as LV, 68; accelerated). Trolley in foreground moves further to right, revealing milling crowd in background (6 fr.).

73. ECU (same as LV, 71; accelerated). Svilova's eyes look down, then move right (2 fr.).

74. HAMLS (same as LVII, 58; accelerated). Dark theater auditorium with audience facing left; pulsating projector beam positioned from upper center foreground to upper left foreground (6 fr.).

75. ECU (same as LV, 73; accelerated). Svilova's eyes look down (2 fr.).

76. HALS (same as LV, 72; accelerated). Crowd milling in street, no trolley (6 fr.).

77. ECU (same as LV, 75; accelerated). Svilova's eyes look down (2 fr.).

78. ECU (same as LV, 62; accelerated). Pendulum moves back and forth against dark background (6 fr.).

79. ECU (same as LV, 77; accelerated). Svilova's eyes look down (2 fr.).

80. HAELS (same as XVI, 38; accelerated). Dark shot of city intersection with lit building; in upper left pedestrians and traffic move through the streets (6 fr.).

81. ECU (same as LV, 79; accelerated). Svilova's eyes look down (2 fr.).

82. LACU (same as LV, 66; accelerated). Traffic signal silhouetted against sky rotates right, coming into diagonal position, upper left to lower right (4 fr.).

83. ECU (same as LV, 81; accelerated). Svilova's eyes look down, then left, and open wide (2 fr.).

84. HAELS (same as LV, 80; accelerated). Traffic continues to move at intersection (6 fr.).

85. ECU (same as LV, 83; accelerated). Svilova's eyes look left, then right (2 fr.).

86. LAMLS (similar to LV, 54; accelerated). Train moves from right foreground diagonally to lower left background against a light sky (6 fr.).

87. ECU (same as LV, 85; accelerated). Svilova's eyes look right (2 fr.).

88. HAELS (same as LV, 84; accelerated). Traffic continues to move at intersection (5 fr.).

89. ECU (same as LV, 87; accelerated). Svilova's eyes look left (2 fr.).

90. MLS (same as LV, 44). Diagonal projector beam pulsates, with an additional spot of light in upper right just above the beam (9 fr.).

91. ECU (same as LV, 89; accelerated). Svilova's eyes look left, then right (2 fr.).

92. LAMLS–MS (similar to LVI, 54; accelerated). Blurred windows of train moving diagonally from upper right to lower left (6 fr.).

93. LACU (same as LV, 82; accelerated). Traffic signal, silhouetted against

gray sky, rotates from diagonal position on upper right to vertical position on lower left (3 fr.).

94. CU (same as LIV, 2). Blurred small circular pendulum of clock with quatrefoil frame moving back and forth against black background (2 fr.).

95. HAMLS (same as LV, 74; accelerated). Dark auditorium with audience facing left; pulsating projector beam positioned from upper center foreground to upper left foreground; one person is at lower brightly lit left corner (7 fr.).

96. LACU (same as LV, 93; accelerated). Traffic signal silhouetted against gray sky rotates from diagonal position on upper left to lower right, then reenters upper right to lower left with wire on left (5 fr.).

97. HAMLS (same as LV, 8; accelerated). Market, crowd milling about and filling entire frame; four brightly lit rectangles formed by tops of roofs (6 fr.).

98. CU (same as LV, 94; accelerated). Pendulum moves back and forth (5 fr.).

99. LACU (same as LV, 96; accelerated). Traffic signal silhouetted against gray sky rotates clockwise from a vertical to a diagonal position, then from an upper right to a lower left position (4 fr.).

100. HAMS (same as LV, 59; accelerated). Crowded market with people dressed in dark and light clothes move quickly in all directions; trolley wires intersect on left (6 fr.).

101. CU (same as LV, 98; accelerated). Pendulum moves back and forth (5 fr.).

102. LAMLS (similar to LV, 19; accelerated). Railroad ties in center receding toward background; small hills on left and right; sky filling upper part of frame; silhouetted coach appears in the foreground, moves over camera toward background (6 fr.).

103. HALS (same as LIV, 69; accelerated). Street with buildings and shadow upper left; light building in upper center and gray building in upper right; crowd moves quickly in all directions (6 fr.).

104. LACU (same as LV, 100; accelerated). Traffic signal silhouetted against gray sky rotates counterclockwise from an upper right to a lower left position, then from upper left to lower right (4 fr.).

105. LAMLS (similar to LV, 103; accelerated). Railroad ties in center receding toward background, with V-shaped telephone poles on right; railway coach silhouetted against light sky moves toward foreground over camera, revealing horizontal strip of brightly lit sky near top; rest of frame dark (6 fr.).

106. CU (same as LV, 102). Pendulum moves back and forth (2 fr.).

107. LAMLS–MS–CU (same as XXI, 18). Front of trolley with bright sky above and dark trees on both sides; trolley moves forward, then passes over camera, filling entire screen with darkness (frames 12–18); trolley exits, revealing horizontal strip of light with sky and trees at bottom of frame. "No. 812" at bottom of trolley, "No. 7" on top (20 fr.).

108. ECU. Camera lens with human eye superimposed over it, both reflected in a mirror; after twenty frames, the iris begins to close over the staring eye (74 fr.).

Black screen with title "Konets" ("The End") in white Cyrillic letters that slowly fade away (69 fr.).

Structurally, the closing sequence produces a vigorous montage out-burst in which the optical beat is conceived as the finale of a symphonic score. The "auditory" overtones triggered in the "Musical Performance with Spoons and Bottles" sequence (L) are replaced here by the optical sensation achieved through a dynamic exchange of shots figuratively connoting the "dissonance" of city life and parallel-ing the "musical" dynamism of human work (including film editing).

The kinesthetic orchestration of shots in this climactic section is achieved by the juxtaposition of different shot compositions, directions of movement, scales, and durations of shots. Once again, the idea of editing is brought to the fore: the close-ups of Svilova's eyes are so brief that, at the sequence's peak, the eyes are perceived merely as two circular shapes punctuating the series of different graphic patterns. With all its abstraction, the musical connotation of this closing sec-tion does not destroy the thematic denotation of shots depicting (1) traffic (trolley, cars); (2) time (the pendulum); (3) people in everyday life (market and street); (4) machines (the train, tracks); (5) film pro-jection (pulsating beam in the movie theater); and, in the closing im-age, (6) the camera lens and human eye. All thematic and formal aspects of the film are reintegrated by way of its most potent means – montage – which builds the ending into a powerful symphonic struc-ture whose kinesthetic intervals overcome the viewers on both percep-tual and sensorial levels.

Ideology of graphic patterns

Formal examination of key sequences in *The Man with the Movie Cam-era* leads to further interpretation of Vertov's and Svilova's selection of graphic patterns and their interaction. All matching or juxtaposing of shot compositions is intentional and executed with precision, espe-cially in the pivotal sequences wherein the montage beat is a domi-nant aspect of the sequence's structure and an essential source of its kinesthetic impact. By grouping the graphic patterns according to their pictorial components, one can discern appropriate meaning for each of the three basic shot compositions as they appear on the screen in clusters and different contexts.

The most dominant pattern in both shape and movement is *horizon-tal,* appearing in numerous shots that depict streets, traffic, trains, cars, pedestrians, country landscapes, workers in action, and athletes exercising. The next prevailing pattern is *vertical,* expressed most dramatically in wide-angle views of factories, smokestacks, chimneys, steel constructions, elevators, cranes, street lights, and tall buildings in the urban environment. The third is *circular,* a pattern that tends to envelop all forms and shapes by virtue of its visually prominent aestheticism. The circular pattern also appears to be significant on the thematic level, and is underscored throughout the film by close-ups of the camera lens, film reels, human faces and eyes, factory wheels and gears, radio loudspeakers, and steel construction equipment. Symbolically, *film stock as material for filmmaking* appears on the screen in all three forms: horizontally (and diagonally), as strips of film (footage) displayed by Svilova on her editing table; vertically, as the classified and selected piece of film footage hung on the shelves in the editing room; circularly, as round film cans ready for shipment, as strips of film being wound into cores, and as 35mm metal reels used for projection in the movie theater.

The repetition of the three basic graphic patterns in *The Man with the Movie Camera* assumes a broad metaphorical implication, associated with the particular thematic representation of each shot. As part of a cognitive system, the three basic patterns can be related to the film's thematic meaning in the following way:

Pattern	Content	Theme
Horizonal	Cars, trains	Communication
Vertical	Smokestacks, cranes	Industry
Circular	The eye, the lens	Creativity

The ideological implication of these patterns depends upon and varies according to the content of the shot. If one analytically focuses on the specific pattern, it turns out that its recurrent association with the respective diegetic connotation tends to develop into an ideological metaphor. For example, the vertical design of factories, tall buildings, and constructions – especially when reenforced by the low camera angle – parallels industrial construction in the new society. The horizontal movement of traffic (trolleys and trains) is naturally linked with the expansive development of the urban environment in postrevolutionary Russia. Of course, horizontal movement is often combined with topics of opposite significance (e.g., bourgeois women driven in a carriage) and can be read as a dialectical unity of opposites achieved through the disruptive–associative montage principle.

An outstanding example of such a dialectical contradiction of movements is demonstrated at the end of the "Death, Marriage, and Birth" sequence (XX), generating a philosophical association with the photographed event – the funeral procession shown only in three shots. In the first shot, the camera shoots from a high angle as it moves parallel to the limousine, which carries an open casket and is progressing *laterally* (Fig. 97). The two parallel movements neutralize the hearse's progression by focusing on the dead man's face in the open casket – the stasis of death. After two disruptive inserts, the stationary shots of a bride (Fig. 98) and a woman experiencing the pain of child labor (Fig. 99), there is another tracking shot of the same funeral procession, this time photographed by the camera withdrawing directly in front of the limousine carrying the open casket (Fig. 100). The position of the camera and its movement in this shot contribute to the viewer's (sensory and emotional) participation in the funeral procession.

In the sequences of street traffic, the lateral movement of the trolleys and pedestrians crossing from left to right or from right to left either obliterates actual physical movement, implying dislocation, or pokes fun (through split-screen, multiple exposure, oblique angles, acceleration) at the hectic tempo of a great city. The positioning of the camera on trains, convertibles, fire engines, motorcycles, carousels, carriages, or simply held by the Cameraman, emphasizes the dynamism of movement through space.

The dislocation of movement that results from the "battle" between different graphic patterns implants itself in the viewer's consciousness, especially when it occurs in composite shots, as in those with flocks of pedestrians moving in various directions and graphically "colliding" with each other within the same frame (Plate 3). Generally, the crowd's movement is perceived as horizontal, whereas the composite shots combine the horizontal movement of the trolleys and pedestrians with the vertical rectangular forms of different vehicles as seen from the rear. In addition, the horizontal traffic movement is counteracted by vertical telephone and electric poles, traffic signs, tall buildings, and the Cameraman himself as he stands above and superimposed over the crowded streets (Fig. 211).

There are also other graphic patterns – diagonal, elliptical, amorphous, zigzag, intercrossing, and disconnected lines – all of which are incorporated within the three basic patterns and movements. The accumulation of particular shapes and movements prompts the viewer to link graphic repetitions with their ideological meanings and thereby arrive at a broader metaphoric reading of the sequences in which the specific pattern dominates. Furthermore, various complementary optical changes on the level of lighting and tonality set off the basic graphic configurations underscored by the dominant patterns in each shot.

The consistent repetition of various graphic patterns as well as their optical integration creates a visual metaphor that comments on the three major themes of *The Man with the Movie Camera:* communication, industry, and the creative process (filmmaking). The three basic graphic patterns are always related to the actual objects and movements occurring in the concrete environment: the sequential progression of the traffic, the verticality of the industrial constructions, and the concentric and/or circular motion of human work/creation. Through their optical mutation, these three graphic patterns generate abstract imagery with symbolic implication:

Graphic pattern	Metaphorical implication
Horizontal (sequential movement)	Environmental progress (traffic)
Vertical (upward development)	Industrial construction (technology)
Circular (concentrated activity)	Human work (creativity)

In the climactic moments of the major sequences, the interaction of the three basic graphic patterns produces a unique "overtonal impact" that, according to Eisenstein, provides the most powerful and uniquely cinematic experience.[26] After an in-depth analysis of the film's structure and repeated viewing of the film, one begins to perceive Vertov's intervals as overtones that kinesthetically fuse the horizontal, vertical, and circular movements perceived in the juxtaposed shots. The metaphorical significance of these overtones exceeds the representational denotation of the photographed objects, and reaches a point where the recorded "life-facts" are not merely "the sum of the facts," but "a higher mathematics"[27] of the recorded, deconstructed, and reconstructed events.

A cornerstone of world cinema

The Man with the Movie Camera demonstrates more than any other film of its time the uniqueness of cinematic language. With its complex formal structure and many levels of thematic meaning, it is Vertov's "theoretical declaration on the screen" of his "Film-Eye" method and the "Film-Truth" principle. It is the outcome of his belief that the "human eye armed with a camera" can reveal the true social reality, the essence of "Life-As-It-Is."[28]

The final shot of *The Man with the Movie Camera* fuses the photographic lens with the eye, reiterating the unification of human perception and technology. As in many other instances, the metaphorical meaning of this merger transcends its obvious signification: it functions as a proposal to the viewer to look at life in a new way, to perceive more than the "naked human eye" can detect, to think not only about what appears right before us but also about what is hidden beneath the surface of reality. That was the *kinoks'* chief objective as Vertov defined it in his early proclamations, following constructivist, futurist, and formalist tenets.[29]

In discussing Mayakovsky, Vertov referred to the poet's work as "a lighthouse"[30] in the world of poetry. The same metaphor can be applied to *The Man with the Movie Camera:* it was, it is, and will continue to be a lighthouse illuminating the path that leads cinema toward a revolutionary art form. As a reaction against fictional (staged) as well as conventional documentary films, produced during a period of intensified political suppression of avant-garde movements and "ideologically inappropriate" experimentation, misunderstood by its contemporaries, proclaimed "inaccessible" to the masses, ideologically attacked from both left and right, rejected by domestic and foreign critics as a "formalist visual attraction," Vertov's film has at last received the recognition it deserves as the most innovative achievement in the silent era. Moreover, it has become increasingly evident that *The Man with the Movie Camera* marks a cornerstone in the entire history of world cinema; as such, it can also be viewed as a monument to the freedom of artistic expression, a tribute to the necessity of personal creativity and the right to experiment in any art or medium. Like an underground current, persistent though inconspicuous, it has paved the path all true artists should take in their struggle against the commercial and political exploitation of any art, cinema in particular.

APPENDIX 1

Introductory statement of the film

THE CREDITS

The Man
with the Movie Camera

A Record on Celluloid
IN 6 REELS

Produced by VUFKU
1929

(An Excerpt from the Diary
of a Cameraman)

(365 fr.)

ATTENTION

VIEWERS:

THIS FILM

Represents in Itself

AN EXPERIMENT
IN THE CINEMATIC COMMUNICATION

Of Visible Events

(629 fr.)

WITHOUT THE AID OF
INTERTITLES

(A Film Without Intertitles)

(765 fr.)

WITHOUT THE AID OF
A S C E N A R I O
(A Film Without a Script)

(901 fr.)

WITHOUT THE AID OF
T H E A T E R
(A Film Without Sets,
Actors, etc.)

(1,061 fr.)

THIS EXPERIMENTAL
WORK WAS MADE WITH THE INTENTION OF
CREATING A TRULY
INTERNATIONAL
ULTIMATE LANGUAGE OF
CINEMA ON THE BASIS OF ITS
T O T A L S E P A R A T I O N
FROM THE LANGUAGE OF THEATER
A N D L I T E R A T U R E

(1,476 fr.)

Author-Supervisor of
the Experiment
Dziga VERTOV

(1,580 fr.)

Chief Cinematographer
M. KAUFMAN

(1,660 fr.)

Assisting Editor
E. SVILOVA

(1,776 fr.)

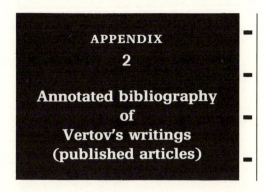

Since the sources of information for this bibliography were often secondary, the original dates of publication as well as pagination are not always included. When not indicated, the place of publication is Moscow. Complete citations for Vertov's unpublished articles can be found in Seth Feldman's *Dziga Vertov: A Guide to References and Resources*, pp. 183–220.

1922

"We. A Variant of a Manifesto" [*My. Variant manifesta*]. *Kinofot*, no. 1 (August 25–31), 11–12.

The first published theoretical statement in which Vertov outlines his concept of the newsreel and documentary cinema, emphasizing his directorial method, "Film-Eye," his principle of shooting, "Film-Truth," and his concept of presenting material on the screen, "Life-As-It-Is." According to Vertov, the first variant of this manifesto was written in 1919 under the title "A Manifesto About the Disarmament of Theatrical Cinema" [*Manifest o razoruzhenii teatral'noi kinematografii*].

"He and I" [*On i ia*]. *Kinofot* (September), 9–10.

A sequel to the first statement. It reviews the first nine issues of the *Film-Truth* series, emphasizing the distinction between the *kinoks'* montage style of presenting everyday events and the theatrical style and content of films produced in America.

"The Fifth Number of Film-Truth" [*Piatyi nomer "kinopravdy"*]. *Teatral'naya Moskva*, no. 50.

An analysis of *Film-Truth No. 5*, emphasizing Vertov's concept of abstracting events (often geographically unrelated) by means of montage.

1923

"Kinoks. Revolution" [*Kinoki. Perevorot*]. *LEF*, no. 3 (June–July), 135–43.

A manifesto about the new concept of "unstaged" documentary film, its aesthetic and social function. The text includes a brief introduction written earlier

203

(January 1923) and signed by the "Council of Three" (Vertov, Kaufman, Svilova).

"Commemorating the Anniversary of Film-Truth" [*K godovshchine "Kinopravdy"*]. *Kino,* nos. 3–7.

In an interview, Vertov surveys the first sixteen issues of the *Film-Truth* series, the material that was chosen to be filmed, as well as the montage structure, and the use of intertitles.

"A Project for Reorganizing Film-Truth" [*Proekt reorganizatsii "Kinopravdy"*]. *Kino,* nos. 3–7.

Vertov's report to Goskino on the first Experimental Unit of newsreel production, suggesting its reorganization into a "Documentary Experimental Station" within Goskino production.

"Film-Truth" [*Kinopravda*]. *Kinofot,* no. 6, 13.

Vertov characterizes the *Film-Truth* series as the most popular "cinema-newspaper" in Russia and outlines the need for the state support of its further development.

"A New Current in Cinema" [*Novoe techenie v kinematografii*]. *Pravda* (July 15), 7.

Vertov summarizes the *kinoks'* declaration that unstaged cinema is the only appropriate response to the commercial photoplay and announces the *kinoks'* plan to produce a "Radio Newsreel."

1924

"Directors on Themselves" [*Rezhissery o sebe*]. *Kino* (March 4).

Vertov answers several questions about his work, emphasizing the unique social function of documentary film.

" 'Film-Eye': A Newsreel in Six Issues" [*Kinoglaz: kinokhronika v 6 seriyakh*]. *Pravda* (July 19), 6.

Vertov summarizes the principles to be applied in shooting the six issues of the *Film-Eye* series. Only the first part, also known as "Life Caught Unawares" [*Zhizn' vrasplokh*], was realized.

"An Answer to Five Questions" [*Otvet na piat' voprosov*]. *Kino,* no. 43 (October 21).

Vertov presents the theoretical platform of the *kinoks,* followed by a critical evaluation of the first issue of *Film-Eye.* It also contains Vertov's critical comments on Gan's film *The Island of the Young Pioneers* [*Ostrov pionerov*] (1924).

1925

"The Basics of Film-Eye" [*Osnovnoe "Kinoglaza"*]. *Kino,* no. 6 (Feb. 3).

Vertov discusses the "Film-Eye" method as the most appropriate means of presenting "Life-As-It-Is," proclaiming it the only way for cinema to "march toward October."

" 'Film-Eye' about 'Strike' " [*"Kinoglaz" o "Stachke"*]. *Kino* (March 24), Kharkov.

Vertov analyzes Eisenstein's *Strike,* characterizing it as the first attempt to apply the "Film-Eye" method and the "Film-Truth" principle to staged (acted) cinema.

" 'Film-Truth' and 'Radio-Truth' " [*"Kinopravda" i "radiopravda"*]. *Pravda* (July 16), 8.

Vertov discusses the possibility of "campaigning with facts," not only in the sphere of viewing ("Film-Truth" and "Film-Eye") but also in the realm of listening ("Radio-Eye"). This article also calls for the application of the "Leninist Proportion" to film production, that is, the balance between the distribution of entertainment movies and the documentary film and newsreel.

"For the Film Newsreel!" [*Za kinokhroniku!*]. *Kino* (December 25).

Vertov delimits the role of the newsreel within "agit-prop" activity as defined in the party's proclamation "About Political Activity in the Villages."

1926

"The Factory of Facts" [*Fabrika faktov*]. *Pravda* (July 24), 6.

Vertov discusses the success of the "Film-Eye" method in documentary films and its influence on staged (acted) films, among which are Eisenstein's *Strike* and *Battleship Potemkin.*

" 'Film-Eye' and the Battle for the Film Newsreel: Three Stages of the Battle" [*"Kinoglaz" i bor'ba za kinokhroniku: tri etapa bor'by*]. *Sovetskii ekran*, no. 15.

Vertov identifies *Film-Weekly, Film-Truth*, and *Film-Eye* as three stages in the process of establishing the creative and ideological platform of the *kinoks.*

"One Sixth of the World" [*Shestaia chast' mira*]. *Kino* (August 17).

In an interview, Vertov describes his film *One Sixth of the World* as the victory of the "Film-Eye" method over the methods used in staged films with actors, sets, and literary plots.

"The Fight Continues" [*Srazhenie prodolzhaetsia*]. *Kino* (October 30).

In a sequel to the previous interview, Vertov discusses the organizational problems faced by the *kinoks* in completing *One Sixth of the World.*

"Film-Eye" [*Kinoglaz*]. In *On the Paths of Art* [*Na putiakh iskusstva*], a collection of articles published by Proletkult, pp. 210–29.

Vertov analyzes the first issue of the *Film-Eye* series; his analysis is followed by a description of the *kinoks'* future projects and of their struggle against staged cinema.

1927

"Dziga Vertov Refutes" [*Dziga Vertov opovergaet*]. *Nasha gazeta* (Jan. 19).

Vertov replies to the Sovkino criticism, explaining the "ideological motives" upon which his theoretical concepts are based. After this article, Vertov was dismissed from Sovkino.

"Letter to the Editor" [*Pis'mo v redaktsiiu*]. *Kinofront*, no. 4.

Vertov replies to the review of *One Sixth of the World* by Ippolit Sokolov printed earlier in the same periodical (no. 2), refuting Sokolov's remark that the film suffers from the inclusion of too many facts and too broad a perspective for the subject.

1928

"Film-Eye and The Eleventh Year" [*"Kino-oko" i 11–i*]. In *The Eleventh Anniversary* [*Odinnadtsatyi*], a collection of articles in Ukrainian, published by VUFKU (Kiev), pp. 17–18.

Vertov characterizes his film *The Eleventh Year* as an experiment in the expression of ideas in "pure film language."

"Film Has Begun to Shout. Our Filmmakers for Sound Film" [*Kino zakrichalo. Nashi fil'mari pro tonfil'm*]. *Kino*, no. 12 (Kiev).

In Ukrainian. Vertov makes a statement on the specific nature and future development of sound film, characterizing it as a new step in the history of cinema.

"The Eleventh Year – Excerpts from the Shooting Diary" [*Odinadtsatyi – Otryvki iz s'emochnogo dnevnika*]. *Sovetskii ekran*, no. 9.

Illustrating five sequences from the script, Vertov comments on the process of shooting the film.

"The First Soviet Film Without Words" [*Pervaia sovetskaia fil'm bez slov*]. *Pravda* (December 1), 7.

Vertov's brief statement describes the structure and goals of the film *The Man with the Movie Camera*, which was then in production.

1929

"Film Is Threatened by Danger" [*Fil'me grozit opasnost'*]. *Novyi zritel'*, no. 5.

In an interview, Dziga Vertov assesses *The Man with the Movie Camera* in the context of the general tendencies of Soviet film production and characterizes it as a "theoretical *exposé* expressed in cinematic terms."

"The Man with the Movie Camera - Ultimate Film-Writing and Radio-Eye" [*"Liudina z kinoaparotom" – Absolutnii kinopis i radio-oko*]. *Nova generatsiia*, no. 1 (Kharkov). In Ukrainian.

Vertov's second and major commentary on his last silent film. This statement, part of which was distributed to the audience during openings in Kiev and Moscow, was published in its entirety only posthumously in Drobashenko's collection of Vertov's articles, *Articles. Diaries. Projects* (1966). The Russian title of the statement reads "*The Man with the Movie Camera* – A Visual Symphony" [*Chelovek s kinoapparatom – zritel'naia symfoniia*].

"A Letter to the Editor" [*Pis'mo v redaktsiiu*]. *Kino* (March 19).

Vertov expresses his indignation at the inappropriate distribution of *The Man with the Movie Camera* in Soviet commercial theaters.

"Watching and Understanding Life" [*Bachiti i chuti zhittia*]. *Kino* (Oct. 1), Kiev. In Ukrainian.

A theoretical discussion of the function of sound in documentary film is related to Vertov's forthcoming project, the production of the first Soviet sound film *Enthusiasm,* or *Symphony of the Don Basin* (1930).

1930

"The March of Radio-Eye" [*Mart radioglaza*]. *Kino i zhizn',* no. 20 (June), 14.
Vertov presents his ideas on using sound in cinema and its relationship to the image in his film *Enthusiasm (Symphony of the Don Basin).*
"The Paths of Documentary Film" [*Puti dokumental'noi fil'my*]. *Kinofront* (May 11).
Vertov replies to the questions formulated by the editors of *Kinofront.* Summarizing his theoretical and artistic goals, he emphasizes the differences between "staged" and "unstaged" cinema, the function of sound in film, and the experimental concept of *The Man with the Movie Camera.*
"Radio-Film-Eye" [*Radio-kino-oko*]. *Vechirnia robitnitsa gazeta* (June 17), Kiev. In Ukrainian.
In an interview, Dziga Vertov outlines the concepts and problems related to the realization of his film *Enthusiasm (Symphony of the Don Basin).*
"Agitational-Propaganda, Scientific-Educational, Instructional Film and the Film-Newsreel" [*Agitatsionno-propagandistskyi, nauchno-uchebnyi, instruktivnyi fil'm i fil'm-khronika*]. In *For Film in the Reconstruction Period,* Moscow, pp. 107–11.
In his statement presented at the First Moscow Conference on nonfeature film, Vertov proposes the formation of a special "film factory" dedicated to the exploration of documentary film and the expansion of the newsreel.

1931

"Film-Eye, Radio-Eye, and So-called 'Documentarism'" [*Kinoglaz, radioglaz i tak nazyvaemyi "dokumentalizm"*]. *Proletarskoe kino,* no. 4.
Vertov contrasts the creative principles of the *kinoks* to the conventional documentary film, presenting a brief survey of the evolution of "Film-Eye" as a new method for capturing reality with a movie camera.
"Let's Discuss the First Sound Film of Ukrainfil'm, *Symphony of the Don Basin"* [*Obsuzhdaem pervuiu zvukovuiu fil'mu Ukrainfil'm, "Simfoniia Donbassa"*]. *Sovetskoe iskusstvo* (Feb. 27).
Vertov explicates the audiovisual structure of his first sound film.
"Whom Does Proletarian Cinema Need?" [*Kto nuzhen proletarskomu kino?*]. *Kino* (March 8).
Vertov replies to questions, asked by editors of the journal, concerning the education of young filmmakers in the areas of documentary and newsreels.
"The First Steps" [*Pervye shagi*]. *Kino* (April 16).
Vertov further elaborates the sight-and-sound counterpoint used throughout *Enthusiasm (Symphony of the Don Basin).*
"In Lieu of a Reply to the Critics" [*Vmesto otveta kritikam*]. *Kino* (June 16).

Vertov defends the ideological meaning and cinematic significance of his film *Enthusiasm*, criticizing the official distributors who insisted on a montage revision of the film for foreign markets.

"Libretto for *Enthusiasm*" [*Libretto za Entuziazm*]. Vertov's statement about sound film has been published only in English in *The Film Society Programme*, performance No. 49 (Nov. 15). Translated by Thorold Dickinson.

Summarizes the three "movements" of the film and characterizes it as the "ice-breaker-in-chief" of the sound newsreel. The original version exists only in manuscript form and is kept in the Dziga Vertov Archive.

1932

"Charlie Chaplin, the Workers of Hamburg, and a Review of Dr. Wirth" [*Charlie Chaplin, gamburgskie rabochie i prikaz doktora Virta*]. *Proletarskoe kino*, no. 3.

Vertov reports on the success of *Enthusiasm* in Germany, expressing his fascination with various aspects of life abroad, concluding with favorable responses from René Clair, Moholy-Nagy, Charlie Chaplin, and the German critic, Wirth.

"Once Again about So-called 'Documentarism'" [*Eshcho o tak nazyvaemom "dokumentalizme"*]. *Proletarskoe kino*, no. 3.

Vertov criticizes "documentarism" as the mechanical application of the "Film-Eye" method to staged films.

"Approaching the Second Five-Year Plan" [*Na podstupakh k vtoroi piatiletka*]. *Proletarskoe kino*, no. 4 (April 19), 5–6.

In reply to the periodical's questionnaire, Vertov outlines his plans for filmmaking over the next five years.

"The Complete Capitulation of Nikolai Lebedev" [*Polnaia kapituliatsiia Nikolaia Lebedeva*]. *Proletarskoe kino*, no. 5 (May), 12–18.

Vertov's rebuttal to Lebedev's criticism of Vertov's "formalism" and "eccentricity" in *The Man with the Movie Camera*, presented in his article "For Proletarian Film Journalism" [*Za proletarskuiu kinopublitsistiku*] published in the journal *Literatura i iskusstvo*, no. 9–10 (Nov. 1931). Vertov condemns Lebedev's orthodox view of art, especially his contention that socialist realism must be applied to, or rather imposed upon, cinema.

"Creative Growth" [*Tvorcheskii pod'em*]. *Kino* (May 24).

Vertov suggests the reorganization of Soviet film production in light of the decree "On the Reorganization of Literary-Artistic Institutions" [*O perestroike literarno-khudozhestvennykh organizatsii*] issued by the Central Committee of the Communist Party.

"In the Front Ranks" [*V pervuiu sherengu*]. *Kino* (Sept. 6).

Vertov's report (which he gave as the representative of Mezhrabpomfil'm Production Company to the International Anti-War Congress in Amsterdam) proposes that cinema be used against warmongers.

1934

"How We Made a Film about Lenin" [*Kak my delali fil'm o Lenine*]. *Izvestiya* (May 24), 6.

Vertov discusses the production of *Three Songs about Lenin,* revealing that during his research, several unknown shots of Lenin were discovered.

"Without Words" [*Bez slov*]. *Rotfil'm* (August 15).

Vertov explains the visual structure of the film *Three Songs about Lenin,* and the difficulties he encountered in the process of shooting.

"Three Songs about Lenin" [*Tri pesni o Lenine*]. *Kommuna* (Nov. 2), Voronezh.

A revised version of the article published in *Rotfil'm.*

"Three Songs about Lenin" [*Tri pesni o Lenine*]. *Sovkhoznaia gazeta* (Nov. 7).

In an interview, Vertov explains that his goal in the film was to emphasize the human aspect of Lenin's character.

"How I Worked on the Film 'Three Songs about Lenin' " [*Kak ia rabotal nad fil'mom "Tri pesni o Lenine"*]. *Proletarskaia pravda* (Nov. 7), Kalinin.

In an interview conducted by the chief editor of the newspaper, Vertov emphasizes the structural nature of the sound track in the film.

"My Report" [*Moi raport*]. *Izvestiya* (Dec. 15), 4.

Vertov gives a short summary of his creative evolution, issued in connection with the fifteenth anniversary of Soviet cinema. It also mentions the success of *Three Songs about Lenin* in New York.

"Three Songs about Lenin" [*Tri pesni o Lenine*]. *Ogonek,* no. 17–18.

Vertov briefly explains the thematic composition and the ideological message of the film.

"Film-Truth" [*Kinopravda*]. *Sovetskoe kino,* no. 11–12, 155–64.

This article was issued in connection with the fifteenth anniversary of Soviet cinema. Vertov elaborates extensively on his creative ideas, emphasizing the fact that *Three Songs about Lenin* expands his "Film-Eye" method and the "Film-Truth" principle.

1935

"Creative Report" [*Tvorcheskii raport*]. *Rabochaia Moskva* (Jan. 5).

Vertov's assessment of his own contribution to cinema. This article was issued in connection with the fifteenth anniversary of Soviet cinema.

"The Last Experiment" [*Poslednii opyt*]. *Literaturnaia gazeta* (Jan. 18).

Vertov explains the making of *Three Songs about Lenin* and its relation to his earlier films.

"Three Songs about Lenin" [*Tri pesni o Lenine*]. *Vechernaia Moskva* (Jan. 21).

Three sequences from the literary script of the film.

"Soviet Directors Speak" [*Govoriat Sovetskie rezhisery*]. *Kino* (March 1).

Vertov gives his impressions of the First International Film Festival in Moscow.

"We Must Arm Ourselves" [*Nado vooruzhat'sia*]. *Sovetskoe kino,* no. 4, 30.

Vertov outlines his ideas on future technical and organizational improvements in Soviet cinema.

"Dziga Vertov on Kino-Eye" and **"Dziga Vertov on Film Technique."** *Filmfront,* no. 2 (Jan. 7), 6–8 and *Filmfront,* no. 3 (Jan. 28), 7–9.

Partial translation from the French (by Samuel Brody) of Vertov's speech delivered during his visit in Paris (January 1929). The original Russian text, under the title "From 'Film-Eye' toward 'Radio-Eye': from the Alphabet of *Kinoks"* [*Ot 'kinoglaza' k radioglazu': iz Azbuki kinokov*], in four parts, containing an explanation of Vertov's "Theory of Intervals," was printed in *Iskusstvo kino,* no. 6 (June), 95–9.

1936

"Lenin in Film" [*Lenin v kino*]. *Sovetskoe iskusstvo* (Jan. 22), no. 22.

Vertov comments on his film dedicated to Lenin and recounts Lenin's remarks made after he saw *Film-Truth No. 13,* entitled *Yesterday, Today, Tomorrow* (1923).

"The Truth about the Struggle of Heroes" [*Pravda o bor'be geroev*]. *Kino* (Nov. 7).

Vertov surveys the first eight issues of the newsreel *On Events in Spain* [*K sobytiiam v Ispanii*], directed by Vladimir Makasseev and Roman Karmen. The same material was used by Esther Shub in her compilation film *Spain* [*Ispaniia*, 1936–1937].

"Song about a Woman" [*Pesnia o zhenshchine*]. *Krest'ianskaia gazeta,* no. 88.

In an interview, Vertov describes preparations for his new project, later to become the sound film *Lullaby* (1937).

1937

"First Encounter" [*Pervaia vstrecha*]. *Krokodil,* no. 29–30.

Vertov's reminiscences of the history and function of the first "campaign-train" *(agit-poezd),* named after M. I. Kalinin and used during the Russian Civil War period.

"Beautiful Life in the Soviet Land" [*Prekrasnaia zhizn' v sovetskoi strane*]. *Kino* (Oct. 12).

Vertov wrote this impressionistic political pamphlet for an occasion – the elections for the Supreme Soviet of the USSR. It also contains a brief reference to the film *Lullaby* (1937).

1938

"Woman in Defense of the Country" [*Zhenshchina v oborone strany*]. *Kurortnaia gazeta* (June 14), Sochi.

Vertov explains his new (unrealized) project with a short summary of a script-in-progress dealing with female Soviet soldiers.

1940

"From the History of the Newsreel" [*Iz istorii kinokhroniki*]. *Kino* (May 5).
Vertov recollects his earliest cinematic experience, focusing on *Film-Weekly* (1919), *Anniversary of the Revolution* (1919), *History of the Civil War* (1922), and *Film-Truth* (1922–5). This is the last article Vertov published before his death in February, 1954.

1957

"From the Working Notebooks of Dziga Vertov" [*Iz rabochikh tetradei Dzigi Vertova*]. *Iskusstvo kino,* no. 4, 112–26.
A lengthy selection of Vertov's writings, all later reprinted in the book *Articles, Diaries, Projects.*

1958

"About the Love for a Real Person" [*O livbvi k zhivomu cheloveku*]. *Iskusstvo kino,* no. 6 (June), 95–9. A draft of this article was written in 1940.
An essay written in the form of an imaginary discussion between Vertov and a film critic on the question of documentary film, emphasizing the "Film-Truth" principle as being the most appropriate means by which documentary cinema can achieve its goals "within the full spectrum of cinematic possibilities."

1959

"Autobiography" [*Avtobiografiia*]. *From the History of Film: Materials and Documents* [*Iz istorii kino: materialy i dokumenty*], II (Moscow: Akademiia nauk SSSR), pp. 29–31.
Dated May 1949. A brief resumé of Vertov's life and achievements.
"Recollections of Filming V. I. Lenin" [*Vospominanie o s'emkakh V. I. Lenina*]. *From the History of Film,* II, pp. 32–5.
Vertov's recollections about filming Lenin and several of Lenin's collaborators.
"The Creative Activity of G. M. Boltyansky" [*Tvorcheskaia deiatel'nost' G. M. Boltyanskogo*]. *From the History of Film,* II, pp. 63–7.
Vertov's eulogy on the work of Boltyansky, an established documentary filmmaker, delivered on March 3, 1945.

1960

"The Song of Two Hundred Million" [*Pesn' dvukhsot millionov*]. *Komsomol'skaia pravda* (March 22).
Vertov's comments on the composition of the film *Three Songs about Lenin,* briefly citing several sequences from the original script.

"Three Songs about Lenin – The Last Experiment" [*Tri pesni o Lenine – poslednii opyt*]. *Sovetskii ekran,* no. 8.

A reprint of the 1935 article "The Last Experiment."

1966

Articles, Diaries, Projects [*Statii, dnevniki, zamysly*], ed. S. Drobashenko. Moscow: Iskusstvo, 1966.

A collection of Vertov's major writings, accompanied by a critical introduction with explanatory comments, sources, and notes.

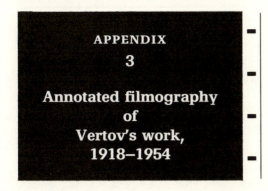

APPENDIX

3

Annotated filmography of Vertov's work, 1918–1954

Duration of films is marked according to the original prints produced in 35mm.

Film Weekly [*Kinonedelia*] 1918–19
A newsreel series each consisting of one reel (12 min. each)
Produced by the Moscow Film Committee of Narkompros (Ministry of Culture)
Director: Dziga Vertov and Mikhail Koltsov
Camera: staff of M.F.C.
The forty-five segments of the series appeared weekly.

The Exhumation of the Remains of Sergei of Radonezh [*Vskrytie moschei Sergeia Radonezheskogo*] 1919
An antireligious agit-film in two reels (25 min)
Produced by the M.F.C.
Author-director and supervisor: Dziga Vertov
Camera: staff of M.F.C.
St. Sergei is one of Russia's most important saints. In the fifteenth century he founded the Trinity-Sergei Monastery in Zagorsk (near Moscow). It became a major center of religious life and still functions today in the Soviet Union.

Anniversary of the Revolution [*Godovshchina revolutsii*] 1919
Compilation film made from *Film-Weekly* material, in twelve reels (150 min.)
Produced by the M.F.C.
Director: Dziga Vertov

Battle of Tsaritsyn [*Boi pod Tsaritsynom*] 1920
Compiled documentary in three reels (40 min.)
Produced by "Revvoensovet" [Revolutionary Military Soviet] and the M.F.C.
Author-director: Dziga Vertov
Camera: staff of M.F.C.
This newsreel covers the famous Soviet offensive against General Anton Denikin and his White Army at Tsaritsyn (December 1919).

The Trial of Mironov [*Protsess Mironova*] 1920
Political newsreel in one reel (12 min.)
Produced by "Revvoensovet" and the M.F.C.
Author-director: Dziga Vertov
Camera: staff of M.F.C.
Filipp Mironov (1872–1921) was a Cossack commander who fought on the

213

side of the Soviets during the Russian Civil War against generals Krasnov and Wrangel.

Instructional Steamer "The Red Star" [*Instruktorskii parokhod "Krasnaia zvezda"*] 1920
A propaganda newsreel in two reels (25 min.)
Produced by VOFKO [All-Russian Photo-Cinema Department in the Ministry of Culture]
Director: Dziga Vertov
Camera: Alexandr Lemberg and Peter Ermolov
The "instructional steamers" were converted river boats that carried artists, Soviet officials, and other propaganda workers down the Volga to spread culture and propaganda among the workers, peasants, and citizens immediately after the revolution.

The All-Russian Elder Kalinin [*Vserosiiskii starosta Kalinin*] 1920
A documentary in two reels (25 min.)
Produced by the M.F.C.
Director: Dziga Vertov
Camera: staff of the M.F.C.
M. I. Kalinin, Soviet president between 1938–46, is shown visiting the agit-train "October Revolution" in early 1920. The agit-trains were sent all over the Soviet Union with propaganda workers who distributed posters, showed films, and performed agit-plays [*agitki*].

The Agit-Train VTSIK [*Agit-poezd VTSIK*] 1921
A travelogue in one reel (12 min.)
Produced by VTSIK [All-Russian Central Executive Committee] and the M.F.C.
Author-director: Dziga Vertov

Commander of the XIII Army, Comrade Kozhevnikov [*Komanduiushchii XIII armiei Tv. Kozhevnikov*] 1921
A compilation film in one reel (12 min.)
Produced by the M.F.C.
Director: Dziga Vertov. Editor: Elizaveta Svilova
Camera: Petr Ermolov
Kozhevnikov was a Soviet general who fought against the White Army at Tsaritsyn. The film compiles footage from various issues of *Film-Weekly* and *Battle of Tsaritsyn.*

History of the Civil War [*Istoriia grazhdanskoi voiny*] 1921
A newsreel in thirteen reels (175 min.)
Produced by VFKO [Central State Film Committee]
Author-director: Dziga Vertov
Camera: staff of VFKO

The Trial of the S.R.s [*Protsess Eserov*] 1922
A political newsreel in three reels (40 min.)
Produced by VFKO
Author-director: Dziga Vertov
Camera: staff of VFKO
The Social Revolutionary Party ("Eser") began in 1901 and was concerned with the interests of the peasants, while the Social Democrats advanced the cause

of the workers in the city. The members of S.R. committed a series of terrorist acts, and one member killed Grand Duke Sergei, uncle of the Tsar and governor-general of Moscow, in 1905. They became the largest political party in Russia. After the October Revolution in 1917, the Bolshevik party and the Social Democrats, suppressed the S.R.s. In Russia, a member of the S.R.s is referred to as an "eser."

Department Store [*Univermag*] 1922
An advertising film, a "film sketch" in two reels (25 min.)
Produced by VOFKO
Director: Dziga Vertov
Camera: staff of VOFKO
"Univermag" stands for *"universal'nyi magazin"* (universal store); one such store, called GUM (State Department Store), had opened in Moscow in Red Square.

Film-Truth [*Kinopravda*] 1922–5
Newsreel series (12 min. each)
Produced by Goskino
Supervisor, author of intertitles: Dziga Vertov
Editor: Elizaveta Svilova
Camera: staff of Goskino
Twenty-three segments were distributed under the general title *Film-Truth*, except for the following segments carrying their own titles:

Yesterday, Today, Tomorrow [*Vchera, segodnia, zavtra*] 1923
    A cinematic poem dedicated to the heroic events of October.
    (*Film-Truth*, no. 13) in three reels (35 min.)
Spring Film-Truth [*Vesennaia Kinopravda*] 1923
    A lyrical newsreel about everyday events (*Film-Truth*, no. 16) in three reels
    (40 min.)
Black Sea–Arctic Ocean–Moscow [*Chernoe more–Ledovityi okean–Moskva*]
1924
    A cinematic visit to a youth camp (*Film-Truth*, no. 20) in one reel (12 min.)
    Camera: Mikhail Kaufman
Leninist Film-Truth [*Leninskaia kinopravda*] February 6, 1924
    A cinematic poem about Lenin (*Film-Truth*, no. 21), in three reels (35 min.)
    Camera: G. Giber, A. Levitsky, A. Lemberg, N. Novitsky, M. Kaufman, E.
    Tisse, and others.
In the Heart of the Peasant Lenin Is Alive [*V serdtse krest'yanina Lenin zhiv*]
March 13, 1925
    A cinematic tale (*Film-Truth*, no. 22) in two reels (25 min.)
    Camera: M. Kaufman, A. Lemberg, and I. Beliakov
Radio Film-Truth [*Radio-Kinopravda*] 1925
    A special experimental newsreel about radio (*Film-Truth*, no. 23), in one
    reel (12 min.)
    Camera: M. Kaufman, I. Beliakov, and A. Bushkin
Five Years of Struggle and Victory [*Piat' let bor'by i pobedy*] 1923
A compilation film in six reels (75 min.)
Produced by VOFKO

Director: Dziga Vertov
Camera: staff of VOFKO
The film was dedicated to the fifth anniversary of the Red Army.

Today [*Segodnia*] 1923
An animated short in one reel (8 min.)
Produced by Goskino
Author-supervisor: Dziga Vertov
Animation: Ivan Beliakov and Boris Volkov
Camera: Mikhail Kaufman
The film depicts in graphic symbols the struggle between the emerging fascist forces and the communists. This is considered the first example of Soviet animation, although Vertov included animated segments earlier in some of his *Film-Truth* series.

Calendar of Goskino Film [*Goskinokalendar*] 1923–5
A newsreel appearing weekly (12 min.)
Produced by Goskino [State Film Enterprise] and Kult'kino [Cultural Film]
Author-director: Dziga Vertov
To distinguish it from *Film-Weekly*, Vertov called this series "Newsreel-Lightning" [*Khronika-molniia*]. Kult'kino was the newly formed documentary section of Goskino, headed by Vertov.

Soviet Toys [*Sovetskie igrushki*] 1924
An animated political satire in one reel (12 min.)
Produced by Kul'tkino
Author-supervisor: Dziga Vertov
Animation: Ivan Beliakov, Alexandr Ivanov, and Alexandr Bushkin
Camera: Alexandr Dorn
The film was commissioned by the State Bank and drew upon the styles of both political newspaper cartoons and ROSTA posters, ridiculing the "Nepman" (speculator taking advantage of the NEP's economic policies).

Humoresques [*Iumoreski*] 1924
An animated propaganda film in one segment (3 min.)
Produced by Kul'tkino
Author-director: Dziga Vertov
Art director: Alexandr Bushkin
Animation: B. Volkov and B. Egerev
Camera: Ivan Beliakov
The film consists of three brief sections, each dealing with the current political and economic situation in the Soviet Union and abroad.

You Have Given Us the Air [*Daesh' vozdukh*] 1924
A special issue of the Calendar in one reel (12 min.)
Produced by Goskino
Author-director: Dziga Vertov
Camera: Mikhail Kaufman
An educational film about aviation.

Film-Eye – Life Caught Unawares [*Kinoglaz – Zhizn' vrasplokh*] 1924
First issue in six reels (82 min.)
Produced by Goskino

Director: Dziga Vertov
Editor: Elizaveta Svilova
Camera: Mikhail Kaufman
The other five issues planned were never produced. The original credits list Vertov as a "Cinema Scout," an allusion to the Soviet youth organization called "Pioneers."

The Foreign Cruise of the Ship of the Baltic Fleet, the Cruiser "Aurora," and the Training Ship "Komsomolets" [*Zagranichnyi pokhod sudov Baltiiskogo flota kreisera "Avroara" i uchebnogo sudna "Komsomolets"*] 1925
A newsreel in one reel (12 min.)
Director: Dziga Vertov
Camera: M. Pertsovich and A. Andiukhin
This is one of the first Soviet films in which footage of foreign countries was shot by Soviet cameramen.

Forward March, Soviet! – The Moscow Soviet in the Present, Past, and Future [*Shagai, Sovet! – Mossovet v nastoiashchem, proshlom i budushchem*] 1926
A film newsreel in seven reels (83 min.)
Produced by Sovkino (formerly Goskino)
Supervisor: Dziga Vertov
Assistant director: Elizaveta Svilova
Camera: Ivan Beliakov
The final intertitle of the film reads: "Socialist Russia will emerge from the Russia of the NEP."

One Sixth of the World – Export and Import of "Gostorg" in the U.S.S.R. – A Race of Film-Eye Around the U.S.S.R. [*Shestaia chast' mira – Eksport i import Gostorga SSSR – Probeg "kinoglaza" po SSSR*] 1926
A cinematic poem in six reels (77 min.)
Produced by Sovkino
Author-supervisor: Dziga Vertov
Assistant: Elizaveta Svilova
Chief cameraman: Mikhail Kaufman
Cameramen: I. Baliakov, S. Bendersky, P. Zotov, N. Konstantinov, A. Lemberg, N. Strukov, and Ia. Tolchen
Film Scouts: A. Kagarlitsky, I. Kopalin, and N. Kudinov
The final title reads: "Into the Current – Of the Common – Socialist – Economy." "Gostorg" stands for the Ministry of Trade.

The Eleventh Year [*Odinnatsatyi*] 1928
An anniversary newsreel in six reels (80 min.)
Produced by VUFKU (All Ukrainian Photo and Film Committee) in Kiev
Author-supervisor: Dziga Vertov
Assistant: Elizaveta Svilova
Camera: Mikhail Kaufman

The Man with the Movie Camera [*Chelovek s kinoapparatom*] 1929
A visual symphony in six reels (95 min.)
Produced by VUFKU
Author-supervisor of the experiment: Dziga Vertov
Montage editor: Elizaveta Svilova
Chief cameraman: Mikhail Kaufman

This is Vertov's last silent film. In his "Summary: Creative Autobiography" [*Konspekt: Tvorcheskaia autobiografiia*], written in 1934 but unpublished, Vertov mentions an earlier version of the film. Seth Feldman, in *Dziga Vertov: A Guide to References and Resources* (p. 110) states that there is no evidence of such a version and that it was not distributed.

Enthusiasm – Symphony of the Don Basin [*Entuziazm – Simfoniia Donbassa*] 1930
A documentary in six reels (96 min.)
Produced by Ukrainfilm VUFKU
Author-director: Dziga Vertov
Assistant: Elizaveta Svilova
Camera: B. Tzeitlin and K. Kuialev
Sound Engineer: Petr Shtro
Music: Timofeyev ("The Don Basin March") and Shostakovich ("The First of May Symphony")
This is Vertov's first sound film. The opening intertitle reads: "The shooting of the audio-visual material has been accomplished by means of the Shorin Recording System on location in mines, factories, etc."

Three Songs about Lenin [*Tri pesni o Lenine*] 1934
A documentary in six reels (70 min.)
Produced by Mezhrabpomfilm (International Workers' Film Enterprise)
Author-director: Dziga Vertov
Assistant: Elizaveta Svilova
Camera: D. Surensky, M. Magidson, and B. Monastirsky
Sound Engineer: P. Shtro
Composer: Yuri Shaporin
In the Vertov Archive in Moscow, there is a montage list and a detailed description of a silent variant of this film. However, Feldman (p. 103) states that there is no evidence of such a version and that it was not distributed. The existing version of the film is approximately sixty minutes long, as restored by Gosfil'mfond in 1970.

Lullaby [*Kolybel'naia*] 1937
A song to the liberated Soviet woman (60 min.)
Produced by Soiuzkinokhronika (Soviet Film Newsreel Production Company)
Author-director: Dziga Vertov
Codirector: Elizaveta Svilova
Camera: Dimitri Surensky and Staff of Soiuzkinokhronika
Composers: Dimitri and Daniil Pokrass
Sound Engineer: I. Renkov
Text of songs: V. Lebedev-Kumach
Commentary: Dziga Vertov

In Memory of Sergo Ordzhonikidze [*Pamiati Sergo Ordzhonikidze*] 1937
A compilation film in two reels (24 min.)
Produced by Soiuzkinokhronika
Directed and edited by Dziga Vertov and Elizaveta Svilova
Camera: Staff of Soiuzkinokhronika
Sergo Ordzhonikidze was commander of the Soviet batallion that routed Denikin's troops in the Ukraine.

Sergo Ordzhonikidze [*Sergo Ordzhonikidze*] 1937
A compilation film in five reels (60 min.)
Produced by Soiuzkinokhronika
Directors: Dziga Vertov, Iakov Bliokh, and Elizaveta Svilova
Camera: M. Oshurkov, I. Beliakov, V. Dobronitsky, Solovev, Adzhibeliashvili
Composer: I. Duanevsky
Sound Engineer: I. Renkov, Semenov
Note: In his "Diaries" (May 1945), Vertov stated that the completed version of this film was subsequently "changed against my will" (*Articles,* p. 263).
The existing version of this film is, actually, an expansion of the newsreel *In Memory of Sergo Ordzhonikidze,* made a few months earlier.

Glory to Soviet Heroines [*Slava sovetskim georiniam*] 1938
A documentary report in one reel (15 min.)
Produced by Soiuzkinokhronika
Author-director: Dziga Vertov
Codirector: Elizaveta Svilova
Camera: staff of Soiuzkinokhronika

Three Heroines [*Tri geroini*] 1938
A documentary portrait in seven reels (90 min.)
Produced by Soyuzkinokhronika
Script: Dziga Vertov and Elizaveta Svilova
Director: Dziga Vertov
Asst. Director: S. Somov
Chief cameraman: S. Semenov
Cameramen: Moskovskoi, Khabarovsky, and staff of the Documentary Studio in Novosibirsk
Cameraman in the train: M. Troyanovsky
Composers: Dimitri and Daniil Pokrass
Sound Engineers: A. Kamponsky, Fomin, Korotkevich
The film depicts the activities of Marina Raskova (a military officer), Valentina Grizodubova (an engineer) and Polina Osipenko (a pilot). Although completed, the film was never released.

In the Region of Peak A [*V raione vysoty A*] 1941
A film chronicle from the front during World War II (known as the War for the Fatherland in Russia), 15 min.
Produced by the Central Studio of the Army Film Chronicle
Director: Dziga Vertov
Camera: T. Buminovich and P. Kasatkin
Part of the regular Army Film Newsreel, no. 87.

Blood for Blood – Death for Death – Crimes of the German Fascist Invaders on the Territory of the U.S.S.R. [*Krov' za krov', smert' za smert' – Zlodeianiia nemetskofashistkikh zakhvatchikov na territorii S.S.S.R*] 1941
A compilation film newsreel (15 min.)
Produced by the Central Studio of the Army Film Newsreel
Author-director: Dziga Vertov
Camera: staff of the Central Studio
A propaganda newsreel.

In the Line of Fire – Newsreel Cameramen [*Na linii ognia – Operatory kino-khroniki*] 1941
A report from the front during World War II, in one reel (12 min.)
Produced by the Central Studio of the Army Film Newsreel
Director: Dziga Vertov
Asst. Director and Editor: Elizaveta Svilova
Aerial shots: N. Vikhirev
Part of the regular "Soiuzkinozhurnal," no. 77.

To You, Front – Kazakhstan's Tribute to the Front [*Tebe, front – Kazakhstan frontu*] 1942
Documentary in five reels (60 min.)
Produced by the Alma-Ata Film Studio
Author-director: Dziga Vertov
Codirector and Editor: Elizaveta Svilova
Camera: K. Pumpiansky
Composers: G. Popov and V. Velikanov
Sound Engineer: K. Bakk
Text of poems: V. Lugovsky
A dramatized documentary with the well-known folksinger Nurpeis Baiganin, who tells the story of a Kazakhstan soldier fighting against the Germans.

In the Mountains of Ala-Tau [*V gorakh Ala-Tau*] 1942
Documentary in two reels (25 min.)
Produced by Alma-Ata Film Studio
Author-director: Dziga Vertov
Codirector and Editor: Elizaveta Svilova
Camera: B. Pumpiansky

The Oath of Youth [*Kliatva molodykh*] 1944
Documentary in three reels (40 min.)
Produced by the Central Studio of the Army Film Newsreel
Directors and Editors: Dziga Vertov and Elizaveta Svilova
Camera: I. Beliakov, G. Amirov, B. Borkovsky, B. Dementev, S. Semenov, V. Kositsyn, and E. Stankevich
The film deals with the activities of the Komsomol during World War II.

News of the Day [*Novosti dnia*] 1944–54
Daily film newsreel (12 min. each)
Produced by the Central Studio of the Army Documentary Film
Dziga Vertov supervised the following issues:
1944: no. 18
1945: nos. 4, 8, 12, 15, 20
1946: nos. 2, 8, 18, 24, 34, 42, 67, 77
1947: nos. 6, 13, 21, 30, 37, 48, 51, 65, 71
1948: nos. 8, 19, 23, 29, 34, 39, 44, 50
1949: nos. 27, 43, 45, 51, 55
1950: nos. 7, 58
1951: nos. 15, 33, 43, 56
1952: nos. 9, 15, 31, 43, 54
1953: nos. 18, 27, 35, 55
1954: nos. 31, 46, 60

This chronology includes the facts important to Vertov's career, with emphasis on those cultural and sociopolitical events that, directly or indirectly, affected his work.

1896	Jan. 2	Denis Arkadievich Kaufman is born in the town of Bialystok, which belonged to the Russian Empire until 1918, and now is part of Poland. His father was a bookdealer and bibliophile.
1905		Writes a poem, "Masha," dedicated to his aunt, a school teacher with liberal views.
1912–1914		Attends high school [realnoe uchilishche] and music school (studying piano and violin) in Bialystok. Reads novels by James Fenimore Cooper, Jack London, and Arthur Conan Doyle.
1915	Spring	Moves with his family to Moscow.
1916	Spring	Moves to Petrograd and enrolls in the Psychoneurological Institute (special interest in human perception). Writes a pamphlet against the monarchist ideology promoted by the Russian government.
	Summer	Organizes the "Laboratory of Hearing," in which he performs experiments with sound recording. Writes a novel, The Iron Hand [Zheleznaya ruka], which has been lost.
1917	Fall	Returns to Moscow.
	Winter	Meets Alexandr Lemberg, a cameraman who introduces him to cinema, in the "Poets' Cafe" [Kafè poetov], the famous meeting place of Moscow artists.
1918	Spring	Meets Mikhail Kol'tsov, a writer who works on the Moscow Film Committee and who gives Vertov a job with the committee (Kol'tsov was also born in Bialystok).
	Summer	Becomes the secretary of the Newsreel Section of the Film Committee (also known as the All-Russian Photo-Film Department). Takes the pseudonym Dziga Vertov (which has been interpreted in many ways, including

		"Spinning Top," "Spinning Gypsy," and "Rotating Film Reel," but the closest translation would be "Spinning Top That is Turning").
1918	Summer	Develops his "Film-Eye" method, a unification of the camera and the filmmaker's eye. First experiment: he photographs in slow motion his jump from the second floor of the Moscow Film Committee.
	June	First appearance of the *Film-Weekly* [*Kino-nedelia*] series.
1919	Spring	Formation of the *kinok*s group (also known as *"kinoglazovtsy"*).
	Summer	Completes his first manifesto, "About the Disarmament of Theatrical Cinema," later expanded into "We. A Variant of a Manifesto."
	Oct. 20	Appointed as the director of the Film Section of the Revolutionary Military Service for the South-Eastern Front.
	Nov.	Release of *The Battle of Tsaritsyn*.
1919	Dec. 5	Supervises the shooting of the Seventh Congress of the Soviets in which Lenin also participated.
		Reads the famous collection of film essays *Cinematography* [*Kinematograf*], focusing on the article "The Social Struggle and the Screen" [*Sotsialnaia bor'ba i ekran*] by Platon Kerezhentsev.
1920		Travels across the country in the agit-train "VTSIK."
		Writes "A Synopsis for a Film About an Agit-Train Passing Through the Soviet Caucasus" (unrealized project).
1921	Summer	Release of *The Agit-Train VTSIK*.
	Fall	Release of *The History of the Civil War*.
		Submits a report on the work of the Film Section (led by Vertov) to Alexandr Lemberg, Chief of the Agitational-Instructional Agit-Train project.
1922	April	Begins to work on the first issue of *Film-Truth* [*Kinopravda*].
	August	Publishes "We. A Variant of a Manifesto."
	Nov. 28	Koltsov's article "On the Screen" [*U èkrana*] published in *Pravda*. Koltsov praises Vertov's method used in the *Film-Truth* series.
	Dec. 8	Publishes the article "The Fifth Number of *Film-Truth*," in which he disagrees with Kol'tsov about the language of cinema as defined in Kol'tsov's article "The Alphabet" [*Azbuka*].
1923		Polemicizes with Aleksei Gan, a leading Soviet Constructivist theorist and the editor of *Kinofot*.
	Feb.	Completes a report to the directorial board of Goskino Film Company, "About the Significance of the Newsreel," later revised and published as "A Project for the Reorganization of the 'Council of Three.'" Vertov,

		Kaufman, and Svilova form the "triumvirate" that runs the *kinoks* workshop.
	Spring	Completes the synopsis of "The Adventures of the Delegates Arriving in Moscow to Attend the Comintern Meeting" (unrealized project).
	July 15	Vertov's first article published in *Pravda*, "A New Current in Cinema." The editor's note states that the article "is printed with the preservation of the author's style."
	Summer	Publishes the manifesto "*Kinoks*. Revolution," initially planned to be the introduction to an unpublished book.
1924	Jan.	Writes a synopsis of the article "The Birth of Film-Eye," later revised and published as " 'Film-Eye': A Newsreel in Six Issues."
	Jan. 21	Supervises the shooting of Lenin's funeral.
	Feb. 6	Release of *Lenin's Burial*, a newsreel edited by Vertov.
	April 9	Presents the paper "Film-Eye" at the *kinoks'* annual meeting.
	Summer	Eisenstein attacks Vertov's method of montage as "impressionistic" and "primitively provocative" in his article "On the Question of a Materialist Approach to Form."
	July 15	Completes the paper "Concerning the Artistic Dramatic Performance and Film-Eye."
	Fall	Release of *Film-Eye – Life Caught Unawares*, the only issue produced out of the projected six series.
1925	Jan.	Participates in the meeting "Left Front," and along with Aleksei Gan defends unstaged cinema and newsreels.
	Jan. 21	Release of *Leninist Film-Truth*, dedicated to the memory of Lenin.
	Feb. 3	*Kinoks* discuss the article "The Basics of Film-Eye" as the creative program of their workshop.
	March	Publishes the article " 'Film-Eye' about *Strike*," which initiates a lengthy controversy between Vertov and Eisenstein.
	March 13	Release of *In the Heart of a Peasant, Lenin is Alive*, which in addition to *Leninist Film-Truth*, commemorates the first anniversary of Lenin's death.
	July 16	Publishes the article "Film-Truth and Radio-Truth" in which the concept of the "Leninist proportion" is discussed.
1926	Jan.	Completes the article "Film-Eye" [*Kinoglaz*], which reiterates his theoretical ideas.
	April	Release of *Forward March, Soviet*!
	Summer	Vertov's ideological conflict with Nikolai Lebedev: he flatly rejects all critical remarks by this staunch defender of socialist realism.
		Completes a proposal for reorganization of the *kinoks'* productions according to the "Film-Eye" method.

	July 24	Publishes the article "The Factory of Facts," in which Vertov claims that Eisenstein used the "Film-Eye" method in his *Battleship Potemkin*.
	Oct.	The release of *One Sixth of the World*. Receives a prize for this film at the World Exposition in Paris.
	Nov.	Completes the script for *Ten Years of October* (unrealized project).
	Fall	The pressure against Vertov in Sovkino increases. His colleagues write a letter to *Pravda*, claiming that the current criticism of Vertov and his *kinoks* is an attack on the entire Soviet documentary film and newsreels.
1927	Jan.	Vertov is dismissed from Sovkino.
	Spring	Vertov accepts an invitation to work for VUFKU, Kiev.
	Dec.	A discussion organized by *Novyi LEF* entitled "*LEF* and Film" declares Vertov's and Shub's concepts of filmmaking identical to what the *LEF* group considered to be a "cinema of facts."
1928	Feb.	Release of *The Eleventh Year*.
	Feb. 16	Public discussion and acclaim of the film *The Eleventh Year*.
	Feb. 20	Writes the first version of his statement on *The Man with the Movie Camera* (published after his death).
	March 19	Writes the second version of his statement "*The Man with the Movie Camera*" (later published in Ukrainian).
	April	Panel discussion entitled "The *LEF* Ring" (organized by *Novyi LEF*) proclaims Vertov's film *The Eleventh Year* "superior" to Eisenstein's *October*.
	Summer	Writes the article "To All Workers of Unstaged Film" in which he criticizes both Eisenstein for making "mediated films" [*promezhutochnyi fil'm*] and Esther Shub for making "imitative films" [*podrazhatel'nyi fil'm*] (published posthumously).
	Fall	Publishes the article "Film Has Begun to Shout," which discusses the inclusion of sound in film as a progressive step in the evolution of the medium.
	Sept.	Previews of *The Man with the Movie Camera* in Kiev and Moscow.
	Nov. 8	Sends a letter to Aleksandr Fevralsky (the editor of *Pravda*), asking him to publish in *Pravda* the declaration "*The Man with the Movie Camera* − The Ultimate Film Writing and 'Radio-Eye.' " The text was rejected.
	Nov. 18	Issues the short statement "The First Soviet Film Without Words," published in *Pravda* (December 1), also known as "The First Statement on *The Man with the Movie Camera*."
1929	Jan. 8	Release of *The Man with the Movie Camera* in Kiev.

	Feb.	Travels to Paris to show a selection from the *Film-Eye* series and deliver a lecture on sound film, the coming of television, and his directorial method.
	Feb.	Attends the International Exhibition "Film and Photography" in Stuttgart. The Soviet section of the exhibition was organized by El Lissitzky and his wife, Sofia Lissitzka-Kupper. Gives a series of lectures on documentary film in Germany (Berlin, Dessau, and Essen).
		Publishes the Russian version of his Paris lecture "From 'Film-Eye' to 'Radio-Eye': From the Alphabet of *Kinoks*," in which his "Theory of Intervals" is elaborated.
		Publishes the second version of his statement, "*The Man with the Movie Camera* – The Ultimate Film Writing and 'Radio-Eye' " in Ukrainian (published only posthumously in Russian). Also known as "The Second Statement."
	Feb.	Writes a series of drafts for statements explaining the montage structure of *The Man with the Movie Camera*. Only one of them is published at this time (in Kharkov), while others, including "*The Man with the Movie Camera* – A Visual Symphony," are published posthumously.
	March 19	Writes a letter to the editor of *Kino* protesting the unsatisfactory distribution of *The Man with the Movie Camera*.
	June	Refuses to participate in the conference on so-called documentarism, organized by the Moscow Association of Film Workers.
	Fall	Participates in the International Anti-War Congress in Amsterdam, representing Mezhrabpomfil'm Company.
1931	Nov. 17	Meets Charles Chaplin (in London), who praises *Enthusiasm* as "one of the most exciting symphonies I've ever heard."
1933	March	Writes the synopses of "She" and "Night of Miniatures" (unrealized projects).
	Summer–Fall	Travels to Northern Russia, the Caucasus, the Ukraine, and Central Asia.
		Undertakes research for the film *Three Songs about Lenin*; looks for documents, records, popular songs, and collects footage of related events.
1934	Spring–Summer	Writes a series of articles explaining the process of shooting and producing *Three Songs about Lenin*; in most of them Vertov emphasizes his intention to expand the *kinoks*' early concepts of nonstaged cinema.
	Oct. 27	A series of discussions concerning *Three Songs about Lenin*, with Vertov's participation, following the film's preview organized by the Moscow Association of Film Workers.

	Nov. 1	Release of *Three Songs about Lenin* in Moscow; masses go to the theater carrying flags with the slogan "We Are Going to See *Three Songs about Lenin*."
	Nov.	Writes several articles on Mayakovsky, emphasizing the relationship between the "Film-Eye" method and Mayakovsky's poetry (published posthumously).
	Nov.–Dec.	Acclamation of *Three Songs about Lenin* in America and various countries in Europe.
	Dec. 15	Publishes the article "Film-Truth," summarizing his creative experience from 1918 up until this point.
1935	Jan. 11	Receives the order of "The Red Star" for his artistic achievements in cinema.
	Jan. 18	Publishes the article "The Last Experiment," which explains the meaning and structure of *Three Songs about Lenin*.
1936	Jan. 22	Publishes the article "Lenin in Film" which discusses the use of sound in *Three Songs about Lenin*.
	Oct. 2	Completes the article "About the Organization of a Creative Laboratory," never published in its original form; the revised parts of the text are later incorporated into different articles.
	Nov.	Plans to make a monumental compilation film about the Spanish Civil War.
1937	Nov.	Release of *Lullaby* in Moscow, in honor of the Twentieth Anniversary of the October Revolution.
1938	Fall	Completes the synopses of "When You Will Go To War" and "Women in Defense of the Country" (unrealized projects).
1939	Feb.	Delivers the paper "In Defense of the Newsreel" at the Symposium Dedicated to Twenty Years of Soviet Cinema.
	April 15	Completes the synopsis of "Gallery of Soviet Women" (unrealized project).
	April 17	Completes the synopsis of "Farmers in Kandybina Kolkhoz" (unrealized project).
	Sept. 19	Completes the synopsis of "Mama, Mama is Falling from the Sky" (unrealized project).
	Sept. 21	Completes the synopsis of "Ukrainian Village" (unrealized project).
	Oct. 4	Completes the synopsis of "A Girl Plays the Piano" (unrealized project).
	Dec. 21	Completes the synopsis of "Outside the Program" (unrealized project).
	Dec. 23	Completes the synopsis of "In My Hometown" (unrealized project).
1940	April 17	Completes the script "Tale About a Giant," written in collaboration with Elena Segal and M. Il'ina (unrealized project).

	June 15	Drafts the article "About the Love for A Real Person," which contains information on the difficulties Vertov encountered in completing his projects. Published posthumously in 1958.
	Summer	Release of the documentary "Our Cinema," directed by Fedor Kiselev, in which Vertov and his work are not mentioned.
1941	Feb.	Completes the synopsis of "Flying Man" (unrealized project).
	April 23	Completes the synopsis of "Meadowlands" (unrealized project).
		Coauthors the synopsis of "Elizaveta Svilova – Editor."
	July 12	Completes the synopsis of "Film Correspondent" (unrealized project).
	July 15	Completes the synopsis of "The Letter From a Girl Tractor Operator" (unrealized project).
	August	Completes the synopsis of "About Wounded Heroes and Their Friends" (unrealized project).
	Sept.	Evacuates to Alma-Ata (Kazakhstan), together with other Soviet artists, after Moscow has been threatened by the advancing German army.
1942	Spring	Works on the script "To You, Front."
	Summer	Works on the realization of the film *To You, Front*.
	Fall	Completes the synopsis of "The New Year's Present" (unrealized project).
	Winter	Organizes a series of screenings of *To You, Front* for the soldiers on the battle lines.
1943	Spring–Summer	Works on an army newsreel series in the Alma-Ata Film Studio.
	Aug. 25	Writes the synopsis of "Gallery of Film Portraits" (unrealized series of documentary films).
	Sept. 4	Writes the synopsis of "The Love for a Real Person" (unrealized project).
	Sept. 8	Writes the synopsis of "Little Anna" (unrealized project).
	Sept. 15	Writes the synopsis of "Germany," a compilation film (unrealized project).
	Oct. 10	Writes the synopsis of "Forward, Only Forward," a short film about a young female pig-herder from the Volga, Alexandra Lyskova (unrealized project).
	Nov.	Returns to Moscow from Alma-Ata.
1944–5		Works on "News of the Day" [*Novosti dnia*], a newsreel produced by the Central Studio of Documentary Film. Continues to work for the Central Studio, completing fifty-four newsreels, with his function defined in the credits as "Author-Director."
	Spring	Prepares a book on his artistic work. The unfinished manuscript consists of fifty-seven pages of notes, enti-

		tled "About the Creative Path" (published posthumously).
1946	Summer	Writes the synopsis of "The Honest Fool" (unrealized project).
	Fall	Delivers a speech during the meeting of the Documentary Section of the Home of Film in Moscow.
1949	Spring	Works on his autobiography; the unfinished manuscript consists of thirty-three pages of notes and recollections covering the period 1917–49.
	March 15	Delivers a speech at the opening of the party meeting discussing the creative aspects of documentary filmmaking.
1952	Spring	Delivers the speech "About Scripts in Documentary Film" during the meeting of the Conference of Cameramen in Moscow.
1954	Feb. 12	Dies in Moscow.
1956	July 15	An evening dedicated to the memory of Vertov in the Film Section of the Academy of Sciences, Moscow.
1959	March 27	An evening dedicated to screenings of Vertov's films in the Central Palace of Film, Moscow.
1966		Publication of the book *Dziga Vertov: Articles, Diaries, Projects*, edited and with an introduction by Sergei Drobashenko.
1967	April	Release of *Life Without Game* [*Zhizn' bez igry*], a documentary about Vertov and his work. Script by Sergei Drobashenko. Directed by Leonid Makhnach.
1974	Summer	Peter Konlechner, director of the Austrian Film Museum, compiles a 60-minute film, *Dziga Vertov*, which includes footage of Elizaveta Svilova commenting on the excerpts from Vertov's films and their cooperation.
1975	Nov. 12	Elizaveta Svilova Ignat'evna dies in Moscow (born September 5, 1900, in Moscow).
1976		Publication of the book *Dziga Vertov in the Reminiscences of His Contemporaries*, edited and with an introduction by Elizaveta Vertova-Svilova and Anna Vinogradova.
1980	March 21	Mikhail Kaufman dies in Moscow (born September 5, 1897, in Bialystok).
	June 24	Boris Kaufman, Vertov's youngest brother, dies in New York (born March 11, 1904, in Bialystok).

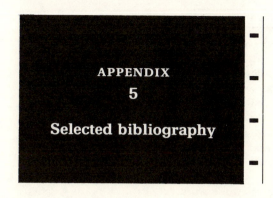

APPENDIX
5

Selected bibliography

Books and studies

Abramov, Nikolai. *Dziga Vertov* (Moscow: Akademiia Nauk, 1962), 165 pp.
Monograph with a thematic interpretation of Vertov's work.

Dobin, E.S., ed. *Razmyshleniia u ekrana* [Reflections on the Screen] (Leningrad: Iskusstvo, 1966), 367 pp.
The chapter entitled "Nasledie Dzigi Vertova i iskaniia 'cinéma-vérité' " [The Heritage of Dziga Vertov and the Aspirations of "Cinéma-Vérité"] by Tamara Selezneva (pp. 337–67) critically compares Vertov's work and method with that of Rouch, Leacock, and Pennebaker.

Drobashenko, Sergei. *Fenomen dostovernosti* [Phenomenon of Authenticity] (Moscow: Akademiia Nauk, 1972), 184 pp.
A theoretical study with a chapter dedicated to Vertov's theory of documentary film.

Feldman, Seth. *Evolution of Style in the Early Work of Dziga Vertov* (New York: Arno Press, 1977), 233 pp.
A facsimile reprint of the author's Ph.D. dissertation dedicated to Vertov's artistic development.

—*Dziga Vertov. A Guide to References and Resources* (Boston: G.K. Hall, 1979), 232 pp.
A chronological survey of Vertov's career with annotated filmography and bibliography.

Freeman, Joseph. *Voices of October: Art and Literature in Soviet Russia,* edited by Joseph Freeman, Joshua Kunitz, and Louis Lozowick (New York: Vanguard Press, 1930), 317 pp.
The chapter entitled "The Soviet Cinema" (pp. 217–64) focuses on Vertov's work on *One Sixth of the World.*

Gan, Aleksei. *Da zdravstvuet demonstratsiia byta!* [Long Live the Demonstration of Everyday Life] (Moscow: Goskino, 1923), 16 pp.
An ideological pamphlet emphasizing the political significance of Vertov's *Film-Truth* series.

Ginzburg, Sergei, ed. *Iz istorii kino: materialy i dokumenty* [From the History of Film: Materials and Documents], Vol. 2 (Moscow: Akademiia Nauk SSSR, 1959), 191 pp.

229

The chapter entitled "Montazhnye listy filmov Vertova" [Montage Lists of Vertov's Films] by Sergei Ginzburg discusses Vertov's early work, with particular attention to his development of the use of title cards (pp. 35–62). The chapter called "Arkhiv Dzigi Vertova" [Dziga Vertov's Archive] by L. M. Listov and Elizaveta I. Vertova-Svilova describes the material contained in Vertov's Archive in Moscow (pp. 132–55).

Herlinghaus, Hermann. *Dziga Wertow: Publizist und Poet des Dokumentarfilms* [Dziga Vertov: Journalist and Poet of the Documentary Films] (Berlin: Club der Filmshaffender der DDR und Deutsche Zentralstelle fur Filmforschung, 1960), 72 pp.

A historical/thematic survey of Vertov's work.

Marshall, Herbert. *Masters of the Soviet Cinema: Crippled Creative Biographies* (Boston: Routledge & Kegan Paul, 1983), 252 pp.

The chapter entitled "Dziga Vertov" (pp. 61–97) is dedicated to Vertov's career and his major films.

Marsollais, Gilles. *L'Aventure du Cinema Direct* [The Adventure of the Direct Cinema] (Paris: Seghers, 1974), 495 pp.

The essay called "Dziga Vertov" (pp. 31–9) discusses the relationship between Vertov and the modern documentary filmmakers.

Montani, Pierro. *Dziga Vertov* (Florence: La Nuova Italia, 1975), 139 pp.

A monograph on Vertov's life and work.

Pisarevskii, D., ed. *Iskusstvo millionov: Sovetskoe kino, 1917–1937* [The Art of Millions: Soviet Cinema, 1917–1937] (Moscow: Iskusstvo, 1958), 624 pp.

The chapter entitled "Fil'my Dzigi Vertova" [The Films of Dziga Vertov] by Il'ya Kopalin examines Vertov's early achievements.

Roshal', Lev. *Dziga Vertov* (Moscow: Iskusstvo, 1982), 264 pp., illust.

A critical survey of Vertov's work with emphasis on the thematic aspects.

Sadoul, Georges. *Dziga Vertov*, edited by Bernard Eisenschitz (Paris: Editions champs libre, 1971), 171 pp.

A critical study of Vertov, with a preface by Jean Rouch entitled "Cinq regards sur Vertov" (pp. 11–14).

Schnitzer, Luda and Jean. *Vertov* (Paris: Avant Scene du Cinema, 1968), 47 pp.

An essay on the ideological significance of Vertov's work.

Selezneva, Tamara. *Kinomysl' 1920'kh godov* [Film Thought of the 1920s] (Leningrad: Iskusstvo, 1972), 184 pp.

In the chapter called "Neigrovaia" [The Unstaged Film], Selezneva discusses Vertov's concept of nonfiction cinema.

Shklovskii, Viktor. *Ikh nastoiaschee* [Their Genuineness] (Leningrad: RSFSR Kinopechat, 1927), 85 pp.

The chapter called "Dziga Vertov" (pp. 61–7), which analyzes Vertov's cinematic concepts in the context of the early Soviet film, is translated in German in *Schriften zum Film* (Frankfurt am Main: Suhrkamp, 1968), pp. 63–8.

Vertova-Svilova, Elizaveta and Anna L. Vinogradova, comps. *Dziga Vertov v vospominaniiakh sovremennikov* [Dziga Vertov as Remembered by His Contemporaries] (Moscow: Iskusstvo, 1976), 277 pp.

A collection of articles dedicated to various aspects of Vertov's career and work.

Articles and essays

Abbott, Jere. "Notes on Movies," *Hound and Horn*, no. 2 (December 1928), pp. 159–62.

Abramov, Nikolai P. "Chelovek v dokumental'nom fil'me" [The Man in the Documentary Film], *Voprosi kino iskusstva*, no. 10 (Summer 1967), pp. 173–89.

—"Dziga Vertov: Poet and Writer of the Revolution," *Soviet Film*, no. 11 (1968), p. 11.

—"Dziga Vertov és a dokumentumfilm müvészete" [Dziga Vertov as a Documentary Filmdirector], *Filmkultura*, no. 5 (Budapest: January 1960), pp. 3–24.

—"Dziga Vertov i iskusstvo dokumental'nogo filma" [Dziga Vertov and the Art of the Documentary Film], *Voprosy kino iskusstva*, no. 4 (1960), pp. 276–308.

Amengual, Barthemlemy. "Vertov et Eisenstein: deux conceptions du cinéma politique" [Vertov and Eisenstein: Two Concepts of the Political Film], *Jeune Cinéma*, no. 93 (March 1976), pp. 10–19.

Anoshchenko, A. "Kinokoki" [A verbal parody of the word *kinok*s], *Kinonedelia* (February 19, 1924), p. 2.

Apon, A. and G. Verhage. "Sergei Eisenstein versus Dziga Vertov," *Skrien* (ed., Apon), 33 (March–April 1973), pp. 3–4.

Aristarco, Guido. "Le fonti di due 'Novatori,' Dziga Vertov e Lev Kuleshov," [The sources of two "Innovators," Dziga Vertov and Lev Kuleshov] *Cinema Nuovo*, no. 137 (Turin: January–February 1959), pp. 31–7.

Armes, Roy. "Vertov and Soviet Cinema," *Film and Reality* (London: Harmondsworth: Penguin Books, 1975), pp. 38–43.

Barrot, Billard, Brunelin, Decaudin, Delmas, Michel, Wyn, and Vrillac. "Le debat est ouvert sur 'L'Homme a la camera' " [Debate on "The Man with the Movie Camera" has opened], *Cahiers du Cinéma*, no. 22 (April, 1953), pp. 36–40 (a transcript of "Tribune de la F.F.C.C.")

Boltianskii, Grigorii M. "Teoriya i praktika kinokov" [Theory and Practice of the *Kinok*s], *Sovetskoe kino*, no. 415 (1926), pp. 10–11.

Bordwell, David. "Dziga Vertov: An Introduction," *Film Comment*, VIII/1 (Spring 1972), pp. 38–45.

Brik, Osip, Viktor Shklovskii, Esther Shub, and Serqei Tretiakov. "LEF i kino" [LEF and Cinema], *Novyi LEF*, nos. 11–12 (November–December 1927), pp. 5–70; English translation in *Screen*, no. 4 (Winter 1971–2), pp. 83–4.

Britton, Lionel. "Kino-Eye: Vertoff and the Newest Film Spirit of Russia," *The Realist*, no. 2 (October–December 1929), pp. 126–38.

Brunius, Jacques Bernard. "Le ciné-art et le ciné-oeil" [The Film-Art and the Film-Eye], *Le Revue de Cinéma*, no. 4 (October 15, 1929), pp. 75–6.

Burch, Noel. "Film's Institutional Mode of Representation and the Soviet Response," *October*, no. 11 (Winter 1979), p. 94.

Casaus, Victor. "Dziga Vertov: Notas sobre su Actualidad" [Notes on Reality], *Cine Cubano*, no. 76–77 (1972), pp. 106–11.

Cornand, J. "Sur deux films de Dziga Vertov 'Kino glaz' and 'L'homme a la camera" [About Two of Vertov's Films, "Film-Eye" and "The Man with the Movie Camera"], *Image et Son*, no. 207 (Paris: June–July 1975), pp. 55–62.

Croft, Stephen and Olivia Rose. "An Essay Toward *Man with the Movie Camera," Screen,* no. 1 (1977), p. 19.

Denkin, Harvey. "Linguistic Models in Early Soviet Cinema," *Cinema Journal,* no. 1 (Fall 1977), pp. 1–13.

Dickinson, Thorold. *"Enthusiasm,* or *The Symphony of the Don Basin,"* The *Film Society Programme* (London: November 15, 1931), unpaginated.

Drobashenko, Sergei. "... und eines Tages flog er durch die Luft" [...One Day He Flew Through the Air], *Film und Fernsehen* (Berlin, DDR), no. 2 (February 1974), pp. 36–8 (interview with Svilova).

Durgnat, Raymond. *"Man With a Movie Camera," American Film,* no. 1 (1984), pp. 78–9, 88–9.

Durus. "Der Mann Mit Der Kamera," [*The Man with the Movie Camera*], Die Rote Fahne, no. 116 (July 5, 1929), unpaginated.

Eisenschitz, Bernard. "Mayakovski, Vertov." *Cahiers du Cinéma,* nos. 220–1 (May–June, 1970), pp. 26–8.

Enzensberger, Marsha. "Dziga Vertov," *Screen,* XIII/4 (Winter 1972–3), pp. 90–107.

Fargier, Jean-Paul, Claude Menard, Alain Leger, Simon Luciani, and Jean-Louis Perrier. " 'Ne copiez pas sur les yeux' disait Vertov" ["Do not Copy with the Eyes," said Vertov], *Cinéthique,* no. 15 (Spring 1973), pp. 55–92.

Feldman, Konstantin. "V sporakh o Vertove" [Disputes about Vertov], *Kino i kultura,* no. 5–6 (1929), pp. 12–13.

Feldman, Seth. "Cinema Weekly and Cinema Truth: Dziga Vertov and the Leninist Proportion," *Sight and Sound,* XLIII/1 (Winter 1973–4), pp. 34–7.

Ferguson, Otis. "Artists Among the Flickers," *The New Republic,* no. 1044 (December 5, 1934), pp. 103–4.

Fevral'skii, Aleksandr V. "Tendentsii iskusstva i 'Radio Glaz' " [Artistic Tendencies in the "Radio-Eye"], *Molodaia gvardiia,* no. 7 (July 1925), p. 167.

—"Tri pesni o Lenine" [*Three Songs about Lenin*], Literaturnaia gazeta (July 16, 1934) (review).

—"Dziga Vertov i 'Pravdisti' " [Dziga Vertov and "Pravdists"], *Iskusstvo kino,* no. 13 (December 1965); French translation "Dziga Vertov et les Pravdisty" in *Cahiers du Cinéma,* no. 229 (May 1971), pp. 27–32.

Fisher, Lucy. *"Enthusiasm:* From Kino-Eye to Radio-Eye," *Film Quarterly,* no. 2 (Winter 1977), pp. 25–34.

Ganly, Raymond. *"Man with a Movie Camera," The Motion Picture News* (October 26, 1929), p. 2.

Gansera, Rainer. "Dziga Wertow," *Filmkritik,* no. 11–12 (Munich 1972), p. 567.

Giercke, Christoph. "Dziga Vertov," *Afterimage,* no. 1 (April 1970), p. 6.

Gorenko, O. "Maister kino-ob'ektiv" [*The Master of the Film Lens*], *Kino,* no. 54 (Kharkov: March 1929), p. 10.

Gusman, Boris. "O kino-glaze" [About Film-Eye], *Pravda* (October 15, 1924), p. 7.

—"Kino-pravda" [Film-Truth], *Pravda* (February 9, 1923), p. 5.

Hall, Mordaunt. "Floating Glimpses of Russia," *The New York Times* (September 17, 1929), p. 36.

Hamilton, James Shelley. *"Three Songs about Lenin,"* National Board of Review Magazine, no. 9 (December 14, 1934), p. 8.

Helman, A. "Dziga Wiertow albo wszechobecnosc kamery. Nasz Iluzjon" [Dziga Vertov, or the Omnipresence of the Camera. Our Illusion], (Poland) *Kino,* VIII/5 (May 1973), pp. 62–4.

Herlinghaus, Hermann. "Wertow-Mayakowski – Futurismus; eine Praemisse Georges Sadouls" [Vertov-Mayakovsky – Futurism; One Premise of George Sadoul], *Filmwissenschaftliche Beitraege,* No. 14 (Berlin, D.D.R., 1963), pp. 154–71.

Herlinghaus, Ruth. "Dsiga Wertow" [Dziga Vertov], *Film,* no. 1 (D.D.R. 1964), pp. 220–30.

Herring, Robert. "Enthusiasm?" *Close Up,* no. 1 (March 1932), pp. 20–4.

—"Three Songs about Lenin," *Life and Letters Today,* no. 13 (December 1935), p. 188.

Hill, Steven P. *"The Man with the Movie Camera,"* Film Society Review, (September 1967), pp. 28–31.

Hughes, J. Pennethorne. "Vertov ad Absurdum," *Close Up,* no. 3 (September 1932), pp. 174–6.

Ivanov, V. V. "Funktsii i kategorii iazyka kino" [The Categories and Functions of Film Language], *Trudy po znakovym sistemam,* no. 7 (1975), pp. 170–92.

Kaufman, Mikhail. "Cine analysis," *Experimental Cinema,* no. 4, pp. 21–3.

—"Le troisieme frere" [The Third Brother], *Cinéma 68,* no. 123 (February 1968), pp. 33–5 (interview with Marcel Matthieu).

—"Vertov tam zhe" [Vertov Is There], *Sovetskii ekran,* no. 45 (1928), p. 15.

—"An Interview," (with Annette Michelson and Naum Kleiman), *October,* no. 11 (Winter 1979), pp. 55–76.

Kaufman, Mikhail, and Elizaveta Vertova-Svilova. "U kinokov" [With the *Kinoks*], *LEF,* no. 4 (August 1924), pp. 220–1.

Khersonskii, Kh. "U istokov" [At the Source], *Sovetskii ekran,* no. 4 (1962), p. 9.

Kol'tsov, Mikhail. "U ekrana" [On the Screen], *Pravda* (November 28, 1922), p. 1.

Kopalin, Il'ia P. "A Life Illuminated by the Revolution," *Soviet Film,* no. 1 (1971), pp. 13–17.

Koster, Simon. "Dziga Vertoff," *Experimental Cinema,* no. 5 (1934), pp. 27–8.

Kracauer, Siegfried. *"Der Man Mit Dem Kinoapparat,"* Frankfurter Zeitung (May 19, 1929), p. 2.

Krautz, A. "Chaplin: Die Professoren sollten von ihm lernen" [Chaplin: The Professors Should Learn from Him], *Film und Fernsehen,* II/2 (February 1974), pp. 41–2.

Kubelka, Peter. "Restoring 'Enthusiasm,'" *Film Quarterly,* no. 2 (Winter 1977), pp. 35–6.

"LEF i kino" [LEF and Film], *Novyi LEF,* no. 11–12 (1927), pp. 50–70; also in *Screen,* no. 4 (Winter 1971–2), pp. 74–80.

Lawton, Anna. "Dziga Vertov: A Futurist with a Movie Camera," *Film Studies Annual,* Part 1 (West Lafayette: Purdue University, 1977), pp. 65–73.

—"Rhythmic Montage in the Films of Dziga Vertov: A Poetic Use of the Language of Cinema," *Pacific Coast Philology,* Vol. XIII (October 1978), pp. 44–50.

Lebedev, Nikolai. "Za proleterskuiu kinopublitsistiku" [For the Proletarian Cinema Journalism], *Literatura i iskusstvo,* no. 9–10 (1931), pp. 3–47.

Lenauer, Jean. "Vertoff, His Work and His Future," *Close Up,* Vol. V, no. 6 (December 1929), pp. 446–68.

Leyda, Jay. *"Three Songs about Lenin,"* The Film Society Programme (London: October 27, 1935), unpaginated; reprinted in *The Film Society Programmes 1925–1939* (New York: Arno Press, 1972), pp. 331–4.

Listov, Viktor. "Na puti k 'Kinopravdy' " [On the Road to "Film-Truth"], *Iskusstvo kino,* no. 2 (1971), pp. 112–19.

—"Kak nachinalas' 'kino-pravda'?" [How Did "Film-Truth" Begin?], *Iskusstvo kino,* no. 7 (July 1972), pp. 96–106.

—"O pis'me iz Petrograda, 'Avtokino' i Vertove" [About a Letter from Petrograd, "Autocinema" and Vertov], *Iskusstvo kino,* no. 1 (January 1975), pp. 109–18.

Macdonald, Dwight. "Eisenstein and Pudovkin in the Twenties," *The Miscellany,* no. 6 (March 1931), p. 24.

Malevich, Kazimir. "I likuiut liki na ekrane" [And Images Triumph on the Screen], *Kunozhurnal ARK,* no. 10 (1925), pp. 7–9; also in *Essays on Art, 1915–1928,* edited by Troels Andersen and translated by Xenia Glowacky-Prus, Vol. 1 (Copenhagen: Bongen 1971), pp. 226–32.

Malcovati, F., ed. "Nikolai Abramov su Vertov" [Nikolai Abramov on Vertov], *Bianco e Nero,* nos. 1–2 (January–February 1973), pp. 65–6 (interview).

— ed. "Cinque domande a Naum Klejman" [Five Questions to Naum Klejman], *Bianco e Nero,* nos. 1–2 (January–February 1973), pp. 63–5 (interview).

Mayne, Judith. "Kino-Truth and Kino-Praxis: Vertov's *Man with a Movie Camera,"* Cine-Tracts, no. 2 (Summer 1977), pp. 81–91.

Michelson, Annette. *"The Man with the Movie Camera:* From Magician to Epistemologist," *Art Forum,* no. 7 (March 1972), pp. 60–72.

—"Dr. Craze and Mr. Clair," *October,* no. 11 (Winter 1979), pp. 31–53.

Moussinac, Leon. "Dziga Vertov et le Kino Pravda" [Dziga Vertov and the Film-Truth], *Miroir du Cinéma,* no. 3 (Paris, October 1962), pp. 38–41.

—"Dziga Vertov on Film Technique," *Filmfront,* no. 3 (January 28, 1935), pp. 7–9.

Oille, Jennifer, " 'Konstruktivizm' and 'Kinematografiya'," *Artforum,* no. 9 (May 1978), pp. 44–9.

Pertsov, Viktor. "Igra i demonstratsiia" [Play and Demonstration], *Novyi LEF,* nos. 11–12 (November–December 1928), pp. 33–44.

Petrić, Vlada. "Dziga Vertov and the Soviet Avant-garde of the 20's," *Soviet Union/Union Sovietique,* 10, Part 1 (1983), pp. 1–58.

—"Dziga Vertov as Theorist," *Cinema Journal,* no. 1 (Fall 1978), pp. 29–44.

—"The Difficult Years of Dziga Vertov: Excerpts From His Diaries," *Quarterly Review of Film Studies,* no. 6 (Winter 1982), pp. 7–21.

Pleynet, Marcellin. "Sur les avant-garde revolutionaires" [About the Revolutionary Avant-Garde], *Cahiers du Cinéma,* nos. 226–7 (January-February 1971), pp. 6–13.

Richter, Erica. "Dsiga Wertow: Publizist und Poet des Dokumentarfilms" [Dziga Vertov: Journalist and Poet of Documentary Films], *Filmwissenschaftliche Mitteilungen* (Berlin), no. 1 (March 1961), pp. 24–5.

Rouch, Jean. "Le film ethnographique" [The Ethnographic Film], *Ethnologie Generale* (Paris: Gallimard, 1968), pp. 37–44.

—"The Cinema of the Future," *Studies in Visual Communication,* no. 1 (Winter 1985), pp. 30–5.

Sadoul, Georges. "Dziga Vertov: Poet du ciné-oeil et prophète de la ciné-oreille" [Dziga Vertov: Poet of the "Film-Eye" and the Prophet of the "Film-Ear"], *Image et Son,* no. 183 (April 1965), pp. 8–18.

—"La notion d'intervalle" [The Notion of the Intervals], *Cahiers du Cinéma,* nos. 220–221 (May–June 1970), pp. 23–4.

—"Kinopravda no. 9," *Cahiers du Cinéma,* nos. 220–1 (May–June 1970), pp. 24–5.

—"Dziga Vertov e i futuristi italiani, Apollinaire e il montaggio delle registrazioni" [Dziga Vertov and the Italian Futurists, Apollinaire and the montage of Sound-Recording], *Bianco e Nero,* no. 7 (July 1964), pp. 1–27. The article is translated as "Les futuristes italianes et Vertov" [The Italian Futurists and Vertov] in *Cahiers du Cinéma,* nos. 220–1 (May–June 1970), pp. 19–22, and in Sadoul's *Dziga Vertov* as "Le montage des enregistrements" [The Montage of Sound-Recording], pp. 15–46.

—"*Kinok*s. Revolution," in *Historie generale du cinéma muet* (Paris: Denoel, 1964); reprinted in revised form in Sadoul's *Dziga Vertov,* pp. 70–100.

—"Bio-Filmographie de Dziga Vertov" [Bio-Filmography of Dziga Vertov], *Cahiers du Cinéma,* no. 146 (August 1963), pp. 21–9; reprinted in Sadoul's *Dziga Vertov,* pp. 145–71.

—"Actualité de Dziga Vertov" [The Reality of Dziga Vertov], *Cahiers du Cinéma,* no. 144 (June 1963), pp. 23–31; translated in Spanish in *Cine Cubano,* no. 72, pp. 98–105.

Sauzier, Bertrand. "An Interpretation of *Man with the Movie Camera,*" *Studies in Visual Communication,* 11, no. 4 (Fall 1985), pp. 30–53.

Seton, Marie. "*Three Songs about Lenin,*" *Film Art* (London: Winter 1934). Reprinted in Lewis Jacobs' *The Documentary Tradition* (New York: Hopkinson and Blake, 1971), p. 100.

Shklovskii, Viktor. "I 'kinoki' di Dziga Vertov" [The *Kinok*s of Dziga Vertov], *Filmcritica,* no. 201 (Rome: October 1969), pp. 314–15. The article is a translation of the material in *Ikh nastoiashche* [Their Genuineness].

—"Kinoki i nadpisi" [The *Kinok*s and the Intertitles], *Kino,* no. 44 (October 30, 1926), pp. 8–9.

—"Kuda shagaet Dziga Vertov" [Where is Dziga Vertov Marching], *Sovetskii ekran,* no. 32 (1926), pp. 6–7.

Shorin, A. "O tekhnicheskoi baze sovetskogo tonkino" [About the Technological Basis of Soviet Sound Film], *Kino i zhizn',* no. 14 (1930), pp. 10–11.

Shutko, K. I. "Odinadtsatom" [For the Eleventh Year], *Odinnadtsatyi* (Moscow: Teakinopechat' 1928), pp. 25–7.

—"Chelovek s kinoapparatom" [*The Man with the Movie Camera*], Pravda (March 23, 1929), p. 6.

Skwara, J. "Dziga Wiertow wczoraj i dzis" [Dziga Vertov, Yesterday and Today], *Kino,* no. 8 (August 1973), pp. 41–3. Also in *Young Cinema* (Prague) no. 4 (Winter 1974), pp. 35–41.

Sokolov, Ippolit. "Shestaia chast' mira" [*One Sixth of the World*], *Kinofront,* no. 2 (1927), p. 9.

Svilova, Elizaveta. "V soviete troikh" [In the Council of Three], *LEF,* no. 4 (August 4, 1924), pp. 220–1.

Toti, Gianni. "La 'produttività' dei materiali di Ejzenstejn e Dziga Vertov" [The Productivity of the Materials of Eisenstein and Dziga Vertov], *Cinema e Film,* no. 3 (Summer 1967), pp. 281–7.

Vaughan, Dai. "*The Man with the Movie Camera,*" *Films and Filming,* no. 7 (November, 1960), pp. 18–20.

Verdone, Mario. "Dziga Vertov nell' avanguardia" [Dziga Vertov in the Avant-garde], *Filmcritica,* nos. 139–40 (Rome: November-December 1963), pp. 661–76.

Viazzi, Glauco. "Dziga Vertov et la tendenza documentaristica" [Dziga Vertov and the Documentary Tendency], *Ferrania* (Milano: August-September 1957), pp. 8–9.

"Vystavka Dzigi Vertova v Berline" [The Dziga Vertov Exhibit in Berlin], *Kino,* no. 3 (March 1974), p. 191.

Wienberg, Herman. "*The Man with the Movie Camera,*" *Film Comment,* no. 1 (Fall 1966), pp. 40–2.

Wibom, Anna-Lena. "Tre sanger om Lenin – Dziga Vertov 1934" [*Three Songs about Lenin* – Dziga Vertov 1934], *Chaplin* no. 6, (Stockholm) pp. 217–23.

Williams, Alan. "The Camera-Eye and the Film: Notes on Vertov's Formalism," *Wide Angle,* no. 3 (1979), pp. 12–17.

Notes

Chapter I

1. Elizaveta Svilova, "In the Council of Three" [*V Soviete troikh*], *LEF*, no. 4 (August 4, 1924), p. 220. The article is preceded by a brief statement entitled "With the *Kinoks*" [*U kinokov*], signed by Mikhail Kaufman who refers to Svilova as "the first *kinok*-editor."

2. Svilova, ibid., p. 221.

3. Dziga Vertov, "The Basics of 'Film-Eye' " (1924) [*Osnovnoe 'kinoglaza'*], in *Articles, Diaries, Projects* [*Stat'i, dnevniki, zamysli*], ed. Sergei Drobashenko (Moscow: Iskusstvo, 1966), p. 81. Later cited as *Articles*. The translation of the quotations in this book is mine. The existing collection, *Kino-Eye: The Writings of Dziga Vertov* (Berkeley: University of California Press, 1984), trans. Kevin O'Brien, edited and with an introduction by Annette Michelson, is available for additional information about Vertov's work.

4. Vertov, ibid., p. 82.

5. Dziga Vertov, "*Kinoks*. Revolution" (1923), [*Kinoki. Pereverot*], *Articles*, pp. 55–6. Originally published in *LEF* (June 1923), no. 3, p. 135, along with Eisenstein's statement, "Montage of Attractions" [*Montazh attraktsionov*], p. 70. Although the above translation of the title of Vertov's article is widely accepted, it would be more correct to translate it as "*Kinoks*. Overthrow."

6. Naum Gabo and Antonin Pevsner, "A Realist Manifesto," trans. Naum Gabo, in Teresa Newman, *Naum Gabo* (London: Tate Gallery, 1967), p. 34.

7. Vertov, "We. A Variant of a Manifesto" [*My. Variant manifesta*], *Articles*, p. 45. Originally published in *Kinofot* (no. 1, 1922), one of the first Soviet film journals edited by Aleksei Gan and designed by Alexandr Rodchenko.

8. Vertov, ibid., p. 47.

9. Michael Kirby, *Futurist Performance* (New York: Dutton Press, 1971), p. 213.

10. Lev Kuleshov, "Americanitis" [*Amerikanshchina*], (1922), *Kinofot*, no. 1 (1922), pp. 14–15. Also in *Kuleshov on Film*, trans. Ronald Levaco (Los Angeles: University of California Press, 1974), p. 129. The English translation is mine.

11. Vertov, "We. A Variant of a Manifesto," *Articles*, p. 47.

12. Annette Michelson, *Kino-Eye: The Writings of Dziga Vertov*, pp. XLII–XLIII (Introduction).

13. Richard Kostelanetz, ed., *Moholy-Nagy* (New York: Praeger, 1970), pp. 118–23. Also in Istvan Nemeskurty, *Word and Image* (London: Clematis Press, 1961), pp. 62–7. The script "Dynamics of a Metropolis: A Film Sketch" was originally drafted in Berlin, 1921–22.

14. Vertov, *"Kinoks.* Revolution," *Articles,* p. 55.

15. Alexandr Rodchenko, "Warning" [*Predosterezhenie*], *Novyi LEF,* no. 11, 1928, pp. 36–7.

16. See Herbert Eagle, *Russian Formalist Theory* (Ann Arbor: University of Michigan Press, 1981), p. 4–20.

17. See John Bolt, "Alexander Rodchenko as Photographer," *The Avant-Garde in Russia 1910–1930,* eds. Stephani Barron and Maurice Tuchman (Los Angeles County Museum of Art, 1980), p. 55.

18. Camila Gray, *The Russian Experiment in Art: 1863–1922* (New York: Harry N. Abrams, 1962), p. 271.

19. Viktor Shklovsky, "Alexandr Rodchenko: Artist-Photographer" [*Aleksandr Rodchenko: khudozhnik-fotograf*], *Prometei,* no. 1 (1966), p. 402.

20. Unsigned, "The Constructivists" [*Konstruktivisty*], *LEF,* no. 1, (March, 1923), p. 251.

21. Vertov, "We. A Variant of a Manifesto," *Articles,* p. 46.

22. El Lissitzky and Ilya Ehrenburg, "Veshch' / Gegenstand / Objet," an editorial in the journal *Veshch' / Gegenstand / Objet,* no. 1–2 (March–April, 1922), p. 1. After three issues, the journal folded.

23. Vertov, "From 'Film-Eye' to 'Radio-Eye' – From the Alphabet of *Kinoks*" (1929). [*Ot "kinoglaza" k "radioglazu" – Iz azbuki kinokov*], *Articles,* p. 112.

24. Vertov, *"Kinoks.* Revolution." The quotation appears in the paragraph entitled "The Council of Three" [*Soviet troikh*], p. 55.

25. Vertov, "About the Importance of Unstaged Cinema" (1923), [*O znachenii neigrovoi kinematografii*], *Articles,* p. 71.

26. Gan, "Constructivism in Cinema" [Konstruktivizm v fil'me], as quoted in *The Tradition of Constructivism,* ed. Stephen Bann (New York: Viking Press, 1974), p. 130.

27. Vertov, "About the Importance of Unstaged Cinema," *Articles,* p. 71.

28. Vertov, "The Artistic Drama and 'Film-Eye,' " (1924) [*Khudozhestvennaia drama i 'kinoglaz'*], *Articles,* p. 81.

29. Vertov, "About 'Film Truth' " (1924) [*O 'kinopravde'*], *Articles,* p. 78.

30. Gan, "Constructivism in Cinema," *The Tradition of Constructivism,* p. 129.

31. Gan, *Long Live the Demonstration of Everyday Life* [*Da zdravstvuet demonstratsiia byta*] (Moscow: Goskino, 1923), p. 14.

32. Jennifer Oille, " 'Konstruktivizm' and 'Kinematografiya' " *Artforum,* 16, no. 9, May 1978, p. 45.

33. Gan, "Film-Truth: The Thirteenth Essay" [*Kinopravda: trinadtsatyi opyt*], *Kinofot,* no. 5 (December 10, 1922), pp. 6–7.

34. Tamara Selezneva, *Film Thought of the 1920s* [*Kinomysl' 1920kh godov*] (Moscow: Iskusstvo, 1972), p. 32.

35. Gan, *Constructivism,* pp. 2, 3, 18.

36. Vertov, "About Film-Truth," *Articles,* p. 79.

37. Vertov, "Artistic Drama and 'Film-Eye,' " *Articles,* p. 81.

38. Vertov, *"Kinoks.* Revolution," *Articles,* p. 58.

39. Vertov, *"Kinoks.* Revolution," *LEF,* no. 3 (1923), p. 135. In addition to some minor discrepancies, the reprinted essay in *Articles* does not contain the epigraph printed in *LEF* above the text in different letter style.

40. Viktor Pertsov, " 'Play' and Demonstration" *["Igra" i demonstratsiia],* *Novyi LEF,* no. 11–12 (1927), pp. 35–42.

41. Sergei Ermolinskii, "At the Pushkin Monument" *[U pamiatnika Push-kinu],* in *Dziga Vertov in Recollections of His Contemporaries* [Dziga Vertov v vospominaniiakh sovremenikov], eds. E. I. Vertova-Svilova and A. L. Vinogra-dova (Moscow: Iskusstvo, 1976), p. 202.

42. Ermolinskii, ibid.

43. *"LEF* and Film" *[LEF i kino], Novyi LEF,* no. 11–12 (1927), p. 51. Also in *Screen,* London, no. 4 (Winter 1971–2), p. 74–7.

44. *"LEF* and Film," ibid., p. 52.

45. Vladimir Mayakovsky, "About Film" (1927) *[O kino],* in *Complete Works* *[Polnoe sobranie sochinenii],* XII, (Moscow: Gosudarstvennoe izdatel'stvo khu-dozhestvennoi literatury, 1959), p. 147.

46. "The *LEF* Ring" *[Rink Lefa], Novyi LEF,* no. 4, (1928), p. 29. Partially translated in *Screen,* no. 4 (Winter 1971–2), pp. 83–8. The characteristic subti-tle of this discussion reads: "Comrades! Hit Each Other With Ideas!" *[Tovar-ishchi! Shibaites' mneniami!].*

47. Sergei Tretyakov, "The New Tolstoi" *[Novyi Tolstoi], Novyi LEF,* no. 1 (1928), p. 1. Following the futurist habit, Tretyakov coined the "verb" from the names Mary Pickford and Henry Ford ("pickfordization").

48. Anatoly Lunacharsky, "Lenin and Film" *[Lenin i kino]* in *Lenin About Culture and Art [Lenin o kul'ture i iskusstve],* a collection of articles (Moscow: IZOGIZ, 1938), p. 311.

49. See "Filmography of American Silent Films in Soviet Distribution" *[Fil-mografiia amerikanskykh nemykh fil'mov byvshikh v sovetskom prokate], Kino i vremia* [Film and Time], (Moscow: Vsesoiuznyi gosudarstvennyii fond kinofi-l'mov inisterstva kul'tury SSSR, 1960), pp. 212–324; "French Silent Films in Soviet Distribution *[Frantsuzkie nemye fil'my v sovetskom prokate],* and "Ger-man Silent Films in Soviet Distribution" *[Nemetskie nemye fil'my v sovetskom prokate], Kino i vremia* (Moscow: Gosfil'mfond, 1965), pp. 348–79, 380–476. Overall, the filmography of the foreign silent films distributed in the Soviet Union between 1922 and 1933 includes 956 American, 535 German, and 202 French commercial titles.

50. Vertov, " 'Film-Truth' and 'Radio-Truth' " *["Kinopravda" i "radiopravda"]* (1925), *Articles,* pp. 84–5.

51. Vertov, "In Defense of Newsreels" (1939), *[V zashchitu khroniki], Articles,* p. 153. For a definition of "typage," see note 126.

52. Vertov, "In Defense of Newsreels," *Articles,* p. 153. The sentence reads: "Bor'ba protiv igrovoi fil'my v khronikal'nykh shtanakh."

53. Vertov, *ibid.*

54. Vertov, "About My Illness" (1935), *[O moei bolezni], Articles,* p. 190.

55. Vertov, ibid., p. 189.

56. Herbert Marshall, *Masters of the Soviet Cinema: Crippled Creative Bio-graphies* (Boston: Routledge & Kegan Paul, 1983), p. 86.

57. Vertov, "The Latest Experiment" (1935) [*Poslednii opyt*], *Articles,* p. 145.

58. Vertov, "Notebooks" (September 7, 1938), *Articles,* p. 219

59. Vertov, "Autobiography" [*Autobiografiia*], *From the History of Cinema: Materials and Documents* [*Iz istorii kino: materialy i dokumenty*] (Moscow: Akademiia nauk SSSR, 1959), p. 29.

60. Vertov, ibid., p. 85.

61. Mayakovsky, "How to Make Verses" (1926) [*Kak delat' stikhi*], *Complete Works,* XII, p. 85.

62. Mayakovsky, "Morning," (1912) [*Utro*], *Complete Works,* I, p. 34.

63. Edward Brown, *Mayakovsky: A Poet in the Revolution* (Princeton: Princeton University Press, 1973), p. 75.

64. Vertov, "From 'Film-Eye' to 'Radio-Eye,' " *Articles,* p. 114.

65. Osip Brik, "Contributions to the Study of Verse Language," *Readings in Russian Poetics: Formalist and Structuralist Views,"* eds. L. Matejka and K. Pomorska (Cambridge: MIT Press), pp. 117–25.

66. Vertov, "Mouths are Gaping through the Window" [*Rty u vitrin*], as cited by Lev Roshal' in *Dziga Vertov,* p. 35. For more information about the "Great Famine," see note 79.

67. Mayakovsky, "Conversation with a Tax Inspector About Poetry" (1926), [*Razgovor s fininspektorom o poezii*], *Complete Works,* VII, pp. 120–1. The translation is by Herbert Marshall, *Mayakovsky,* pp. 352–53. The original lines read: "Govoria po-nashemu, / rifma – / bochka. / Bochka s dinamitom. / Strochka – fitil'. Stroka dodymit, / vzryvaetsia strochka, –/ i gorod / na vozdukh / strofoi letit."

68. As cited by Mikhail Kaufman in *Dziga Vertov in Recollections of His Contemporaries,* p. 71.

69. Kaufman, ibid.

70. Mayakovsky, "Film and Film" (1922) [*Kino i kino*], *Complete Works,* XII, p. 29. The poem was originally published in *Kinofot,* no. 4, (October 4–12, 1922), p. 2.

71. Mayakovsky, "Theater, Cinematography, Futurism" (1913) [*Teatr, kinematograf, futurizm*], *Complete Works,* I, p. 295. The article was originally published in *Kinozhurnal,* no. 14 (July 14, 1913), p. 3.

72. Mayakovsky, "The Destruction of the 'Theater' by Cinematography as a Sign of a Rebirth of the Theatrical Art" (1913) [*Unichtozhenie kinematografom 'teatra' kak priznak vozrozhdeniia teatral'noga iskusstva*], *Complete Works,* I, p. 278. The article was initially published in *Kinozhurnal,* no. 16, August 1913, p. 5.

73. Mayakovsky, "The Relationship of Today's Theater and Cinematography to Art" (1913) [*Otnoshenie segodniashnego teatra i kinematografii k iskusstvu*], *Complete Works,* I, p. 281. The article was initially published in *Kinozhurnal,* no. 17, Sept 8, 1913, p. 7.

74. Mayakovsky, ibid., p. 284.

75. A note scribbled by Tsar Nikolai II on the margin of a police report, in 1913. Quoted by I. S. Zilbershtein in "Nikolai II About Film" [*Nikolai II o kino*], *Sovetskii ekran,* April 12, 1927, p. 10.

76. Vertov, "We. A Variant of a Manifesto," *Articles,* p. 49.

77. Vertov, "About the Importance of Newsreels" [*O znachenii khroniki*] (1923), *Articles*, p. 67.

78. Tretyakov, *"LEF* and NEP" [*LEF i NEP*], *LEF*, no. 2, April 1923, pp. 70–8.

79. Vertov, "Mouths are Gaping through the Window," as cited by Lev Roshal' in *Dziga Vertov*, p. 35. Although not dated, the poem was probably written between 1932 and 1933, at the time the Ukraine, Kazakhstan, and Volga regions were struck by famine; almost six million people, mostly peasants and their children, died – a fact the official Soviet press has never admitted. See Maksudov, "Geography of the Hunger of 1933" [*Geografiia goloda 1933 goda*], *SSSR: vnutrennie protivorechiia*, no. 7 (New York: Chalidze Publications, 1983), pp. 5–17. Herbert Marshall, on the other hand, doubts that Vertov would have written about the famine of 1932–3, which was man-made by Stalin to crush the Ukrainian "kulaks" and therefore was forbidden to be discussed. In contrast, the natural famine of 1922, was extensively covered by the press, and many artists dealt with it, including Mayakovsky who dedicated to it his 1922 poem "Volga."

80. Mayakovsky, as cited in *Speeches Delivered at a Dispute About the Politics of Sovkino* [*Vstupleniia na dispute o politike Sovkino*] (1927), (Moscow: Iskusstvo, 1954), pp. 441–2.

81. Mayakovsky, "Film and Wine" [*Kino i vino*] (1928), *Complete Works*, IX, p. 64. Originally published in *Kinofot*, no. 4–5 (1922), p. 2. The poem opens with the lines: "Skazal / filosof iz Sovkino: / "Rodnie sestry / kino i vino."

82. Mayakovsky, "Film and Film," *Complete Works*, XII, p. 29. Ivan Mozhukin (1890–1939) was the famous Russian actor who moved to Paris after the revolution. He was known for highly emotional performances in conventional melodramas before 1918. Kuleshov used one of Mozhukin's old close-ups for his montage experiment known as the "Kuleshov effect."

83. Annette Michelson, *Kino-Eye: The Writings of Dziga Vertov*, p. LXI (Introduction).

84. Vertov, "About Mayakovsky" (1934) [*O Maiakovskom*], *Articles*, p. 182.

85. Vertov, "More About Mayakovsky" (1935) [*Eshche o Maiakovskom*], *Articles*, p. 184.

86. Vertov, "Notebooks" (February 29, 1936), *Articles*, p. 200.

87. Sergei Eisenstein, "Word and Image," *The Film Sense*, ed. and trans. Jay Leyda (New York, Harcourt, Brace & World, 1947), p. 63. The original essay, "Montage 1938" [*Montazh 1928*], was changed in the English translation to "Word and Image." Actually, this was the second part of Eisenstein's extensive study on montage which at that time had not yet been published. The other part, written in 1937, was published for the first time in Eisenstein's *Selected Works* [*Izabrannye proizvedeniia*] (Moscow: Iskusstvo, 1964), pp. 329–484.

88. Vertov, *"The Man with the Movie Camera* – A Visual Symphony" [*Chelovek s kinoapparatom – zritel'naia symfoniia*] (1929), *Articles*, pp. 278–9. Vertov wrote several articles explaining this film and its structure. This is the first manifesto related to the film, published in February 1929, at the time of the film's release in Kiev and Moscow. The second manifesto, entitled *"The Man with the Movie Camera:* The Absolute Film and the Radio-Eye" [*Ludina z kinoaparatom: absolutnii kinopis i radio-oko*], was published two months later in the Ukrainian journal *Nova generatsiia* (Khrakov). A detailed analysis of both manifestos can be found in my article, "Dziga Vertov as Theorist," *Cinema Journal*, XVIII, no. 1 (Fall, 1978), pp. 29–47.

89. Mayakovsky, "Conversation with a Tax Inspector about Poetry," in *Complete Works*, VII, p. 121. The translation is by Herbert Marshall, *Mayakovsky*, p. 353. The original lines read: "Poeziia – / ta zhe dobycha / radiia / v gramm dobicha, / v god trudy. / Izvodish' / edinoga slova radi / tysiachi toni / slovesnoi rudy."

90. Vertov, "About Mayakovsky," *Articles*, p. 185.

91. Vertov, ibid., p. 289.

92. Vertov, "A Girl Plays the Piano," *Articles*, p. 291.

93. Vertov, ibid., p. 297.

94. Vladimir Markov, *Russian Futurism: A History* (Berkeley: University of California Press, 1968), p. 185. Another study related to this topic is *Russian Formalism: History – Doctrine*, by Victor Erlich (The Hague, Netherlands: Mouton, 1955), with an excellent chapter entitled "Marxism versus Formalism."

95. Anna Lawton, "Dziga Vertov: A Futurist with a Movie Camera," *Film Studies Annual* (West Lafayette: Purdue University, 1977), Part One, p. 66.

96. For more information about the technique used in the ROSTA posters, see Roberta Reeder, "The Interrelationship of Codes in Mayakovsky's ROSTA Posters," *Soviet Union / Union Sovietique*, no. 7 (1980), p. 41.

97. Mayakovsky, "Cloud in Trousers" (1914–15) [*Oblako v shtanakh*], *Complete Works*, I, p. 185. The translation is by Max Hayward and George Reavey, *The Bedbug and Selected Poetry*, p. 79. The original lines read: "A za poètami – / ulichnye tyshchi: / studenty, / prostitutki, / podriadchiki. / . . . "Pomogi mne!" / Molit' o gimne, / ob oratorii! / My sami tvortsy v goriashchem gimne – / shume fabriki i laboratorii."

98. Mayakovsky, "Brooklyn Bridge" (1925) [*Bruklinskii most*], *Complete Works*, VII, p. 83. The translation is by Herbert Marshall, *Mayakovsky*, p. 337.

99. For an extensive discussion of "Zaum," see Marsha Enzensberger, "Dziga Vertov," *Screen*, no. 4 (Winter 1972–3), p. 55; or *Screen Reader* I (London: SEFT, 1977), p. 37. In Russian, the best explanation of "Zaum poetry" [*zaumnaia poezia*], as well as other formalist terms, is provided by Boris Thomashevsky, *Theory of Literature* [*Teoriia literatury*] (Leningrad: Gosudarstvennoe izdatel'stvo, 1925), p. 68.

100. Vertov, "My Last Experiment," *Articles*, p. 144.

101. Fevralsky, "The Forwardlooking" [*Vperedsmotriashchii*], Dziga Vertov in Recollections of His Contemporaries, p. 140–3.

102. Vertov, "More About Mayakovsky," *Articles*, p. 184.

103. Vertov, "The Importance of Newsreels," *Articles*, p. 67.

104. Vertov, "More About Mayakovsky," *Articles*, p. 184.

105. Raymond Durgnat, *"Man with a Movie Camera,"* *American Film*, Oct. 1984, p. 79.

106. Vertov, "About the Importance of Unstaged Cinema," *Articles*, p. 71.

107. Esfir Shub, "My Life – Cinema" [*Zhizn' moia – kinematograf*] (Moscow: Iskusstvo 1972), p. 380.

108. As quoted by Boris Thomson, *Lot's Wife and the Venus of Milo* (Cambridge: University of Cambridge Press, 1978), p. 73.

109. Viktor Shklovsky, *Their Genuineness* [*Ikh nastoiashchee*] (Moscow: Kino-pechat', 1927), pp. 24–5. Also in *For Forty Years* [*Za sorok let*] (Moscow: Is-kusstvo, 1956), pp. 71–2.

110. Viktor Shklovsky, *Eisenstein* [*Èizenshtein*] (Moscow: Iskusstvo, 1973), p. 142.

111. Sergei Drobashenko, *Phenomenon of Authenticity* [*Fenomen dostover-nosti*] (Moscow: Akademiia nauk SSSR, 1972), pp. 48–50.

112. Tamara Selezneva, *Film Thought of the 1920s*, pp. 26–56.

113. Vertov, "In Defense of Newsreels," *Articles*, p. 153.

114. Eisenstein, "Toward the Question of a Materialist Approach to Form" (1925) [*K voprosu o materialisticheskom podkhode k forme*], *Selected Works*, I, (Moscow: Iskusstvo, 1964), pp. 113–14. Translation is mine. Also, in *The Avant-Garde Film*, ed., P. Adam Sitney (New York: New York University Press, 1978), pp. 15–22, translated by Roberta Reeder.

115. Vertov, "The Factory of Facts" (1926) [*Fabrika Faktov*], *Articles*, p. 89. Originally published in *Pravda* (July 24, 1926), p. 6.

116. Shub, in *Close-Up* [*Krupnym planom*] (Moscow: Iskusstvo, 1959), p. 113.

117. Svilova, "A Recollection about Vertov" [*Pamiat' o Vertove*], *Dziga Vertov in Recollections of His Contemporaries*, p. 69.

118. Svilova, ibid.

119. Drobashenko, *Phenomenon of Authenticity*, p. 65.

120. Drobashenko, ibid., p. 66.

121. Drobashenko, ibid., p. 67.

122. Eisenstein, "Through Theater to Cinema" [*Srednaia iz trekh, 1924–9*], in *Film Form*, ed., trans. Jay Leyda (New York: Harcourt, Brace & World, 1949), p. 5.

123. Eisenstein, "Toward the Question of a Materialist Approach to Form," *Selected Work*, I, pp. 113–14.

124. Eisenstein, ibid., p. 114.

125. Shklovsky, "Where is Dziga Vertov Marching?" [*Kuda shagaet Dziga Vertov?*], *Sovetskii ekran* (August 10, 1926), p. 4.

126. "Typage" is a term with several definitions since its meaning evolved in the process of Eisenstein's own theoretical development. In the most general sense, typage implies "casting" both actors and nonprofessionals, insisting on the characteristic facial features that are capable of instantly revealing (on the screen) the typical psychology, social background, and behavior of the characters. Later, Eisenstein emphasized that "*typage* must be understood as broader than merely a face without make-up or substitution of 'naturally expressive' types for professional actors," and proposed that the definition of typage be understood as "the filmmaker's specific approach to the events embraced by the content of the film." *Film Form*, p. 8.

127. Osip Brik, "The *LEF* Ring," *Novyi LEF*, no. 4 (1928), p. 29. In his discussion, Brik also stated: "In order to show Lenin's figure in *October*, Eisenstein opted for the most shameful method, a method below any cultural standard. He found a person, i.e., the worker Nikandrov, resembling Lenin, to play Lenin. The result turned out to be a presumptuous fake which can satisfy only those spectators who are totally insensitive to historical truth" (p. 30). Soon after the discussion about *October*, the practice of impersonating Lenin, Stalin, and other Soviet leaders became common in Soviet cinema. Certain actors even specialized in playing exclusively the roles of important Soviet leaders in feature films.

128. Mayakovsky, "About Film," *Complete Works*, XII, p. 147.

129. Eisenstein, "Montage of Attractions" [*Montazh attraktsionov*], LEF (June 1923), no. 3, p. 70. Also, *The Film Sense*, pp. 230–1.

130. Vertov, "*Kinok*s. Revolution," *LEF* (June 1923), no. 3, p. 135. Also, *Articles*, p. 55.

131. Eisenstein, "On the Question of a Materialist Approach to Form," *Selected Works*, I, pp. 113–14.

132. Eisenstein, ibid., p. 114.

133. Eisenstein, ibid., p. 115.

134. Vertov, "The Factory of Facts" (1926), *Articles*, pp. 88–9.

135. Vertov, ibid., p. 89.

136. Eisenstein, "The Fourth Dimension in Film" (1929), [*Chetvertoe izmerenie v kino*], *Selected Works*, II, p. 51. Also, *Film Form*, p. 73. In the original, the quotation reads: "Podobnym otritsatel'nym primerom mozhet sluzhit *Odinnatsatyi* Dzigi Vertova gde metricheskii modul' nastol'ko matematicheski slozhen, chto ustanovit' v nem zakonomernost' mozhno tol'ko s arshinom v rukakh, to est' ne vospriiatiem a izmereniem."

137. Eisenstein, "Behind the Frame" (1929) [*Za kadrom*], *Selected Works*, II, p. 295. Also, *Film Form*, p. 43. In the original, the quotation reads: "Eshche chasche – prosto formal'nye biriul'ki i nemotirovannoe ozornichan'e kameroi (Chelovek s kinoapparatom)."

138. Eisenstein, "What Lenin Gave To Me" (1932) [*Chto mne dal V. I. Lenin*], *Selected Works*, V, p. 530. The article was published for the first time in *Iskusstvo kino*, no. 4 (1964), pp. 2–8.

139. Andrei Zhdanov (1896–1948) one of the most orthodox party officials, was responsible for the appropriate realization of socialist realism in Soviet art. Throughout the 1930s and 1940s, Zhdanov's power became so strong that this entire period of artistic suppression is known as "Zhdanovism." Already at the First Congress of the Union of Soviet Writers, in 1934, he launched a "practical definition" of the doctrine by demanding that all Soviet artists "produce a truthful and historically concrete representation of reality, combined with the task of ideological transformation and political education in the spirit of Socialist Realism." As quoted by Slonim, *Soviet Russian Literature*, p. 151.

140. Eisenstein, "Pantagruel is Born" (1933) [*Roditsia Pantagruel*], *Selected Works*, II, p. 300.

141. Eisenstein, "Film Directing. The Art of Mise-en-Scène" (1933–4) [*Rezhissura. Iskusstvo mizanstseny*], *Selected Works*, IV, p. 67. It appeared for the first time in Eisenstein's *Selected Works*, published in 1956.

142. Eisenstein, ibid. In the original, the sentence reads: "Gipertrofiiu montazha v ushcherb izobrazheniiu-kadru mozhno bylo by rassmatrivat' kak svoeobraznuiu kinoshizofreniiu."

143. Eisenstein, "The Most Important of the Arts" (1935) [*Samoe vazhnoe iz iskusstv*], *Selected Works*, V, p. 530.

144. Eisenstein, "Twenty" (1940) [*Dvadtsat'*], *Selected Works*, V, p. 104.

145. Eisenstein, "About Stereoscopic Film" (1948) [*O stereokino*], *Selected Works,* III, p. 130.

146. Eisenstein, Pudovkin, Yutkevich, "Statement," *Film Form,* p. 257. The original title was "The Future of the Sound Film − A Statement" [*Budushchee zvukovoi fil'my − Zaiavka*], *Selected Works,* II, p. 315.

147. Vertov, "Let Us Discuss First Sound Film, *Symphony of the Don Basin* Produced by Ukrainfilm" (1931) [*Obsuzhdaem pervuiu zvukovuiu fil'mu Ukrainfil'm, "Simfoniia Donbassa"*], *Articles,* p. 125.

148. Vertov, ibid.

149. As stated by Elizaveta Svilova in an interview with Seth Feldman in *Evolution of Style in the Early Work of Dziga Vertov,* a facsimile reprint of Feldman's dissertation (New York: Arno Press, 1977), p. 166. The interview took place in Stockholm (March 7, 1974).

150. Ippolit Sokolov, "The Potentialities of Sound Cinema" [*Vozmozhnosti zvukovogo kino*], *Kino,* no. 45 (February, 1929), p. 5.

151. Sokolov, ibid.

152. Sokolov, ibid.

153. Eisenstein et al., "Statement," *Film Form,* p. 258.

154. Vertov, "The First Steps," *Articles,* p. 129.

155. Vertov, ibid.

156. Vertov, "In Defense of Newsreels," *Articles,* p. 154.

157. Vertov, "Notebooks" (February 12, 1940), *Articles,* p. 229.

158. Vertov, "Notebooks" (February 12, 1940), *Articles,* p. 228.

159. Vertov, ibid.

160. Vertov, "Stanzas About Myself" [*Stikhi o sebe*], as quoted by Lev Roshal', *Dziga Vertov,* p. 6. The poem is not dated.

161. Vertov, "Notebooks" (February 4, 1940), *Articles,* p. 228. In the original the sentence reads: "Mozhno li umeret' ne ot fizicheskogo a ot tvorcheskogo goloda? Mozhno!"

162. Cited by Osip Brik in his essay "Mayakovsky and the Literary Movement of 1917–30" (1936). The quotation is taken from the English translation of the essay by Diana Matias, published in *Screen,* London, no. 3 (Autumn 1974), pp. 59–81.

163. Brik, ibid., pp. 70–8.

164. Vertov, "Notebooks" (May 17, 1934), *Articles,* p. 178.

165. Vertov, "Notebooks" (May 19, 1934), *Articles,* p. 178.

166. Vertov, "Notebooks" (December 25, 1939), *Articles,* p. 227.

167. Vertov, "Notebooks" (August 14, 1939), *Articles,* p. 224.

168. K. I. Shutko, *"The Man with the Movie Camera"* [*Chelovek s kinoapparatom*], *Pravda,* March 23, 1929, p. 4.

169. Ermolinskii, "At the Pushkin Monument," *Dziga Vertov in Recollections of His Contemporaries,* p. 200.

170. As quoted by Lev Roshal' in his book *Dziga Vertov* (Moscow: Iskusstvo, 1982), pp. 200–4.

171. Roshal', ibid.

172. Roshal', ibid.

173. Roshal', ibid.

174. A. Fedorov-Davydov, "Toward a Realist Art" [*K realisticheskomu iskusstvu*], *Kino,* March 30, 1936, p. 4.

175. Vertov, "In Defense of Newsreels" (1939), *Articles,* p. 154.

176. Vertov, "Notebooks" (March 26, 1944), *Articles,* p. 253.

177. Vertov, "Notebooks" (August 14, 1943), *Articles,* p. 246.

178. See Georges Sadoul, " 'Film-Truth' and 'Cinéma-Vérité' " [*Kinopravda et cinéma vérité*], Dziga Vertov (Paris: Edition Champ libre, 1971), pp. 108–38; "From Dziga Vertov to Jean Rouch" [*De Dziga Vertov a Jean Rouch*], pp. 139–44.

179. Nikolai Abramov, *Dziga Vertov* (Moscow: Izdatel'stvo Akademii Nauk, 1962), p. 98.

180. Jean Rouch, "Five Views On Vertov" [*Cinq regards sur Vertov*], in Georges Sadoul, *Dziga Vertov,* pp. 11–14.

181. Jean Laude, *L'année 1913* [*The year 1913*] (Paris: Klicksieck, 1971), p. 205. For more information about Zhdanov, see note 139.

182. Gerard Conio, "Debate on the Formal Method," *The Futurists, the Formalists, and the Marxist Critique,* ed., Cristopher Pike (Highland N.J.: Humanities Press, 1979), p. 45.

183. Vertov, "Notebooks" (March 26, 1944), *Articles,* p. 252.

184. Vertov, "Notebooks" (October 11, 1944), *Articles,* p. 261.

185. Vertov, "Notebooks" (September 7, 1938), *Articles,* p. 218.

Chapter II

1. Vertov, *"The Man with the Movie Camera"* (1928), *Articles,* p. 109.

2. Vertov, "From the History of the *Kinoks*" (1929), *Articles,* p. 118.

3. Annette Michelson, *Kino Eye: The Writings of Dziga Vertov,* p. XXIV (Introduction).

4. Stephen Croft and Olivia Rose, "An Essay Toward *Man with the Movie Camera,*" *Screen,* no. 1 (1977), p. 19.

5. Mikhail Kaufman, "Interview," *October,* p. 69.

6. Vertov, *"The Man with the Movie Camera," Articles,* p. 109.

7. Vertov, *"Kinoks* and Montage," p. 97, is a segment of the article "Temporary Instructions to the 'Film-Eye' Groups" (1929), *Articles,* pp. 94–104.

8. Croft and Rose, "An Essay Toward *Man with the Movie Camera,*" *Screen,* no. 1, pp. 15–16.

9. Bertrand Sauzier, "An Interpretation of *Man with the Movie Camera," Studies in Visual Communication,* 11, no. 4 (Fall 1985), pp. 34–53.

10. Vertov, "About 'Film-Truth,' " (1924), *Articles,* p. 78.

11. Eisenstein, see Chapter I, note 137.

12. Vertov, "About the Importance of Unstaged Cinema," *Articles,* p. 71.

13. See, for example, Annette Michelson, *"The Man with the Movie Camera:* From Magician to Epistemologist," *Artforum,* no. 7. (March 1972), pp. 60–72.

14. Noel Burch, "Film's Institutional Mode of Representation and the Soviet Response," *October,* no. 11 (Winter 1979), p. 94. This is the authorized translation of the original paper delivered in French at the International Federation of Film Archives (FIAF) in Varna (Bulgaria), March 1977.

15. Vertov, *"The Man with the Movie Camera* – A Visual Symphony," *Articles,* p. 280.

16. Vertov, "A Letter from Berlin" (1929) [*Pis'mo iz Berlina*], *Articles,* p. 121.

17. Vertov, *"The Man with the Movie Camera* – A Visual Symphony" (1929), *Articles,* p. 280.

18. Vertov, ibid., p. 278.

19. Vertov, "About 'Film-Truth' " (1924), *Articles,* p. 78.

20. Vertov, *"The Man with the Movie Camera* – A Visual Symphony," *Articles,* p. 279.

21. Vertov, *"The Man with the Movie Camera," Articles,* p. 106.

22. Friedrich Engels, *Anti-Dührung,* as quoted in *Marxism and Art,* ed., M. Solomon (New York: Vintage Books, 1974), p. 28.

23. Eisenstein, "Methods of Montage," *Film Form,* p. 83.

24. Eisenstein, "The Filmic Fourth Dimension," *Film Form,* p. 69.

25. Vertov, "About the Importance of Unstaged Cinema," *Articles,* p. 71.

26. Vertov, "The Artistic Drama and 'Film-Eye,' " *Articles,* p. 81.

27. Pudovkin, "Close-Up of Time," *Film Technique and Film Acting,* p. 146.

28. Vertov, "The Birth of 'Film-Eye' " (1924), *Articles,* p. 15.

29. Vertov, *"Kinoks.* Revolution," *Articles,* p. 55.

30. Vertov, "Notebooks" (September 6, 1936), *Articles,* p. 198.

Chapter III

1. Arthur Knight, *The Liveliest Art* (New York: Mentor Books, 1979), p. 86.

2. Vertov, *"The Man with the Movie Camera," Articles,* p. 109.

3. Anna Lawton, "Rhythmic Montage in the Films of Dziga Vertov: A Poetic Use of the Language of Cinema," *Pacific Coast Philology,* Vol. XIII (October 1978), p. 44.

4. Vertov, "About Mayakovsky," *Articles,* p. 183.

5. Vertov, "From 'Film-Eye' to 'Radio-Eye,' " *Articles,* p. 114.

6. Vertov, "We. A Variant of a Manifesto," *Articles,* p. 49.

7. Vertov, ibid.

8. Vertov, ibid., p. 48.

9. Vertov, ibid., p. 47.

10. Eisenstein, "The Cinematographic Principle and Ideogram," *Film Form,* pp. 38–40. The original title of this essay, "Behind the shot" [*Za kadrom*], was published in 1929, the year *The Man with the Movie Camera* was completed.

11. Eisenstein, ibid., p. 47.

12. Vertov, "We. A Variant of a Manifesto" (1922), *Articles,* p. 47.

13. Vertov, "About Love for a Living Man" (1958), *Articles,* p. 159.

14. Vertov, *"The Man with the Movie Camera," Articles,* p. 109.

15. Wilson Bryan Key, *Subliminal Seduction: Ad Media's Manipulation of a Not So Innocent America* (New York: New American Library, 1974), p. 51.

16. Key, ibid., p. 52.

17. N. F. Dixon, *Subliminal Perception: The Nature of a Controversy* (London, McGraw-Hill, 1971), p. 166.

18. Vertov, "Notebooks" (March 26, 1944), *Articles,* p. 254.

19. Maya Deren's statement at "A Symposium With Maya Deren, Arthur Miller, Dylan Thomas, Parker Tyler" (1963), *Film Culture Reader*, ed., P. Adams Sitney (New York: Praeger, 1970), p. 174.

20. Eisenstein, *Film Form*, pp. 67–72, 78–81.

21. Vertov, "From the History of the *Kinoks*," *Articles*, p. 119.

22. Vertov, "We. A Variant of a Manifesto," *Articles*, p. 48.

23. Vertov, "From the History of the *Kinoks*," *Articles*, p. 119.

24. Eisenstein, Pudovkin, Yutkevich, "Statement," *Film Form*, p. 258.

25. Vertov, "From 'Film-Eye' to 'Radio-Eye,' " *Articles*, p. 112.

26. Eisenstein, *Film Form*, pp. 67–72, 78–81.

27. Vertov, *"The Man with the Movie Camera"* (1928), *Articles*, p. 109.

28. Vertov, ibid.

29. Vertov, ibid.

30. Vertov, "About Mayakovsky," *Articles*, p. 182.

Frame enlargements

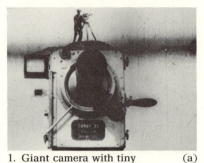

1. Giant camera with tiny   (a)   (b)
Cameraman

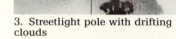

2. Building with drifting   3. Streetlight pole with drifting
clouds   clouds

4. Empty movie theater   (a)   (b)

(c)   5. Chandeliers and light   (a)
       fixtures

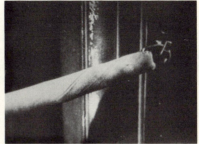

(b)   6. Cordon at theater entrance

7. Cameraman entering movie
theater

8. Projector (a)

(b)   9. Projector and projectionist

10. Film can

11. Projector's reel and
projectionist

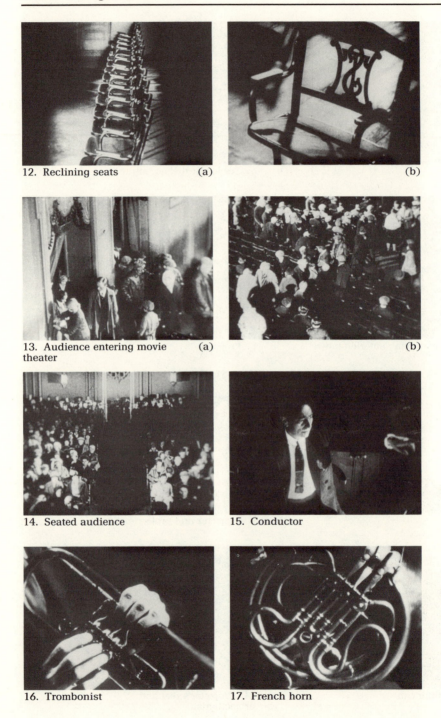

12. Reclining seats (a) (b)

13. Audience entering movie (a) (b)
theater

14. Seated audience 15. Conductor

16. Trombonist 17. French horn

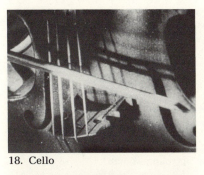

18. Cello

19. Orchestra

20. Trumpet

21. Drummer

22. Violinist

23. Numeral "1"

24. Window   (a)

(b)

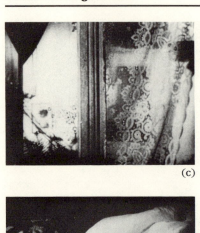

(c)

25. Sleeping young woman (a)

(b)

26. *The Awakening of a* (a)
Woman movie poster

(b)

27. Terrace with trees, and (a)
giant bottle

(b)

28. Derelicts and vagrants (a)

(b)

(c)

(d)   29. Hosing of streets   (a)

(b)   30. Bolshoi Theater

31. Store windows   (a)   (b)

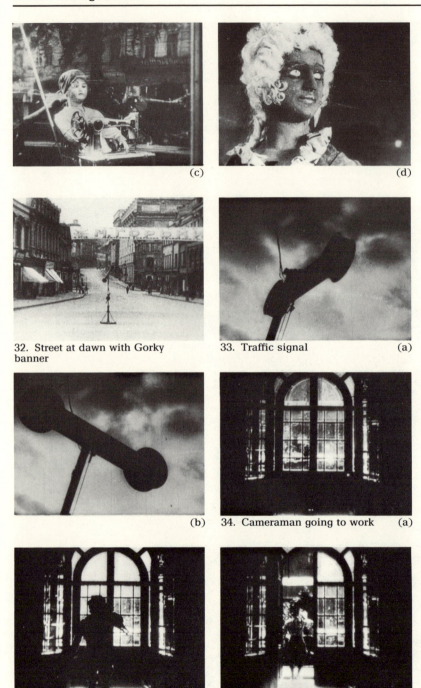

(c)   (d)

32. Street at dawn with Gorky banner   33. Traffic signal   (a)

(b)   34. Cameraman going to work   (a)

(b)   (c)

(d)

35. Steel bridge construction
with Cameraman below

36. Pigeons flying backwards – (a)
reverse motion

(b)

37. Cameraman shooting train (a)

(b)

38. Speeding train – acceler- (a)
ated motion

(b)

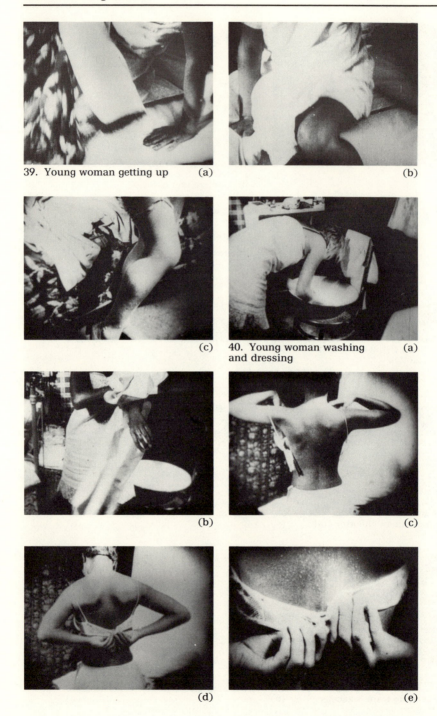

39. Young woman getting up   (a)

(b)

(c)   40. Young woman washing   (a)
and dressing

(b)

(c)

(d)

(e)

41. Railway tracks – pixilla-   (a)   (b)
tion, flickering-effect

42. Cameraman leaving rail-   (a)   (b)
way tracks

43. Cameraman in convertible   44. Convertible and telephone
crossing railway tracks   poles – accelerated motion

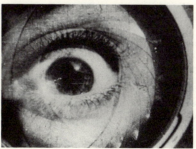

45. Cameraman's hands put-   46. Camera lens with superim-   (a)
ting on telescopic lens   posed eye and reflected
                                     Cameraman

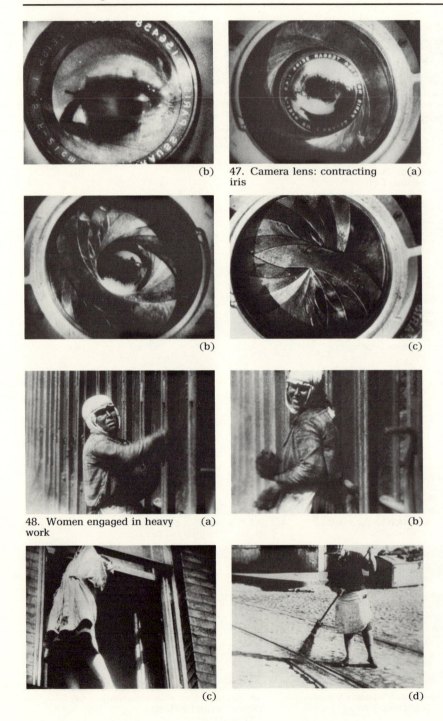

(b) 47. Camera lens: contracting   (a)
iris

(b)   (c)

48. Women engaged in heavy (a)   (b)
work

(c)   (d)

(e)

49. Blinking   (a)

(b)

50. Venetian blinds –   (a)
pixillation

(b)

51. Blooming flowers   (a)

(b)

52. *The Sold Appetite* movie
poster

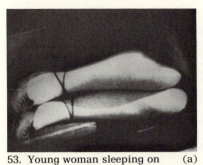

53. Young woman sleeping on   (a)
bench

(b)

(c)

54. Passing trolleys   (a)

(b)

56. Workers pulling carts   (a)

55. Cameraman and movie
poster

(b)

(c)   57. Cameraman lying on
          ground

58. Cameraman walking over   59. Peasant women   (a)
steel bridge

(b)   60. Cameraman walking   (a)
          through crowd

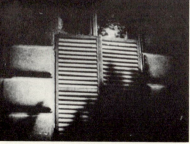

(b)   61. Opening of window shut- (a)
          ters – overlapping

(b)

(c)

(d)

62. "Glasses – pince nez," opti-   (a)
cian's shop

(b)

63. Reflection in revolving door   (a)

(b)

(c)

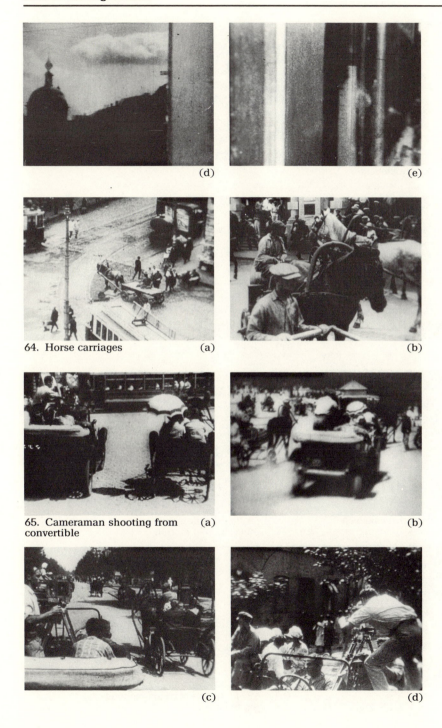

(d)

(e)

64. Horse carriages (a)

(b)

65. Cameraman shooting from (a)
convertible

(b)

(c)

(d)

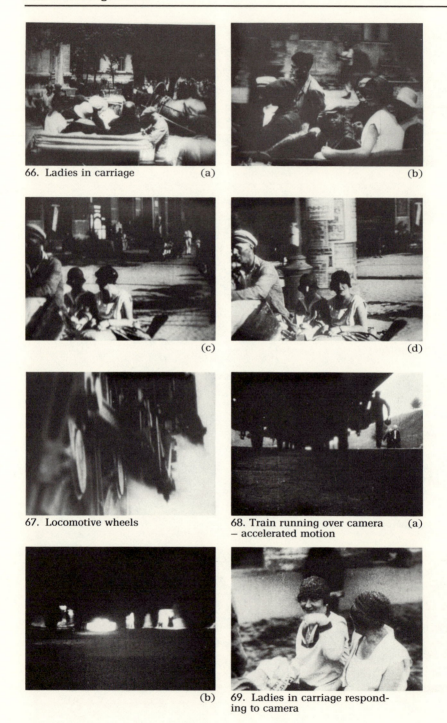

66. Ladies in carriage   (a)   (b)

(c)   (d)

67. Locomotive wheels

68. Train running over camera   (a)
– accelerated motion

(b)   69. Ladies in carriage responding to camera

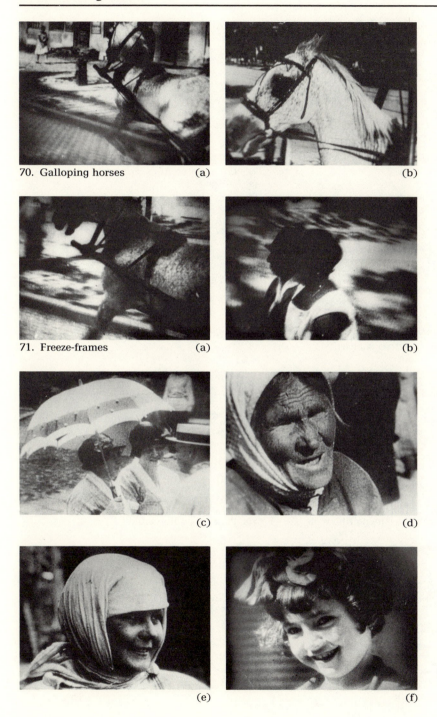

70. Galloping horses   (a)   (b)

71. Freeze-frames   (a)   (b)

(c)   (d)

(e)   (f)

(g)

72. Little girl on filmstrip with perforations

73. Film reels on shelves (a)

(b)

74. Editor classifying film reels

75. Plate on editing table

76. Filmstrips (a)

(b)

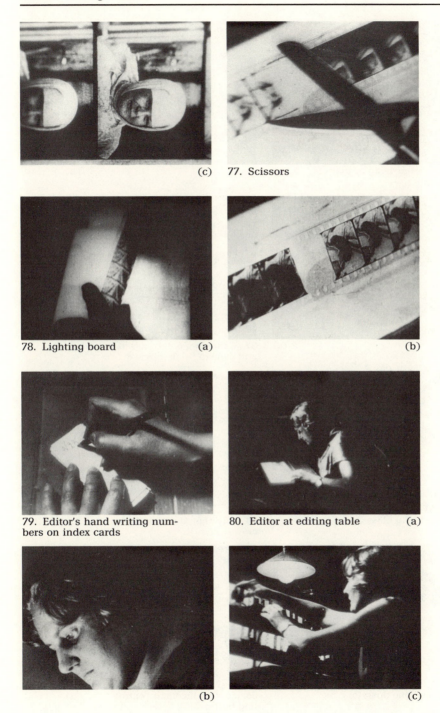

(c)   77. Scissors

78. Lighting board   (a)   (b)

79. Editor's hand writing num-   80. Editor at editing table   (a)
bers on index cards

(b)   (c)

81. Ladies arriving at their   (a)   (b)
apartment

                             (c)   82. Cameraman carrying tripod
and camera – lateral view

83. Revolving glass door   84. City square with milling
crowd

85. Telephone operator   86. Street policeman control-   (a)
ling traffic

(b)

(c)

(d)

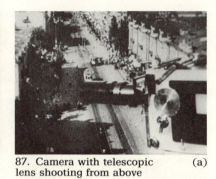

87. Camera with telescopic (a)
lens shooting from above

(b)

88. Marriage Bureau: first cou- (a)
ple with clerk

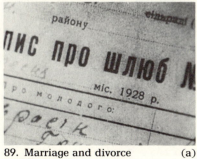

(b)

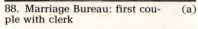

89. Marriage and divorce (a)
certificates

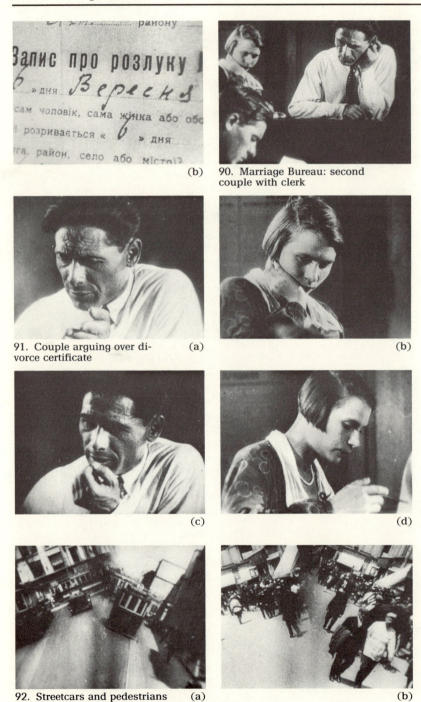

(b)

90. Marriage Bureau: second couple with clerk

91. Couple arguing over divorce certificate   (a)

(b)

(c)

(d)

92. Streetcars and pedestrians   (a)
falling apart – split screen

(b)

93. Marriage Bureau: third (a) (b)
couple with clerk

(c) 94. Trolleys passing each other

95. Old woman weeping over
tombstone

96. Old woman crying over
grave

97. Open casket funeral – lateral view

98. Bride getting out of
carriage

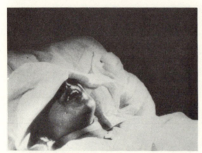

99. Woman in labor

100. Open casket funeral –
frontal view

101. Bride and groom climbing
into carriage

102. Bride and groom carrying
icons

103. Funeral cortege – frontal
view

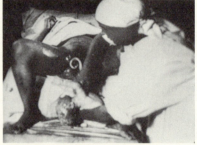

104. Woman giving birth

105. Cameraman shooting over
tall buildings – superimposition

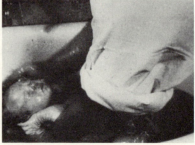

106. Washing baby

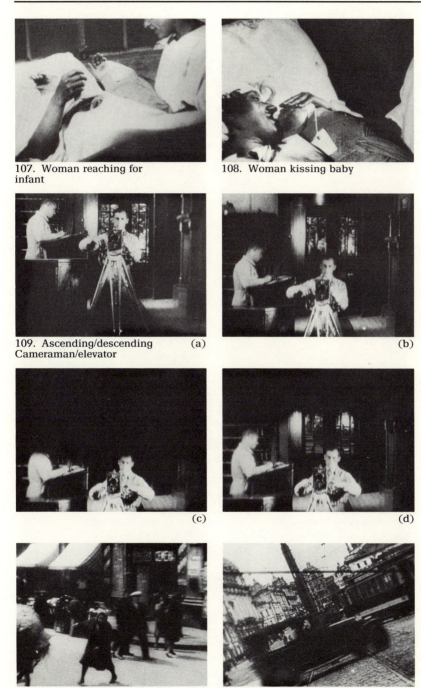

107. Woman reaching for infant

108. Woman kissing baby

109. Ascending/descending (a)
Cameraman/elevator

(b)

(c)

(d)

110. Pedestrians "threatened" by moving camera

111. Oblique view of traffic – (a) accelerated motion

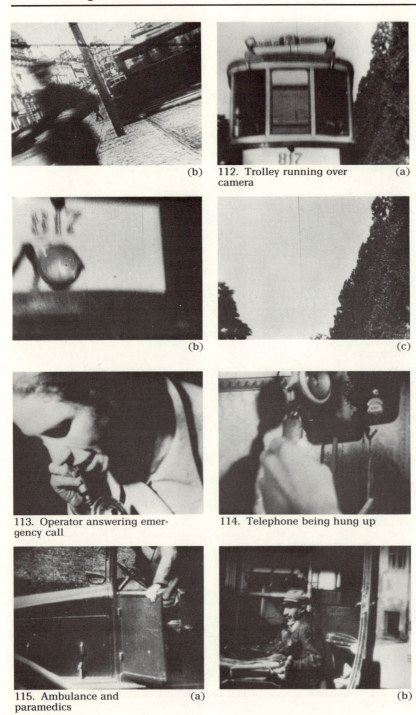

(b)

112. Trolley running over (a)
camera

(b)

(c)

113. Operator answering emergency call

114. Telephone being hung up

115. Ambulance and (a)
paramedics

(b)

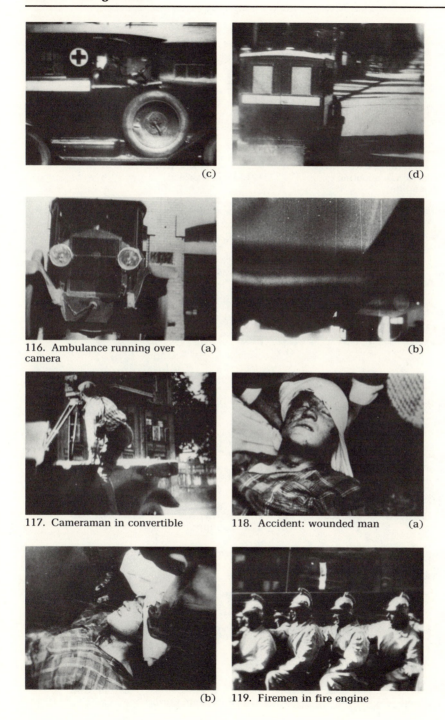

(c)

(d)

116. Ambulance running over (a)
camera

(b)

117. Cameraman in convertible

118. Accident: wounded man (a)

(b) 119. Firemen in fire engine

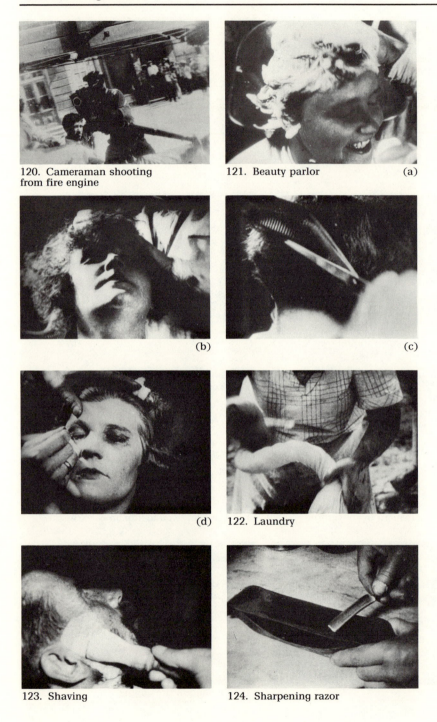

120. Cameraman shooting
from fire engine

121. Beauty parlor   (a)

(b)

(c)

(d)   122. Laundry

123. Shaving

124. Sharpening razor

125. Sharpening ax

126. Cameraman reflected in shoemaker's sign

127. Shoeshining

128. Fingernail polishing

129. Splicing film

130. Stitching

131. Sewing

132. Packing cigarettes — ac- (a) celerated motion

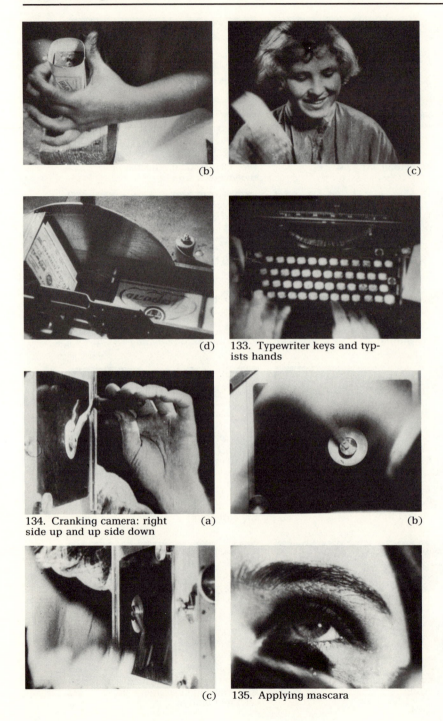

(b)

(c)

(d)   133. Typewriter keys and typists hands

134. Cranking camera: right   (a)   (b)
side up and up side down

(c)   135. Applying mascara

136. Hand calculating on abacus

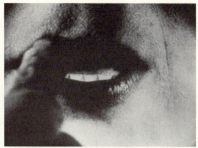

137. Putting on lipstick

138. Hand operating cash register

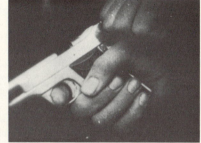

139. Loading pistol

140. Hand plugging in electric cord

141. Hand pressing telephone buzzer

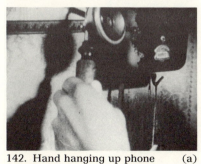

142. Hand hanging up phone and depressing receiver hook

(a)

(b)

143. Operator's hands at work   (a)
on switchboard

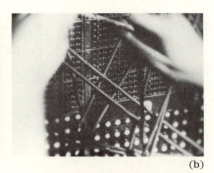

(b)

144. Hands playing piano

145. Cameraman shooting in
mine

146. Cameraman warned by
worker

147. Cameraman at work   (a)

(b)

(c)

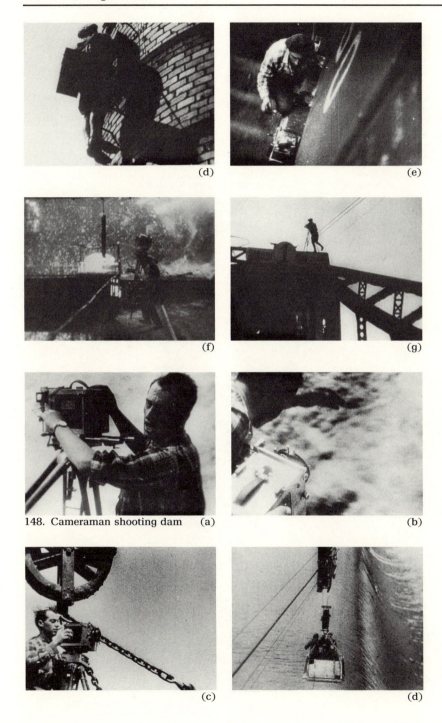

(d)

(e)

(f)

(g)

148. Cameraman shooting dam   (a)

(b)

(c)

(d)

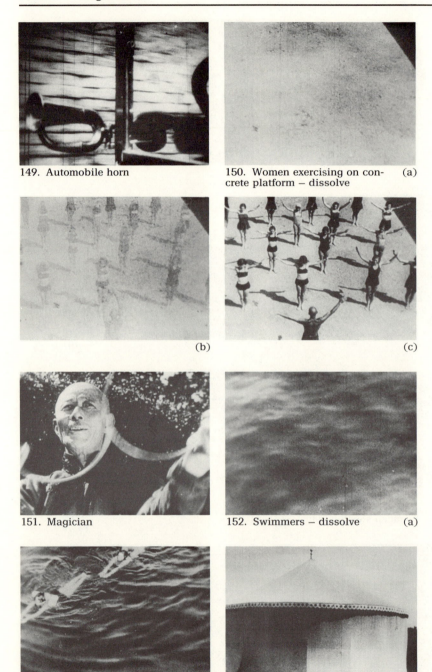

149. Automobile horn

150. Women exercising on concrete platform – dissolve (a)

(b)

(c)

151. Magician

152. Swimmers – dissolve (a)

(b)

153. Carousel – dissolve (a)

(b)

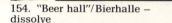

154. "Beer hall"/Bierhalle – (a)
dissolve

(b)

155. Moving sticks – animation (a)

(b)

156. Wall newspaper (a)

(b)

157. "About Sports," newspaper article

158. Sports fans   (a)   (b)

(c)   (d)

(e)   (f)

159. Sports events – slow   (a)   (b)
motion

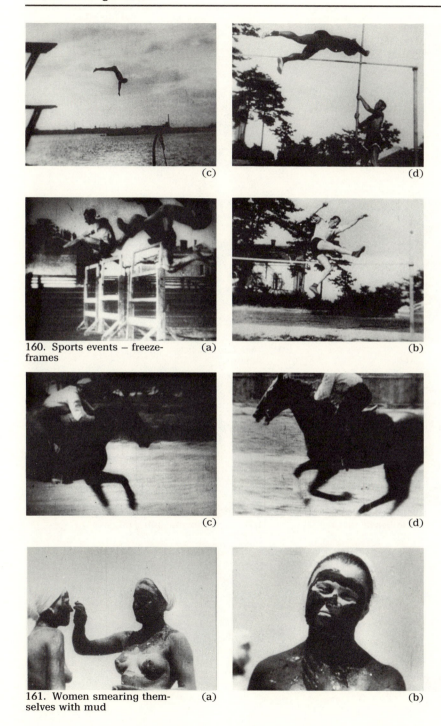

(c)

(d)

160. Sports events – freeze-frames (a)

(b)

(c)

(d)

161. Women smearing them-selves with mud (a)

(b)

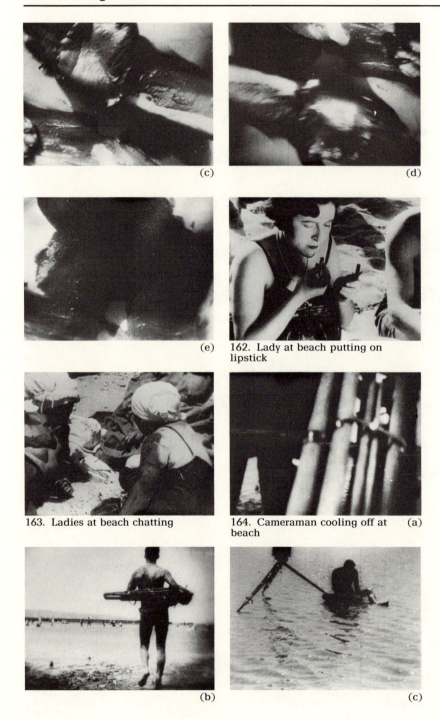

(c)

(d)

(e)

162. Lady at beach putting on lipstick

163. Ladies at beach chatting

164. Cameraman cooling off at (a) beach

(b)

(c)

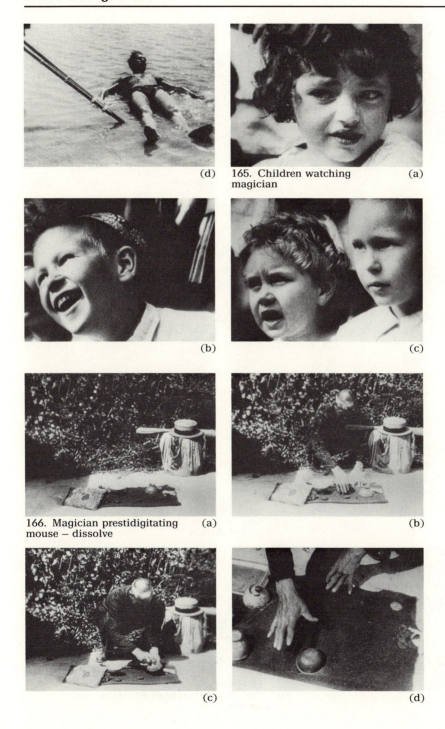

(d)

165. Children watching magician (a)

(b)

(c)

166. Magician prestidigitating mouse – dissolve (a)

(b)

(c)

(d)

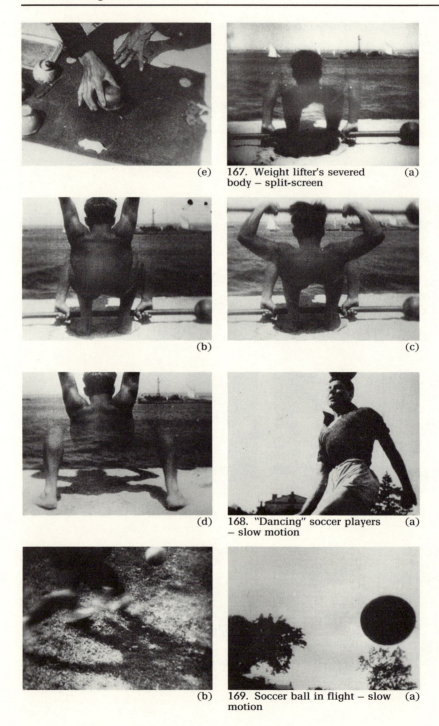

(e)

167. Weight lifter's severed (a)
body – split-screen

(b)

(c)

(d)

168. "Dancing" soccer players (a)
– slow motion

(b)

169. Soccer ball in flight – slow (a)
motion

(b)

170. Athlete throwing javelin

171. Goalkeeper catching ball
– slow motion

172. Somersaulting runner –   (a)
reverse motion

(b)

(c)

(d)   173. Motorcycles   (a)

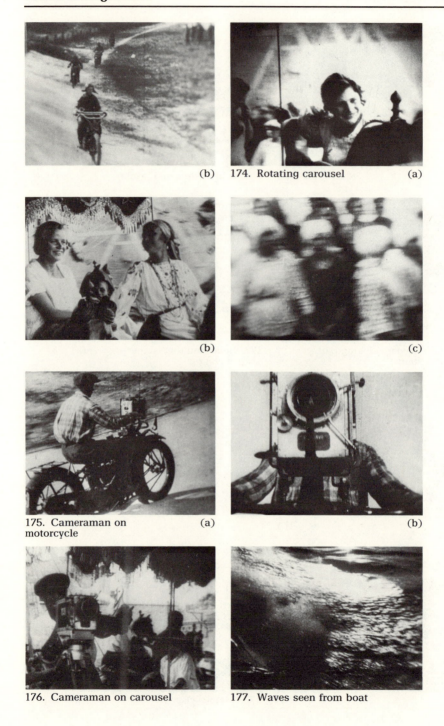

(b)   174. Rotating carousel   (a)

(b)   (c)

175. Cameraman on   (a)   (b)
motorcycle

176. Cameraman on carousel   177. Waves seen from boat

178. Torn filmstrip – freeze-frame, split-screen

179. *Green Manuela* movie (a)
poster – swish pan

(b)

(c)

(d)

(e)

180. Cameraman above city – composite shot

181. People in beer hall (a)

(b)

182. Cameraman "drowned" in (a)
beer mug – superimposition

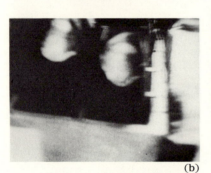

(b)

183. Beer bottles carried on (a)
tray

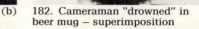

(b)

184. Churches turned into (a)
workers' club – swish pan

(b)

(c)

(d)

185. Checkers arranging them- (a)
selves – reverse motion

(b)

186. Chess figures arranging (a)
themselves – reverse motion

(b)

187. Workers relaxing in work- (a)
ers' club

(b)

(c)

188. Dziga Vertov playing   (a)   (b)
chess in workers' club

189. Young woman holding rifle   190. "Swastika" target

191. "Death to Fascism" target   192. Young woman shooting

193. Beer bottle targets –   (a)   (b)
animation

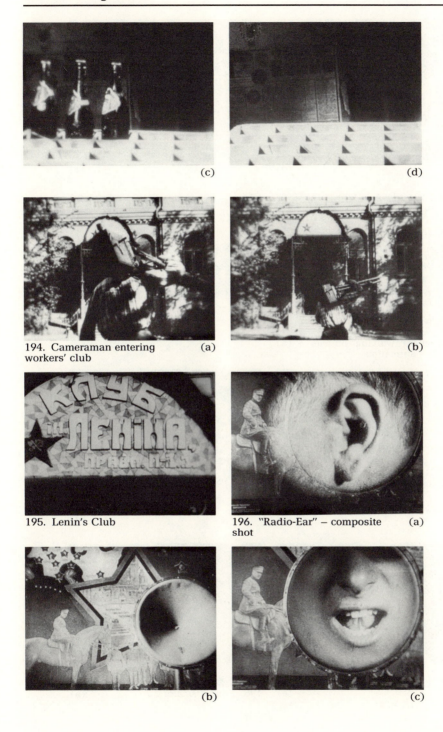

(c)

(d)

194. Cameraman entering (a)
workers' club

(b)

195. Lenin's Club

196. "Radio-Ear" – composite (a)
shot

(b)

(c)

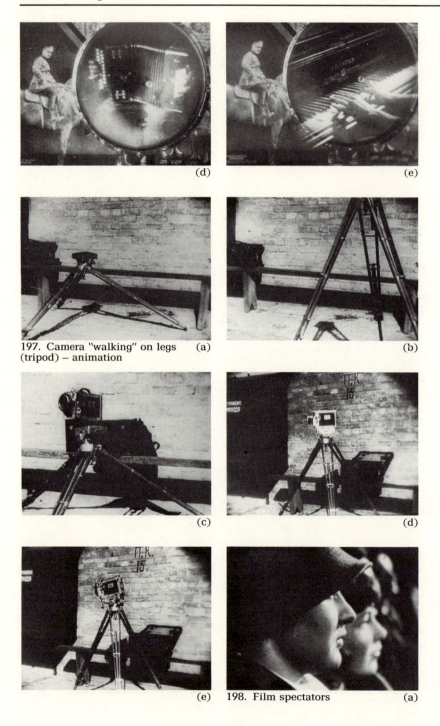

(d)

(e)

197. Camera "walking" on legs (a)
(tripod) – animation

(b)

(c)

(d)

(e)   198. Film spectators   (a)

(b)

(c)

(d)

(e)

(f)

(g)

(h)

199. Seated audience – high
angle view

(a)

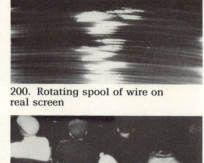

(b) 200. Rotating spool of wire on real screen

201. Rotating spool on screen-within-screen – composite shot

202. Conversing movie audience

203. Dancers and piano – su-perimposition, composite shots (a)

(b)

(c) 204. Audience – frontal view

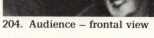

205. Glittering threads and spinning looms – superimposition

206. Spinning wheel and fe- (a)
male worker – superimposition

(b)

207. Locomotive wheels on real screen

208. Locomotive on screen-within-screen – composite shot

209. Well-dressed women (a)

(b)

(c)

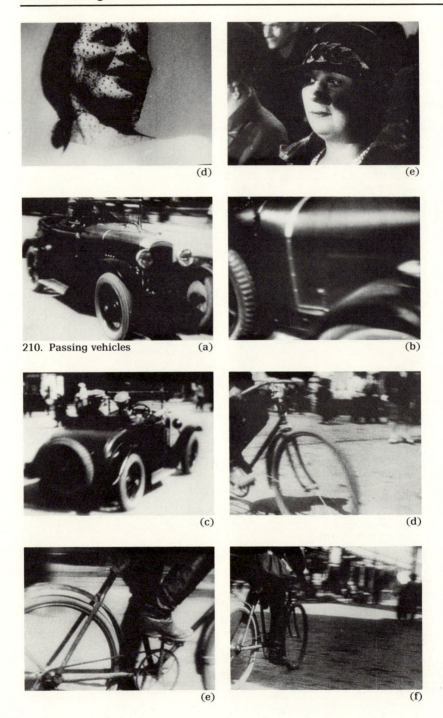

(d)

(e)

210. Passing vehicles (a)

(b)

(c)

(d)

(e)

(f)

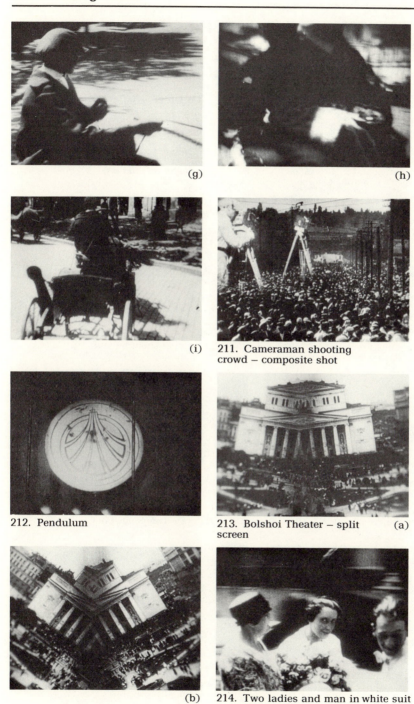

(g)

(h)

(i)

211. Cameraman shooting crowd – composite shot

212. Pendulum

213. Bolshoi Theater – split screen (a)

(b)

214. Two ladies and man in white suit in carriage on real screen

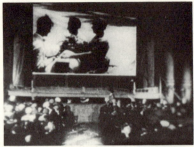

215. Two ladies and man on screen-within-screen – composite shot

216. Cameraman on motorcycle on real screen

217. Cameraman on screen-within-screen – composite shot

218. Cameraman shooting with telescopic lens

219. Cameraman shooting planes

220. Trolleys – composite shot

221. Typists – composite shot

222. Switchboard operators – composite shot (a)

(b)

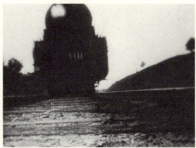

223. Train on real screen

224. Train on screen-within-
screen – composite shot

225. Cameraman in convertible
on real screen

226. Cameraman on screen-
within-screen – composite shot

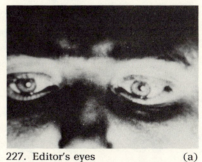

227. Editor's eyes (a)

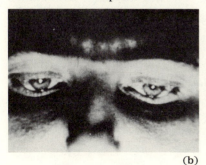

(b)

228. Movie theater and projec- (a)
tion beam

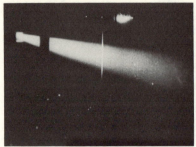

(b) 229. Traffic signal (a)

(b) 230. Car mounted on tracks, (a)
running over camera

(b) 231. Swinging pendulum (a)

(b) 232. Trolley running over (a)
camera

(b)

233. Lens with eyeball: iris (a)
closing – superimposition

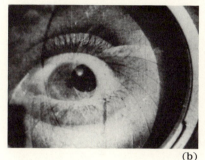

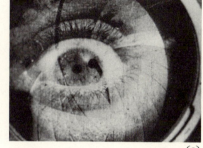

(b)

(c)

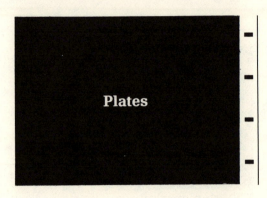

Plates

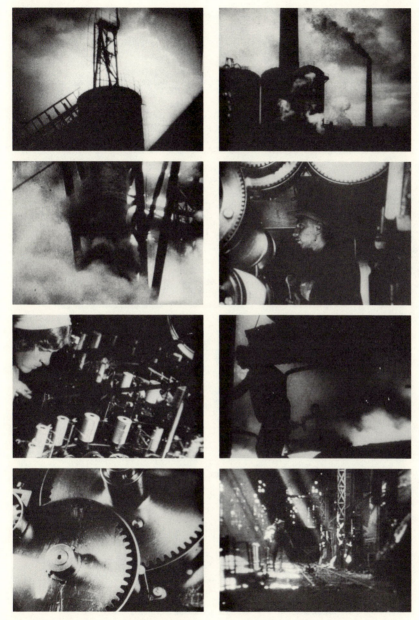

Plate 1. Factory and machines

Plate 2. Traffic and means of communication

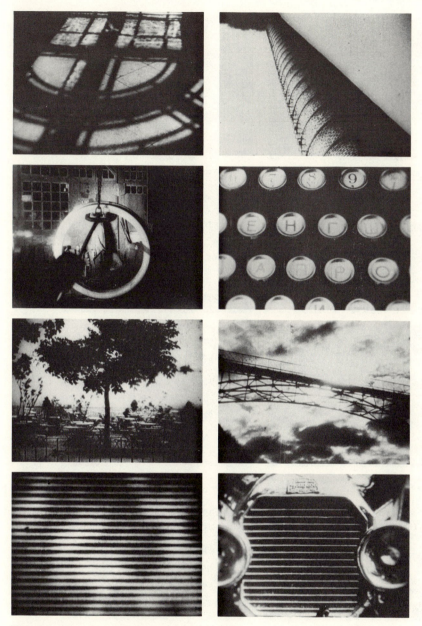

Plate 3. Graphic shot composition

Plate 4. Visual abstraction

Plate 5. "Awakening" sequence (VII)

Plate 6. "Street and Eye" sequence (XXII)

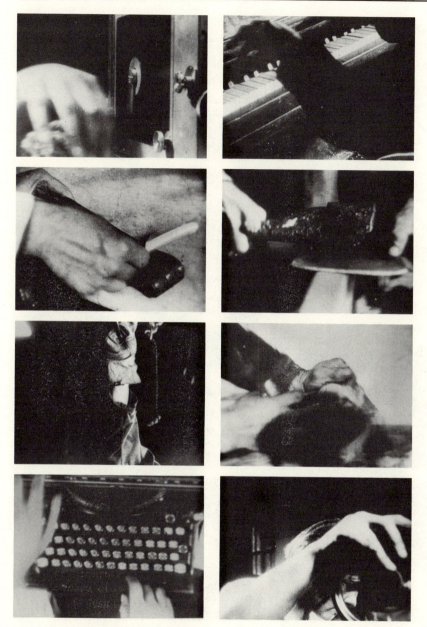

Plate 7. "Working Hands" sequence (XXVII)

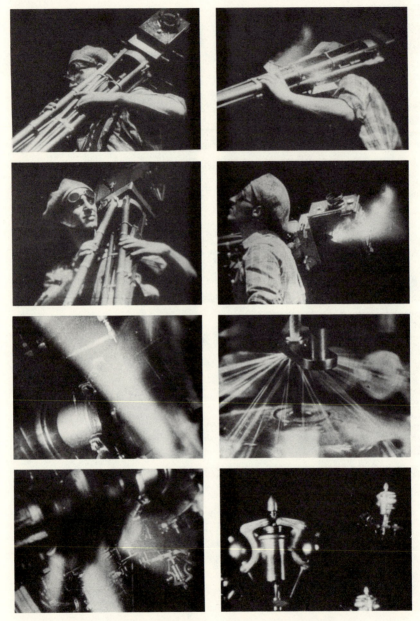

Plate 8. "Cameraman and Machines" sequence (XXXI)

Plate 9. "Musical Performance with Spoons and Bottles" sequence (L)

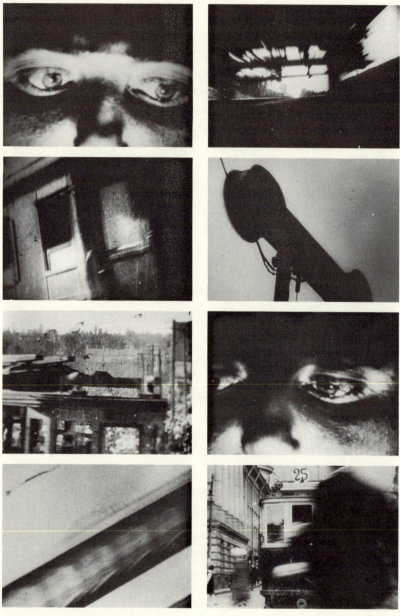

Plate 10. "Editor and the Film" sequence (LV)

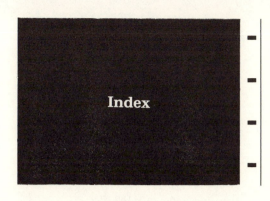

Index

So sacred Scripture and the texts of the sacred liturgy are made to seem to have meant *males only* every time "peace to men of good will" was wished, a claim that is not only absurd, but also quite simply an enormous injustice to all the good people who have preceded us, from the first translators into English, through all the generations that repeated this supposedly male-only vision of the divine.

When these games are played at Mass, the unity, peace, and serenity are damaged, with the excuse, "You cannot imagine how alienated the women have been all these years," despite the evidence that the overwhelming majority of Catholics frequenting the sacraments were and are precisely women, and that the vast majority of them experienced no such alienation, and still do not despite the efforts of the feminists to make them feel oppressed.

Bishops and priests who keep trying to satisfy people infected by this kind of hatred through little concessions are being totally unrealistic. They succeed only in producing an atmosphere of constant change and experiment, rife with tension, while feminists are never satisfied. As Monsignor Sokolowski commented in *L'Osservatore Romano*,[208] the liturgy should convey a sense of permanence. It is only twenty-five years since the most abrupt liturgical upheaval in the history of the Church. If now the American bishops get by with a rewrite of the very sense of what has been handed down from the evangelists as Holy Scripture, the sense will be conveyed that all is change, everything is up for grabs, anything can be toyed with to convey the "sensibilities" of some group or other at any time, in keeping with power games.

What most seek from the liturgy, in the midst of a world swept up in frenetic change and massive dislocation, is a haven of repose, of familiarity, where one can find orientation and a reminder that at the core of the many truths there remains the One Eternal Truth, "as it was in the beginning, is now and shall be forever. Amen." Leave the political turmoil at the door. This kind of insistent demand for ongoing change is driven by an agenda that has no intention of stopping with the ordination of women.[209]

208. Ibid.
209. The resentment mentality never has any logical stopping place. Its drive for "equality" may be about justice on the surface (and sincerely, too, on the part of many who are swept along), but at its driving core all too often it is about *self-hatred*. The greatest pity in such sociopsychological dynamics is that the extremes become an easy (and totally unjustifiable) excuse for the "moderate" to avoid addressing burning *concrete* issues of real abuse.

Another sad thing that happens: those who simply ignore such "progress" come to look "reactionary," simply by remaining faithful to what they know required no fixing. In the process, those driving for "modernization" widen a split; then at a certain point, some of the less balanced "reactionaries" do indeed *react*, summoning up the hatred in their own psyches, and so actually do lurch toward the extremes, as they seek refuge from unnecessary unpleasantness. The "progressivists" then point to the most extreme and unfortunate positions of these "loonies of the right" as characteristic of all

While one hears horror stories of irreverent liturgies, dancing in the sanctuary, and wild improvisations in the canon—recently, a seminarian in a religious order told me of an invalid mass, so changed was everything, even the words of consecration. I am happy to report—from Toronto, from Tulsa, from Paris and southern France and Vienna, the parts of the world I frequent—no such irreverence has been observed. On the contrary, from that small sample one would conclude, incorrectly, I suppose, that, the period of experimentation gone-by, the Church has settled into a period of calm, prayerful, and canonically correct vernacular liturgies, with even the more dignified of the new liturgical music percolating to the fore.[210]

I promised earlier to say a word about sermons as a weakness in the liturgy. In many daily Masses in Toronto there is no comment on the Scripture readings at all. This flaunts the *ordo*—the sermon is an integral part of the Liturgy of the Word—and priestly duty, and what an apostolic occasion missed to inspire and instruct! This negligence, not only of the sermon but also of offertory petitions, contributes to the Mass being too short. Sunday sermons are often carefully prepared, and sometimes are quite good.

When I look back over the whole question of Catholic practice such as I see it among my students and friends, one factor stands out starkly: the rapport between personal prayer life and solidity in the faith. What I am about to claim is based on anecdotal experience, and I may be badly reading attitudes in the small random sample provided by my direct acquaintances. With certain exceptions there seems to be a high correlation between contemplative prayer life and solidity of understanding and practice of the faith.[211] I believe that routine Mass attendance, where one just sits there and lets others do the praying, is deadly,[212] and often leads eventually to not-so-regular Mass attendance, and little understanding of what the faith is all about. (It is certainly not from the run-of-the-mill Sunday sermons one is going to find out!) Only through contemplative prayer does one adequately "internalize" what is objectively happening in the Eucharist. I cannot think of a single instance of someone whom I know to be a real

---

who simply want quietly to continue the form of adoration and of language they have always known and loved. (*Diabolein* means "to divide.")

210. Daily Mass attendance is strong in Toronto. To a recent Gallup Survey 18 percent of those who responded that they were "Catholic" said they try to go to Mass more than once a week. Three or four hundred people, many of them young, attend our College Church (St. Basil's) daily, and last Lent more than fifty thousand communions were distributed.

211. Not the least troubling among these exceptions are a few priests and religious who, despite quite regular prayer lives, remain disagreeable human beings negative about the "institutional Church." On the other side are some loyal Catholics who are people of utmost integrity and generosity, who put little stock in personal prayer and what they call "piety."

212. *Ex opere operato* means Christ is truly present in the Eucharist, and in the communicant when he consumes it, but if we do not open ourselves to respond, nothing "automatic" happens.

apostle who has not learned somewhere along the line how to pray. My Opus Dei friends insist, for instance, that while daily Mass is vital, "it is not enough . . . a half hour, minimum, of contemplative prayer a day is essential," and so it is a "norm" required of members of "the Work." I recall going on a business trip with one of my Opus Dei friends, and the call he left for six A.M., so that he could say his prayers. In seventeen years of formal Catholic education by exemplary religious and priests, I was never taught how to pray, which, incidentally, is not easy, either to learn or to carry out faithfully, nor did I in my youth acquire a love for Scripture, but a passion for the more rationalistic dimensions of Saint Thomas's thought was successfully communicated. Prayerful liturgies help. So do retreats (we never had any in high school, and I went to two while at a Jesuit university). Charismatics, who are good at prayer, manage to convey in their prayer meetings a sense of personal prayer, as "talking with Jesus." But Catholic education should make this a central consideration, starting with family prayer, continuing with Bible-study and prayer groups, and culminating in personal interior prayer.

Before leaving the question of liturgy, a word of lament about the state of Church architecture, art, and music. In the Archdiocese of Toronto about forty-eight churches have been constructed in the last twenty-five years. A Christianity and Culture student from St. Michael's College, Mrs. Joe Ann Boyle, did an architectural "liturgical suitability" appraisal of most of these churches, and concluded that more attention had gone into keeping costs low than in building "for the ages" a house suitable for showing forth the glory of God. Many of these churches are reasonably dignified liturgical spaces. Not one can be defended as a work of architecture meriting a detour to go admire. Few present a space suggesting the mystery that ought to surround the greatest of all miracles.

Every building reflects its time. When at the turn of this century imitation Romanesque and Gothic (and even some moderate baroque) churches were constructed from a mixture of modern and ancient materials, they reflected at once the zeal of poor immigrant populations wishing to have grand churches, their pedestrian taste and lack of artistic creativity, and the eclecticism that results from mindlessly marrying old forms to new materials and construction materials.

The new churches are more honest—and architectural honesty is not a bad symbol for a Church! Forms, sometimes at least,[213] grow out of a search for proper liturgical function, with a tendency to break with Roman-basilica style toward a gathering of the people around the table of the Lord. This shift from austere crucifixion on a far-away Golgotha toward disciples gathered about the Lord at His Last Supper certainly invites a more "folksy" liturgy,

213. Msgr. Alan McCormack, the former chancellor of the archdiocese of Toronto who tried in vain to get a liturgical building code developed for the Archdiocese, told me of architects coming to ask, "In what part of this building do you do your liturgical thing?"

but, as has been proved in many parishes, a supernatural banquet can also be conducted with dignity and courtesy. The effort to mold form in order to focus light has rarely worked well—even when full, many of these churches feel empty, as the light is too blaring, like too much of the amplified electric-instrument music. The tensile strength of modern materials permits soaring spaces at low cost, but often the materials used for decoration look cheap: rare is the window of quality, and the stonework too often looks like veneer.[214] A sign of a good architect is his care in details. Bad architects fill churches with absurd light fixtures, slap on acoustical tile ceilings, and ugly door handles.

The art objects to be found in these churches of course vary wildly in quality. Even when they are rather good, they all too rarely marry properly with the architecture. Again, the exceptions prove it can still be done. In the French village chapel of Beaumont, an artist who lives nearby painted on the six concrete panels on each side, set at an angle to direct light from the windows toward the altar, three-quarter life-size figures sparsely representing personages in the life of Christ, with suggestions of motif. These constitute, along with an enormous, plain massive-timber cross in the chancel, the sole decoration. They are contemplative, dignified, and modestly decorative, not crying out for attention to themselves. But when one knows, first, that the artist is a person of exceptional talent,[215] and, second, that she is intensely prayerful and orthodox, and finally, that she spent four years worrying and praying over the iconography, the design, the colors of these few figures, consulting many persons of both taste and theological acumen, literally pouring a portion of her life into this work, executed without pay, one begins to understand what must go into a genuine work of religious art.

Those who despair of recent liturgical music would be resuscitated if they could experience a Taizé evening of prayer like the one described earlier—again, the work of a composer of obvious talent who has prayerfully devoted much of his life to developing this music for the express purpose of creating a certain kind of atmosphere of prayer. He has drawn skillfully on the tradition, East and West, for both words (often in Latin) and musical forms and tonalities. The resulting four-part harmonic mantras of Gregorian inspiration, easily learned, are capable of enveloping those who seek to enter into a meditative atmosphere in a perfumed air of

214. When the stone wall actually bears, it can be better, and when, as in olden times, masons are invited to put their love into constructing a wall, superb results can still be obtained. In the small village church of Beaumont-Villages (Indre et Loire, France), the basically concrete structure includes a curving absidial wall (the curve itself is exquisitely designed) sweeping high, for the construction of which the architect solicited the best mason he could command, with quite lovely results. So it can be done, even today!

215. Mlle. Jacqueline Mesnet, the artist, tied seven times for the prestigious Prix de Rome after the jury debated for seven hours—she was beaten once by Brazillier who has gone on to world fame.

quiet loveliness. This is contemporary liturgical music that works! More thought needs to be given to just what one ought to be trying to achieve at each moment music is introduced. Choir directors should remember such important matters as this: hymns need to be repeated often enough so that the congregation gets to know them; when one admits music during the Offertory, the participating public will not hear the important words of the offering itself. I have seen that important part of the Mass pass unperceived all too often. Shut up during the canon! Do not position the "musicians" so that they and their performance become the center of attention.

PART 2: THE AUTHENTIC CATHOLIC ATTITUDE *(VERHALTEN)*

Neither "Left" nor "Right" but "There Where Mother Teresa Stands, the Center"

Now that I have come to the end of this incomplete and necessarily superficial survey of the contemporary landscape of the Church—mostly of the Europe and North America I know—I propose to step back and examine what, in the face of all these challenges and in the light of the depths of truth kept alive by this ancient tradition, I believe should be a Catholic's basic attitude today.

At the end of *Being and Truth,* I explored the notion of the appropriate attitude for one seeking wisdom. The German word for attitude, *verhalten,* borrowed from Heidegger, brings out well the sense of a way of comporting (*halten* means "holding") oneself in the lifelong quest for truth[216]—what I there argued should be "a non-ideological attitude." When Balthasar describes the *Christlicher Stand* (the Christian state of life),[217] he reminds us that the Christian must achieve a consistency of character and personhood by responding to the mission sent by God, a consistent and sustained activity of following in the Way of Christ. The "Stand" is not a static standing, rather, it is like our Father of faith, Abraham, accepting the challenge to leave all security and set out for the promised land by comporting ourselves in a way formed by the supernatural virtues of faith, hope, and charity. Believing, hoping, loving together constitute the Christian way of standing out toward the future, of "ek-sisting," as we take up our stand as regards what we believe has been handed us by the past, and pro-ject a future, directing ourselves, or rather allowing ourselves to re-spond to Christ's call directing us. It is only by letting the gift of the supernatural virtues form our lives (produce the character called for by our *Stand* and attitude) that we can follow authentically in the way brought to us by this tradition and the institution, the Church and the social structures it makes available to us.

There are two issues that merit consideration in this regard: What is the proper attitude of Catholics in the face of the tensions within the Church? He

216. See *BT,* chap. 10, sec. B., "Toward a Non-Ideological Attitude."
217. See Balthasar, *The Christian State of Life.*

can be tempted to become depressed by the massive indifference of many European and North American so-called Catholics (not to speak of the masses who have left the Church), fruits of that sinfulness of its members, lay and priestly, which leads to mediocrity and poor witness. And then, finally, how should the proper attitude relate to our understanding of the overall situation of the Church in the present planetary situation, the context for Catholic action?

Regarding the tensions,[218] it is not enough simply to remember this is nothing new, that the apostles had themselves portrayed in the Gospels with glaring human failings, recognizable as failures of love (John is the exception—he and Mary were at the foot of the cross), and that the Acts of the Apostles and the letters of Saint Paul show severe tensions in the early Church. We see every day those failures, preventing God from being able to answer as fully as He would like the prayers for unity and peace in the Church. In every eucharistic sacrifice we pray, "Lord Jesus Christ, . . . look not on our sins but on the faith of your Church, and grant us the peace and unity of your kingdom where you live for ever and ever."

The authentic attitude of one who is becoming genuinely a self by

---

218. A good sample of the tensions: On June 29, 1996, retired Archbishop John Quinn, in a major address at Oxford University, on the occasion of the one hundredth anniversary of Campion Hall, provided us with a neat catalog of the issues high on the priority list of what we might call the "anti-Roman" wing of the Church. (I am quoting from a report in *The Catholic Register*, Toronto [July 8, 1996]: 1 and 2.) Just as do the "entrenched Romans" he first drapes his enterprise in the prestige of the holy father, invoking the pope's call for leaders and theologians of Christian churches to help him "find a way of exercising the primacy which, while in no way renouncing what is essential to its mission, is nonetheless open to a new situation." (encyclical *Ut Unum Sint*, para. 95.) The archbishop's major complaint: the curia interposes itself between the pope and bishops, acting as though it were a *tertium quid*, subordinated to the pope, but over the bishops. Archbishop Quinn does not mention that the same problem exists on the level of National Bishops Conferences, where committees often of only three or so bishops take stands in the name of all the bishops that enrage many of them. He does not point out that without a curia the pope could in no way exercise any meaningful authority over thirty-five hundred bishops and a billion laymen. That neither the curia nor the National Bishops Committees' bureaucrats will ever consult and dialogue sufficiently to satisfy every bishop that he has been truly heard for the simple reason that this is physically impossible, he fails to mention. Of course, there is a real problem at both levels, and it will not go away. In each concrete case what is called for is more charity all around. But the thrust of Archbishop Quinn's desires is toward greater "subsidiarity," that is, massive decentralization, which of course makes it easier for local churches to go their own ways. He listed issues the new council he is calling for should address: priestly celibacy, women's role in the church, the role of bishops' conferences, general absolution, liturgical inculturation, ordination of women, contraception, and reception of the sacraments by the divorced and remarried. He ignores the pope's firm setting aside of ordination of woman as impossible in view of Christ's institution of the priesthood as male; he ignores the vast effort of the encyclical *Evangelium Vitae* to give a proper setting to the Church's upholding the natural-law principle that interfering with an act's prime purpose is immoral. And this pope has done everything he can to strengthen commitment to the ideal of priestly celibacy. So while professing loyalty to the Holy See, Archbishop Quinn seems prepared to ignore the pope in fact.

allowing himself to be expropriated by Christ—the saint—is the fruit of grace, and we should daily pray for it, "give me a pure heart . . . not my will but thine be done!" I was talking recently to a seminarian who was lamenting that one of his colleagues "had gone over to the other church." By this he meant the majority group in his seminary (they are only seven!) who are influenced by a certain "Protestantizing" agenda for the Church,[219] driven by an antiauthority attitude, often today labeled "political correctness." I cautioned him against thinking in terms of "the other church." For one thing, by falling into such rhetoric, he contributes, if ever so little, to furthering the very schism he would lament. (He is not one of those superorthodox without love who would welcome a split.) He does make it sound as though he himself is devoid of the supernatural virtue of hope, which requires charity, too.

The Church is the only human social reality for the survival of which we possess a divine guarantee. It has already survived two thousand years of tensions, even schisms. However, the promise of survival obviously does not guarantee against tensions developing into ever more debilitating splits (we were warned from the beginning by Christ Himself: "Do not suppose that I have come to bring peace to the earth: it is not peace I have come to bring but a sword. For I have come to set a man against his father" [Matt. 10:34]); nor does it guarantee against the loss of important parts of the Christian world to non-Christian forces (as of much of the Middle East and all North Africa to Islam, temporarily, the Christian hopes, but that time has already lasted more than a millennium). One should never forget that the Church remains the principal object of Satan's relentless destructive wiles— it is he who turns Christ's presence into division—and the visible institution the "stumbling block" to the secular, humanist dream of uniting mankind without God, through a vague universal reason.[220] Balthasar explains:

> The institution of the Church, which from within is the presence of Christ and a liberating sphere, acts as a scandal to the world. And this very scandal is indispensable to prevent the incarnation of God, and even his crucifixion, from dissolving into idealistic vapor and shallow morality. The Church, in her rigid structure, will always be a "stumbling block" for people. . . . Attempts will be made to decimate the Church, perhaps with success; but she will never be completely demolished. In our times "an assault on Christianity by the world religions" has been noted. . . . Why? After they have drained off what seemed to be assimilable, the indigestible, annoying institution is left; it must disappear. But, as is well known, precisely the rind of a fruit contains the most vitamins. "Life is in the living form itself." But, more importantly, do not forget either that the Church is the first love of the infinitely powerful Holy Spirit.

Whenever a party arises within the Church with a fairly clear agenda for serious transformations (as *ecclesia semper reformanda*, this can be a good thing), then prayerful discernment, not just by the ecclesial authorities,

---

219. See Archbishop Quinn's catalog in the previous note.

220. This, with every passing day, appears less reasonable. See Balthasar, *New Elucidations*, 102.

but by each Catholic as he becomes subject to the siren's call, becomes an unavoidable responsibility. One must ask these questions of the would-be reformers: Does their agenda show signs of being informed by a spirit of fundamental regard, not only for the stated positions of the magisterium in general, but also for the good faith of those who are the voices of the actual teaching authority of the Church? Is the magisterium considered by the proponents of change or correction as authoritatively representative of "the center," or are pope and the vast majority of the bishops treated as, effectively, *extreme*, perhaps even heretical (they will be "giving too much away")? Are the reformers treating them as "reactionaries" and "too rigid," and "unfaithful to the apostolic tradition" (or both!)?

I am not forgetting that one can have total respect for the good faith of pope and bishop and loving obedience to their authority yet put forward suggestions for better doctrinal formulation, reform of a discipline, innovations in institutional form. A Teilhard de Chardin, a de Lubac, a Rahner, or a Balthasar can even take Catholic thinking off in horizon-opening new directions, while being adamant about their intention of remaining faithful to the entire tradition and obedient to the magisterium. They may subsequently be judged off on one position or another without their basic fidelity to the Church being legitimately in doubt.

Between an attitude of respect for the authority, while advocating a correction or even a great innovation, and the kind of mistrust and resentment one finds in some camps of reformers there is all the difference in the world—genuine new creation and fraternal correction, while acknowledging the charism of office, and incipient revolt against authority are hardly the same thing. When anger and resentment dominate, one will see the movement of "reform" move from discussion into public acts of disobedience. When that happens, such a party is indeed playing with schism. In our own time, we have seen the Lefevrites do just that: so they ended fairly rapidly in a break "with Rome." (They deny they are in schism.) The old saying "more Catholic than the pope" applies literally to these followers of then retired (and since deceased) Archbishop Marcel Lefevre, who denounced the Second Vatican Council as a corruption of true Catholic teaching.[221]

But what is more serious in the long run than particular acts to show frustration (*rage* is the latest Marxist "in" word) is the attitude, quite apparent on our local scene, driving a certain agenda that is undeniably having a

---

221. At the other end of the spectrum, about the worst acts to have occurred on the part of what we might call "the woman's ordination party," so far as I have heard, have been a few mock Masses (one in Toronto, in a church, with laughing priests present and a nun presiding at the "mass") and refusal by some nun-representatives to participate in a Eucharist with the local "patriarchal" bishops in a Midwestern U.S. meeting to discuss the women's issues. Both events were either not widely broadcast, in the case of the "mass," or private, in the case of the meeting. The first was blasphemous, the second an act of momentary schism. It is not clear whether the sisters intended to call into question the validity of the bishops' orders, or simply to lodge an on-the-spot protest. Both actions were without disciplinary consequences.

considerable effect. In some seminaries and in some religious congregations the present "Polish pope," the "Panzer Kardinal" Ratzinger, "the bishops" in general ("patriarchal hierarchs"), and important elements in the official teaching of the Church are treated with undisguised scorn. Students are being taught to mistrust if not indeed to loathe "the hierarchy," a concept that has become in a libertarian democratic society per se a dirty word. An increasingly well-organized movement of reformers calls itself "We are the Church." (Are the bishops still allowed in?) A well-orchestrated program of "sensibilization of women" is being carried out through a network of "lay ministers" in the parishes, who themselves have been formed in the dissenting theological faculties.

In general, most bishops do little about this, standing by while this influence spreads at the lowest level, among the students, and through the seminary and theological faculties, risks deforming future priests, so that it will soon be felt at a higher level. If a bishop were to take the slightest disciplinary action, he is guaranteed vilification in the press, even international attention. Even to dare to raise an eyebrow in the face of this concerted program is to get oneself classified as a hateful "reactionary." Many bishops are in fact afraid to act. Some await the occasion when positions come open to move more faithful people into key positions.

I return to my question, now more concretely: what should be the individual Catholic's attitude, not in general, but in the face of this kind of provocation?

First, as I said to the seminarian, avoid the temptation to overreact. Panic, anger, temptation to anathemas are usually signs of sins against hope and charity, and hence against faith itself. Second, there is, fortunately, much one can do positively. To start with, in the present situation, where many laypeople in the West—and not as before just some clerics—possess at least a fairly good basic education, and where libraries and opportunities for instruction abound, the individual has a personal responsibility and a real possibility to get himself informed about the Church's actual teaching (read the papal documents—you might be surprised to find them more "liberal" than you imagined, and certainly inspiring), and *prayerfully* work first to understand it himself, in a spirit of docility (assuming he does indeed believe in the divine reality of the Church). Further, the individual must be aware he may be victim himself of a program of misinformation, a distortion of what the Church actually teaches and of what is happening; he cannot avoid his personal responsibility for equipping himself *at the true source* to recognize such distortions when they are presented. The issuance of the new *Catechism of the Catholic Church* provides him with the most authoritative and up-to-date handbook of what the Church does in fact teach, rendering it unnecessary to suffer a moment longer from disinformation on this level at least.[222]

222. Those with unorthodox agendas are, of course, furious at this "authoritarian" initiative, for it ruins all games of confusing people about what the Church actually

Third, and most important, he must strive to be open to allowing God to install within him "a pure heart" (Mother Teresa's daily prayer), a heart that is not self-serving but open to whatever the Holy Spirit demands of it. Knowledge of what the Church actually teaches is indispensable. But without a life of interior prayer and frequenting of the sacraments, without prayerful meditation on the Holy Scriptures, one cannot expect to stand up against the assault; one will inevitably be drawn into answering ideology with ideology.

Finally, always one must seek "the center," found in the chair of Peter, and manifested in works of charity. I would repeat here the words of Balthasar, responding to the question of where in an age of polarization in the Church, does one find hope:

> Where should one look to see a dawn? One should look to where in the tradition of the Church something truly spiritual appears, where Christianity does not seem a laboriously repeated doctrine but a breathtaking adventure. Why is all the world suddenly looking at the wrinkled but radiant face of the Albanian woman in Calcutta? What she is doing is not new for Christians. Las Casas and Peter Claver did something similar. But suddenly the volcano that was believed extinguished has begun to spit fire again. And nothing in this old woman is progressive, nothing is traditionalist. She embodies effortlessly the center, the whole.[223]

How to Balance One's Effort between Personal, Family, Local, National, International, and Civilizational-Intellectual Demands?

The responsible, educated individual, as *Tradition and Authenticity* pointed out, faces a challenge balancing his concern and efforts between

---

does teach. The council documents are much easier to twist into different senses than this *Catechism*, to fabricate a spurious "Spirit of the Council." And *The Catechism* is authoritative. It was requested by a general synod, the drafts were scrutinized by all the bishops, whose corrections—twenty-five thousand suggested amendments were taken into account—were included, and the final version was gone over, line by line, first by Cardinal Ratzinger, then, I have been told, by Pope John Paul himself. (The archbishop of Tours, Jean Honoré, member of the drafting committee, commenting on Cardinal Ratzinger's work of correction, had this to say: "Cardinal Ratzinger acted as final arbiter in the debates. Most of the texts are due, not to the literal drafting of Cardinal Ratzinger—that is not his style—but to his arbitration, always realized in a very intelligent and precise manner by the Cardinal. I have an esteem for him that approaches admiration. I frankly believe that he does not enjoy the reputation that is his due" [*Catholic World Report* (December 1992): 50]. The same interview contains many more details of how *The Catechism* was composed.) An auxiliary bishop of London, Ontario, welcoming the new text, responded, no doubt prodded by the journalist, that, yes, he feared extremists of the right would misuse *The Catechism* as a "litmus paper test of orthodoxy." Without any doubt, a cavalier or, worse, hostile attitude to *The Catechism* will indeed be a bad sign for anyone who believes in the divine teaching authority of the Church. "Litmus test" is severe, but it does seem to me *The Catechism* will help make it clear just to what extent someone is faithful to or dissenting from the genuine teaching of the whole Catholic Church. That reflects, I believe, an attitude, not of "reaction" but of *fidelity*.

223. Hans Urs von Balthasar, *A Short Primer for Unsettled Laymen* (San Francisco: Ignatius Press, 1985), 17.

the pressing needs of his family, of his work, and the more diffuse but nonetheless real needs of his city, his province, and his nation. In addition, the Christian has a responsibility for the Church; he must find time, and then within it balance concern for his parish, perhaps for a movement that may be important to his spiritual formation, the local church as a whole, the church in his country, and the worldwide Christian family.

This demands not just a lifelong balancing act for the individual, seeking to organize his time—a necessary HTX sport!—but it poses a strategic problem for the leaders of the Church, the bishop, the pope, and the curia, and heads of religious congregations and spiritual organizations such as Communione e Liberazione and Opus Dei. Every bishop is expressly told that his mandate is not only for the local church but also for the universal body of Christ, and indeed many bishops find themselves spending much time on affairs of Rome and of the national bishops' conference.

The very immensity of the Catholic Church poses managerial and communications problems. These are made worse by the fact that the average Catholic has difficulty relating humanly with such a vast entity. That may be one reason Catholics show relatively poor solidarity with their brothers in Christ who are suffering horrible repression. (To suggest there is no compassion at all would be unfair; the universal Church helps considerably in many instances, but relative to its potential strength, what it does—what we Catholics do—for instance, for the embattled small Christian population of Lebanon being strangled by an increasingly militant and ruthless Islamic sect and Syrian secular power, is ludicrous compared to what, if properly mobilized, Catholics could do.)

But already on the local level the challenges are daunting. "The vineyard is vast and the laborers few." When I think of how I have been pulled between family (a Bible discussion group for my children and friends when they were teens), parish (helping Sundays to run a catechetical Mass, and in recent years nothing but a minimal financial contribution), diocese (chairing the Commission on the Family, and now initiating a Communio Circle to bring together young people serious about their theological and spiritual formations, and two other groups to study *The Catechism*), Catholic college (major effort to start and operate the Christianity and Culture program, and to complete the six volumes of a philosophical work, which is obviously part of my Catholic life, and of my professional work), national (presiding over the Catholic Civil Rights League for all of Canada, which is time- and attention-absorbing), and international concerns (Communio editorial boards, the International Family Congresses, Hospiz Sonntagberg [an enormous international family-evangelization center in Lower Austria] and yearly meetings in Rome to discuss strategies of evangelization), I am impressed more by what remains undone than by what little appears to have been accomplished by all this frenetic activity, especially (to cite an urgent item on an unfinished agenda) the need to do something as a senior educator to mobilize the forces within the vast Catholic school system to save its Catholic character. All of this while trying to be present to my sixteen children and grandchildren.

When I lament both this distraction of being pulled in so many directions and my failure to address initiatives that cry out for some leadership and much presence, I am admonished sharply by my spiritual director not to think that it is up to me to save the world all by myself. I could not agree more. But still I ask him, every time we get into this, "Well, which of these do you advise me to drop?" He never has any viable suggestions; he is rather like the overworked physician who says "Get some rest!" The truth is there is pathetically little leadership apparent in our community and nation. I know for sure the other people giving active leadership in the many excellent Catholic initiatives that are going on are likewise overburdened and pulled in too many directions. Everyone tells them also, "Slow down"!

In the midst of this unending balancing act, one thing is clear regarding priorities: attend first to demands of family, job, and personal spiritual development. My spiritual director points out that on the success of the last will depend the extent to which God will be able to use one effectively as an instrument of what is, after all, his work. Just as Balthasar has so clearly shown the disaster that occurs to the Church when theology is pursued independently of, or, so to speak, "out ahead of," holiness, so, in the guise of gaining the whole world for the Church, we can be losing our own souls. (Recall Cardinal Ratzinger's biting remark: "One can spend his whole life running from Church committee to Church committee and not even be Christian"!)

Just these three demand considerable time: family, job, and spiritual development. If one allows oneself to become absorbed in a job—especially if it is a career position, with open-ended responsibility—the temptations to shortchange the family and to leave no serious time exclusively for God are overwhelming. (Offering to God each morning all one's works, joys, and sufferings is fine but doth not a sustained contemplation make.) Two major reasons that members of the good bourgeoisie today are so often such spiritual failures is just this: never have so many been so enticed to become so completely absorbed in their work as in today's high-tech, competitive world. Second, the dispersion of family means one must formally plan visits and then travel great distances to see one another (two hours' travel time for our nearest child and travel to the Alaskan outback for the farthest).[224]

In the face of this reality, one can understand the practical wisdom of the rigidity of the "norms" Opus Dei asks its members to take on voluntarily: daily Mass, rosary, half an hour of interior prayer and meditation, monthly evenings of reflection, annual retreat, with a sense of urgency to make these

224. Forget the daughter working in the backwoods of Alaska out of telephone reach! Well, we did not. We flew in a bush plane through the fog-filled canyon of the Taku River to go see her. "Nearer my God to Thee." This midrash illustrates well the dilemmas of yuppie HTX families.

the first responsibility, even at the risk of some tension in the family. That means not only at least two hours a day to fulfill them correctly, but also countless little moments of "quickie" reflection to recall that what one is doing at work or in the kitchen is only for the greater glory of God, and that He is present in all these tiring tasks.

Yet, my Opus Dei friends recall, one can do all that and still risk "just going though the motions." What counts most, they insist, is the quality of contemplative presence in "interior prayer" that one is supposed to achieve every day. Amid the wild distractions of my life, I really wonder how they can possibly manage to achieve contemplative intensity every day.

This brings home what is to me on the personal level the central point in this endeavor to appropriate the Catholic tradition, which has brought home the force of the truth that truth is to be lived. While the communal dimension of appropriation, the help from sources, the critique from friends is vital, the appropriation is in the final analysis a personal act and a personal responsibility. That personal point then is this: Jesus Christ came to save *my* soul, and while I know that I am not to be saved without the others being saved, for we were created essentially social,[225] and altogether we form the body of Christ, the health of the individual cell is essential to the health of the whole body. So I am doing the best I can for strengthening the Church when I respond in love to those who need my help the most and when I receive the sacraments and when I plunge into prayer and contemplation,[226] which provide the sustenance for this love.

The Church faces similar problems when, on the strategic level, the responsible parties reflect on where to put resources. Should the Church concentrate on penetrating nominal so-called practicing Catholics with proper doctrine and spiritual life, or expend energy instead on bringing

---

225. Eve being taken from Adam's rib, and all men coming forth thereafter from the wombs of women.

226. When I acknowledge how poorly I cooperate with the graces God gives us to permit such prayer to be possible I am appalled at the ineffectiveness (assigning no blame on others) of twelve years of Catholic school (good at communicating doctrine and for the example of fidelity that the Benedictine Sisters of Oklahoma and the priests of the Diocese of Tulsa gave in those days, but with little success in teaching me to pray), a B.A. and an M.A. in Catholic philosophy at a Jesuit university (excellent at communicating a love for Catholic culture, but again making no dent on getting me past the "never miss Mass on Sunday" stage), a Ph.D. at the Institut Catholique de Paris (again, wonderful example of priestly life in austerity and obedience, but no effect on personal spiritual development), and after teaching in secular universities and being married to a Catholic wife and then teaching these last fourteen years in a Catholic college of, let us say, very mixed fidelity, neither Sunday sermons nor all that formal education led me to make the slightest progress in prayer—nor did twenty years of sitting on the editorial board of *Communio*. I have received much more care and help in understanding and beginning the practice of a bit of prayer from a charismatic friend met through the International Family Congresses and from a couple of rather persistent Opus Dei missionaries than from all the remaining influences in my life! I justify the personal nature of this note by the fact that, after all, this is where the bottom line is to be found.

the message to those outside who have never heard it, both the masses in Europe and the United States who require "new evangelization" and the even more immense and distant masses in the non-Christian world who require elementary missionary activity? None of these dimensions of the overall mission can responsibly be neglected. The question is one of priorities, where the secondary priorities are still to receive resources and necessary attention.

Applying what was said above, the answer seems to me clear: first comes deepening the spiritual lives of the core of the Church, for quite simply, from whatever "success" God grants with turning casual Catholics into zealous Christians will come the instruments for the "new evangelization" so desperately needed in Europe and America, and the missionaries who will continue to stream out to the entire world.[227]

PART 3: REFLECTIONS ON THE CHURCH'S POSITION IN THE EMERGING WORLD SYSTEM

Three Competitors for Men's Hearts

I close with a consideration of the position of the Catholic Church today in a world formed by what has become a planetary force. I want to do this not because I am suddenly forgetting the faith's experience of the Church as a supernatural reality complete in the perfect sanctity of its Mother Mary, or just because I do believe that humanly its leaders must reflect on the Church's strategic situation at this moment in history. Rather, I seek to sharpen our sense of the principal spiritual combat that is being waged both on the level of the individual soul—yours, mine, our neighbors—and on the cultural level.

We should begin such an overview of the planetary Kulturkampf by considering the three actual great competitors against the Church for the hearts of men.

First, we confront secular materialistic humanism. Open societies permit the Church to operate unfettered, but their success-oriented consumerist cultures erode the religious sense. A question for all citizens of these dynamic societies, not just the Church: are these open societies not spinning out of control, is not confusion growing in the hearts of men, and nihilism rising? Are these societies long to survive as "open"? The most creative, most hard-driving, most pathologically workaholic elements in them set an impossible pace everyone is supposed to catch up to.[228] The

227. In a recent address to the Spanish Episcopal Conference, the cardinal archbishop of Madrid, Angel Suquia, argues convincingly for the need to center the new evangelization on Catholics themselves (*Communio* [winter 1992]: 515–40).

228. One evening at a cocktail party in her Palm Beach home, overlooking the dusky Atlantic, a handsome Catholic middle-aged woman, vice president of one of the largest international corporations, told me that she, and many of her lady executive friends,

strain on families resulting from every manner of challenge, erosion of religious sense generally, the sexual revolution, and so on, results not only in below-replacement birthrates (every European people is dying out—indeed, fifty-one nations—and sperm count is down 50 percent in a century), but also in ever increasing numbers of children disturbed by family breakups and abuse from addicted, violent parents. Meanwhile, the growing elitist tyranny of "political correctness" shows signs of a disturbing need for some kind of socially accepted "truth" psychologically linked to pitting one group against another (ostracization of the white male is the latest rather large target—but then so was the "capitalist bourgeoisie" a large target for the Bolsheviks, and they were wiped out in Russia and China! Are males not being effectively neutered?). Marxism is grounded in a profound psychological understanding of resentment-driven hatred. Marx could hate with the best![229]

Second, we consider evangelical Christianity, a strong competitor in North and South America and Africa, and in the Philippines. And *competitor* is the word, for vast numbers of evangelicals work nonecumenically to spread anti-Catholic feelings, to take Catholics—and not just indifferent ones—away from the Church.

Third, we must consider Islam, which is quite militant in general, and extremely so in the form of the various fundamentalist movements.[230] The most solidly Muslim countries do not allow any Christian missionary activity: the Arabian peninsula, the Persian Gulf, Iran, North Africa, the Turkish lands, Afghanistan, Pakistan; the pockets of ancient Christianity in the Middle East and Egypt are under severe pressure.[231] The big question

---

rushed back home from across the continent to their weekend homes to engage in forty-eight hours of meditation, serious religious endeavor, as condition for retaining their equilibrium.

229. China and North Korea bear huge psychological grudges, rather like the sick, racist motives that led to "ethnic cleansing" in Bosnia: China, resentment against both Japan and the conquering Europeans, North Korea against Japan as well, but especially against the U.S.-sponsored economic success of South Korea. China, remember, is a country embracing a quarter of mankind still closed to the Church; and North Korea is completely closed. Is it the Church's best hope to be able to evangelize China and North Korea that they be opened up to the "open world" of free circulation not only of goods but also of ideas? The Church followed the dastardly Conquistadores to South America, and the colonial powers to Africa, with success in implanting itself in both instances. This time it may be obliged to follow the democratizers to China. The Communist regime is already persecuting it on the grounds that it is part of the democratizing force.

230. This is an alliance of religion and politics that my (believing) Muslim friends in North Africa condemn as a frightening monstrosity. As one Tunisian put it, it is "a pre-empting of religion by tyrannical political movements."

231. The largest group, the Egyptian Christians, the "Copts" (from Greek, *Egyptos*), are best able to resist, for they are perhaps 9 million, making up as much as a sixth of the population. But they are under intense persecution at this time. Little headway is being made either among Muslims in India or Hindus, but the Church is making great progress in Indonesia, where Islam is lukewarm (it is now being reawakened by

for the future is rather like that of China, whether as Westernization brings secularization to Islamic countries, will they eventually open up a bit to Catholic and evangelical missionary activity? Or will they all succumb to the fundamentalist counterattack against all "Westernization"?[232] Should Catholics be working harder to present the gospel to Muslims who have moved to Western countries?[233]

From the high pinnacle of philosophy the reasons advanced in *Tradition and Authenticity* for cooperative search for an ecumenical wisdom by thinkers in all the great traditions still seem valid: appreciation for one another is far better than war; and caring persons in each tradition can learn from the other traditions, whose distinctive truth claims formulate the fruits of rich experiences, witnessed to in practice by the saints of each tradition. Since first formulating these noble considerations a quarter of a century ago I have become gloomier about the intractability of ideological engagement—the mistrust of the other and the insecurity about one's own little ideological house of cards, a defensiveness inherent in fallen human nature, and maintained by the ignorance of most regarding even their own traditions.

I am not, however, losing hope, not, anyway, supernatural hope in God's transforming healing and in His mercy for all His children, at the end time for all peoples and in the meantime for individual souls who respond to His prompting to ask for mercy. I have never thought that Christ envisioned some kind of convergence of the family of man through ecumenical dialogue between the great traditions, and gradual strategic improvement of social, political, and economic structures. As these would be ways of peace, I am sure He is happy with any progress we may manage to achieve. But I doubt that "salvation of the nations" will come that way. Five years as a leader in Canadian Professors for Peace in the Middle East (a Zionist organization), as many years in Islamic-Christian Encounter,[234] and a couple of intense years in fruitful collaboration with an "open" and profoundly thinking evangelical academic, while despairing sometimes at the paucity of "ecumenical openness" among Catholics of different tendencies, and the self-centered indifference of many in the "movements" toward what is happening to fellow Catholics outside their community, have not overall enhanced my optimism about the possibility for any great advance in mutual understanding and appreciation between

---

fundamentalists); upward of one hundred thousand conversions a year to the Church are reported.

232. I would not underestimate the power of these simple reactionary forces in what are still underdeveloped countries.

233. The myth that Muslims do not convert has been laid to rest by the fascinating account of a great variety of conversions in the words of the converts themselves, collected and commented by Jean-Marie Gaudeul, *Appelés par le Christ: Ils viennent de l'Islam* (Paris: Cerf, 1991).

234. *Rencontre* is a French word meaning some good times together in Tunis while remaining intellectually "ships passing in the night"!

"communities." My evangelical friend is having little luck getting the leaders of his own local church to respond to important theological and pastoral problems he has submitted to them in writing. Between loving individuals, determined to grow in friendship, yes, fruitful dialogue does take place; while rare, it exists, as our local Communio group will testify in our intense work together with the theologically sophisticated evangelical friend I mentioned. We have learned many wise things from him, and we have been inspired by his love for and complete trust in Jesus as he goes through a difficult period in his family life.

On the other hand, I have witnessed the tenacity of pathological blocks and the devastation wreaked by a fundamental option for evil (the sinister dynamics of which are nowhere better portrayed in literature than in Bernanos's characters, Mouchette and Monsieur Ouine).[235] Meanwhile, the tendency of the HTX to functionalize everyone—flattening us out into "consumers" and "personnel," producing unisex, faceless masses whose opinions register wild and erratic swings but whose depth of wisdom and penetration impresses no one—pressing forward relentlessly toward a standardized world in which all the old traditions progressively lose their grip, except for islets of increasingly intense fanatic "fundamentalists," and a few sound, loving communities. The fanatics and the seriously engaged lovers in community both in their opposite ways resist the planetary homogenization.

From his many pronouncements in this vein, clearly it is the view of the supreme pontiff[236] that the most urgent task of the Church strategically is the "new evangelization" of the traditional European heartland. This is not just a nice-sounding slogan, but rather it signals a strategy emerging from the Church's profound thinkers—the pope at their head.[237] Start, says Angel Cardinal Suquia, archbishop of Madrid,[238] with those nominal Catholics who may still be reached and converted to become living witnesses to the one and only true way to the unification of mankind. *Life* in that Jesus Christ sent by the Creator to call us into loving union with Him is *the only way possible,* because it alone is in keeping with how He, who is *agape,* has

235. In *Sous le soleil de Satan,* "la nouvelle Mouchette" in *Joie,* and Monsieur Ouine (inspired by the person of André Gide) in *Monsieur Ouine.*

236. For a catalog of his pronouncements on this, see E. Caparros, "La rechristianisation de la société: Le rôle de laics dans la perspective du Canon 225," *Fidelium Iura de Derechos y Deberes Fundamentales del Fiel* 3 (1993): 38 n. 2.

237. See the papers from the *Communio* conference on the new evangelization, in *International Theological Review: Communio* 19 (winter 1992).

238. Angel Cardinal Suquia, "The New Evangelization: Some Tasks and Risks of the Present," an important speech to the Spanish Bishops' Conference, published in *Communio: International Catholic Review* 19:4 (winter 1992): 515–40. Also in the same number, see Bishop Karl Lehmann (president of the German Bishops' Conference), "The Meaning of the New Evangelization in Europe," and David Schindler's address to a large gathering of bishops in Texas, "Towards a Eucharistic Evangelization," a profound and challenging piece.

made us.[239] The goal is well put by Bishop Karl Lehmann: "to promote the infectious testimony of every individual Christian."[240]

The dynamic rush of planetary secular society not only toward splendid advances in knowledge and liberation of increasing numbers from numbing poverty and the oppression that goes with it but also toward nihilism confronts Christians with the most daunting task of evangelization ever faced, arguably worse than what faced the tiny band as it confronted a vigorous Augustan Empire. But take courage from the fact of the severity of the challenges taken on by earlier Christians: first starting with a few thousand to survive in a "worldwide" ecumenical empire, a high civilization with a civic theology of great flexibility, absorption power, and coherence. (What drove the empire to try to destroy the little Christian band was its outrage at their unabsorbable, uncompromising Absolute.)

Then the Christian remnant of the seventh and eighth centuries with the empire collapsed all about it and engulfed by the barbarian flood, confronted with the task of civilizing coherent brutal tribes, with only fragments of the destroyed old culture remaining. A few Christians built a new survival culture: the monasteries. And see what the Holy Spirit eventually accomplished through them: the building of European civilization, forged from remnants of classical culture over three hundred incredibly hard years, which historians still call the Dark Ages.[241] Or imagine the seventeenth-century Christians surveying the smoking ruins of "Christendom" after the devastating Thirty Years' War that produced a huge outbreak of atheism, including the de-Christianization of entire countrysides; or the bourgeois Christians of Kierkegaard's time, faced with the onslaught of industrialization and a Church so demoralized by the ravages of the Revolution—the truly most radical of all, the French Revolution—so demoralized there was question around 1840 of whether it would even be worthwhile electing another pope. (But when they did, they soon had the great Leo XIII on their hands, one of the most effective popes in the Church's entire history, and no mean strategic head himself!)

239. In one critical respect that life has already begun, in Christ's reconciliation of us all with the Father on the cross, unique foundation of our forgiving reconciliation with one another.

240. Bishop Karl Lehmann, in "Pastoral Letter," *Communio* 19:4 (winter 1992): 546.

241. This was not conceived by them as a strategy. But our times are different: to be sure, the driving strategy of the new evangelization is the Holy Spirit's; but one of the forms His gifts take in our situation is that set of notions, fruit of the Spirit's good influence on the HTX, that produce a sense of management, strategic planning, a more critical and empirically based gazing into the future. Yes, I am serious—what is good in these evolving mentalities is of the Holy Spirit. They are, to be sure, also full of traps of their own (the Other Spirit is not on vacation!); yet at their base they respond to something demanded by the actual situation. The unprecedented strategic effort of the present pontificate, contemplatively responsive to the guidance of the Holy Spirit, has been prompted not only to "read the signs of the times" but also to respond to what this situation demands. For the best glimpse of the power of this strategic thinking read the pope's letter *On Preparing the Third Millennium (Tertio Millennio Adveniente).*

The ecumenical empire of planetary proportions, the dense weave of institutions, traditions, symbols systems making up the HTX, provides resources of knowledge, communications, transportation, organizing skills that Christians can use, not just "to survive in virtual reality"[242] but also to cooperate in the Holy Spirit's penetration of it with the only lasting reality. Contemporary Christians too face barbarians, both from without—whole civilizations and cultures little touched as yet by any direct Christian influence—and, more menacing, from within, symbolized by those nihilistic mobs in East Germany the blind terrorism of which Cardinal Ratzinger recently identified as a singular menace of our time.[243] Not only do our "atheist humanist" elites have no idea how to deal with this threat, but they are also inadvertently aiding and abetting it through their simpleminded liberalism, which undermines discipline and thus is damaging of genuine freedom.[244] And this ill-conceived pseudoliberalism is penetrating sectors of the Church.

Yes, the task is daunting. But that is the Catholics' apostolic challenge, so let them "be wise as serpents and patient as doves" (Matt. 10:16) and get on with it. Archbishop Karl Lehmann reminds us that they do not have to start from scratch. "Many of the things that will be necessary for survival of the Church are already in place in individual communities, such as groups, associations, parishes, and religious communities. There are already advanced parties in the field."[245]

This "European heartland" of the *catholica* that must be the first target of the new evangelization divides into five zones, each with its own particular species of challenge: Western Europe, Eastern Europe, North America, Latin America, and Oceania. All are suffering in advanced form from the same basic secularizing pressures, which began four hundred years ago. This tiresomely repeated truism hides an important principle: to understand how the Catholic Church failed to hold faithful the majority of its European peoples, one needs a long perspective to appreciate how deep are the roots of its failure. For many of those factors are still operative, with

242. *HTX: Learning to Survive in Virtual Reality* examines the challenge the HTX poses to everyone, and the peculiar opportunities and dangers for Christians. It will not be enough for isolated Christians and little bands to "hook up on the Net," and to create "virtual universities." These can easily become just so many more distractions burning up time that could be spent in contemplation and in deepening real friendships, maybe even with our own families.

243. On the week of September 5, 1994, eighteen hundred "punks" descended on the old center of Hanover for no apparent reason and trashed it. After an all-night battle the police arrested three hundred. Senseless violence for its own sake. I saw no mention of this in the Toronto press. There is much more of this to come. No society can put one-third of the population in prison and survive.

244. The same day a German priest told me of the trashing of Hanover, a youngish woman from the black center of Washington with whom I walked from the Metro, opined, "Something has to be done about these ne'er-do-wells. But the lawyers ruin everything. A friend of mine had her purse snatched, two men apprehended the criminal, and he sued them with help of Legal Aid, and won!"

245. Lehmann, "Pastoral Letter," 547.

the result that the Church today is still losing a scandalous and unacceptable portion of those "raised Catholic." It is failing to get virtually anything of its supernatural message across to masses of people in these Europeo-American countries, beyond those elements of its revealed teaching that have long since been "drained off" into the secular culture in sanitized versions.[246]

One possible awakener: the "life issues," not just abortion, but euthanasia. One does not have to be sharp to start to worry, "What happens to *me* when I am old and deemed 'useless'?" And not just death but lack of birth menaces the survival of the "post-Christian" societies: high-tech contraception and materialistic lifestyles have resulted in the fated disappearing of every European and American people, much more rapidly than almost anyone is prepared to admit. Even Latin America, says a recent U.N. conference, is starting to see a population decline that is already causing economic concern!

But whence will come local guidance and strategy to combat the civilization of death at levels below the world-spanning vision of the papacy? The ineffectiveness of the regional Church's official think tanks, located in the bureaucracies of the bishops' conferences, and of overloaded chancery offices, causes resources to be wasted on bureaucracies, which for the most part—especially in the case of bishops' conferences—are often captive of elements less interested in "new evangelization" and in sound strategic thinking than in pushing an agenda of reform in the Church and in society generally, based, at least in many flagrant instances, in a liberal democratic conviction that the people must be "empowered."[247] This results in the neglect of catechizing of the society; and it nurtures tension in the Church, for "traditional" Catholics do not share a concept of Christianity as being about "empowering" at all, but grateful obedience to the Word. To the extent a kind of neo-Marxist "class struggle" resentment model gets taken over from secular humanism into these currents within the Church bureaucracies, they instinctively tend to pit identifiable groups against one another. And so instead of converting hearts and fostering *communio*, they in fact incite divisive struggle.

Many good people get absorbed in the pursuit of empowerment, seeking to redress social wrongs (such as the plight of the homeless) and sometimes pseudowrongs (like the "right" of women to ordination) on the "structural" level (which too often ends up being the abstract, ideational, and hence inevitably ideological), mixed with partisan secular politics, rather than in concentrating on concretely addressing actual real situations, patiently, person by person and community by community. (This modest slugging

246. "At times, it seems that in the future Christians will only be people who personally and decidedly profess a faith that is also reflected by the conduct of their daily lives" (ibid., 544–45). The end of "ethnic Christianity"? Is that perhaps not all bad?

247. New evangelization does not call for "expanding the institutional dimensions of the Church; rather, all the structures must be evaluated in accordance with their ability to enable the faithful to witness to their faith" (ibid., 546).

it out in the real world lacks the "sex appeal" of dreaming about "altering great unjust structures.") In the process, real economic problems get demonized into class-conspiracy theories. To be fair, many of the "activists" driving these strategies are also charitable people who give of themselves in helping individuals and needy small groups. But to the extent they grind their emotional axes while carrying out such activity, they run the risk of disorienting the people they truly mean to help. I might add the extreme reactions from the emotional opposite pole that they irritate takes a further toll on the community's charity.

These remarks are not intended to distract from the responsibility of all Christians to work for the *political* redressing of injustices. But those are proper *secular political* activities in which all citizens should be involved, each according to his possibilities, without mobilizing the prestige of the Church for necessarily ambiguous partisan solutions. Christians of goodwill are free to disagree about these. In any event, personal and familial spiritual development must come first. Turning the episcopal conference or the local parish into an instrument for political agitation is bound to obfuscate and interfere with the proper, immense work of the Church, which is evangelization of all and spiritual development of her people,[248] and corporeal works of mercy to all, without partisan political discrimination.

The Evangelical Challenge

The holy father expressed beautifully the most basic need in our time, not of the Church only, but of mankind: "We need heralds of the Gospel, experts in humanity . . . and who, at the same time, are contemplatives in love with God. That is why we need new saints. *The great evangelizers of Europe were*

---

248. Of course, in countries where the political situation permits no structures through which people can work to redress wrongs, it may well be that Church structures are the only hope. That is a different case from that to be found in any of the democratic open societies. I might add that efforts, from here, to effect big changes in other countries, like Guatemala (which for some reason has become the present obsessive target of leftist forces throughout North America) can produce serious mischief in the lucky targeted country and a diversion of energy here at home that strikes me as pathological in origin. The charge of mischief abroad is thoroughly documented in case after case, but that mass of facts has no impact on those who insist on playing such games. And that is what they are—games—but not innocent ones. I shall not myself fall into the same trap by here pontificating about how in countries like Guatemala the Church forces and politics should interact. I shall limit myself to one general comment: when in Brazil, Honduras, Guatemala, and I do not know where else evangelicals make tremendous inroads among what are supposed to be centuries-old Catholic peoples, something is drastically wrong in the local churches. It has been suggested (and this appears reasonable to me) that efforts on the part of the already thin church forces to work centrally in political consciousness raising at the expense of a sound catechesis that brings the poor closer to the eucharistic reality of Our Lord leaves these people thirsting for a simple religious experience they can grasp. The evangelicals are operating in the vacuum. But that is not our problem in Europe and North America.

saints. We need to beg the Lord to send us new saints to evangelize the world of today."[249]

Only "contemplatives who are in love with God" can penetrate to the simple heart of this, the oldest and most sophisticated of traditions: that the carpenter's son, Jesus of Nazareth, is Lord and Christ, the wisdom of God, living among us eucharistically as the Church. Nothing "Catholic," neither seminary nor school, no movement, not even a parish is really Catholic unless the persons forming it, because of their true love, are sound and zealous in the faith. Only to the extent that they are responding to God's grace can they bring the antiquity, complexity, and modern sophistication of this tradition to bear fruitfully on the unbelievably complex problems of our HTX time.

New organizations like Legatus, a spiritual movement for CEOs of large corporations, will continue to talk of forging a strategic sense of evangelizing the society. I would not have undertaken this book if I did not agree that Christians must show the way forward societally. But they can do so validly only to the extent their hearts are Marian. The essence of the Catholic tradition radiates from the sacred heart of Jesus, and is "beauty ever old, ever new," as Saint Augustine sang. The Christian who does not seek daily to allow himself to be centered there will not advance the new evangelization, and indeed distorts the tradition. The Eucharist, received in an atmosphere of interior prayer, producing the fruits of mercy and love, manifests the ever surprising essence of the Incarnate Word. Every strategy, whether global, local, or personal, must be centered in Him. Only those who devote themselves to knowing and loving the Word can radiate it to the society.

249. John Paul II, Discourse to the Sixth Symposium of European Bishops, in *L'Osservatore Romano* (October 22, 1985): 1, 4, 5.

INDEX

Adoptionism: and the creedal formulations, 162; mentioned, 56, 65, 67–70 *passim*, 86, 136

Albertus Magnus, Saint: and Saint Thomas Aquinas, 155, 158

Alexander, Bishop of Alexandria, 58

Alexander VI, Pope [Rodrigo Borgia], 25, 160

Anabaptists: and Islam, 42; and baptism, 233–34, 443; mentioned, 209

Anglican Protestantism: high Anglicanism, 209; synodal structure, 323–24n26, 324n27, 473; and Catholicism, 449–50

Anicetus, Bishop of Rome, 108

Apostolein: "sent forth," 40, 105; in the New Testament, 241

Apparitions, 493–501 *passim*

Aquinas, Saint Thomas: the influence of his thought, 145–47; and phenomenology, 164–65; existence/essence distinction, 165, 167; and modern ontology, 174; mentioned, 25, 155–57, 160–62, 167–68, 177, 399, 401, 402

Architecture, 150–51, 189–90, 509–10

Arianism, 56–59, 132, 198n52, 410n2

Aristotle: and Saint Thomas Aquinas, 157–58; and phenomenology, 164–65; categories, 346n10; mentioned, 151–52, 162–67 *passim*, 298n145

Athanasius, Saint, 108

Atheism, 255–66 *passim*

Augustine, Saint: and Western secular attitudes, 119, 123, 134; and the Trinity, 135–37; and "the seminal reasons," 150; and Scholasticism, 152, 154–55n22; *imago Dei*, 153; and Saint Thomas Aquinas, 156, 162; mentioned, 17, 25, 26, 147, 177, 249, 371, 384–85, 390

Balthasar, Hans Urs von: and experience [*Erfahrung*], 32, 61; and the Eucharist, 38; and Islam, 40; and love, 102, 313; and apostolic succession, 116, 320, 412; and secularization, 250–51; and female ordination, 299n148; and Communio, 435–36; ecumenical dialogue between East and West, 502–3; mentioned, 5, 6n9, 10n17, 23–24n4, 25n5, 29–30, 33n20, 40, 46n40, 46n42, 154n19, 275n78, 279n89, 279n90, 288–89, 290n127, 314–15, 324–25n28, 361n11, 372, 373, 384n27, 467–68, 513

Baptism: and the Jews, 58; of Jesus, 69, 88; of pagans by Peter, 101; Luther's position, 224–25, 314; and Anabaptists, 233–34, 443; and the charismatics, 443–44, 480; and hope, 449; mentioned, 42, 44, 48, 84, 356

Basil, Saint, 132

Bernard of Clairvaux, Saint, 201

Bible: and revelation, 37

Birth Control. *See* Sexuality; Feminist issues

Black Death (plague), 142, 181

Blondel, Maurice, 169

Bonaventure, Saint: and Saint Thomas Aquinas, 158; mentioned, 171

Bouyer, Louis, 99, 102, 107, 129n31, 487–89

Brown, Lawrence E., 203

Buddhism, 31, 43, 178

Calvinism, 216n33

Cathari: central tenets of their beliefs, 199–200; and Docetism, 200–201; mentioned, 193–94

Catherine of Siena, Saint, 187

Catholicism: as universal, 4, 14n26, 23–24, 42, 62; defined, 15, 22–23; and papal authority, 97–112 *passim*; and Communion, 190n32; and original sin, 213; and a corporeal notion of "virtue," 214; based on presence of Christ, 391

Chalcedon. *See* Ecumenical Councils

Chantraine, Georges, 472, 475

Charisms: Luther's position, 225; of the